THEORY AND APPLICATION
OF
DIGITAL
SIGNAL PROCESSING

THEORY AND APPLICATION OF DIGITAL SIGNAL PROCESSING

Lawrence R. Rabiner

Bell Laboratories

Bernard Gold

MIT Lincoln Laboratory

PRENTICE-HALL, INC. Englewood Cliffs, New Jersey

Library of Congress Cataloging in Publication Data

RABINER, LAWRENCE R
    Theory and application of digital signal processing.

    Includes bibliographical references and index.
    1. Digital electronics. 2. Signal theory (Telecommunication) I. Gold, Bernard, joint author. II. Title
TK7868.D5R32   621.3819′58′2   74-22332
ISBN 0-13-914101-4

10 9 8 7 6 5

Printed in the United States of America

PRENTICE-HALL INTERNATIONAL, INC., *London*
PRENTICE-HALL OF AUSTRALIA, PTY. LTD., *Sydney*
PRENTICE-HALL OF CANADA, LTD., *Toronto*
PRENTICE-HALL OF INDIA PRIVATE LIMITED, *New Delhi*
PRENTICE-HALL OF JAPAN, INC., *Tokyo*

Contents

3   THE THEORY AND APPROXIMATION OF FINITE DURATION IMPULSE RESPONSE DIGITAL FILTERS   **75**

6 SPECTRUM ANALYSIS AND THE FAST FOURIER TRANSFORM 356

7 AN INTRODUCTION TO THE THEORY OF TWO-DIMENSIONAL SIGNAL PROCESSING 438

11 GENERAL-PURPOSE HARDWARE FOR SIGNAL PROCESSING FACILITIES   627

Preface

Any undertaking as ambitious as this book requires the assistance and aid of several individuals for successful completion. The authors would like to acknowledge the assistance provided by Bell Laboratories in all phases of the preparation of this book. In particular we should like to thank Henry Raupp of the drafting department, and Mrs. Richards and Mrs. Olczvary of the typing department for their invaluable assistance. Mrs. Beverly MaSaitis also provided typing assistance throughout the preparation of the various drafts of this book, and Mrs. Kathy Shipley provided programming assistance which was used to obtain several of the figures which appear in the book. The authors also wish to thank the Bell Laboratories book review board for overseeing the preparation of the manuscript.

Almost concurrently with the publication of this book, a complementary book by Oppenheim and Schafer will be published. Their book is directed at graduate electrical engineering students while ours is oriented more towards advanced courses and the practicing engineer. Thus while their book accents careful treatment of the fundamental theoretical questions, ours stresses more advanced topics such as in-depth treatments of filter design techniques, hardware, and applications. A useful graduate sequence would be a semester based on the Oppenheim and Schafer text followed by an advanced course using this book. This is a good place to acknowledge our debt to these gentlemen for the many stimulating technical interactions we have had with them. In particular, one of us (L. R. R.) has interacted closely with Ron Schafer for almost seven years, and this relationship has been a deeply satisfying and a professionally rewarding one. All of us entered this field in its early stages and have grown with it together. It is interesting to note that although both books were written independently, some of the material treated is common to both books; this is unavoidable in books which strive for some sense of completeness. In this connection we gratefully acknowledge the strong pedagogical influence by Oppenheim and Schafer on our presentation of some of the theoretical material in the early chapters; perhaps this influence is most noticeable in Chapter 2, on the theory of discrete-time linear systems and in parts of Chapter 5, on finite wordlength arithmetic. However, as we are sure our colleagues would agree, our gratefulness for such help is very minor compared to the debt we owe them for a warm and lasting technical relationship and, more important still, for their friendship. We also want to express our gratitude to two outstanding contributors to digital signal processing, Charles Rader of Lincoln Laboratory and James Kaiser of Bell Laboratories, whom we have had the good fortune to know well. Our technical interaction with these two gentlemen has also been most rewarding. We should like to acknowledge the direct contributions of Charles Rader in writing Section 6.19 on Convolution and Correlation Using Number Theoretic Transforms.

Much of the material in Chapter 5 on quantization effects in digital filters relies heavily on the valuable contributions of Leland Jackson of Rockland Systems Corporation, and Cliff Weinstein of MIT Lincoln Laboratory. In particular Jackson provided several of the figures used in this chapter and in Chapter 9. The photographs used in Section 7.21 were supplied by Professor Thomas Stockham of the University of Utah and by Professor Leon Harmon and Alan Strelzoff of Case Western Reserve University.

Much of the material in Chapter 8 on hardware for digital signal processing was obtained from published works and with the aid of several colleagues at MIT Lincoln Laboratory. Of the published material, a good deal was borrowed from specifications and data manuals published by Motorola Corporation. Paul McHugh, Joe Tierney, Alan McLaughlin, Peter Blankenship, Albert Huntoon, and Charles Rader, of MIT Lincoln Laboratory, all aided in the preparation of this chapter. In addition, the work of Stylianos Pezaris on array multipliers greatly influenced the presentation on multipliers. A special debt of gratitude is owed to Peter Blankenship, both for allowing us to incorporate his work on dividers and floating hardware into this chapter, and also for his useful corrections and detailed criticism of our first draft of this chapter.

The authors wish to acknowledge the very useful guidance, criticism and contributions to Chapter 13 on applications to radar of Edward Hofstetter, Peter Blankenship, C. E. Muehe, and Ted Bially of MIT Lincoln

Laboratory. In particular Hofstetter kindly reviewed and corrected the first draft of this chapter.

Finally a large note of appreciation goes to Charles Rader for an extremely careful and thorough review of the entire manuscript. In addition to finding numerous errors in equations and logic, Rader made valuable suggestions which hopefully helped to clarify several important issues in the book. We are greatly indebted to him for his help.

A personal note of gratitude goes to James L. Flanagan of Bell Laboratories for his guidance and inspiration throughout the years of one of us (L. R. R.) at Bell Laboratories. His understanding of the trials and tribulations which went into the preparation of this book have made working for him a personal pleasure.

Perhaps the deepest appreciation and thanks must go to our respective families for their patience, tolerance and understanding which far exceeded that normally required in a usual family relationship. Since this book was prepared, for the most part, out-of-hours, i.e., in the spare time which is generally allocated to family life, the loss of a period of $2\frac{1}{2}$ years is no small contribution to this work. To our wives and children, for all their help we are especially grateful.

LAWRENCE R. RABINER and BERNARD GOLD

THEORY AND APPLICATION
OF
DIGITAL
SIGNAL PROCESSING

1

Introduction

1.1 Brief Historical Introduction and Some Comments

Since World War II, if not earlier, electronics engineers have speculated on the applicability of digital hardware techniques to the many problem areas in which signal processing plays a role. Thus, for example, Laemmel (1948) reports a lunchtime conversation among Shannon, Bode, and several other Bell Telephone Laboratories scientists on the possibility of employing digital elements to construct a filter. Needless to say, the conclusion then was not favorable. Cost, size, and reliability strongly favored analog filtering and analog spectrum analysis techniques. In the 1950's, Stockham (1955) reports that Linville, at the time an MIT professor, discussed digital filtering at graduate seminars. By then, control theory, based partly on Hurewicz's (1945) work, had become established as a discipline; the concepts of sampling and its spectral effects were well understood and the mathematical tools of z-transform theory, which had existed since Laplace's time, were propagating into the electronics engineering community. Technology at that point, however, was only able to support practical efforts to be directed toward either low-frequency control problems or low-frequency seismic signal processing problems. While seismic scientists made notable use of digital filter concepts to solve many interesting problems, it was not until the mid 1960's that a more formal theory of digital signal processing began to emerge. By then the potential of integrated circuit technology was appreciated and it

1

was not unreasonable to imagine complete signal processing systems that could best be synthesized with digital components.

The first major contributions to the field of digital signal processing were by Kaiser (at Bell Laboratories) in the area of digital filter design and synthesis. Kaiser's work showed clearly how to design useful digital filters using the bilinear transform. At about that time tremendous impetus was given to this emerging field by the Cooley–Tukey (1965) paper on a fast method of computing the discrete Fourier transform, a method that was subsequently popularized and extended via many papers in the *IEEE Transactions of the Group on Audio and Electroacoustics* and other journals. This set of techniques has come to be known as the fast Fourier transform (FFT). Its value lies in the reduction (by one to two orders of magnitude for most practical problems) in computing time for the discrete Fourier transform (DFT).

At the time of the Cooley–Tukey paper, the development of a formal and quite comprehensive theory of digital filters was well under way. The great importance of the FFT was that it showed quite strikingly how digital, as opposed to analog, methods could be intrinsically more economic to employ for spectrum analysis. This resulted in accelerated activity that by now has led to a wide variety of applications for signal processing problems extending from the low-frequency spectrum of seismology through the acoustic spectrum of sonar and speech into the video spectrum of radar systems.

Perhaps the most interesting aspect of the development of the field of digital signal processing is the changing relationship between the roles of FIR (finite impulse response) and IIR (infinite impulse response) digital filters. Initially Kaiser analyzed FIR filters using window functions, which indicated that IIR filters were much more efficient than FIR filters. However, Stockham's work on the FFT method of performing convolution, or more specifically FIR digital filtering, indicated that implementation of high-order FIR filters could be made extremely computationally efficient; thus comparisons between FIR and IIR filters are no longer strongly biased toward the latter. These results also inspired significant research for efficient designs for FIR filters.

Paralleling this research activity has been a formidable amount of pedagogy, exemplified by the incorporation into many graduate school curricula of courses on digital signal processing and the entering of much of this material into undergraduate electrical engineering courses.

The book by Gold and Rader (1969) was the first attempt at a comprehensive theory of digital signal processing. Being an early book, it has had to serve the double purpose of graduate text and guide for the practicing engineer. On both counts it inevitably had shortcomings; it is axiomatic that a well-rounded graduate text should be grounded in at least several years of teaching plus a good collection of problems, while the practicing engineer desires more design data and more highly developed synthesis procedures than were available at that time.

Given the brief history above, we can now describe the purpose of this book. We have addressed ourselves primarily to the hardware engineers and computer programmers, that is, to those engineers who are interested in designing a system via either hardware or software. In order to present a sufficiently complete picture, we feel it is necessary to connect theory, on the one hand, with a variety of applications from radar, sonar, speech, music, seismic, and medical signal processing, and, on the other hand, with the digital component technology that is the main driving force for progress in this field as well as the overall field of computer design. In addition, much new digital filter design information is now available and the field of digital filter synthesis, while still (we feel) in its early states, is beginning to take some shape. Similarly, while FFT algorithms have been explored quite thoroughly, the design and synthesis of digital spectrum analyzers has not been sufficiently organized and formulated; one of our aims is to do this. By surveying the table of contents, the reader can view in more detail how we intend to fulfill our purpose.

A final general comment—from one point of view, digital signal processing is a collection of computer algorithms and thus can be thought of as simply another branch of computational mathematics. We feel, however, that the content of digital signal processing theory has much in common with classical network and filter theory and transform theory as presently taught in standard engineering curricula and that it is well worth maintaining this formal structure. Thus, while we intend to accent the practical design and synthesis questions, we shall try not to do this at the expense of theoretical unity.

1.2 An Overview of Digital Signal Processing

The field of digital signal processing has grown enormously in the past decade to encompass and provide firm theoretical backgrounds for a large number of individual areas. Figure 1.1 illustrates one view of how the field has emerged and spread out. Since digital signal processing, for the most part, relies on the theory of discrete-time linear time-invariant systems, we show this as a major unifying influence for the entire field.

The major subdivisions of the field of digital signal processing are digital filtering and spectrum analysis. The field of digital filtering is further divided into finite impulse response (FIR) filters and infinite impulse response (IIR) filters. The field of spectrum analysis is broken into calculation of spectra via the discrete Fourier transform (DFT) and via statistical techniques as in the case of random signals—e.g., quantization noise in a digital system. As discussed in Sec. 1.1, the fast Fourier transform (FFT) and the related area

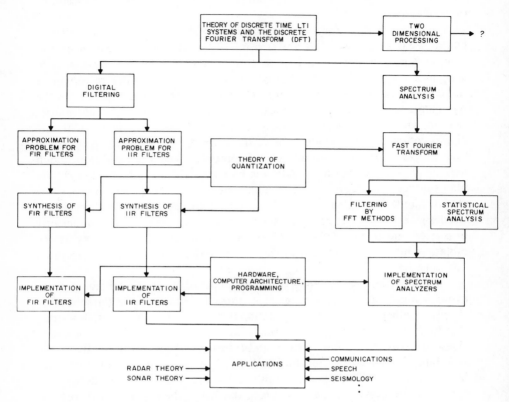

Fig. 1.1 Overview of digital signal processing.

of fast convolution are almost exclusively used in practical spectrum analysis techniques. Two-dimensional signal processing is a relatively new area. Thus in Fig. 1.1 there is indicated a question mark concerning its direction.

The remaining aspects of digital signal processing, as shown in Fig. 1.1, are the important topics of implementations of digital systems and application areas. Almost all the theoretical considerations involved in the design of digital filters and spectrum analyzers would be of little value if a good understanding of the issues involved in practical implementations of such systems in finite precision software or hardware were not also available. Thus the area of implementation is first concerned with the mathematical basis of quantization in discrete systems; then, depending on whether a software or hardware implementation is desired, it is important to understand the inherent limitations and advantages in both these implementations.

The reader should note that all the applications areas listed in Fig. 1.1 are established fields that have traditionally relied on analog components for their signal processing. Many questions have been raised as to the desirability of applying digital signal processing technology when so much progress has been and continues to be made with analog methods. The answer, we believe, goes to the very core of scientific and engineering methods. Whereas the *formulation* of engineering problems is often as vague as those of the "softer" sciences (such as anthropology, psychology, etc.), the *execution* of these problems appears to depend on greater and greater accuracy and reproducibility. Witness, for example, the dogged attempts to measure even more accurately the velocity of light, Planck's constant, and other universal constants. Consider the enormous efforts that have gone into obtaining frequency and time standards. Thus, although many signal processing systems will be most economically implemented with analog methods, it is the capability of digital systems to achieve a *guaranteed* accuracy and essentially perfect reproducibility that is so appealing to engineers. This, in turn, will help motivate components manufacturers to emphasize ways of improving digital technology that should eventually bias the economics toward the digital implementation. To summarize, the importance of digital signal processing should eventually surpass that of analog signal processing for the same reasons that digital computers have surpassed analog computers.

1.3 Organization of the Book

Figure 1.1 provides a good introduction into the organization of this book. Chapter 2 provides a general discussion of the theoretical framework of discrete-time linear systems. It is not intended to be overly theoretical but instead to provide a comprehensive summary of the major areas of digital signal processing that will be relied on throughout the book. It also provides an introduction to the DFT and the concept of fast convolution.

Chapters 3 and 4 discuss the design of digital filters. Chapter 3 is concerned with the theory of FIR filters and Chapter 4, the theory of IIR filters. Since the digital filter is one of the essential building blocks of most digital systems, these chapters endeavor to provide a discussion as complete as possible of the nature of the various design algorithms. In Chapter 3 the methods of windowing, frequency sampling design, and minimum peak error design are discussed in detail and compared whenever possible. Since it has been shown that the minimum peak error filters are optimum in the Chebyshev sense and since algorithms (including FORTRAN programs) have been published for designing them, a majority of Chapter 3 has been devoted to a discussion of this class of filters and their specific properties.

In Chapter 4 two classes of IIR design procedures are considered. The first procedure is the classical closed form approach of digitizing an analog filter using an appropriate filter transformation. The techniques of mapping of differentials, impulse invariant transformation, bilinear transformation, and matching poles and zeros all fall into this class of design procedures. The second class of IIR design procedures is the open form approach using modern optimization algorithms to design filters that are optimum according to some specified criteria. Included among these approaches are minimum mean square methods, minimum absolute error methods, equiripple methods, and time domain optimization to match a prescribed time response. It is not possible to select any one IIR design procedure and say that it will be applicable in *all* cases of interest. Thus it is especially important that the reader fully understand the nature of the various IIR design procedures and know their advantages and limitations. Chapter 4 also includes some simple comparisons between one class of FIR filters and an "equivalent" class of IIR filters. Such comparisons allow the reader to gain an understanding of the tradeoffs in complexity between these two broad classes of filters.

Chapters 5, 6, and 7 provide theoretical discussions of the mathematics of quantization in digital filters, the topics of spectrum analysis and the FFT, and the theory of two-dimensional systems including two-dimensional filter design techniques. This group of chapters provides, for the most part, the necessary theoretical background for the discussion of digital hardware and practical applications that comprise the remainder of the book.

Chapter 5 deals with the effects of quantization in digital systems. The concepts of roundoff error, analog-to-digital conversion noise, and coefficient sensitivity are introduced. A key result is the relation between dynamic range and roundoff noise in fixed point implementations of recursive digital filtering structures such as the direct or cascade forms. These dynamic range roundoff noise interactions lead to the ideas of section ordering and pole–zero pairing in cascade realizations of filters in order to maximize attainable signal-to-noise ratio of the filter. For the case of FIR filters we present a method for analyzing roundoff noise effects and scaling the filter variables to prevent overflow within the filter. Finally, the concept of correlated roundoff noise or limit cycles for IIR filters is presented.

Chapter 6 provides a fairly general discussion of the FFT—perhaps the single most important algorithm in digital signal processing. The well-known radix 2 decimation-in-time and decimation-in-frequency versions of the FFT are derived in fairly simple terms. A unified theory of the FFT is presented in which a one-dimensional transform is converted into a two-dimensional transform of lower dimensions. This unified treatment leads to a fairly simple explanation of the concepts of digit reversal, twiddle factors, in-place, etc. Following the discussion of the FFT, a fairly general discussion of discrete spectrum analysis is given. The concepts of sliding and hopping spectra are introduced. It is shown how an FFT spectrum analyzer can be made equivalent to a spectrum analyzer composed of a bank of bandpass filters. Finally, the concept of spectrum analysis of random signals is quantified and several methods for performing such statistical spectral analyses are discussed. This chapter concludes with a discussion of the application of number theoretical techniques to the computation of convolutions.

Chapter 7 gives an introduction to the theory of discrete-time, linear, two-dimensional systems along with some discussion as to digital filter design methods in two dimensions. Most of the theory is similar to that presented in Chapter 2 for one-dimensional systems; however, several important concepts do not extend into more than one dimension and these concepts are stressed in Chapter 7. Although filter design techniques in two dimensions are fairly rudimentary, one excellent mapping technique from one to two dimensions for FIR filters is discussed and is claimed to be the best currently available way of designing a two-dimensional filter.

The next group (Chapters 8 to 11) discusses the application of special-purpose digital hardware to digital signal processing problems. In particular, Chapter 8 provides an introduction to digital hardware; Chapter 9 discusses the use of digital hardware for digital filter implementation and for special-purpose hardware devices; Chapter 10 considers digital hardware for implementing the FFT; and Chapter 11 deals with programmable signal processing computers.

Chapter 8 lays down the foundations for building hardware to implement any signal processing function. The concepts of memory, arithmetic elements, and control logic are presented as the key building blocks of a digital system. Various alternatives for memory and processing elements are discussed. The construction of adders, subtracters, and multipliers is discussed in terms of speed of operation, cost, and complexity.

Chapter 9 provides a thorough description of how special-purpose digital hardware can be used to realize a general-purpose digital filter as well as a dedicated piece of hardware such as a digital signal generator. Hardware realizations of FIR and IIR filters are explained and then several systems are presented in which a digital filter is only one of several system components. Included in this class are a digital touch tone receiver and a digital time-division-multiplexed to frequency-division-multiplexed translator. The concept of multiplexing, or sharing of an arithmetic unit among several digital systems, is discussed in terms of the cost effectiveness of such shared or multiplexed utilization of an expensive piece of hardware. Examples of digital hardware for generating a sine wave and a random noise signal are also discussed in this chapter.

Chapter 10 provides a thorough description of how digital hardware can provide several orders of magnitude speedup in performing FFT's over conventional software algorithms. The concepts of arithmetic parallelism, pipelining, and memory-processor overlap, etc., are discussed as ways of speeding up the overall throughput rate of the system.

Chapter 11 deals with the application of digital hardware in the design of a programmable, high-speed, specially designed signal processing computer.

The case in point here is the Fast Digital Processor (FDP), which was designed and built at Lincoln Laboratories. The theory behind the architecture of such a high-speed processor, as well as programming considerations for optimal use of the capabilities of such a machine, is given in detail. A brief discussion of a general-purpose signal processing facility is also included.

Chapters 12 and 13 provide examples of how the ideas discussed throughout the book have found application in the specific areas of speech and radar processing. The ideas presented in these chapters serve to illustrate typical applications of the theory and in no way are meant to serve as an exhaustive treatment of all the possible approaches to these problem areas or, in fact, even to the best approaches. It is felt that the engineering approach to problems in the areas of speech and radar, as presented here, may serve to stimulate further thought as to better and more powerful algorithms for signal processing in other areas of interest.

In summary, we have endeavored to present a large body of advanced ideas in the area of digital signal processing that is aimed at the graduate or practicing engineer level.

2

Theory of Discrete-Time Linear Systems

2.1 Introduction

The theory of discrete-time linear systems deals with the representation and processing of sequences, both in time and in frequency. Throughout this chapter and, for the most part, throughout this book we shall handle sequences as though there were no quantization of the amplitudes of the members of the sequence. This assumption of infinite bit precision on both the sample values of sequences and the coefficients of linear systems leads to a fairly general body of theory on discrete-time (continuous amplitude) systems. Later on we shall discuss the various finite word length effects in digital systems where the amplitudes are quantized to some specified accuracy.

The processing to be discussed throughout the book may be carried out via simulation on a digital computer or in special-purpose digital hardware. In later chapters we shall discuss computer structures that are especially appropriate for signal processing applications, as well as hardware organization for efficient implementation of digital systems.

2.2 Sequences

Discrete-time signals are defined only for discrete values of the independent variable, time. Generally time is quantized uniformly; i.e., $t = nT$, where T is the interval between time samples. Discrete-time signals are represented mathematically as sequences of numbers whose amplitude may take on a continuum of values. The notation that is used to describe sequences may be any of the following:

$$\{h(n)\} \qquad N_1 \leq n \leq N_2 \qquad\qquad (2.1a)$$

$$\{h(nT)\} \qquad N_1 \leq n \leq N_2 \qquad\qquad (2.1b)$$

$$h(n) \qquad N_1 \leq n \leq N_2 \qquad\qquad (2.1c)$$

$$h(nT) \qquad N_1 \leq n \leq N_2 \qquad\qquad (2.1d)$$

Expressions (2.1a) and (2.1c) may apply to nonuniformly spaced samples; whereas Expressions (2.1b) and (2.1d) explicitly assume uniform spacing in time.

A sequence may be obtained in several ways. One particularly simple manner is to generate a set of numbers and order them into a sequence. For example, the numbers $0, 1, 2, \ldots, N - 1$ form the ramp-like sequence $h(n) = n$, $0 \leq n \leq N - 1$. Another way is from some numerical recursion relation; i.e., $h(n) = h(n - 1)/2$ with initial condition $h(0) = 1$ gives the sequence $h(n) = (1/2)^n$, $0 \leq n \leq \infty$. A third way is by uniformly sampling a continuous-time waveform and using the amplitudes of the samples to form a sequence; i.e., $h(nT) = h(t)|_{t=nT}$, $-\infty < n < \infty$, where T is the sampling interval. Generally an analog-to-digital (A/D) converter is used to obtain a sequence by sampling a continuous-time waveform. [Further discussion of A/D converters and digital-to-analog (D/A) converters is given in Chapter 5.] The first two methods of obtaining a sequence are

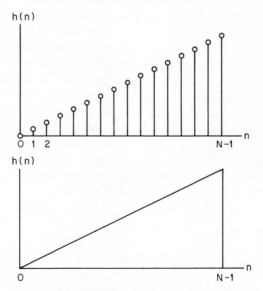

Fig. 2.1 Representations of a sequence.

independent of any time frame, whereas the third method is critically dependent on time. Thus we see that, in some sense, all the expressions of Eq. (2.1) are valid for describing sequences.

It is often useful and informative to be able to depict a particular sequence graphically. For this purpose one of the dual representations of Fig. 2.1 will be used throughout this text. [As a representative example, the sequence $h(n) = n$, $0 \leq n \leq N - 1$ is used in Fig. 2.1.] The upper figure shows a line of proper amplitude at each value of n_0 to give the sample value at $n = n_0$. In many cases it is impractical to draw each individual sample and it suffices to draw only the envelope of the sequence amplitudes as shown at the bottom of Fig. 2.1.

Some important sequences frequently used in digital signal processing are given below and sketched in Fig. 2.2. Figure 2.2(a) shows $u_0(n)$, a digital

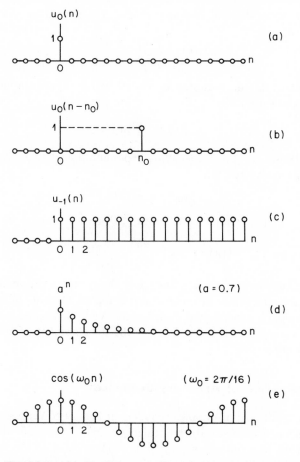

Fig. 2.2 Important sequences used in digital signal processing.

impulse or unit sample defined by the relation

$$u_0(n) = \begin{cases} 1 & n = 0 \\ 0 & n \neq 0 \end{cases} \tag{2.2}$$

This sequence plays much the same role in discrete-time systems that the analog impulse or Dirac delta function $\delta(t)$ plays in continuous-time systems. The one important difference is that the digital impulse is a plausible signal, whereas the analog impulse only exists in the sense of a generalized function or distribution. Figure 2.2(b) shows $u_0(n - n_0)$, an impulse delayed by n_0 samples and defined as

$$u_0(n - n_0) = \begin{cases} 1 & n = n_0 \\ 0 & n \neq n_0 \end{cases} \tag{2.3}$$

Figure 2.2(c) shows the unit step sequence $u_{-1}(n)$ defined as

$$u_{-1}(n) = \begin{cases} 1 & n \geq 0 \\ 0 & n < 0 \end{cases} \tag{2.4}$$

It is easily seen that the unit step is related to the digital impulse by the summation relation

$$u_{-1}(n) = \sum_{l=-\infty}^{n} u_0(l) \tag{2.5}$$

Figures 2.2(d) and (e) show a decaying exponential $g(n)$ and a sinusoid $h(n)$ defined, respectively, as

$$g(n) = \begin{cases} a^n & n \geq 0 \\ 0 & n < 0 \end{cases} \tag{2.6}$$

$$h(n) = \cos\left(\frac{2\pi n}{n_0}\right) \qquad \text{all } n \tag{2.7}$$

Another particularly important sequence is the complex exponential $e^{j\omega n} = \cos(\omega n) + j\sin(\omega n)$. Since this sequence is complex, separate plots of the real and imaginary parts are required to depict it. We shall see later that many of the sequences above play major roles in the development of the theory of digital signal processing.

2.3 Representation of Arbitrary Sequences

Using the basic digital impulse sequence it is easy to represent arbitrary sequences in terms of delayed and scaled impulses. Consider the sequence generated from the numbers $\ldots, a(0), a(1), a(2), a(3), \ldots$, where $a(n)$ represents the amplitude of the nth member of the sequence. A simple

way of representing this is

$$\{a(n)\} = \sum_{m=-\infty}^{\infty} a(m)u_0(n - m) \tag{2.8}$$

From this basic representation of arbitrary sequences and from the properties of discrete, linear, time-invariant (LTI) systems, we now derive a way of characterizing LTI systems.

2.4   Linear, Time-Invariant Systems

A discrete-time system is essentially an algorithm for converting one sequence (called the *input*) into another sequence (called the *output*). A simple representation of discrete-time systems is given in Fig. 2.3. The input

Fig. 2.3 Representation of a discrete-time system.

is called $x(n)$ and the output is called $y(n)$. The output is functionally related to the input by the relation

$$y(n) = \phi[x(n)] \tag{2.9}$$

where $\phi(\cdot)$ is determined by the specific system.

A *linear* system is defined in the following manner. If $x_1(n)$ and $x_2(n)$ are specific inputs to a linear system and $y_1(n)$ and $y_2(n)$ are the respective outputs, then if the sequence $ax_1(n) + bx_2(n)$ is applied at the input, the sequence $ay_1(n) + by_2(n)$ is obtained at the output, where a and b are arbitrary constants.

In a *time-invariant* system, if the input sequence $x(n)$ produces an output sequence $y(n)$, then the input sequence $x(n - n_0)$ produces the output sequence $y(n - n_0)$ for all n_0.

We now show that in the case of linear, time-invariant systems a convolutional relation exists between the input and output sequences. Let $x(n)$ be the input to an LTI system and $y(n)$ be the output of the system. Let $h(n)$ be the response of the system to a digital impulse. [The sequence $h(n)$ is called the *impulse response* or *unit sample response*]. From Eq. (2.8) we can write $x(n)$ as

$$x(n) = \sum_{m=-\infty}^{\infty} x(m)u_0(n - m) \tag{2.10}$$

Since $h(n)$ is the response to the sequence $u_0(n)$, then by time invariance we can say that $h(n - m)$ is the response to the sequence $u_0(n - m)$. Similarly, by linearity, the response to the sequence $x(m)u_0(n - m)$ must be

$x(m)h(n - m)$. Thus, the response to $x(n)$ must be

$$y(n) = \sum_{m=-\infty}^{\infty} x(m)h(n - m) \qquad (2.11a)$$

which is the desired convolution relation. Equivalently, by a simple change of variables, Eq. (2.11) can be converted to the form

$$y(n) = \sum_{m=-\infty}^{\infty} h(m)x(n - m) \qquad (2.11b)$$

Thus in the case of LTI systems, the sequence $h(n)$ completely characterizes the system as shown in Fig. 2.4.

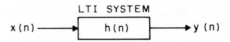

Fig. 2.4 Representation of an LTI system.

Figure 2.5 shows a simple pictorial representation of how convolution (Eq. 2.11) is carried out in practice. Figure 2.5(a) shows a typical input sequence $x(n)$ that is nonzero in the range $0 \le n \le 4$. Figure 2.5(b) shows a typical impulse response sequence $h(n)$ that is nonzero in the interval $0 \le n \le 7$. Figures 2.5(c) to (f) show simultaneous plots of $x(m)$ and $h(n - m)$ for $n = 0, 2, 10$, and 11. Clearly for $n < 0$ and for $n > 11$, there is no overlap between $x(m)$ and $h(n - m)$; therefore $y(n)$ is exactly 0. Finally Fig. 2.5(g) shows $y(n)$, which is the desired convolution.

2.5 Causality, Stability

An LTI system is said to be causal or realizable if the output at $n = n_0$ is dependent only on values of the input for $n \le n_0$. For LTI systems this implies that the impulse response $h(n)$ is zero for $n < 0$. As we shall see in Chapters 3 and 4, there are several important systems that are nonrealizable, e.g., the ideal lowpass filter or the ideal differentiator. Thus a large part of filter theory is concerned with methods of approximating nonrealizable systems by realizable ones.

An LTI system is said to be *stable* if every bounded input produces a bounded output. A necessary and sufficient condition on the impulse response for stability is

$$\sum_{n=-\infty}^{\infty} |h(n)| < \infty \qquad (2.12)$$

The necessity and sufficiency of Eq. (2.12) may be easily shown as follows.

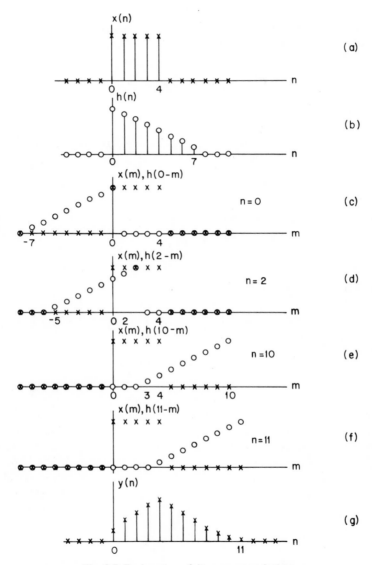

Fig. 2.5 Explanation of discrete convolution.

First we assume Eq. (2.12) is not true; i.e.,

$$\sum_{n=-\infty}^{\infty} |h(n)| = \infty \tag{2.13}$$

Consider the bounded sequence $x(n)$ defined by

$$x(n) = \begin{cases} +1 & \text{if } h(-n) \geq 0 \\ -1 & \text{if } h(-n) < 0 \end{cases} \tag{2.14}$$

Then from Eq. (2.11) the output at $n = 0$ is

$$y(0) = \sum_{m=-\infty}^{\infty} x(m)h(-m) = \sum_{m=-\infty}^{\infty} |h(-m)|$$

$$= \sum_{m=-\infty}^{\infty} |h(m)| = \infty \tag{2.15}$$

Thus, $y(0)$ is not bounded, showing Eq. (2.12) to be a necessary condition for stability. To prove sufficiency, we assume Eq. (2.12) is valid and that we have a bounded input sequence $x(n)$; i.e.,

$$|x(n)| \leq M \tag{2.16}$$

From Eq. (2.11) we find

$$|y(n)| = \left| \sum_{m=-\infty}^{\infty} x(m)h(n-m) \right|$$

$$\leq \sum_{m=-\infty}^{\infty} |x(m)| \, |h(n-m)|$$

$$\leq M \sum_{m=-\infty}^{\infty} |h(n-m)| < \infty \tag{2.17}$$

Thus, $y(n)$ is bounded, completing the proof of the stability conditions. Figures 2.6(a) and (b) show examples of the impulse response of a stable and an unstable system, respectively. The impulse response of Fig. 2.6(a) is of the form $h(n) = \alpha^n u_{-1}(n)$, with $0 < \alpha < 1$. Thus Eq. (2.12) is satisfied and the system is stable. The impulse response of Fig. 2.6(b) is of the same form as that of Fig. 2.6(a) but $\alpha > 1$. Thus Eq. (2.12) is not satisfied and the system is unstable.

2.6   Difference Equations

A subset of the class of linear time-invariant discrete systems are those where the input and output sequences $x(n)$ and $y(n)$ are related via a constant coefficient linear difference equation. The difference equation representation of LTI systems is extremely important because it often offers insights into

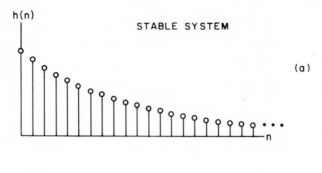

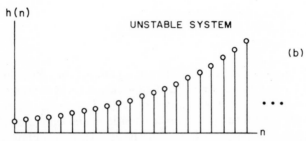

Fig. 2.6 The impulse responses of a stable and an unstable system.

efficient ways of realizing these systems. Furthermore, from the appropriate difference equations one can learn a great deal about the characteristics of the specific system under investigation, e.g., the natural frequencies of the system and their multiplicity, the order of the system, the frequencies at which there is zero transmission, etc.

The most general expression for an Mth-order linear, constant coefficient, difference equation of a causal system is of the form

$$y(n) = \sum_{i=0}^{M} b_i x(n-i) - \sum_{i=1}^{M} a_i y(n-i) \qquad n \geq 0 \qquad (2.18)$$

where $\{b_i\}$ and $\{a_i\}$ characterize the appropriate system and $a_M \neq 0$. (The system order M characterizes the mathematical properties of the difference equations in the manner shown below.) Equation (2.18) is written in an especially useful form for solution by direct substitution. Given the set of initial conditions [e.g., $x(i), y(i), i = -1, -2, \ldots, -M$] and the input sequence $x(n)$, the output sequence $y(n)$ for $n \geq 0$ may be computed directly from Eq. (2.18). For example, the difference equation

$$y(n) = x(n) - 3y(n-1) \qquad (2.19)$$

with initial condition $y(-1) = 0$ and input $x(n) = n^2 + n$ may be solved directly to give

$$
\begin{aligned}
y(0) &= x(0) - 3y(-1) = & 0 \\
y(1) &= x(1) - 3y(0) &= & 2 \\
y(2) &= x(2) - 3y(1) &= & 0 \\
y(3) &= x(3) - 3y(2) &= & 12 \\
y(4) &= x(4) - 3y(3) &= & -16 \\
y(5) &= x(5) - 3y(4) &= & 78 \\
y(6) &= x(6) - 3y(5) &= & -192
\end{aligned}
$$

.

.

.

etc.

Although direct solution of difference equations may be appropriate in some cases, it is far more useful to obtain closed form expressions for the solution. Techniques for obtaining such closed form solutions are well described in the literature on difference equations and only a brief review of these techniques will be given here. The basic idea is to obtain two sets of solutions to the difference equation, a homogeneous solution and a particular solution. The homogeneous solution is obtained by setting terms involving the input $x(n)$ to zero and finding outputs that are possible with zero inputs. It is this class of solution that essentially characterizes the specific system. The particular solution is obtained by guessing a sequence $y(n)$ that would be obtained with the given input sequence $x(n)$. The initial conditions are used to determine the arbitrary coefficients in the homogeneous solution. To illustrate this procedure we solve Eq. (2.19) in this manner. The homogeneous equation is

$$ y(n) + 3y(n - 1) = 0 \tag{2.20} $$

It is readily shown that solutions of the form $A\alpha^n$ are characteristic solutions of the homogeneous equation for linear, constant coefficient difference equations. Thus, substituting $A\alpha^n$ for $y(n)$ in Eq. (2.20) gives

$$ A\alpha^n + 3A\alpha^{n-1} = 0 $$

$$ A\alpha^{n-1}(\alpha + 3) = 0 \tag{2.21} $$

$$ \alpha = -3 $$

$$ y_h(n) = A(-3)^n $$

For the particular solution the sequence

$$ y_p(n) = Bn^2 + Cn + D \tag{2.22} $$

is guessed to be the probable output sequence in response to $x(n) = n^2 + n$.
From Eq. (2.19) the equation

$$Bn^2 + Cn + D + 3B(n-1)^2 + 3C(n-1) + 3D = n^2 + n \quad (2.23)$$

is obtained. Since coefficients in like powers of n must match, B, C, and D
are solved for as

$$B = \tfrac{1}{4}, \qquad C = \tfrac{5}{8}, \qquad D = \tfrac{9}{32} \quad (2.24)$$

Thus, the total solution is

$$y(n) = \frac{n^2}{4} + \frac{5n}{8} + \frac{9}{32} + A(-3)^n \quad (2.25)$$

From the initial condition $y(-1) = 0$, A is solved for as $A = -\tfrac{9}{32}$. The
final solution is

$$y(n) = \frac{n^2}{4} + \frac{5n}{8} + \frac{9}{32}[1 - (-3)^n] \quad (2.26)$$

A casual check of Eq. (2.26) for $n \geq 0$ shows complete agreement with the
direct solution presented earlier. The obvious advantage of Eq. (2.26) over
direct solution is that one can evaluate $y(n)$ for any particular $n = n_0$ in a
simple manner.

In digital signal processing, difference equations are important because
of the simplicity with which one can realize the system from them. Thus,
the most general first-order difference equation

$$y(n) = -a_1 y(n-1) + b_0 x(n) + b_1 x(n-1) \quad (2.27)$$

can be realized in the form shown in Fig. 2.7. The box labeled DELAY

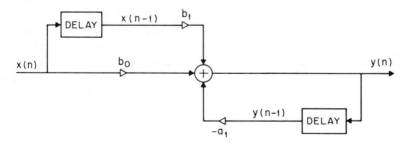

Fig. 2.7 Realization of a simple first-order difference equation.

consists of a one-sample delay. This form of realization where separate delays
are used for both the input and output has been called direct form 1. Later
we shall discuss different methods of realization for this and other digital
systems. The most general second-order difference equation, which is of the
form

$$y(n) = -a_1 y(n-1) - a_2 y(n-2) + b_0 x(n) + b_1 x(n-1) + b_2 x(n-2) \quad (2.28)$$

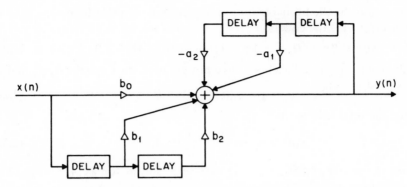

Fig. 2.8 Realization of a second-order difference equation.

can be realized as shown in Fig. 2.8. Again separate delays are used for both the input and output sequences.

As seen later in this chapter, the specific cases of first- and second-order systems are very important in the realization of higher-order systems because these higher-order systems can be decomposed into a cascade or a parallel combination of first- and second-order systems.

2.7 Frequency Response

Until now we have been concerned with the response of LTI systems to arbitrary input sequences. In this section we restrict ourselves to a special class of inputs, $x(n) = e^{j\omega n}$, in order to investigate the frequency domain representation of LTI systems. As will be shown, this class of inputs is the set of eigenfunctions of discrete-time, LTI systems; i.e., the output sequence is the input sequence multiplied by a complex weighting factor that depends solely on ω.

Consider a class of input sequences of the form

$$x(n) = e^{j\omega n} \qquad -\infty < n < \infty \qquad (2.29)$$

If this input is applied to an LTI system with impulse response $h(n)$, then from Eq. (2.11a) the output is

$$y(n) = \sum_{m=-\infty}^{\infty} h(m)e^{j\omega(n-m)} \qquad (2.30)$$

$$= e^{j\omega n} \sum_{m=-\infty}^{\infty} h(m)e^{-j\omega m} \qquad (2.31)$$

$$= x(n)H(e^{j\omega}) \qquad (2.32)$$

Thus for this special class of inputs, we see from Eq. (2.32) that the output is identical to the input to within a complex multiplier $H(e^{j\omega})$, which is defined from the impulse response as

$$H(e^{j\omega}) = \sum_{n=-\infty}^{\infty} h(n)e^{-j\omega n} \qquad (2.33)$$

Since an input of the form $e^{j\omega n}$ is functionally equivalent to a sampled sinusoid of frequency ω, the multiplier $H(e^{j\omega})$ is called the *frequency response* of the system since it gives the transmission of the LTI system for every value of ω.

As an example of a frequency response calculation, consider an LTI system with impulse response $h(n) = a^n u_{-1}(n)$ ($|a| < 1$). Then the frequency response is

$$H(e^{j\omega}) = \sum_{n=0}^{\infty} a^n e^{-j\omega n}$$

$$= \sum_{n=0}^{\infty} (ae^{-j\omega})^n \tag{2.34}$$

Since $|a| < 1$, the geometric series of Eq. (2.34) can be summed to give

$$H(e^{j\omega}) = \frac{1}{1 - ae^{-j\omega}} \tag{2.35}$$

Figure 2.9 shows plots of $h(n)$ and the magnitude and phase of $H(e^{j\omega})$ versus ω for frequencies in the range $0 \le \omega \le 2\pi$.

Several properties of the frequency response are worth noting. It is readily seen that the frequency response is a periodic function of ω with period of

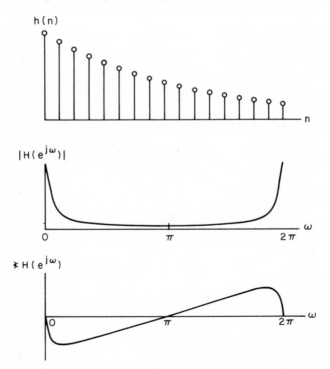

Fig. 2.9 Time and frequency responses of a first-order system.

2π. This periodicity is related to the character of sampled waveforms. Thus an input sequence of frequency $(\omega + 2m\pi)$ $(m = \pm1, \pm2,\ldots)$ is indistinguishable from an input sequence of frequency ω; i.e.,

$$\hat{x}(n) = e^{j(\omega+2m\pi)n} = e^{j\omega n} = x(n) \tag{2.36}$$

Since $H(e^{j\omega})$ is periodic, any interval of length 2π is sufficient to completely describe this function. Generally the interval $0 \leq \omega < 2\pi$ is used.

Another important property of the frequency response is that in the case of real $h(n)$ (as usually is the case) the magnitude of $H(e^{j\omega})$ is symmetric and the phase of $H(e^{j\omega})$ is antisymmetric over the interval $0 \leq \omega < 2\pi$. Similarly, the real part of $H(e^{j\omega})$ is symmetric and the imaginary part of $H(e^{j\omega})$ is antisymmetric over the same interval. Thus the interval of ω of interest is generally further reduced to $0 \leq \omega \leq \pi$ for real impulse responses.

2.8 Frequency Response of First-Order Systems

Consider the difference equation of a first-order system

$$y(n) = x(n) + Ky(n - 1) \tag{2.37}$$

with initial condition $y(-1) = 0$. Then $h(n)$, the impulse response, is readily obtained as

$$h(n) = \begin{cases} K^n & n \geq 0 \\ 0 & n < 0 \end{cases} \tag{2.38}$$

Thus using Eq. (2.33) the frequency response of the first-order system is

$$H(e^{j\omega}) = \frac{1}{1 - Ke^{-j\omega}} \tag{2.39}$$

Representing $H(e^{j\omega})$ as

$$H(e^{j\omega}) = |H(e^{j\omega})|\, e^{j \sphericalangle H(e^{j\omega})} \tag{2.40}$$

gives

$$|H(e^{j\omega})| = \frac{1}{(1 + K^2 - 2K \cos \omega)^{1/2}} \tag{2.41}$$

$$\sphericalangle H(e^{j\omega}) = \omega - \tan^{-1}\left(\frac{\sin \omega}{\cos \omega - K}\right) \tag{2.42}$$

Figure 2.10 shows plots of $\log |H(e^{j\omega})|$ and $\sphericalangle H(e^{j\omega})$ for various values of K. In all cases, $H(e^{j\omega})$ is a lowpass characteristic. In Sec. 2.18, we see how $|H(e^{j\omega})|$ and $\sphericalangle H(e^{j\omega})$ can be obtained from geometric considerations.

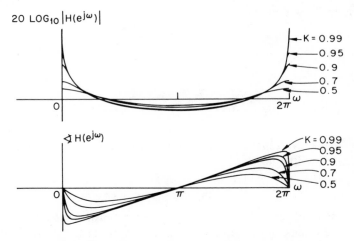

Fig. 2.10 Frequency responses of several first-order systems.

2.9   Frequency Response of Second-Order Systems

The difference equation of a second-order system may be written in the form

$$y(n) = x(n) + a_1 y(n-1) + a_2 y(n-2) \qquad (2.43)$$

[A more general second-order section includes terms of the form $b_1 x(n-1)$ and $b_2 x(n-2)$ but for the sake of simplicity we shall omit such terms.] If we again assume the initial conditions $y(-1) = 0$ and $y(-2) = 0$, then the impulse response is readily shown to be one of two types. The impulse response may be either of the form[1]

$$h(n) = \alpha_1 (p_1)^n + \alpha_2 (p_2)^n \qquad \text{(I)} \qquad (2.44)$$

where p_1 and p_2 are real, or of the form

$$h(n) = \alpha_1 r^n \sin(bn + \varphi) \qquad \text{(II)} \qquad (2.45)$$

The impulse response of Eq. (2.44) represents two first-order systems and thus the impulse response decays as p_1^n and p_2^n. The impulse response of Eq. (2.45) represents a second-order system in which the impulse response is a damped sinusoid. The necessary condition on the coefficients of the difference equation [Eq. (2.43)] for the impulse response to be a damped sinusoid is

$$a_2 < -\frac{a_1^2}{4} \qquad (2.46)$$

---

[1] We are assuming that the roots of the homogeneous equation are distinct. Trivial modifications can be made if this is not the case.

which implies $a_2 < 0$. In the case of when the condition of Eq. (2.46) holds, it is easy to show that

$$r = \sqrt{-a_2} \tag{2.47}$$

$$\cos b = \frac{a_1}{2\sqrt{-a_2}} \tag{2.48}$$

$$\varphi = b \tag{2.49}$$

$$\alpha_1 = \frac{1}{\sin b} \tag{2.50}$$

The frequency response corresponding to the impulse response of Eq. (2.45) can be written as

$$H(e^{j\omega}) = \frac{1}{1 - 2r(\cos b)e^{-j\omega} + r^2 e^{-2j\omega}} \tag{2.51}$$

The log magnitude and phase responses of second-order systems corresponding to a fixed value of b $(\pi/4)$ and varying r are shown in Fig. 2.11. From these plots it is clear that a second-order system represents a simple digital resonator.

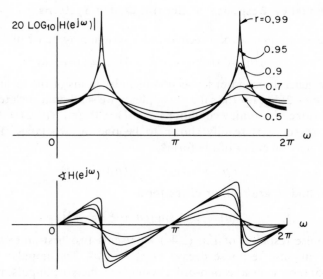

Fig. 2.11 Frequency responses of several second-order systems.

2.10   Discrete Fourier Series

Since the frequency response is a periodic function of ω, Eq. (2.33) can be viewed as a Fourier series representation of $H(e^{j\omega})$ where the impulse response coefficients are also the Fourier series coefficients. From the theory of Fourier series the coefficients $h(n)$ can be derived from $H(e^{j\omega})$ as

$$h(n) = \frac{1}{2\pi} \int_{-\pi}^{\pi} H(e^{j\omega})e^{j\omega n} \, d\omega \tag{2.52}$$

Equations (2.33) and (2.52) thus form a discrete Fourier series pair of relations. Equation (2.52) shows that $h(n)$ can be considered to be essentially a superposition of sinusoids $e^{j\omega n}$ of amplitude $H(e^{j\omega n})$. Since the relations of Eq. (2.33) and (2.52) are valid for any sequence where Eq. (2.33) can be summed, an arbitrary input sequence can also be represented in the form

$$x(n) = \frac{1}{2\pi} \int_{-\pi}^{\pi} X(e^{j\omega})e^{j\omega n} \, d\omega \tag{2.53}$$

where

$$X(e^{j\omega}) = \sum_{n=-\infty}^{\infty} x(n)e^{-j\omega n} \tag{2.54}$$

Since the response to the input $e^{j\omega n}$ is $H(e^{j\omega})e^{j\omega n}$ [from Eq. (2.31) and (2.32)], it is easy to determine the response $y(n)$ to Eq. (2.53) as

$$y(n) = \frac{1}{2\pi} \int_{-\pi}^{\pi} X(e^{j\omega})H(e^{j\omega})e^{j\omega n} \, d\omega \tag{2.55}$$

where we have used linearity to sum up the individual responses. Using the relation

$$Y(e^{j\omega}) = X(e^{j\omega})H(e^{j\omega}) \tag{2.56}$$

we see that Eq. (2.55) is part of a Fourier series pair for $y(n)$. Therefore, as we have just shown, convolution in the time domain is converted to multiplication in the frequency domain. Thus even though $H(e^{j\omega})$ represents the response of the system to the restricted set of inputs $e^{j\omega n}$, $0 \le \omega < 2\pi$, Eq. (2.53) shows that an arbitrary input sequence $x(n)$ may be represented as a continuous superposition of these inputs. In this way, the frequency response is an important indicator of the system's response to almost any input sequence.

2.11 Comments on Units of Frequency

It is often desirable to express the frequency response of a sequence $h(nT)$ in terms of units of frequency that involve the sampling interval T. In this case Eq. (2.33) and (2.52) are modified to the form

$$H(e^{j\omega T}) = \sum_{n=-\infty}^{\infty} h(nT)e^{-j\omega n T} \tag{2.57}$$

$$h(nT) = \frac{T}{2\pi} \int_{-\pi/T}^{\pi/T} H(e^{j\omega T})e^{j\omega n T} \, d\omega \tag{2.58}$$

$H(e^{j\omega T})$ is periodic in ω with a period of $(2\pi/T)$. For Eq. (2.57) and (2.58) the units of ω are radians per second. Alternatively, one can replace ω by $2\pi f$ to express the frequency response as a function of f where the units of f are hertz.

As an example, if $T = 0.0001$ sec $(1/T = $ sampling frequency $= 10,000$ Hz), then $H(e^{j2\pi fT})$ is periodic in f with a period of $10,000$ Hz, and $H(e^{j\omega T})$ is periodic in ω with a period of $20,000\pi$ rad/sec. An example of a typical frequency response for a real input sequence with this sampling interval is shown in Fig. 2.12. Since the input is real, the frequency response has the symmetry properties discussed earlier.

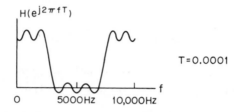

Fig. 2.12 Frequency response for a 10 kHz system.

2.12 Relation Between Continuous and Discrete Systems

As discussed earlier, the sequence $x(nT)$ is often derived by sampling a continuous-time waveform $x(t)$ once every T seconds. In such cases it is important to be able to relate the frequency response of the sequence $X(e^{j\omega T})$ to the Fourier transform $X_A(j\Omega)$ of the continuous-time waveform. In this section we shall outline the necessary steps for arriving at the result and discuss its implications.

The Fourier transform relations for the continuous-time waveform $x(t)$ are

$$X_A(j\Omega) = \int_{-\infty}^{\infty} x(t)e^{-j\Omega t}\,dt \tag{2.59}$$

$$x(t) = \frac{1}{2\pi} \int_{-\infty}^{\infty} X_A(j\Omega)e^{j\Omega t}\,d\Omega \tag{2.60}$$

For the discrete-time waveform the necessary transform relations are

$$X(e^{j\omega T}) = \sum_{n=-\infty}^{\infty} x(nT)e^{-j\omega nT} \tag{2.61}$$

$$x(nT) = \frac{T}{2\pi} \int_{-\pi/T}^{\pi/T} X(e^{j\omega T})e^{j\omega nT}\,d\omega \tag{2.62}$$

Since $x(nT) = x(t)|_{t=nT}$, it is possible to relate $X_A(j\Omega)$ and $X(e^{j\omega T})$ by evaluating Eq. (2.60) at $t = nT$ and performing the infinite integral as an infinite sum of pieces of the integral, each of width $2\pi/T$; i.e.,

$$x(nT) = \frac{1}{2\pi} \sum_{m=-\infty}^{\infty} \int_{(2m-1)\pi/T}^{(2m+1)\pi/T} X_A(j\Omega)e^{j\Omega n T}\, d\Omega \qquad (2.63)$$

Simple manipulation of the integrals in Eq. (2.63) and substituting ω for Ω yields the result

$$x(nT) = \frac{T}{2\pi} \int_{-\pi/T}^{\pi/T} \left[\frac{1}{T} \sum_{m=-\infty}^{\infty} X_A\left(\omega + \frac{2\pi}{T} m \right) \right] e^{j\omega n T}\, d\omega \qquad (2.64)$$

Equating the term inside the brackets of Eq. (2.64) to Eq. (2.62) yields the desired relation

$$X(e^{j\omega T}) = \frac{1}{T} \sum_{m=-\infty}^{\infty} X_A\left(\omega + \frac{2\pi}{T} m \right) \qquad (2.65)$$

Equation (2.65) shows the periodic frequency response of the sequences to consist of a sum of an infinite number of components of the frequency response of the analog waveform. In the case where the analog frequency response is bandlimited to the range $|\Omega| \leq \pi/T$, i.e., $X_A(j\Omega) = 0$, $|\Omega| > \pi/T$, then Eq. (2.65) shows that in the frequency range $|\omega| \leq \pi/T$

$$X(e^{j\omega T}) = \frac{1}{T} X_A(\omega) \qquad (2.66)$$

In this case the digital frequency response is related in a straightforward manner to the analog frequency response. Figures 2.13(a) and (b) illustrate this case. When $X_A(j\Omega)$ is not bandlimited to the range $|\Omega| \leq \pi/T$, then the

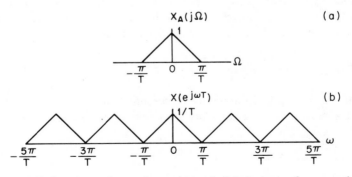

Fig. 2.13 Sampling relations for analog and digital systems for properly sampled inputs.

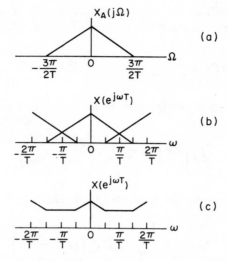

Fig. 2.14 The effects of undersampling on the digital frequency response.

relation between the digital and analog frequency responses is not so straight-forward. Figures 2.14(a), (b), and (c) illustrate one such typical case. In Fig. 2.14(a) it is seen that the analog frequency response is bandlimited to the range $|\Omega| \leq 3\pi/(2T)$. Using Eq. (2.65) we see that the terms for $m = 0$, ± 1 all enter into determining $X(e^{j\omega T})$ in the range $|\omega| \leq \pi/T$, as shown in Fig. 2.14(b). Figure 2.14(c) shows the resulting digital frequency response, which is clearly not simply related to the analog frequency response, as in the previous example. The problem here lies in the fact that the sampling rate $1/T$ was not high enough to prevent the shifting of high-frequency information in $X_A(j\Omega)$ into lower frequencies in $X(e^{j\omega T})$. This shifting of information from one band of frequencies to another one is called *aliasing*. The sequence corresponding to the frequency response of Fig. 2.14(c) is said to be an aliased representation of $x(t)$. Clearly the way to avoid aliasing is to sample the continuous-time waveform at a sufficiently high rate.

2.13   The z Transform

One of the most useful techniques for representing and manipulating sequences is the z transform. Given a sequence $x(n)$, defined for all n, its z transform is defined as

$$X(z) = \sum_{n=-\infty}^{\infty} x(n)z^{-n} \tag{2.67}$$

where z is a complex variable. Clearly, the complex function of Eq. (2.67) is defined only for certain values of z for which the power series converges. Since a detailed discussion of convergence is beyond the scope of this book (and is readily available in several texts), we shall state the general results for cases of interest throughout this book.

1. Finite Duration Sequences

In the case where $x(n)$ is nonzero only in the interval $N_1 \leq n \leq N_2$ $(N_1 < N_2)$ where N_1 and N_2 are finite, then $X(z)$ converges everywhere in the z plane except possibly $z = 0$ or $z = \infty$. If a finite duration sequence is the impulse response of an LTI system, the system is called a *finite impulse response* (FIR) system or, equivalently, an FIR filter. We shall see in Chapter 3 that an important class of filter design procedures is based on finite duration sequences.

h(n)

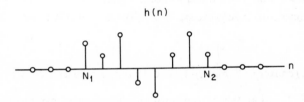

Fig. 2.15 A finite duration sequence.

Figure 2.15 shows a typical FIR sequence $\{h(n)\}$. If every element of $\{h(n)\}$ is finite, it is easy to see that an LTI system with $h(n)$ as its impulse response is *always* stable since the stability test [Eq. (2.12)] involves a finite sum of finite numbers. Furthermore such a system can always be made realizable by delaying the impulse response by an appropriate amount, i.e., $-N_1$ samples if N_1 is less than zero.

In contradistinction to the definition above, a system (filter) is called an *infinite impulse response* (IIR) system (filter) if its impulse response extends to either minus infinity (i.e., $N_1 = -\infty$) or plus infinity (i.e., $N_2 = \infty$) or both. IIR sequences form the basis for another whole class of filter design techniques as shown in Chapter 4.

2. Causal Sequences

In the case when $x(n)$ is nonzero only in the region $0 \leq N_1 \leq n < \infty$, then $X(z)$ converges everywhere *outside* a circle of radius R_1. The value of R_1 depends on the locations of the singularities of $X(z)$ (these singularities are called the *poles* of the system). If $R_1 < 1$, we shall see that the system being represented is stable. Causal sequences are important in that they form the basis for most physical systems.

3. Noncausal Sequences

In the case when $x(n)$ is nonzero in the region $-\infty < n < N_1 \leq 0$, then $X(z)$ converges everywhere *inside* a circle of radius R_1, where R_1 again depends on the singularities of $X(z)$. Noncausal sequences are not generally of practical interest but are useful in several theoretical problems.

At this point it is worthwhile evaluating the z transform of several useful sequences.

EXAMPLE 1.   Find the z transform of a digital impulse.

Solution: Since $x(n)$ is zero except for $n = 0$ where $x(n)$ is 1, we find
$$X(z) = 1$$
Since a digital impulse is a finite duration sequence, $X(z)$ converges everywhere in the z plane.

EXAMPLE 2.   Find the z transform of a digital step.

Solution: Since $x(n)$ is zero except for $n \geq 0$ where $x(n)$ is 1, we find
$$X(z) = \sum_{n=0}^{\infty} z^{-n} = \frac{1}{1 - z^{-1}}$$
where $X(z)$ converges for $|z| > 1$ since the only singularity of $X(z)$ is at $z = 1$.

EXAMPLE 3.   Find the z transform of the complex exponential $x(n) = 0$, $n < 0$; $x(n) = e^{jn\omega}$, $n \geq 0$.

Solution: Solving for the z transform, we find
$$X(z) = \sum_{n=0}^{\infty} e^{jn\omega} z^{-n}$$
$$= \sum_{n=0}^{\infty} (z^{-1} e^{j\omega})^n = \frac{1}{1 - z^{-1} e^{j\omega}}$$
where $X(z)$ converges for $|z| > 1$ since the only singularity of $X(z)$ is at $z = e^{j\omega}$.

EXAMPLE 4.   Find the z transform of the simple exponential sequence $x(n) = 0$, $n < 0$; $x(n) = a^n$, $n \geq 0$.

Solution: Solving for the z transform, we find
$$X(z) = \sum_{n=0}^{\infty} a^n z^{-n} = \sum_{n=0}^{\infty} (az^{-1})^n = \frac{1}{1 - az^{-1}}$$
where $X(z)$ converges for $|z| > a$ since the only singularity of $X(z)$ is at $z = a$.

2.14   Relation Between the z Transform and the Fourier Transform of a Sequence

The z transform of a sequence may be viewed as a unique representation of that sequence in the complex z plane. From Eq. (2.67) we see that if the

z transform is evaluated on a circle of unit radius, i.e., $z = e^{j\omega}$, then we find

$$X(z)|_{z=e^{j\omega}} = X(e^{j\omega}) = \sum_{n=-\infty}^{\infty} x(n)e^{-j\omega n} \qquad (2.68)$$

which is the Fourier transform of the sequence. It will also be shown that in the case where all singularities of $X(z)$ are inside the unit circle, the system whose impulse response is represented by the given sequence is stable. For these reasons the unit circle in the z plane plays a very definite role. For example, there are many important unrealizable systems, such as the ideal lowpass filter and the ideal differentiator, whose z transforms converge *only* on the unit circle; i.e., they have only Fourier transforms and no z transforms.

A typical way of displaying the information that the z transform contains is in terms of the singularities (poles) and the zeros of $X(z)$. Thus, for example, the z transform of Example 4 might be shown as in Fig. 2.16(a)

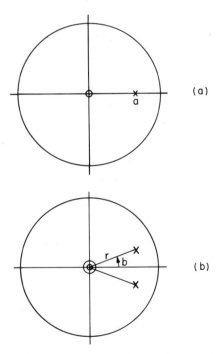

Fig. 2.16 Pole and zero locations for first- and second-order systems.

where an x marks a pole and O, a zero. With the additional assumption of causality, one can uniquely reconstruct the z transform (to within a constant multiplier) from such a pole–zero plot.

EXAMPLE 5.   Find the z transform of the sequence that is represented by the impulse response

$$h(n) = \begin{cases} \dfrac{r^n \sin [(n + 1)b]}{\sin b} & n \geq 0 \\ 0 & n < 0 \end{cases}$$

Solution: Using the definition of the z transform we find

$$H(z) = \sum_{n=0}^{\infty} r^n \left\{ \frac{\sin [(n + 1)b]}{\sin b} \right\} z^{-n}$$

$$= \sum_{n=0}^{\infty} \frac{r^n z^{-n}}{\sin b} \left[\frac{e^{jb(n+1)} - e^{-jb(n+1)}}{2j} \right]$$

$$= \sum_{n=0}^{\infty} (rz^{-1}e^{jb})^n \frac{e^{jb}}{2j \sin b} - \sum_{n=0}^{\infty} \frac{(rz^{-1}e^{-jb})^n e^{-jb}}{2j \sin b}$$

$$= \frac{1}{2j \sin b} \left(\frac{e^{jb}}{1 - rz^{-1}e^{jb}} - \frac{e^{-jb}}{1 - rz^{-1}e^{-jb}} \right)$$

$$H(z) = \frac{1}{1 - 2r (\cos b)z^{-1} + r^2 z^{-2}}$$

which converges for $|z| > r$. The z-plane pole–zero plot for this resonator is shown in Fig. 2.16(b) to consist of a complex conjugate pair of poles at $z = re^{\pm jb}$ and a double zero at $z = 0$.

As mentioned above, one can essentially reconstruct $X(z)$ from knowledge of the positions of the poles and zeros of $X(z)$. Thus if we know that $X(z)$ has N poles at $z = p_1, p_2, \ldots, p_N$ and M zeros at $z = z_1, z_2, \ldots, z_M$, then we can write $X(z)$ in factored form as

$$X(z) = A \frac{\displaystyle\prod_{i=1}^{M} (1 - z_i z^{-1})}{\displaystyle\prod_{i=1}^{N} (1 - p_i z^{-1})} \tag{2.69}$$

where A is an arbitrary constant. By multiplying the individual factors together, it is clear that a fairly general form for $X(z)$ is a rational fraction in z—i.e.,

$$X(z) = \frac{\displaystyle\sum_{i=0}^{M} a_i z^{-i}}{1 + \displaystyle\sum_{i=1}^{N} b_i z^{-i}} \tag{2.70}$$

The form of Eq. (2.70) is often used for general filter design problems.

2.15   The Inverse z Transform

It is important to be able to go from a sequence to its z transform and back to the sequence. The mechanism for transforming back to a sequence is called the *inverse z* transform, which may be formally given as

$$x(n) = \frac{1}{2\pi j} \oint_{C_1} X(z) z^{n-1} \, dz \tag{2.71}$$

Equation (2.71) represents a contour integral in the z plane over any closed path in the region of convergence that encompasses the origin in the z plane. For example, the path of integration could be a circle of radius $C_1 > R_1$, the radius of convergence of the z transform [i.e., we are assuming that $x(n)$ is causal].

The inverse z transform may be practically evaluated in several ways including

1. Direct evaluation of Eq. (2.71) using the residue theorem.
2. Partial fraction expansion of $X(z)$.
3. Long division.
4. Power series expansion.

For method 1 we can use the theorem of complex variable theory that states that a contour integral [Eq. (2.71)] can be evaluated directly as

$$x(n) = \sum [\text{residues of } X(z) z^{n-1} \text{ inside } C_1] \tag{2.72}$$

Consider Example 4 on p. 30 where $X(z) = 1/(1 - az^{-1})$. From Eq. (2.72), for $n \geq 0$, we obtain $x(n) = $ residue of $z^{n-1}/(1 - az^{-1})$ at $z = a$ or $x(n) = a^n$, $n \geq 0$. For $n < 0$ there is a multiple-order pole at $z = 0$. Direct evaluation of the residue in the poles at $z = 0$ shows $x(n) = 0$ for $n < 0$.

For method 2 we take a z transform in the form of Eq. (2.69) and write it as a summation of the form

$$X(z) = \sum_{i=1}^{N} \frac{\alpha_i}{1 - p_i z^{-1}} \tag{2.73}$$

Recognizing that each factor $\alpha_i/(1 - p_i z^{-1})$ has inverse z transform $\alpha_i(p_i)^n$, we obtain

$$x(n) = \begin{cases} \sum_{i=1}^{N} \alpha_i(p_i)^n & n \geq 0 \\ 0 & n < 0 \end{cases} \tag{2.74}$$

Methods 3 and 4 will not be considered here. Instead the reader is referred to the various references on the z transform.

2.16 Properties of the z Transform

The z transform serves many useful purposes in discrete-time, LTI systems. To make full use of the power of z transforms one must understand several of these properties.

The basic properties of z transforms to be discussed include

1. Linearity.
2. Delay.
3. Convolution of sequences.
4. Multiplication of sequences.
5. Delay of causal sequences.

1. Linearity

The z transform is linear; i.e., if $X_1(z)$ is the z transform of $x_1(n)$ and $X_2(z)$ is the z transform of $x_2(n)$, then the z transform of $ax_1(n) + bx_2(n)$ is $aX_1(z) + bX_2(z)$ for all real a and b.

2. Delays

If a sequence $x_1(n)$ has z transform $X_1(z)$, then the sequence $x_1(n - n_0)$ has the z transform $z^{-n_0}X_1(z)$ for all n_0. This property of z transforms is particularly useful for converting between a difference equation representation of an LTI system and its z transform representation. For example, the difference equation

$$y(n) = x(n) - b_1 y(n - 1) - b_2 y(n - 2) \qquad (2.75)$$

has a z transform representation

$$Y(z) = X(z) - b_1 z^{-1} Y(z) - b_2 z^{-2} Y(z) \qquad (2.76)$$

or

$$Y(z) = \frac{X(z)}{1 + b_1 z^{-1} + b_2 z^{-2}} \qquad (2.77)$$

where

$$Y(z) = \sum_{n=-\infty}^{\infty} y(n)z^{-n}$$

$$X(z) = \sum_{n=-\infty}^{\infty} x(n)z^{-n}$$

We have used the delay properties above to express the z transforms of $y(n - 1)$ and $y(n - 2)$ in terms of the z transform of $y(n)$.

3. Convolution of Sequences

If $x(n)$ is the input to a discrete-time LTI system with impulse response $h(n)$ and $y(n)$ is the output, then

$$Y(z) = X(z)H(z) \qquad (2.78)$$

where $X(z)$, $H(z)$, and $Y(z)$ are the respective z transforms of $x(n)$, $h(n)$, and $y(n)$. Thus convolution of sequences is converted to multiplication of their

z transforms. It is easily seen that $H(z)$ can be defined from Eq. (2.78) as

$$H(z) = \frac{Y(z)}{X(z)} \qquad (2.79)$$

Thus in the example of Eq. (2.75), it is clear that one can obtain $H(z)$ or, equivalently, $h(n)$ from the difference equation of the system and vice versa. For the system of Eq. (2.75), $H(z)$ is of the form

$$H(z) = \frac{1}{1 + b_1 z^{-1} + b_2 z^{-2}} \qquad (2.80)$$

The importance of Eq. (2.78) as a practical tool for determining the output of a system from the input and its impulse response, without having to perform a convolution, should not be underestimated. By multiplying two transforms and inverse transforming the result to give $y(n)$, one can often reduce a complex problem to a simple one. As an example, consider the input sequence $x(n) = u_{-1}(n)a^n$ applied to a LTI system with impulse response $h(n) = u_{-1}(n)b^n$. The z transforms of $x(n)$ and $h(n)$ are

$$X(z) = \frac{1}{1 - az^{-1}} \qquad |z| > a$$

$$H(z) = \frac{1}{1 - bz^{-1}} \qquad |z| > b$$

Multiplication of $X(z)$ by $H(z)$ gives

$$Y(z) = \frac{1}{(1 - az^{-1})(1 - bz^{-1})} \qquad |z| > \max [a, b]$$

Assuming $a \neq b$, we can do a partial fraction expansion of $Y(z)$ to give

$$Y(z) = \frac{-a/(b - a)}{1 - az^{-1}} + \frac{b/(b - a)}{1 - bz^{-1}}$$

From Eq. (2.72), $y(n)$ is obtained as

$$y(n) = \left(\frac{-a}{b - a} a^n + \frac{b}{b - a} b^n \right) u_{-1}(n)$$

4. Multiplication of Sequences

If $x_1(n)$ and $x_2(n)$ are two sequences with z transforms $X_1(z)$ and $X_2(z)$, then the sequence $x_3(n) = x_1(n) \cdot x_2(n)$ has a z transform $X_3(z)$ defined as

$$X_3(z) = \frac{1}{2\pi j} \oint_C X_1(v) X_2 \left(\frac{z}{v} \right) v^{-1} \, dv \qquad (2.81)$$

The convergence regions of $X_3(z)$ consists of all z such that if v is in the convergence region of $X_1(z)$, then z/v is in the convergence region of $X_2(z)$.

The contour of integration for Eq. (2.81) is a closed contour inside the overlap convergence regions for $X_1(v)$ and $X_2(z/v)$.

Equation (2.81) is called the *complex convolution theorem* because it expresses the z transform of a product $[x_1(n) \cdot x_2(n)]$ as a complex convolution of the z transform of the respective sequences. By making the substitutions $z = e^{j\omega}$, $v = e^{j\theta}$ in Eq. (2.81), we can evaluate the Fourier transform of the product in terms of the individual Fourier transforms. The resulting relation is

$$X_3(e^{j\omega}) = \frac{1}{2\pi} \int_{-\pi}^{\pi} X_1(e^{j\theta}) X_2(e^{j(\omega-\theta)}) \, d\theta \tag{2.82}$$

Equation (2.82) is the more familiar form of a convolution of two Fourier transforms. It will be important in the discussion of filter design by windowing techniques as well as in understanding various modulation systems.

An important application of Eq. (2.81) is the so-called Parseval equations, which relate the "energy" in a signal to the "energy" in its spectrum. A "generalized" form of Parseval relations can be obtained by defining $y(n)$ as

$$y(n) = x(n)w^*(n)$$

From Eq. (2.81) is is seen that $Y(z)$, the z transform of $y(n)$, is

$$Y(z) = \frac{1}{2\pi j} \oint_C X(v) W^*\left(\frac{z^*}{v^*}\right) v^{-1} \, dv$$

Evaluating $Y(z)$ at $z = 1$ gives

$$Y(z)|_{z=1} = \sum_{n=-\infty}^{\infty} y(n) = \sum_{n=-\infty}^{\infty} x(n)w^*(n) = \frac{1}{2\pi j} \oint_C X(v) W\left(\frac{1}{v^*}\right) \frac{dv}{v}$$

Finally letting the contour of integration be the unit circle (i.e., letting $v = e^{j\omega}$) gives

$$\sum_{n=-\infty}^{\infty} x(n)w^*(n) = \frac{1}{2\pi} \int_{-\pi}^{\pi} X(e^{j\omega}) W^*(e^{j\omega}) \, d\omega$$

For the case when $w(n) = x(n)$, we find the important special case

$$\sum_{n=-\infty}^{\infty} |x(n)|^2 = \frac{1}{2\pi} \int_{-\pi}^{\pi} |X(e^{j\omega})|^2 \, d\omega$$

which is generally referred to as the Parseval relation.

5. Delay of Causal Sequences—The One-Sided z Transform

Since we are generally dealing with causal sequences in most applications, it is often useful to define a "one-sided" z transform as

$$X(z) = \sum_{n=0}^{\infty} x(n)z^{-n} \tag{2.83}$$

where one takes the attitude that the behavior of $x(n)$ prior to $n = 0$ is unknown and hence can be neglected. For many sequences the properties of the one-sided z transform are identical to the properties of the normal z transform. The major exception is the property related to delayed sequences. Consider, for example, a sequence $x_1(n)$ with one-sided z transform $X_1(z)$ and a delayed sequence $x_2(n) = x_1(n - 1)$. The one-sided z transform of $x_2(n)$ may be written as

$$X_2(z) = \sum_{n=0}^{\infty} x_2(n)z^{-n} = \sum_{n=0}^{\infty} x_1(n - 1)z^{-n} \tag{2.84}$$

Letting $m = n - 1$, we obtain

$$X_2(z) = \sum_{m=-1}^{\infty} x_1(m)z^{-m}z^{-1} \tag{2.85}$$

which can be written as

$$X_2(z) = z^{-1}\left[x_1(-1)z + \sum_{m=0}^{\infty} x_1(m)z^{-m} \right] \tag{2.86}$$

$$= z^{-1}[X_1(z)] + x_1(-1) \tag{2.87}$$

The effect of a single delay is still to multiply the one-sided z transform by z^{-1}; however, account must be made of possible values of $x_1(n)$ for $n < 0$; i.e., initial conditions are important.

As a second example, if $x_3(n) = x_1(n - 2)$, then

$$X_3(z) = z^{-2}[X_1(z)] + x_1(-2) + x_1(-1)z^{-1} \tag{2.88}$$

From Eq. (2.87) and (2.88) one can infer the resulting equations for an arbitrary delay of n_0 samples ($n_0 > 0$) as

$$Y(z) = z^{-n_0}X_1(z) + x_1(-n_0) + x_1(-n_0 + 1)z^{-1} + \cdots + x_1(-1)z^{-(n_0-1)} \tag{2.89}$$

where

$$y(n) = x(n - n_0)$$

2.17 Solution of Difference Equations Using the One-Sided z Transform

Since difference equations are generally defined for $n \geq 0$, subject to prescribed initial conditions, it is fairly easy to see how the one-sided z transform can be utilized to solve for the output sequence given the appropriate input sequence. As an example, consider the first-order difference equation

$$y(n) = x(n) + ay(n - 1) \tag{2.90}$$

with initial condition $y(-1) = K$. Let the input be $x(n) = e^{j\omega n}u_{-1}(n)$. The one-sided z transform of Eq. (2.90) is obtained by multiplying both sides of

Eq. (2.90) by z^{-n} and summing from 0 to ∞ to give

$$\sum_{n=0}^{\infty} y(n)z^{-n} = \sum_{n=0}^{\infty} x(n)z^{-n} + a \sum_{n=0}^{\infty} y(n-1)z^{-n} \qquad (2.91)$$

Using the delay property of Sec. 2.16 gives

$$Y(z) = X(z) + az^{-1}Y(z) + ay(-1) \qquad (2.92)$$

or solving for $Y(z)$

$$Y(z) = \frac{X(z) + ay(-1)}{1 - az^{-1}} \qquad (2.93)$$

Since

$$x(n) = e^{j\omega n}, \qquad X(z) = \frac{1}{1 - e^{j\omega}z^{-1}}$$

and

$$Y(z) = \frac{aK}{1 - az^{-1}} + \frac{1}{(1 - az^{-1})(1 - e^{j\omega}z^{-1})} \qquad (2.94)$$

Making a partial fraction expansion of the second term of Eq. (2.94) gives

$$Y(z) = \frac{aK}{1 - az^{-1}} + \frac{a/(a - e^{j\omega})}{1 - az^{-1}} + \frac{-e^{j\omega}/(a - e^{j\omega})}{1 - e^{j\omega}z^{-1}} \qquad (2.95)$$

Inverse z transforming Eq. (2.95) gives

$$y(n) = \left[a^{n+1}K + \frac{a^{n+1}}{a - e^{j\omega}} - \frac{e^{j\omega(n+1)}}{a - e^{j\omega}} \right] u_{-1}(n) \qquad (2.96)$$

The first term within the brackets represents the response due to initial conditions and the second term is the transient response of the system. Both these terms decay exponentially for $a < 1$. The third term represents the system's forced response to the input.

To generalize the discussion above to higher-order systems is straightforward. The general Lth-order difference equation may be written as

$$y(n) = \sum_{i=0}^{L} a_i x(n-i) - \sum_{i=1}^{L} b_i y(n-i) \qquad (2.97)$$

with initial conditions $\{y(-1), y(-2), \ldots, y(-N)\}$. [*Note:* We are assuming $x(n) = 0$, $n < 0$.] Taking the one-sided z transform of Eq. (2.97) gives

$$Y(z) = \sum_{i=0}^{L} a_i z^{-i} X(z)$$

$$- \sum_{i=1}^{L} b_i [z^{-i}Y(z) + y(-i) + y(-i+1)z^{-1} + \ldots + y(-1)z^{-(i-1)}] \qquad (2.98)$$

One can then solve for $Y(z)$ in terms of $X(z)$ and the initial conditions and obtain $y(n)$ by the inverse z transform.

2.18   Geometric Evaluation of the Fourier Transform

As discussed earlier [Eq. (2.69)], the z transform of a sequence can always be written as a ratio of a product of factors representing the zeros of $X(z)$ and a product of factors representing the poles of $X(z)$. Thus $X(z)$ assumes the form

$$X(z) = A \frac{\displaystyle\prod_{i=1}^{M}(1 - z_i z^{-1})}{\displaystyle\prod_{i=1}^{N}(1 - p_i z^{-1})} \tag{2.99}$$

The Fourier transform or system function of the sequence may be obtained by evaluating $X(z)$ on the unit circle, i.e., for $z = e^{j\omega}$. Thus

$$X(e^{j\omega}) = A \frac{\displaystyle\prod_{i=1}^{M}(1 - z_i e^{-j\omega})}{\displaystyle\prod_{i=1}^{N}(1 - p_i e^{-j\omega})} \tag{2.100}$$

Writing the complex quantity $X(e^{j\omega})$ as $|X(e^{j\omega})|e^{j \angle X(e^{j\omega})}$, we find

$$|X(e^{j\omega})| = |A|\frac{\displaystyle\prod_{i=1}^{M}|1 - z_i e^{-j\omega}|}{\displaystyle\prod_{i=1}^{N}|1 - p_i e^{-j\omega}|} = |A|\frac{\displaystyle\prod_{i=1}^{M}|e^{j\omega} - z_i|}{\displaystyle\prod_{i=1}^{N}|e^{j\omega} - p_i|} \tag{2.101}$$

$$\angle X(e^{j\omega}) = \angle A + \sum_{i=1}^{M}\angle(1 - z_i e^{-j\omega}) - \sum_{i=1}^{N}\angle(1 - p_i e^{-j\omega}) \tag{2.102}$$

A geometrical interpretation of Eq. (2.100) to (2.102) is given in Fig. 2.17.

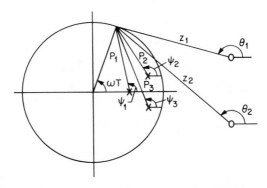

Fig. 2.17 A graphic interpretation of the frequency response measurement.

From the point $z = e^{j\omega}$ (on the unit circle) vectors are drawn to each of the poles and zeros. The magnitudes of the vectors are used to determine the magnitude of the system function at the given frequency (ω) and the angles are used to determine the phase. In the example of Fig. 2.17 there are three poles ($N = 3$) and two zeros ($M = 2$) and $A = 1$ so (with reference to Fig. 2.17)

$$|X(e^{j\omega})| = \frac{Z_1 Z_2}{P_1 P_2 P_3}$$

$$\sphericalangle X(e^{j\omega}) = \theta_1 + \theta_2 - (\psi_1 + \psi_2 + \psi_3)$$

To determine the entire system function (as ω varies from 0 to π) z moves counterclockwise around the unit circle beginning at $z = +1$ and ending at $z = -1$.

2.19   Digital Filter Realizations (Structures)

A digital filter transfer function can often be realized in a variety of ways. Noise and inaccuracies caused by quantization (see Chapter 5) of any practical digital filter implementation are very dependent on the precise digital filter structure. In a broad sense, the methods of realization can be divided into two classes, recursive and nonrecursive. For a *recursive realization*, the functional relationship between the input sequence of the filter $\{x(n)\}$ and the resulting output sequence $\{y(n)\}$ can be described as

$$y(n) = F[y(n-1), y(n-2), \ldots, x(n), x(n-1), \ldots]$$

i.e., the current output sample $y(n)$ is a function of *past* outputs as well as present and past input samples. For a *nonrecursive realization*, the relation between the output and input sequences becomes

$$y(n) = F[x(n), x(n-1), \ldots]$$

i.e., the current output sample $y(n)$ is a function only of past and present inputs. Here and in Sec. 2.20 several possible realizations of a digital filter are shown.

As noted earlier, the z transform of a digital filter can be expressed as a rational polynomial in z^{-1}; i.e.,

$$H(z) = \frac{Y(z)}{X(z)} = \frac{\sum\limits_{i=0}^{N} a_i z^{-i}}{\sum\limits_{i=0}^{N} b_i z^{-i}} \tag{2.103}$$

and

$$b_0 \overset{\Delta}{=} 1$$

(We are assuming the degrees of the numerator and denominator are identical.) A difference equation relating $y(n)$ and $x(n)$ can be derived by cross-multiplying the terms of Eq. (2.103) to give

$$Y(z) \sum_{i=0}^{N} b_i z^{-i} = X(z) \sum_{i=0}^{N} a_i z^{-i} \qquad (2.104)$$

or

$$\sum_{i=0}^{N} b_i z^{-i} Y(z) = \sum_{i=0}^{N} a_i z^{-i} X(z) \qquad (2.105)$$

Interpreting the term $z^{-k} Y(z)$ as the inverse transform of $y(n-k)$ we can inverse z transform Eq. (2.105) to give the difference equation

$$\sum_{i=0}^{N} b_i y(n-i) = \sum_{i=0}^{N} a_i x(n-i) \qquad (2.106)$$

Since $b_0 = 1$, we can rewrite Eq. (2.106) to solve for $y(n)$ explicitly as

$$y(n) = \sum_{i=0}^{N} a_i x(n-i) - \sum_{i=1}^{N} b_i y(n-i) \qquad (2.107)$$

A simple structure for realizing this difference equation is shown in Fig. 2.18.

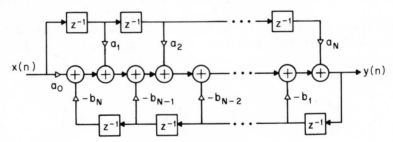

Fig. 2.18 Direct form 1.

This structure is called direct form 1 and uses separate delays for both the numerator and denominator terms of Eq. (2.103). The simplicity of the structure as well as its direct relation to the z transform are some of its attributes. As we shall see in Chapter 5, however, in cases where the poles of $H(z)$ are reasonably close to each other or to the unit circle (i.e., frequency selective filters), there is a severe coefficient sensitivity problem with this structure; hence, it generally is avoided in most practical situations.

If Eq. (2.103) is rewritten in a slightly different form, i.e.,

$$H(z) = \frac{Y(z)}{X(z)} = \underbrace{\left(\frac{1}{\sum_{i=0}^{N} b_i z^{-i}} \right)}_{H_1(z)} \underbrace{\left(\sum_{i=0}^{N} a_i z^{-i} \right)}_{H_2(z)} \qquad (2.108)$$

then a different structure can be obtained for realizing the digital filter. Equation (2.108) gives $H(z)$ as a cascade of a filter containing only poles with transfer function $H_1(z)$ and a filter containing only zeros with a transfer function $H_2(z)$. If we write

$$H_1(z) = \frac{W(z)}{X(z)} = \frac{1}{\sum\limits_{i=0}^{N} b_i z^{-i}} \tag{2.109}$$

and

$$H_2(z) = \frac{Y(z)}{W(z)} = \sum\limits_{i=0}^{N} a_i z^{-i} \tag{2.110}$$

we obtain the pair of difference equations (assuming $b_0 = 1$)

$$w(n) = x(n) - \sum\limits_{i=1}^{N} b_i w(n - i) \tag{2.111}$$

$$y(n) = \sum\limits_{i=0}^{N} a_i w(n - i) \tag{2.112}$$

which can be realized as shown in Fig. 2.19. Since the signal $w(n)$ is delayed identically for both $H_1(z)$ and $H_2(z)$, one set of delays suffices for the entire

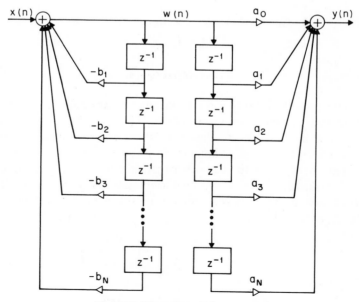

Fig. 2.19 Direct form 2 (noncanonic).

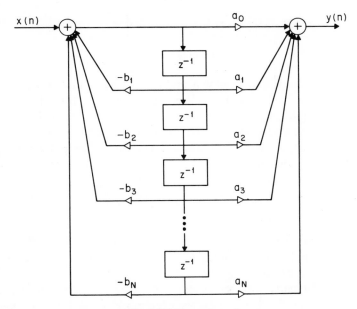

Fig. 2.20 Direct form 2.

filter. Thus Fig. 2.19 can be redrawn as shown in Fig. 2.20. The resulting structure is called direct form 2 and has also been called the canonic form because it has the minimum number of multiplier, adder, and delay elements. (Since other configurations also have this property, the term *canonic form* for the structure of Fig. 2.20 is not recommended.) The desirable and undesirable features of direct form 2 are similar to those of direct form 1 and will be discussed later.

A third structure for realizing digital filters can be obtained by writing Eq. (2.103) in the form

$$H(z) = \frac{Y(z)}{X(z)} = a_0 \prod_{i=1}^{K} H_i(z) \qquad (2.113)$$

where $H_i(z)$ is either a second-order section, i.e.,

$$H_i(z) = \frac{1 + a_{1i}z^{-1} + a_{2i}z^{-2}}{1 + b_{1i}z^{-1} + b_{2i}z^{-2}} \qquad (2.114)$$

or a first-order section, i.e.,

$$H_i(z) = \frac{1 + a_{1i}z^{-1}}{1 + b_{1i}z^{-1}} \qquad (2.115)$$

and K is the integer part of $(N + 1)/2$. The realization of Eq. (2.113) is

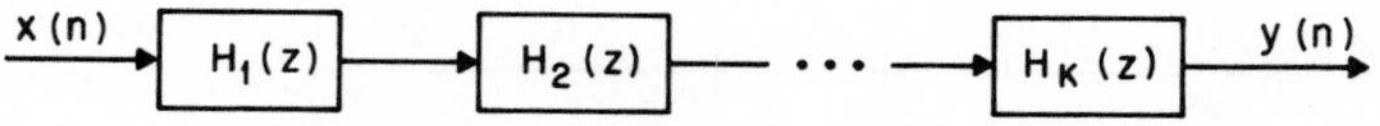

Fig. 2.21 Cascade form.

called the cascade form and is shown in Fig. 2.21. The individual second-order sections of Fig. 2.21 may be realized in either direct form 1 or 2. The reason for choosing a cascade of second-order sections (and possibly a first-order section) is that a second-order section is required to realize a complex pole, or zero, with real filter coefficients. Since not all the poles and zeros are complex, some authors like to define the cascade structure by the z transform

$$H(z) = a_0 \left[\prod_{i=1}^{K_1} H_{1i}(z) \right] \left[\prod_{i=1}^{K_2} H_{2i}(z) \right] \qquad (2.116)$$

where $H_{1i}(z)$ is a first-order system [as defined in Eq. (2.115)], $H_{2i}(z)$ is a second-order system [as defined in Eq. (2.114)], K_2 is the maximum number of complex poles or zeros, and $K_1 = N - 2K_2$.

One general difficulty with the cascade structure is that we must decide which poles to pair with which zeros. An even more complicated problem is deciding the exact order in which to cascade the individual first- and second-order systems. For example, if the desired transfer function is

$$H(z) = \frac{\displaystyle\prod_{i=1}^{5} N_i(z)}{\displaystyle\prod_{i=1}^{5} D_i(z)}$$

then a possible cascade realization is

$$H(z) = \frac{N_1(z)}{D_2(z)} \times \frac{N_3(z)}{D_5(z)} \times \frac{N_4(z)}{D_1(z)} \times \frac{N_5(z)}{D_4(z)} \times \frac{N_2(z)}{D_3(z)}$$

where the pairing is N_1 with D_2, N_3 with D_5, N_4 with D_1, N_5 with D_4, and N_2 with D_3. The implied ordering is N_1/D_2 first, followed by N_3/D_5, N_4/D_1, N_5/D_4, and finally N_2/D_3. In the limit of infinite bit precision for the word lengths of all variables, the questions of pairing and ordering are insignificant. In a practical situation, however, they are quite important. A more complete discussion of this problem is given in Chapter 5. A further difficulty with the cascade structure is the necessity for having scaling multipliers between the individual sections in the cascade to prevent the filter variables from becoming too large or too small. We shall defer discussion of scaling to Chapter 5.

A fourth digital filter structure can be obtained by writing Eq. (2.103) in its partial fraction expansion as

$$H(z) = C + \sum_{i=1}^{K} H_i(z) \qquad (2.117)$$

where $H_i(z)$ is either a second-order section of the form

$$H_i(z) = \frac{a_{0i} + a_{1i}z^{-1}}{1 + b_{1i}z^{-1} + b_{2i}z^{-2}} \qquad (2.118)$$

or a first-order section of the form

$$H_i(z) = \frac{a_{0i}}{1 + b_{1i}z^{-1}} \qquad (2.119)$$

K is the integer part of $(N + 1)/2$, and $C = a_N/b_N$ as defined in Eq. (2.103). Figure 2.22 shows a realization of the structure of Eq. (2.117), which is called

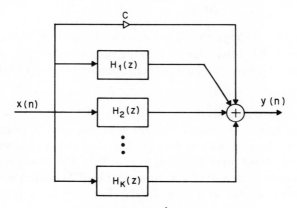

Fig. 2.22 Parallel form.

the parallel form. The individual first- and second-order sections of Eq. (2.118) and (2.119) are realized using one of the direct forms given earlier in this section.

The filter structures given above are by no means the only available structures. They are, however, the most widely used both in simulations and in digital hardware. Additional structures may be obtained in a countless variety of ways. For example, one could define a parallel–cascade structure where the parallel form is used to realize part of the transfer function and the cascade form for the remainder. Additionally, transpose configurations for all the structures above can be obtained by reversing the directions of all signal flow (i.e., by reversing the directions of all arrows) and by interchanging all branch nodes and summing junctions. The resulting structures have the same transfer functions but different properties with respect to finite word length effects.

The choice among the various structures given above is dictated by their economy of implementation, whether it be hardware or software. This, in turn, depends in many instances on properties of the structures when finite word lengths are attached to the filter coefficients and the filter variables. Further discussion on this matter is deferred to Chapter 5.

2.20 Structures for All-Zero Filters

In the important special case where the denominator of Eq. (2.103) is a constant (call it 1 for simplicity), the difference equation describing the system is of the nonrecursive type; i.e., the current output $y(n)$ depends entirely on a finite number of past and present values of the input. In this case it is customary to rewrite Eq. (2.103) explicitly in terms of the filter impulse response $h(n)$ as

$$H(z) = \frac{Y(z)}{X(z)} = \sum_{n=0}^{N-1} h(n)z^{-1} \tag{2.120}$$

where the upper limit on the summation is changed to $N - 1$ in order to describe a realizable filter with an impulse response duration of N samples. The appropriate difference equation corresponding to Eq. (2.120) is

$$y(n) = h(0)x(n) + h(1)x(n - 1) + \ldots + h(N - 1)x(n - N + 1) \tag{2.121}$$

which is of the nonrecursive type.

There are several structures that are generally used to implement filters with finite impulse responses of the type described above. The most common structure is the direct form described in Sec. 2.19. For this special case there is only one direct form and this is shown in Fig. 2.23. Because of the resemblance of this structure to a tapped delay line, this form is often called a

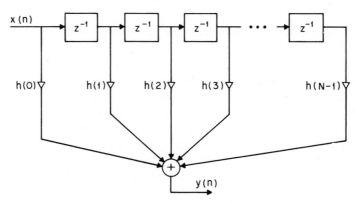

Fig. 2.23 Direct form for finite impulse response filters.

tapped delay line filter or sometimes a transversal filter. As we shall see later, the implementation of the structure of Fig. 2.23 in digital hardware can be particularly simple, requiring only a single multiplier, a single accumulator, and two circulating shift register memories.

The cascade structure is also a particularly useful one for implementing a filter with only zeros. In this case, the z transform of the filter (Eq. 2.120) is factored and expressed as a product of second- and first-order systems; i.e.,

$$H(z) = \prod_{n=1}^{N_M} H_n(z) \tag{2.122}$$

where

$$H_n(z) = a_{0n} + a_{1n}z^{-1} + a_{2n}z^{-2} \quad \text{(second-order system)}$$

or

$$H_n(z) = a_{0n} + a_{1n}z^{-1} \quad \text{(first-order system)}$$

and N_M is the integer part of $(N + 1)/2$.

There are several other structures that are often used to implement these all-zero filters, for which there are no counterparts for arbitrary filters containing both poles and zeros. The most common of these structures is called the fast convolution technique. This technique is based on realizing a convolution as the inverse Fourier transform of the product of the Fourier transforms of the input and impulse response sequences. The details of this method will be described later in this chapter.

Other structures for implementing Eq. (2.120) may be derived through the use of classic interpolation formulas for the representation of a polynomial. For example, the use of the Lagrange interpolation formula leads to the realization

$$H(z) = \prod_{n=0}^{N-1} (1 - z^{-1}z_n) \sum_{m=0}^{N-1} \frac{A_m}{1 - z^{-1}z_m} \tag{2.123}$$

where

$$A_m = \frac{H(z_m)}{\prod_{\substack{n=0 \\ n \neq m}}^{N-1} (1 - z_n z_m^{-1})} \tag{2.124}$$

and $\{z_n\}$, $0 \leq n \leq N - 1$, are N arbitrary points in the z plane at which the z transform [Eq. (2.120)] is evaluated to give the values $H(z_n)$ as used in Eq. (2.124). The resulting structure is seen from Eq. (2.123) to consist of a cascade of N first-order sections (containing zeros at $z = z_n$, $n = 0, 1, \ldots,$ $N - 1$) in cascade with a parallel combination of N first-order sections

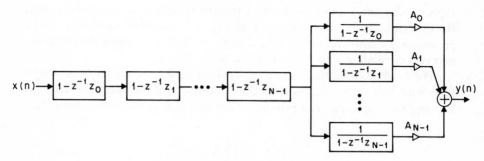

Fig. 2.24. General Lagrange structure.

(containing poles at $z = z_n$, $n = 0, 1, \ldots, N - 1$). The general structure of Eq. (2.123) is shown in Fig. 2.24.

It is easy to verify that the structure of Eq. (2.123) is capable of realizing an arbitrary z transform of the form of Eq. (2.120). First it is seen that the pole of *each* of the parallel paths exactly cancels one of the zeros in the cascade, giving an equivalent filter with $(N - 1)$ zeros. Furthermore at each of the points z_n the value of $H(z)$ is exactly $H(z_n)$ as desired. Since $H(z)$ is an $(N - 1)$st-degree polynomial, it is specified exactly by its values at N independent points. Hence Eq. (2.123) is seen to represent Eq. (2.120) exactly.

It should be noted that the Lagrange structure is noncanonic in the number of delays required to realize the filter as there are N delays for the cascade and one extra delay in each of the parallel paths; i.e., there are a total of $2N$ delays. The importance of this and similar structures lies in special cases where they may be applied (as discussed below). Other possible advantages may be found when the sensitivity of this structure to finite precision representation of filter coefficients is known.

An important special case of the Lagrange structure is when the sequence z_n consists of points equally spaced around the unit circle in the z plane; i.e.,

$$z_n = e^{j(2\pi/N)n} \qquad n = 0, 1, \ldots, N - 1 \qquad (2.125)$$

For this case the product of terms in Eq. (2.123) can be shown to be of the form

$$\prod_{n=0}^{N-1} [1 - z^{-1} e^{j(2\pi/N)n}] = (1 - z^{-N}) \qquad (2.126)$$

and Eq. (2.124) becomes

$$A_m = \frac{1}{N} \{H[e^{j(2\pi/N)m}]\} \qquad (2.127)$$

Equation (2.126) may be easily verified by noting that the roots of the equation $z^{-N} = 1$ are precisely the N principle roots of unity, i.e., those given in Eq. (2.125). Equation (2.127) is obtained by evaluating Eq. (2.124) under the conditions of Eq. (2.125) giving

$$A_m = H(z_m) \lim_{z \to z_m} \left[\frac{1 - z^{-1}z_m}{\prod\limits_{n=0}^{N-1}(1 - z^{-1}z_n)} \right]$$

$$= H(z_m) \lim_{z \to z_m} \left(\frac{1 - z^{-1}z_m}{1 - z^{-N}} \right)$$

$$= \frac{H(z_m)}{N}$$

The resulting filter structure is described by the z transform

$$H(z) = \frac{(1 - z^{-N})}{N} \sum_{n=0}^{N-1} \frac{H[e^{j(2\pi/N)n}]}{1 - z^{-1}e^{j(2\pi/N)n}} \tag{2.128}$$

and is shown in Fig. 2.25. This structure has been called the frequency

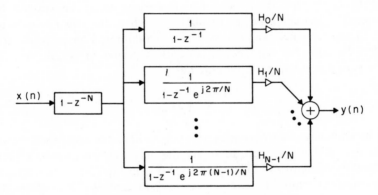

Fig. 2.25. Frequency sampling structure.

sampling structure since its basic coefficients $\{H[e^{j(2\pi/N)n}]\}$ are the values of the filter's frequency response $[H(e^{j\omega})]$ sampled at N points equally spaced around the unit circle.

This frequency sampling structure has several interesting properties. For example, using finite precision arithmetic to represent the parallel part of the structure of Eq. (2.128), the poles of the filter will not exactly cancel the zeros of the structure that are lumped together in the term $(1 - z^{-N})$. Thus, the resulting structure will have both poles and zeros, and its impulse response will not be of finite duration. The importance of this structure lies in the fact that it is an extremely efficient realization of a filter for which the majority of the filter multiplier coefficients {the terms $H[e^{j(2\pi/N)n}]$} are exactly zero.

For those paths where $H[e^{j(2\pi/N)n}] = 0$, the particular parallel paths need not be realized. Thus, for example, in the case of a narrowband filter when only a small number of the filter multiplier coefficients are nonzero, the resulting structure requires very few multiplications per output sample and is thus a very efficient realization.

Other nonrecursive structures may be derived for the all-zero filter based on application of other polynomial interpolation formulas. For instance, a Newton structure may be obtained via the Newton interpolation formula, as well as a Hermite structure and a Taylor structure among others. Since the advantages and disadvantages of such structures are not generally well understood, we shall refrain from any further discussion of these specific structures.

2.21 The Discrete Fourier Transform

We have already discussed several techniques for representing sequences or equivalently discrete-time systems including the convolution summation, the Fourier transform, and the z transform. In the case where the sequence to be represented is periodic (and, as we shall see, when the sequence is of finite duration), the sequence can be represented in a discrete Fourier series. Thus, consider the periodic[2] sequence $x_p(n)$ with a period of N samples. Since $x_p(n)$ is periodic, it can be represented as

$$x_p(n) = \sum_{k=-\infty}^{\infty} X_p(k)e^{j(2\pi/N)kn} \qquad (2.129)$$

where the only possible frequencies of which $x_p(n)$ can be composed are $\omega_k = 2\pi k/N$, $-\infty < k < \infty$, since these are the only frequencies whose periods are integrally related to N. The quantity $X_p(k)$ in Eq. (2.129) represents the amplitude of the sinusoid at frequency ω_k. Equation (2.129) contains much redundant information since the complex exponential at frequency $\omega_k = 2\pi k/N$ is indistinguishable from the complex exponential at frequency $\omega_{k\pm mN} = (2\pi/N)(k \pm mN)$, $0 < m < \infty$, because of the periodicity of the function $e^{j\theta}$ every 2π; e.g.,

$$e^{j(2\pi/N)kn} = e^{j(2\pi/N)(k\pm mN)n} \qquad 0 < m < \infty \qquad (2.130)$$

Thus Eq. (2.129) can be expressed in the form

$$x_p(n) = \sum_{k=0}^{N-1} X_p(k)e^{j(2\pi/N)kn} \qquad (2.131)$$

emphasizing the fact that there are only N distinct exponentials that are

---

[2] All periodic sequences will have the subscript p.

periodic of period N samples. For convenience, and to conform with popular notation, we rewrite Eq. (2.131) as

$$x_p(n) = \frac{1}{N} \sum_{k=0}^{N-1} X_p(k) e^{j(2\pi/N)kn} \tag{2.132}$$

where the division by $1/N$ has no bearing on the form of the representation. To obtain a relation for $X_p(k)$ in terms of $x_p(n)$, we multiply both sides of Eq. (2.132) by $e^{-j(2\pi/N)mn}$ and sum over n giving

$$\sum_{n=0}^{N-1} x_p(n) e^{-j(2\pi/N)nm} = \frac{1}{N} \sum_{n=0}^{N-1} \sum_{k=0}^{N-1} X_p(k) e^{j(2\pi/N)n(k-m)} \tag{2.133}$$

Interchanging orders of summation on the right side of Eq. (2.133) and using the relation

$$\sum_{n=0}^{N-1} e^{-j(2\pi/N)n(k-m)} = \begin{cases} N & \text{if } k = m \\ 0 & \text{otherwise} \end{cases} \tag{2.134}$$

we obtain

$$\sum_{n=0}^{N-1} x_p(n) e^{-j(2\pi/N)nm} = \sum_{k=0}^{N-1} X_p(k) u_0(k - m) \tag{2.135}$$

or (reversing left and right sides of the equation and substituting the index k for m)

$$X_p(k) = \sum_{n=0}^{N-1} x_p(n) e^{-j(2\pi/N)nk} \tag{2.136}$$

Equation (2.136) has been called the discrete Fourier transform (DFT) and Eq. (2.132) is called the inverse discrete Fourier transform (IDFT).

It is clear from Eq. (2.132) and (2.136) that both the sequences $x_p(n)$ and $X_p(k)$ are periodic with period N samples. It is also clear from Eq. (2.136) that $X_p(k)$ may be determined exactly from one period of $x_p(n)$. This brings up the interesting question of what is the relation between the z-transform representation of a finite duration sequence composed of a single period of a periodic sequence and the DFT of the periodic sequence from which the original finite sequence was created. Stated in a different way, consider the finite duration sequence $x(n)$ defined as

$$x(n) = \begin{cases} x_p(n) & 0 \leq n \leq N - 1 \\ 0 & \text{all other } n \end{cases} \tag{2.137}$$

where $x_p(n)$ is periodic and of period N samples; i.e., $x(n)$ is a single period of $x_p(n)$. The z transform of $x(n)$ is

$$X(z) = \sum_{n=0}^{N-1} x(n) z^{-n} \tag{2.138}$$

Evaluation of Eq. (2.138) at the point $z = e^{j(2\pi/N)k}$, i.e., at a point on the unit circle with angle $2\pi k/N$, gives

$$X(z)\big|_{z=e^{j(2\pi/N)k}} = X[e^{j(2\pi/N)k}] = \sum_{n=0}^{N-1} x(n)e^{-j(2\pi/N)nk} \qquad (2.139)$$

Since $x_p(n) = x(n)$ in the interval $0 \leq n \leq N - 1$, a comparison of Eq. (2.139) and (2.136) shows

$$X_p(k) = X[e^{j(2\pi/N)k}] \qquad (2.140)$$

Thus *the DFT coefficients of a finite duration sequence are the values of the z transform of that same sequence at N evenly spaced points around the unit circle.* An even more important observation is that *the DFT coefficients of a finite duration sequence constitute a unique representation of that sequence* because the inverse DFT relations may be used to reconstruct the desired sequence exactly from the DFT coefficients. Thus even though the DFT, IDFT relations are derived on the basis of periodic sequences, they are even more important in their ability to represent finite duration sequences.

To illustrate the ideas above, consider the periodic sequence (of period N) shown in Fig. 2.26(a) and defined as

$$x_p(n) = a^n \qquad\qquad 0 \leq n \leq N - 1$$
$$x_p(n + mN) = x_p(n) \qquad m = \pm 1, \pm 2, \ldots$$

The DFT of $x_p(n)$ may be obtained from Eq. (2.136) as

$$X_p(k) = \sum_{n=0}^{N-1} a^n e^{-j(2\pi/N)nk}$$

$$= \sum_{n=0}^{N-1} [ae^{-j(2\pi/N)k}]^n$$

$$= \frac{1 - a^N}{1 - ae^{-j(2\pi/N)k}} \qquad 0 \leq k \leq N - 1$$

The sequence $X_p(k)$ (both magnitude and angle) is shown in Fig. 2.26(b) for $a = 0.9$, $N = 16$. Figure 2.26(c) shows the finite duration sequence $x(n)$ defined as

$$x(n) = \begin{cases} a^n & 0 \leq n \leq N - 1 \\ 0 & \text{other } n \end{cases}$$

i.e., a single period of $x_p(n)$. The z transform of $x(n)$ is

$$X(z) = \sum_{n=0}^{N-1} a^n z^{-n} = \frac{1 - a^N z^{-N}}{1 - az^{-1}}$$

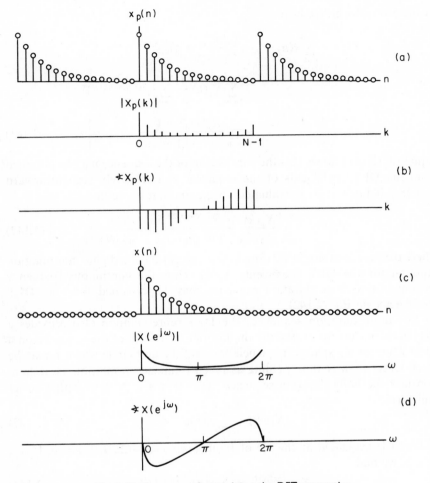

Fig. 2.26. Sequences for studying the DFT properties.

When $X(z)$ is evaluated on the unit circle, we obtain

$$X(e^{j\omega}) = \frac{1 - a^N e^{-j\omega N}}{1 - a e^{-j\omega}}$$

Figure 2.26(d) shows $X(e^{j\omega})$ (both magnitude and angle) for $0 \leq \omega \leq 2\pi$. At the points $\omega = 2\pi k/N$, $X_p(k)$ and $X(e^{j2\pi k/N})$ are clearly identical.

Since the DFT provides a unique representation of a finite duration sequence, it must be possible to obtain the z transform of the sequence in terms of the DFT coefficients. From Eq. (2.137) and (2.132) and the definition of the z transform, we have

$$
\begin{aligned}
X(z) = \sum_{n=0}^{N-1} x(n)z^{-n} &= \sum_{n=0}^{N-1} \frac{1}{N} \sum_{k=0}^{N-1} X_p(k)e^{j(2\pi/N)nk}z^{-n} \\
&= \sum_{k=0}^{N-1} X_p(k) \frac{1}{N} \sum_{n=0}^{N-1} [e^{j(2\pi/N)k}z^{-1}]^n \\
&= \sum_{k=0}^{N-1} \frac{X_p(k)}{N} \left[\frac{1 - z^{-N}}{1 - z^{-1}e^{j(2\pi/N)k}} \right]
\end{aligned}
\tag{2.141}
$$

Equation (2.141) shows that the z transform of the sequence may be obtained from the DFT coefficients of the sequence in a relatively straightforward manner. If Eq. (2.141) is evaluated on the unit circle, we find

$$
X(e^{j\omega}) = \sum_{k=0}^{N-1} \frac{X_p(k)}{N} \frac{e^{-j\omega[(N-1)/2]} \sin(\omega N/2)}{e^{j(\pi k/N)} \sin(\omega/2 - \pi k/N)}
\tag{2.142}
$$

where the functions $\sin(\omega N/2)/\sin(\omega/2 - \pi k/N)$ act as a digital interpolation function for the DFT coefficients $X_p(k)$. Thus the continuous frequency response of a finite duration sequence may be obtained from its DFT coefficients via Eq. (2.142).

For finite duration sequences the DFT representation also provides a convenient means for evaluating the Fourier transform of the sequence on a set of L uniformly spaced frequencies around the unit circle where L may be much larger than N thus yielding as much frequency resolution as desired. Consider the finite duration sequence $\{x(n), 0 \le n \le N - 1\}$ with Fourier transform

$$
X(e^{j\omega}) = \sum_{n=0}^{N-1} x(n)e^{-j\omega n}
\tag{2.143}
$$

If $X(e^{j\omega})$ is evaluated at the set of frequencies $\omega_l = (2\pi/L)l$, $l = 0, 1, \ldots,$ $L - 1$, we find

$$
X[e^{j(2\pi/N)l}] = \sum_{n=0}^{N-1} x(n)e^{-j(2\pi/L)ln}
\tag{2.144}
$$

If we define an L-point sequence $\hat{x}(n)$ as $(L > N)$,

$$
\hat{x}(n) = \begin{cases} x(n) & 0 \le n \le N - 1 \\ 0 & N \le n \le L - 1 \end{cases}
\tag{2.145}
$$

and take the L-point DFT of $\hat{x}(n)$, we find

$$
X(k) = \sum_{n=0}^{L-1} \hat{x}(n)e^{-j(2\pi/L)kn}
\tag{2.146}
$$

Since $\hat{x}(n) = 0$, $n \ge N$, Eq. (2.146) can be written as

$$
\hat{X}(k) = \sum_{n=0}^{N-1} x(n)e^{-j(2\pi/L)kn}
\tag{2.147}
$$

Comparison of Eq. (2.147) and (2.144) shows

$$\hat{X}(k) = X[e^{j(2\pi/L)k}] \tag{2.148}$$

Thus the simple technique of augmenting a finite duration sequence with zero-valued samples allows arbitrary resolution in computing the Fourier transform of the sequence on a set of uniformly spaced points around the unit circle. This simple procedure is one of the most useful techniques in the spectrum analysis of any finite duration sequence.

We have just shown that the DFT provides a unique representation of an N-point finite duration sequence where the DFT coefficients are values of the z transform of the sequence at N uniformly spaced points around the unit circle. Similarly the z-transform of a (possibly infinite) sequence is a unique representation of that sequence. We have also seen that sampling in the time domain causes aliasing in the frequency domain. We now would like to show how sampling in the frequency domain leads to aliasing in the time domain. We now show what sequence is obtained if we use for the DFT coefficients the values of an *arbitrary* z transform evaluated at N uniformly spaced points around the unit circle. To be more explicit, assume the sequence $h(n)$ (possibly infinite in duration) has z transform $H(z)$ defined as

$$H(z) = \sum_{n=-\infty}^{\infty} h(n)z^{-n} \tag{2.149}$$

If we define a set of "DFT coefficients" as

$$H_p(k) = H(z)\big|_{z=e^{j(2\pi/N)k}} = \sum_{n=-\infty}^{\infty} h(n)e^{-j(2\pi/N)nk} \tag{2.150}$$

we can solve for the periodic sequence $h_p(n)$ with the DFT coefficients $H_p(k)$ as

$$h_p(n) = \frac{1}{N}\sum_{k=0}^{N-1} H_p(k)e^{j(2\pi/N)nk} \tag{2.151}$$

Substituting Eq. (2.150) into Eq. (2.151) (and changing the summation index to m) gives

$$h_p(n) = \frac{1}{N}\sum_{k=0}^{N-1}\left[\sum_{m=-\infty}^{\infty} h(m)e^{-j(2\pi/N)mk}\right]e^{j(2\pi/N)nk}$$

$$= \sum_{m=-\infty}^{\infty} h(m)\left[\frac{1}{N}\sum_{k=0}^{N-1}e^{-j(2\pi/N)k(m-n)}\right]$$

$$= \sum_{m=-\infty}^{\infty} h(m)\sum_{r=-\infty}^{\infty} u_0(m-n+rN)$$

$$h_p(n) = \sum_{r=-\infty}^{\infty} h(n-rN) \tag{2.152}$$

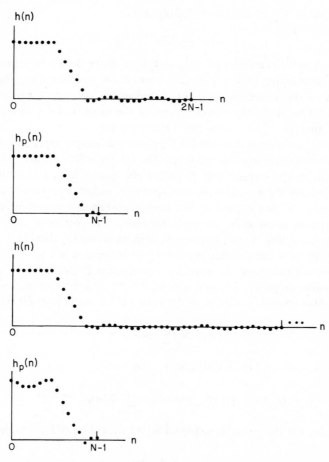

Fig. 2.27. Two sequences [$h(n)$] and their periodic counterparts [$h_p(n)$].

The importance of Eq. (2.152) should not be minimized. It shows that the periodic sequence created by using as its DFT coefficients the values of the z transform evaluated at N points around the unit circle of a nonperiodic sequence, is an *aliased* version of the nonperiodic sequence. Thus if the duration of $h(n)$ is less than or equal to N samples, there is no aliasing in $h_p(n)$. Equation (2.152) also shows that the aliasing effects in representing an infinite sequence by a finite set of DFT coefficients can be reduced by increasing N the size of the DFT. Figure 2.27 shows two sequences $h(n)$ and their "equivalent" N-point periodic counterparts. In the first example the duration of $h(n)$ is approximately N and so there is little aliasing in creating $h_p(n)$. In the second example the duration of $h(n)$ is much greater than N and so the resulting periodic sequence bears less resemblance to $h(n)$.

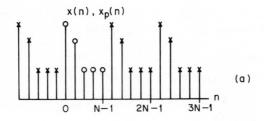

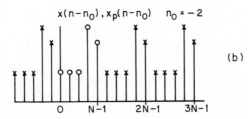

Fig. 2.28. The shifting property of the DFT.

2.22 Properties of the DFT

There are several properties of the DFT that play an important role in practical techniques for processing signals. In this section we shall summarize these properties and only when necessary shall we discuss them in detail.

1. Linearity

If $x_p(n)$ and $y_p(n)$ are periodic sequences (both of period N samples) with DFT's $X_p(k)$ and $Y_p(k)$, then the DFT of $x_p(n) + y_p(n)$ is $X_p(k) + Y_p(k)$. A similar result holds in the case of finite duration sequences.

2. Shifting Property

If $x_p(n)$ is a periodic sequence (period N samples) with DFT $X_p(k)$, then the periodic sequence $x_p(n - n_0)$ has DFT $X_p(k)e^{-j(2\pi/N)n_0 k}$. If we are considering finite duration sequences, special attention must be given to the meaning of a time-shift of the sequence. As shown in Fig. 2.28(a), the finite duration sequence $x(n)$ is of duration N samples. The N-point DFT of $x(n)$ is $X(k)$. Also shown in Fig. 2.28(a) (the x's) are the samples of the "equivalent" periodic sequence $x_p(n)$ with the same DFT as $x(n)$. If we want to define the DFT of $x(n - n_0)$ $(n_0 < N)$, we must consider a shift of the periodic sequence $x_p(n - n_0)$ and the equivalent finite duration sequence [with DFT $X(k)e^{-j(2\pi/N)n_0 k}$] is that part of $x_p(n - n_0)$ in the interval $0 \leq n \leq$

$N - 1$. Thus, as shown in Fig. 2.28(b), as far as the DFT representation is concerned, the sequence $x(n - n_0)$ is created by *rotating* $x(n)$ circularly by n_0 samples.

3. Symmetry Properties

If $x_p(n)$ is a real, periodic sequence of period N samples, then $X_p(k)$ satisfies the symmetry conditions

$$\text{Re } [X_p(k)] = \text{Re } [X_p(N - k)]$$

$$\text{Im } [X_p(k)] = -\text{Im } [X_p(N - k)]$$

$$|X_p(k)| = |X_p(N - k)| \tag{2.153}$$

$$\measuredangle X_p(k) = -\measuredangle X_p(-k)$$

[Similar results hold if $x(n)$ is a real finite duration sequence with the N-point DFT $X(k)$.] If the additional symmetry requirement is made that $x_p(n)$ is symmetrical, e.g.,

$$x_p(n) = x_p(N - n) \tag{2.154}$$

then the additional restriction is imposed that $X_p(k)$ is purely real.

Since one generally deals with real sequences, the symmetry constraints of Eq. (2.153) may be conveniently used to obtain the DFT's of two sequences using a single DFT. Consider the real, periodic sequences $x_p(n)$ and $y_p(n)$, both of period N samples, with N-point DFT's $X_p(k)$ and $Y_p(k)$, respectively. If we define the (complex) sequence $z_p(n)$ as

$$z_p(n) = x_p(n) + jy_p(n) \tag{2.155}$$

then the DFT of $z_p(n)$ is

$$Z_p(k) = \sum_{n=0}^{N-1} [x_p(n) + jy_p(n)]e^{-j(2\pi/N)nk} \tag{2.156}$$

$$Z_p(k) = X_p(k) + jY_p(k) \tag{2.157}$$

Equating real and imaginary parts of Eq. (2.157) gives

$$\text{Re } [Z_p(k)] = \text{Re } [X_p(k)] - \text{Im } [Y_p(k)]$$

$$\text{Im } [Z_p(k)] = \text{Im } [X_p(k)] + \text{Re } [Y_p(k)] \tag{2.158}$$

Since the real and imaginary parts of $X_p(k)$ and $Y_p(k)$ have even and odd symmetry, respectively, it is clear that these components can be recovered by a simple "even–odd" separation; i.e.

$$\text{Re } [X_p(k)] = \frac{\text{Re } [Z_p(k)] + \text{Re } [Z_p(N - k)]}{2}$$

$$\text{Im } [Y_p(k)] = \frac{\text{Re } [Z_p(N - k)] - \text{Re } [Z_p(k)]}{2}$$

$$\text{Re } [Y_p(k)] = \frac{\text{Im } [Z_p(k)] + \text{Im } [Z_p(N - k)]}{2} \qquad (2.159)$$

$$\text{Im } [X_p(k)] = \frac{\text{Im } [Z_p(k)] - \text{Im } [Z_p(N - k)]}{2}$$

Thus a single N-point DFT can effectively transform two N-point real sequences at the same time. If additional symmetry constraints are imposed on the input sequences, further reductions in the necessary number of operations required to obtain the DFT can be achieved.

2.23 Convolution of Sequences

If $x_p(n)$ and $h_p(n)$ are two periodic sequences of period N with DFT's

$$X_p(k) = \sum_{n=0}^{N-1} x_p(n)e^{-j(2\pi/N)nk} \qquad (2.160)$$

$$H_p(k) = \sum_{n=0}^{N-1} h_p(n)e^{-j(2\pi/N)nk} \qquad (2.161)$$

then the N-point DFT of the sequence $y_p(n)$, defined as the circular or periodic convolution of $x_p(n)$ and $h_p(n)$, i.e.,

$$y_p(n) = \sum_{l=0}^{N-1} x_p(l)h_p(n - l) \qquad (2.162)$$

is

$$Y_p(k) = H_p(k) \cdot X_p(k) \qquad (2.163)$$

Since the implications of Eq. (2.163) are so important, we shall show how to derive this equation. First, however, it is important to understand exactly what a circular convolution is. In Fig. 2.29(a) and (b) are shown the periodic sequences $x_p(n)$ and $h_p(n)$. Figure 2.29(c) shows how the circular convolution is evaluated [Eq. (2.162)] for $n = 2$. Because of periodicities we need only consider the sequences $x_p(l)$ and $h_p(n - l)$ in the interval $0 \leq l \leq N - 1$. As n varies, $h_p(n - l)$ slides along $x_p(l)$. As the sample of $h_p(n - l)$ slides past $l = N - 1$, the identical sample appears at $l = 0$. Thus the term

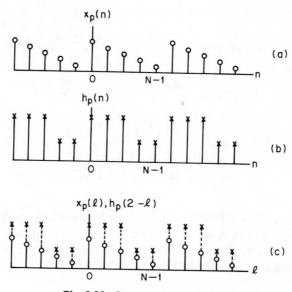

Fig. 2.29. Circular convolution.

circular convolution is a description of the convolution of two sequences defined on a circle.

Equation (2.163) may be obtained by taking the *N*-point DFT of Eq. (2.162) giving

$$
\begin{aligned}
Y_p(k) &= \sum_{n=0}^{N-1} \left[\sum_{l=0}^{N-1} x_p(l) h_p(n-l) \right] e^{-j(2\pi/N)nk} \\
&= \sum_{l=0}^{N-1} x_p(l) \underbrace{\left[\sum_{n=0}^{N-1} h_p(n-l) e^{-j(2\pi/N)(n-l)k} \right]}_{H_p(k)} e^{-j(2\pi/N)lk} \\
&= H_p(k) \underbrace{\sum_{l=0}^{N-1} x_p(l) e^{-j(2\pi/N)lk}}_{X_p(k)} \\
&= H_p(k) \cdot X_p(k) \tag{2.164}
\end{aligned}
$$

In the case when we are working with finite duration sequences, the result above also holds if $x_p(n)$ and $h_p(n)$ are interpreted as the equivalent periodic sequences with the same DFT's as the given finite duration sequences. Since, in the case of finite duration sequences, we are generally interested in performing linear or aperiodic rather than circular convolutions, modifications must be made in the technique above to yield useful results.

2.24 Linear Convolution of Finite Duration
Sequences

Consider two finite duration sequences $x(n)$ and $h(n)$. The duration of $x(n)$ is N_1 samples; i.e., $x(n)$ is nonzero only in the interval $0 \leq n \leq N_1 - 1$. The duration of $h(n)$ is N_2 samples; i.e., $h(n)$ is nonzero only in the interval $0 \leq n \leq N_2 - 1$. *The linear or aperiodic convolution* of $x(n)$ and $h(n)$ yields the sequence $y(n)$ defined as

$$y(n) = \sum_{m=0}^{n} h(m)x(n - m) \tag{2.165}$$

where $h(m)$ and $x(n - m)$ are zero outside the appropriately defined intervals. Figure 2.30 shows typical sequences $x(n)$, $h(n)$, and $y(n)$. Clearly $y(n)$ is a finite duration sequence of duration $(N_1 + N_2 - 1)$ samples.

We have already shown that multiplying the DFT's of two finite duration sequences and then inverse transforming the products is equivalent to circularly convolving the equivalent periodic sequences created from the given sequences. This fact and the example of Fig. 2.29 suggest a simple procedure for linearly convolving two finite duration sequences. When two periodic sequences are convolved, the output sequence is periodic and of the same period as each of the input sequences. Thus from the example of Fig. 2.30, since the period of $y(n)$ is $(N_1 + N_2 - 1)$ samples, in order to obtain this period from a circular convolution requires that both $x(n)$ and $h(n)$ must be

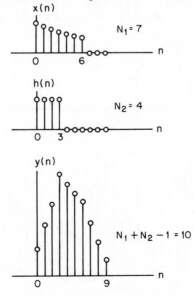

Fig. 2.30. Linear or aperiodic convolution.

$(N_1 + N_2 - 1)$-point sequences. To achieve this result we simply append the appropriate number of zero-valued samples to both $x(n)$ and $h(n)$ to make them $(N_1 + N_2 - 1)$-point sequences; then we take the $(N_1 + N_2 - 1)$-point DFT's of $x(n)$ and $h(n)$, multiply the DFT's and inverse transform to give the correct $y(n)$. This process is illustrated in Fig. 2.31, which shows the equivalent periodic sequences used for the circular convolution. As easily seen, the appending of zeros to each of the sequences $x(n)$ and $h(n)$ creates the correct length sequence to eliminate the wraparound difficulties inherent in circular convolution. Thus a single period of $y_p(n)$ of Fig. 2.31 is identical to $y(n)$ of Fig. 2.30. The technique of convolving two finite duration sequences using DFT techniques has been called fast convolution as opposed to the direct evaluation of Eq. (2.165), which is called direct or slow convolution. The term *fast* is used because the DFT can be evaluated rapidly and efficiently using any of a large class of algorithms called fast Fourier transforms. It can be shown that, even for moderate values of $(N_1 + N_2 - 1)$ (i.e., on the order of 30), fast convolution is more efficient than direct convolution. Hence the procedure above is an important signal processing tool.

In a practical sense, it is important to note that the size of the DFT's in the example above need not be restricted to $(N_1 + N_2 - 1)$-point

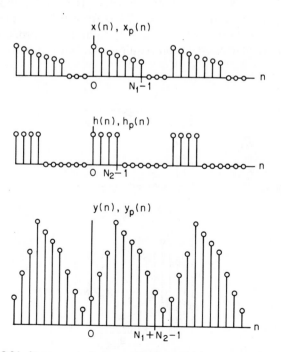

Fig. 2.31. Linear convolution realized using circular convolutions.

transforms. It should be clear that any number L can be used for the transform size, subject to the restriction $L \geq N_1 + N_2 - 1$. If $L > N_1 + N_2 - 1$, the only difference in the procedure described above is that additional zero-valued samples are appended to $x(n)$ and $h(n)$. Furthermore, the equivalent periodic output sequence $y_p(n)$ has $(L - N_1 - N_2 + 1)$ zero-valued samples at the end of the period. Clearly these differences in no way affect the desired result. The advantage in being able to arbitrarily choose L lies in the fact that practical algorithms for computing the DFT are not equally efficient for all values of L. Thus, for example, some algorithms require L to be a power of 2. In such a case, one is forced to choose L as the power of 2 greater than or equal to $N_1 + N_2 - 1$.

2.25 Sectioned Convolutions

In many practical situations, one is interested in computing the convolution of two finite duration sequences where one sequence is much longer than the other sequence; i.e., $N_1 \gg N_2$ or $N_2 \gg N_1$ in the examples above. Of course one can always use a value $L \geq N_1 + N_2 - 1$ but this is generally inefficient and impractical for several reasons. First this means the entire longer sequence must be available before the convolution can be carried out. In the case of practical waveforms, such as speech or radar signals, this is not always the situation. Furthermore since no processing occurs before the entire sequence is available, this implies long delays before the output is obtained. Finally, if $N_1 + N_2 - 1$ is too large, it is impractical to compute the DFT because of the large amounts of memory required and because of other practical considerations associated with fast Fourier transform algorithms. To alleviate these problems, two techniques can be used to section the larger sequence and compute partial results that can be pieced together to form the desired output sequence.

The first technique is called the overlap–add method. The mechanics of this technique are best explained with the use of the example shown in Fig. 2.32. For simplicity we assume $x(n)$ is effectively of infinite duration and the duration of $h(n)$ is N_2 samples. The sequence $x(n)$ is sectioned into pieces, each of duration N_3 samples, as shown in Fig. 2.32. (Considerations in choosing the value for N_3 are fairly complex but a good rule of thumb is to choose N_3 on the order of N_2.) Thus the input sequence $x(n)$ can be represented as

$$x(n) = \sum_{k=0}^{\infty} x_k(n) \tag{2.166}$$

where

$$x_k(n) = \begin{cases} x(n) & kN_3 \leq n \leq (k+1)N_3 - 1 \\ 0 & \text{otherwise} \end{cases} \tag{2.167}$$

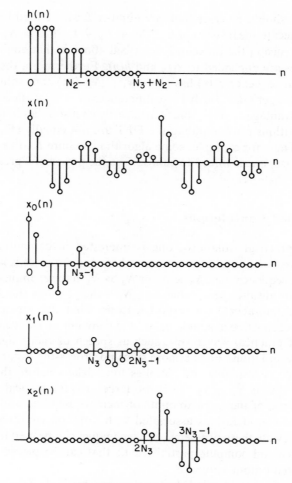

Fig. 2.32. Overlap–add method.

The linear convolution of $x(n)$ and $h(n)$ thus can be written as

$$y(n) = \sum_{m=0}^{n} h(m) \sum_{k=0}^{\infty} x_k(n - m) \qquad (2.168)$$

$$= \sum_{k=0}^{\infty} h(n) * x_k(n) = \sum_{k=0}^{\infty} y_k(n) \qquad (2.169)$$

The duration of each of the convolutions of Eq. (2.169) is $(N_3 + N_2 - 1)$ samples—thus there is a region of $(N_2 - 1)$ samples over which the kth convolution overlaps the $(k + 1)$st convolution and the appropriate output sequences must be added. Figure 2.33 shows the individual output terms $y_k(n)$

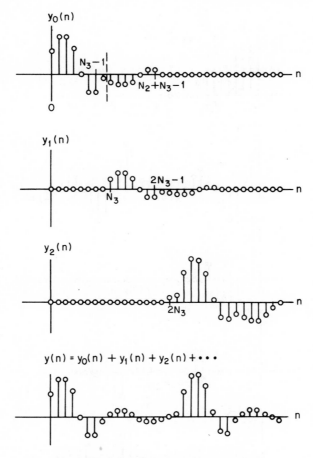

Fig. 2.33. Outputs of overlap–add method.

and how they must be added. Each of the convolutions of Eq. (2.169) is computed via the fast convolution technique noted in Sec. 2.24. This method has been called the overlap–add method because the output sequences overlap and must be added together to produce the correct result.

As stated earlier there is a second method of sectioning that can be used to achieve the desired linear convolution of a very long sequence with a much shorter one. This technique is called the overlap–save method and it involves overlapping input sections rather than output sections and discarding those samples in the circular convolution of sections that are not valid. Again we illustrate this procedure by way of an example. Figure 2.34 shows $h(n)$, an N_2-point sequence, and $x(n)$, which is sectioned into sequences of duration $(N_3 + N_2 - 1)$ samples, each of which overlap by $(N_2 - 1)$ samples. (Note that the overlap between sequences is placed in the second half of the

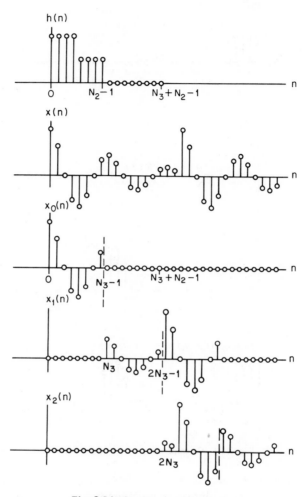

Fig. 2.34. Overlap–save method.

sequence $x_k(n)$ for convenience in computing circular convolutions via the DFT.) The $(N_3 + N_2 - 1)$-point circular convolution of $h(n)$ and $x_k(n)$ is computed for each section yielding the sequences $y_k(n)$ as shown in Fig. 2.35. The last $(N_2 - 1)$ samples of each $y_k(n)$ are discarded [since they are essentially incorrect due to the wraparound of $h(n)$] and the remaining samples of section $y_k(n)$ are abutted to the remaining samples of section $y_{k+1}(n)$, etc., to give $y(n)$. Thus using either the overlap–add or the overlap–save technique, it is relatively easy to compute the convolution of a short sequence with a much longer one in small chunks that can be pieced together in an appropriate manner.

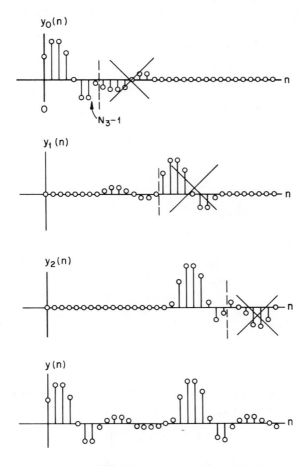

Fig. 2.35. Outputs of overlap–save method.

2.26 The Discrete Hilbert Transform

Until now we have discussed various representations of sequences including z transforms, Fourier transforms, and discrete Fourier series. In this section we show that for a linear, stable, causal digital system the z transform may be evaluated anywhere outside the unit circle explicitly in terms of the real or imaginary part of the sequence's Fourier transform. Consequently, the real and imaginary parts of the Fourier transform of the sequence can be expressed as functions of each other.

Consider the causal sequence $x(n)$ [i.e., $x(n) = 0, n < 0$] with z transform $X(z)$. We assume $X(z)$ is analytic outside the unit circle—i.e., all poles of $X(z)$ are within the unit circle. We let $X_R(e^{j\omega})$ and $X_I(e^{j\omega})$ be the real and imaginary parts of the Fourier transform of the sequence; i.e.,

$$X(e^{j\omega}) = X_R(e^{j\omega}) + jX_I(e^{j\omega}) \tag{2.170}$$

Defining $x_e(n)$, the even part of $x(n)$, as

$$x_e(n) = \tfrac{1}{2}[x(n) + x(-n)] \tag{2.171}$$

then $x(n)$ can be written as

$$x(n) = 2x_e(n) \cdot s(n) \tag{2.172}$$

where

$$s(n) = \begin{cases} 1 & n > 0 \\ \tfrac{1}{2} & n = 0 \\ 0 & n < 0 \end{cases} \tag{2.173}$$

The z transform of $x(n)$ may be evaluated outside the unit circle (for $z = re^{j\omega}, r > 1$) from Eq. (2.172) as

$$\begin{aligned} X(re^{j\omega}) &= \sum_{n=-\infty}^{\infty} x(n)r^{-n}e^{-jn\omega} \\ &= \sum_{n=-\infty}^{\infty} 2x_e(n)s(n)r^{-n}e^{-jn\omega} \end{aligned} \tag{2.174}$$

The right-hand side of Eq. (2.174) is recognized to be the Fourier transform of the sequence $y(n)$ defined as

$$y(n) = [2x_e(n)][s(n)r^{-n}] \tag{2.175}$$

Since $y(n)$ is the product of two sequences, the complex convolution theorem [Eq. (2.81)] can be used to express the Fourier transform of this product as a convolution of the Fourier transforms of the individual sequences; i.e.,

$$\begin{aligned} X(re^{j\omega}) &= X(z)\big|_{z=re^{j\omega}} \\ &= \frac{1}{2\pi j} \oint_C X_R(v)\left(\frac{e^{j\omega} + r^{-1}v}{e^{j\omega} - r^{-1}v}\right)\frac{dv}{v} \end{aligned} \tag{2.176}$$

where the z transform of $x_e(n)$ is $X_R(z)$, the z transform of $s(n)r^{-n}$ is $0.5(1 + r^{-1}z^{-1})/(1 - r^{-1}z^{-1})$, and the contour C is the unit circle.

Equation (2.176) expresses $X(re^{j\omega})$ outside the unit circle in terms of its real part on the unit circle. In the case when $X_R(z)$ is a rational fraction, Eq. (2.176) is easily evaluated by contour integration using residues.

In a similar manner an equation relating $X(e^{j\omega})$ to $X_I(e^{j\omega})$ can be obtained by representing $x(n)$ as

$$x(n) = 2x_0(n)s(n) + x(0)u_0(n) \tag{2.177}$$

where $x_0(n)$ is the odd part of $x(n)$ and is defined as

$$x_0(n) = \tfrac{1}{2}[x(n) - x(-n)] \tag{2.178}$$

The resulting equation for $X(re^{j\omega})$ is of the form

$$X(re^{j\omega}) = X(z)\big|_{z=re^{j\omega}}$$

$$= \frac{1}{2\pi} \oint_C X_I(v)\left(\frac{e^{j\omega} + r^{-1}v}{e^{j\omega} - r^{-1}v}\right)\frac{dv}{v} + x(0) \tag{2.179}$$

where the z transform of $x_0(n)$ is $X_I(z)$ and the contour C is again the unit circle.

In order to be able to evaluate $X_R(e^{j\omega})$ in terms of $X_I(e^{j\omega})$ and vice versa, Eq. (2.176) and (2.179) must be evaluated in the limit as $r \to 1$. In this case care must be taken to evaluate the integrals correctly as a pole is approaching the contour of integration. If the appropriate integrals are interpreted as Cauchy principle values of the integrals, i.e.,

$$\frac{1}{2\pi j} P \oint \frac{f(z)}{z - z_0}\, dz = \begin{cases} f(z_0) & |z_0| < 1 \\ 0 & |z_0| > 1 \\ \tfrac{1}{2}f(z_0) & |z_0| = 1 \end{cases} \tag{2.180}$$

(where $P \oint g(z)\, dz$ is the Cauchy principle value of the integral), then Eq. (2.176) and (2.179) can be evaluated in the limit as $r \to 1$. Rather than using Eq. (2.176) and (2.179) directly, we first let $v = e^{j\theta}$ (i.e., the unit circle is the contour of integration) and let $X(re^{j\omega})$ be represented as

$$X(re^{j\omega}) = X_R(re^{j\omega}) + jX_I(re^{j\omega}) \tag{2.181}$$

Equation (2.176) may be put in the form

$$X_R(re^{j\omega}) + jX_I(re^{j\omega}) = \frac{1}{2\pi} \int_{-\pi}^{\pi} X_R(e^{j\theta})$$

$$\times \frac{1 - r^{-2} + j2r^{-1}\sin(\theta - \omega)}{1 - 2r^{-1}\cos(\theta - \omega) + r^{-2}}\, d\theta \tag{2.182}$$

$$X_I(re^{j\omega}) = \frac{1}{2\pi} \int_{-\pi}^{\pi} \frac{X_R(e^{j\theta})2r^{-1}\sin(\theta - \omega)}{1 - 2r^{-1}\cos(\theta - \omega) + r^{-2}}\, d\theta \tag{2.183}$$

Similarly Eq. (2.179) can be manipulated to give the relation

$$X_R(re^{j\omega}) = -\frac{1}{2\pi} \int_{-\pi}^{\pi} \frac{X_R(e^{j\theta})2r^{-1}\sin(\theta-\omega)}{1-2r^{-1}\cos(\theta-\omega)+r^{-2}}\, d\theta + x(0) \quad (2.184)$$

Taking the limit in Eq. (2.183) and (2.184) as $r \to 1$ [and using Eq. (2.180) to evaluate the integrals] gives

$$X_I(e^{j\omega}) = \frac{1}{2\pi} P \int_{-\pi}^{\pi} X_R(e^{j\theta}) \cot\left(\frac{\theta-\omega}{2}\right) d\theta \quad (2.185)$$

and

$$X_R(e^{j\omega}) = x(0) - \frac{1}{2\pi} P \int_{-\pi}^{\pi} X_I(e^{j\theta}) \cot\left(\frac{\theta-\omega}{2}\right) d\theta \quad (2.186)$$

which is the desired result. Equations (2.185) and (2.186) are referred to as discrete Hilbert transforms. These equations give explicit relations for determining the imaginary (real) part of the frequency response of a causal system from the real (imaginary) part of the frequency response.

Discrete Hilbert transform relations can also be derived relating the log magnitude of the frequency response and the phase for a causal, *minimum phase* system. A minimum phase system is one for which all the poles and zeros of the system transfer function are inside the unit circle. In general these relations are not too useful because the zeros of the transfer function are overly restricted—i.e., no zeros may occur on or outside the unit circles, as is frequently the case.

2.27 Hilbert Transform Relations for Real Signals

One of the most important applications of the Hilbert transform relations is in modulation systems that often process complex, bandpass signals, such as might be used in a single sideband modulation system. Such signals have the property that their frequency response is exactly zero when evaluated on the lower half of the unit circle, i.e., in the region $\pi \le \omega < 2\pi$. If $v(n)$ is such a signal, its Fourier transform $V(e^{j\omega})$ has the property

$$V(e^{j\omega}) = 0 \qquad \pi \le \omega < 2\pi \quad (2.187)$$

Clearly $v(n)$ is a complex signal since the Fourier transforms of real signals have the property

$$V^*(e^{-j\omega}) = V(e^{j\omega}) \quad (2.188)$$

which would imply $V(e^{j\omega}) = 0$ if $v(n)$ were real. Since $v(n)$ is complex, it may be represented as

$$v(n) = x(n) + j\hat{x}(n) \quad (2.189)$$

where $x(n)$ and $\hat{x}(n)$ are real. Equation (2.187) requires

$$V(e^{j\omega}) = X(e^{j\omega}) + j\hat{X}(e^{j\omega}) = 0 \qquad \pi \leq \omega < 2\pi \qquad (2.190)$$

or

$$\hat{X}(e^{j\omega}) = jX(e^{j\omega}) \qquad \pi \leq \omega < 2\pi \qquad (2.191)$$

Since $\hat{x}(n)$ and $x(n)$ are real, it is easily seen that

$$\hat{X}(e^{j\omega}) = -jX(e^{j\omega}) \qquad 0 \leq \omega < \pi \qquad (2.192)$$

Thus the signal $\hat{x}(n)$ may be obtained from $x(n)$ by filtering it with a filter with frequency response $H(e^{j}\omega)$ defined as

$$H(e^{j\omega}) = \begin{cases} -j & 0 \leq \omega < \pi \\ j & \pi \leq \omega < 2\pi \end{cases} \qquad (2.193)$$

In this case $V(e^{j\omega}) = 2X(e^{j\omega})$ in the interval $0 \leq \omega < \pi$, and $V(e^{j\omega}) = 0$ in the interval $\pi \leq \omega < 2\pi$, as desired. The impulse response of the filter with the frequency response of Eq. (2.193) may be obtained as

$$h(n) = \frac{1}{2\pi}\left(\int_0^\pi -je^{j\omega n}\, d\omega + \int_\pi^{2\pi} je^{j\omega n}\, d\omega \right) \qquad (2.194)$$

$$h(n) = \begin{cases} \dfrac{1 - e^{j\pi n}}{\pi n} & n \neq 0 \\ 0 & n = 0 \end{cases}$$

or equivalently

$$h(n) = \begin{cases} \dfrac{2\sin^2(\pi n/2)}{\pi n} & n \neq 0 \\ 0 & n = 0 \end{cases} \qquad (2.195)$$

Equations (2.193) and (2.195) define the ideal digital Hilbert transform network. Figure 2.36 shows the frequency response and impulse response of the Hilbert transform network. It is clear that $h(n)$ is unrealizable since it is doubly infinite in extent. Furthermore the z transform of Eq. (2.195) converges only on the unit circle. Thus the ideal Hilbert transformer is like the ideal lowpass filter and the ideal differentiator in that such filters can never be realized exactly in practice. We shall see in Chapters 3 and 4 how realizable approximations to these ideal filters may be designed in practice.

Since $\hat{x}(n)$ may be obtained from $x(n)$ by a simple filtering operation, a convolutional relation exists between $x(n)$ and $\hat{x}(n)$. Such a relation may be written by inspection as

$$\hat{x}(n) = \frac{2}{\pi} \sum_{\substack{m=-\infty \\ m \neq n}}^{\infty} x(n - m) \frac{\sin^2(\pi m/2)}{m} \qquad (2.196)$$

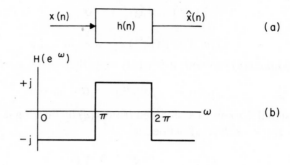

HILBERT TRANSFORMER

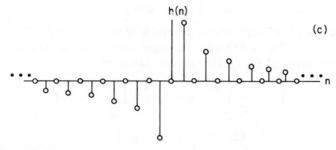

Fig. 2.36. Frequency and time response of the ideal Hilbert transformer.

Equivalently $x(n)$ may be obtained from $\hat{x}(n)$ via a filtering operation using a filter with impulse response the negative of Eq. (2.195). Thus a second relation between $x(n)$ and $\hat{x}(n)$ is

$$x(n) = -\frac{2}{\pi} \sum_{\substack{m=-\infty \\ m \neq n}}^{\infty} \hat{x}(n-m) \frac{\sin^2(\pi m/2)}{m} \qquad (2.197)$$

Equations (2.196) and (2.197) are the Hilbert transform relations between the real signals $x(n)$ and $\hat{x}(n)$.

The signal $v(n) = x(n) + j\hat{x}(n)$ has been called the analytic signal because of its similarity to continuous-time analytic signals in which the frequency response is zero for negative frequencies. An important application of the analytic signal is in the bandpass sampling problem. This application is illustrated in Fig. 2.37. Figure 2.37(a) shows the frequency response of a real bandpass signal $x(n)$. The analytic signal $v(n) = x(n) + j\hat{x}(n)$ has the frequency response shown in Fig. 2.37(b). For the signal $v(n)$, the sampling rate can now be reduced to $(B/2\pi)$ complex samples per second without aliasing of the spectrum. The spectrum of the reduced rate signal $w(n)$ is shown in Figure 2.37(c) (assuming $2\pi/B$ is an integer). To recover the original signal at the original sampling rate, a bandpass interpolation filter

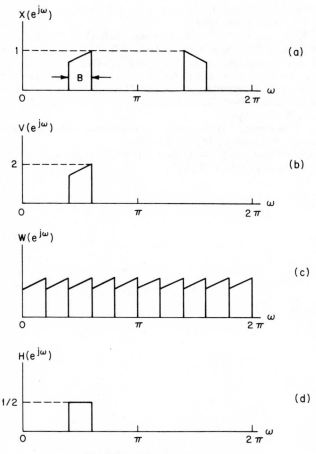

Fig. 2.37. Bandpass sampling.

is used as shown in Fig. 2.37(d). The real part of the interpolation filter output is the desired real signal.

REFERENCES

1. B. Gold and C. M. Rader, *Digital Processing of Signals*, McGraw-Hill Book Co., New York, 1969.
2. H. Freeman, *Discrete Time Systems*, John Wiley & Sons, Inc., New York, 1965.
3. E. I. Jury, *Sampled-Data Control Systems*, John Wiley & Sons, Inc., New York 1958.
4. J. R. Ragazzini and G. F. Franklin, *Sampled-Data Control Systems*, McGraw-Hill Book Co., New York, 1958.

5. E. I. JURY, *Theory and Application of the Z-Transform Method*, John Wiley & Sons, Inc., New York, 1964.

6. F. F. KUO AND J. F. KAISER, *Systems Analysis by Digital Computer*, John Wiley & Sons, Inc., New York, 1966.

7. C. M. RADER AND B. GOLD, "Digital Filter Design Techniques in the Frequency Domain," *Proc. IEEE*, **55**, No. 2, 149–171, Feb., 1967.

8. B. GOLD, A. V. OPPENHEIM, AND C. M. RADER, "Theory and Implementation of the Discrete Hilbert Transform," *Proc. Sym. Computer Proc. in Communication*, 235–250, 1969.

9. T. G. STOCKHAM, "High Speed Convolution and Correlation," *AFIPS Conference Proceedings*, **28**, 229–233, 1966.

10. A. V. OPPENHEIM AND R. W. SCHAFER, *Digital Signal Processing*, Prentice-Hall, Inc., Englewood Cliffs, N.J., 1975.

3

The Theory and Approximation of Finite Duration Impulse Response Digital Filters

3.1 Introduction

As seen in Chapter 2, the class of finite duration sequences possesses certain desirable properties from the point of view of filter design. For example, the questions of stability and realizability never arise since FIR sequences are always stable and, with an appropriate finite delay, can always be made realizable. Furthermore, as shown later in this chapter, FIR sequences can be designed so that their frequency responses have an *exactly* linear phase characteristic. Hence filter design problems in which an arbitrary magnitude response is desired can be tackled using FIR sequences.

It is interesting to note that until the advent of the fast Fourier transform (FFT) algorithm, FIR filters were generally felt to be computationally unfeasible because exceedingly long sequences were required to approximate sharp cutoff filters adequately. The computational efficiency of the FFT made the techniques of fast convolution feasible, however, and currently available FIR filters are in fact competitive even with sharp cutoff IIR filter designs.

3.2 Issues in Filter Design

The general process of designing a digital filter (for realization either in hardware or software) involves the following four basic steps:

1. Solve the approximation problem to determine filter coefficients that satisfy performance specifications.

2. Choose a specific structure in which the filter will be realized and quantize the resulting filter coefficients to a fixed word length.
3. Quantize the digital filter variables, i.e., the input, output, and intermediate variable word lengths.
4. Verify by simulation that the resulting design meets given performance specifications.

The results of step 4 generally lead to revisions in steps 2 and 3 in order to meet specifications.

Although it would be desirable to be able to perform steps 1 to 3 simultaneously, i.e., to be able to solve the approximation problem for arbitrary structures, with arbitrary word lengths, it is not likely that such a design procedure will be available in the foreseeable future. Thus for the time being we must be content to solve each of these problems independently. Thus in this chapter only the approximation problem for FIR filters will be discussed. Steps 2 and 3 will be discussed in Chapter 5 and step 4 is sufficiently straightforward so that no discussion is required.

3.3 Discussion of FIR Filters

There are many reasons for studying how to design FIR filters. Among the advantages of FIR filters are

1. FIR filters with exactly linear phase can be easily designed. This simplifies the approximation problem, in many cases, when one is only interested in designing a filter that approximates an arbitrary magnitude response. Linear phase filters are important for applications where frequency dispersion due to nonlinear phase is harmful—e.g., speech processing and data transmission.
2. Efficient realizations of FIR filters exist as both recursive and nonrecursive structures.
3. FIR filters realized nonrecursively, i.e., by direct convolution, are always stable.
4. Roundoff noise, which is inherent in realizations with finite precision arithmetic, can easily be made small for nonrecursive realizations of FIR filters.

Among the possible disadvantages of FIR filters are

1. A large value of N, the impulse response duration, is required to adequately approximate sharp cutoff filters. Hence a large amount of processing is required to realize such filters when realized via slow convolution.
2. The delay of linear phase FIR filters need not always be an integer number of samples. This nonintegral delay can lead to problems in some signal processing applications.

In the next sections the properties of FIR filters with linear phase are derived and then a discussion of several design techniques for FIR filters is given.

3.4 Characteristics of FIR Filters with Linear Phase

Let $\{h(n)\}$ be a causal finite duration sequence defined over the interval $0 \leq n \leq N - 1$. The z transform of $\{h(n)\}$ is

$$H(z) = \sum_{n=0}^{N-1} h(n)z^{-n} = h(0) + h(1)z^{-1} + \ldots + h(N-1)z^{-(N-1)} \quad (3.1)$$

The Fourier transform of $\{h(n)\}$ is

$$H(e^{j\omega}) = \sum_{n=0}^{N-1} h(n)e^{-j\omega n} \quad (3.2)$$

which is periodic in frequency with period 2π; i.e.,

$$H(e^{j\omega}) = H[e^{j(\omega+2\pi m)}] \qquad m = 0, \pm 1, \pm 2, \ldots \quad (3.3)$$

With the restriction that $\{h(n)\}$ is real, additional constraints on $H(e^{j\omega})$ are obtained by expressing it in terms of its magnitude and phase; i.e.,

$$H(e^{j\omega}) = \pm |H(e^{j\omega})| \, e^{j\theta(\omega)} \quad (3.4)$$

The operator $\pm$ in Eq. (3.4) is necessary since $H(e^{j\omega})$ is actually of the form

$$H(e^{j\omega}) = H^*(e^{j\omega})e^{j\theta(\omega)} \quad (3.5)$$

where $H^*(e^{j\omega})$ is a real function taking on both positive and negative values. From Eq. (3.2) it is seen that the magnitude of the Fourier transform is a symmetric function and the phase an antisymmetric function—i.e.,

$$|H(e^{j\omega})| = |H(e^{-j\omega})| \qquad 0 \leq \omega \leq \pi \quad (3.6a)$$

$$\theta(\omega) = -\theta(-\omega) \quad (3.6b)$$

For many practical FIR Filters, exact linearity of phase is a desired goal. We now determine the constraints on the impulse response, $h(n)$, leading to exact phase linearity. Thus we impose the additional constraint that the phase is linear, i.e., $\theta(\omega)$ is of the form

$$\theta(\omega) = -\alpha\omega \qquad -\pi \leq \omega \leq \pi \quad (3.7)$$

where α is a constant phase delay in samples. Using the results of Eq. (3.4) and (3.7), Eq. (3.2) can be written in the form

$$H(e^{j\omega}) = \sum_{n=0}^{N-1} h(n)e^{-j\omega n} = \pm |H(e^{j\omega})| \, e^{-j\alpha\omega} \quad (3.8)$$

Equating real and imaginary parts of the components of Eq. (3.8) gives the two equations

$$\pm |H(e^{j\omega})| \cos(\alpha\omega) = \sum_{n=0}^{N-1} h(n) \cos(\omega n) \tag{3.9a}$$

$$\pm |H(e^{j\omega})| \sin(\alpha\omega) = \sum_{n=0}^{N-1} h(n) \sin(\omega n) \tag{3.9b}$$

By taking the ratio of Eq. (3.9b) to Eq. (3.9a) (to eliminate $\pm |H(e^{j\omega})|$ from the equations) the relation

$$\frac{\sin(\alpha\omega)}{\cos(\alpha\omega)} = \tan(\alpha\omega) = \frac{\sum_{n=0}^{N-1} h(n) \sin(\omega n)}{\sum_{n=0}^{N-1} h(n) \cos(\omega n)} \tag{3.10}$$

is obtained or, equivalently,

$$\tan(\alpha\omega) = \frac{\sum_{n=1}^{N-1} h(n) \sin(\omega n)}{h(0) + \sum_{n=1}^{N-1} h(n) \cos(\omega n)} \tag{3.11}$$

There exist two possible solutions to Eq. (3.10) [or Eq. (3.11)]. The first possibility is that $\alpha = 0$, which implies [from Eq. (3.11)]

$$0 = \frac{\sum_{n=1}^{N-1} h(n) \sin(\omega n)}{h(0) + \sum_{n=1}^{N-1} h(n) \cos(\omega n)} \tag{3.12}$$

for which the only solution is that $h(0)$ is arbitrary and $h(n) = 0, n \neq 0$—i.e., the impulse response of the filter is an impulse—a result that is not too useful. The only other possible case is when $\alpha \neq 0$, which means that Eq. (3.11) can be rewritten by cross-multiplying terms to give

$$\sum_{n=0}^{N-1} h(n) \cos(\omega n) \sin(\alpha\omega) - \sum_{n=0}^{N-1} h(n) \sin(\omega n) \cos(\alpha\omega) = 0 \tag{3.13}$$

or, equivalently,

$$\sum_{n=0}^{N-1} h(n) \sin[(\alpha - n)\omega] = 0 \tag{3.14}$$

Since Eq. (3.14) is of the form of a Fourier series, if any solution can be found, it is guaranteed to be unique. It is fairly easy to see that one solution to Eq. (3.14) is the set of conditions

$$\alpha = \frac{(N-1)}{2} \tag{3.15}$$

$$h(n) = h(N - 1 - n) \qquad 0 \leq n \leq N - 1 \tag{3.16}$$

The significance of Eq. (3.15) and (3.16) should be emphasized. Equation (3.15) says that for every value of N there is only one value of phase delay α for which linear phase can be obtained exactly. Equation (3.16) says that for the value of α of Eq. (3.15) the impulse response sequence must have a special kind of symmetry.

It is worthwhile examining the implications of Eq. (3.15) and (3.16) for the separate cases where N is odd and even. When N is odd, α is an integer, which means the filter delay is an integer number of samples. A typical impulse response sequence of a linear phase filter is shown in Fig. 3.1 for the case $N = 11$ or $\alpha = 5$. The center of symmetry of the sequence occurs at the fifth sample.

A typical impulse response sequence for a linear phase filter for N even is shown in Fig. 3.2. For this example, $N = 10$; from Eq. (3.15), $\alpha = \frac{9}{2} = 4.5$. Thus the filter delay is $4\frac{1}{2}$ samples, reflecting the fact that the center of symmetry of the impulse response lies midway between two samples—as shown in Fig. 3.2. In Secs 3.36 and 3.37 we shall discuss several important filters that advantageously use this nonintegral number of samples delay property for linear phase filters with even impulse response durations.

The definition of a linear phase filter [Eq. (3.7)] requires the filter to have both constant group delay and constant phase delay. If only constant group

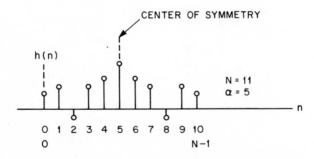

Fig. 3.1. Typical impulse response for N odd, even symmetry.

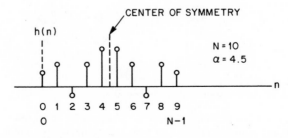

Fig. 3.2. Typical impulse response for N even, even symmetry.

delay[1] is desired (as is often the case), however, a second type of "linear phase" filter can be defined in which the phase of $H(e^{j\omega})$ is a piece-wise linear function of ω; i.e.,

$$H(e^{j\omega}) = \pm|H(e^{j\omega})|\, e^{j(\beta-\alpha\omega)} \tag{3.17}$$

Using an argument similar to the one in Eq. (3.8) to (3.14), it can be shown that the only possible new solutions for $\{h(n)\}$, α and β, are

$$\alpha = \frac{(N-1)}{2} \tag{3.18a}$$

$$\beta = \pm\frac{\pi}{2} \tag{3.18b}$$

$$h(n) = -h(N-1-n) \qquad 0 \le n \le N-1 \tag{3.18c}$$

Filters that satisfy Eq. (3.18) again have a delay of $[(N-1)/2]$ samples but their impulse responses are anti-symmetric around the center of the sequence, as opposed to the true linear phase sequences that are symmetric around the center of the sequence. By way of example, Fig. 3.3(a) and (b) show impulse

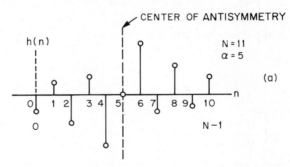

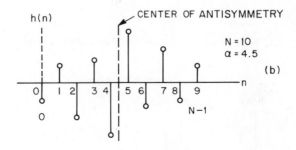

Fig. 3.3. Impulse responses for odd and even values of N, odd symmetry.

[1] The group delay of a filter is defined as the derivative of the phase with respect to frequency, as opposed to the phase delay which is defined as the phase divided by the frequency.

responses of filters that satisfy Eq. (3.18) for both odd and even values of N. It should be noted that for odd values of N, Eq. (3.18c) requires $h[(N-1)/2] = 0$. Application of the conditions of Eq. (3.18) will be given in Sec. 3.36 and 3.37 for the design of wideband differentiators and Hilbert transformers.

In summary, depending on the value of N (odd or even) and the type of symmetry of the filter impulse response sequence (symmetric or antisymmetric), there are four possible types of linear phase FIR filters.

3.5 Frequency Response of Linear Phase FIR Filters

Expressing the frequency response of linear phase FIR filters in the form

$$H(e^{j\omega}) = H^*(e^{j\omega})e^{j(\beta-\alpha\omega)} \tag{3.19}$$

where $H^*(e^{j\omega})$ is purely real and α and β are given by Eq. (3.18), the function $H^*(e^{j\omega})$ may be expressed in terms of the impulse response coefficients for each of the four cases of a linear phase filter. Such formulas are derived in this section for later use in describing various techniques for designing FIR filters to match prescribed frequency response characteristics.

Case 1 Symmetrical Impulse Response, N Odd
For this case $H(e^{j\omega})$ can be put in the form

$$H(e^{j\omega}) = \sum_{n=0}^{(N-3)/2} h(n)e^{-j\omega n} + h\left[\frac{(N-1)}{2}\right]e^{-j\omega(N-1)/2}$$

$$+ \sum_{n=(N+1)/2}^{N-1} h(n)e^{-j\omega n} \tag{3.20}$$

Making the substitution $m = N - 1 - n$ in the last summation in Eq. (3.20) gives

$$H(e^{j\omega}) = \sum_{n=0}^{(N-3)/2} h(n)e^{-j\omega n} + h\left[\frac{(N-1)}{2}\right]e^{-j\omega(N-1)/2}$$

$$+ \sum_{m=0}^{(N-3)/2} h(N-1-m)e^{-j\omega(N-1-m)} \tag{3.21}$$

Since $h(n) = h(N-1-n)$, the first and third summations can be combined and the term $e^{-j\omega(N-1)/2}$ can be factored out to give

$$H(e^{j\omega}) = e^{-j\omega(N-1)/2}$$

$$\times \left[\sum_{n=0}^{(N-3)/2} h(n)\{e^{j\omega[(N-1)/2-n]} + e^{-j\omega[(N-1)/2-n]}\} + h\left[\frac{(N-1)}{2}\right]\right]$$

$$\tag{3.22}$$

or, equivalently,

$$H(e^{j\omega}) = e^{-j\omega(N-1)/2} \left\{ \sum_{n=0}^{(N-3)/2} 2h(n) \cos\left[\omega\left(\frac{N-1}{2} - n\right)\right] + h\left[\frac{(N-1)}{2}\right] \right\}$$

(3.23)

Letting $m = (N-1)/2 - n$, Eq. (3.23) becomes

$$H(e^{j\omega}) = e^{-j\omega(N-1)/2} \left[\sum_{m=0}^{(N-3)/2} 2h\left(\frac{N-1}{2} - m\right) \cos(\omega m) + h\left[\frac{(N-1)}{2}\right] \right]$$

(3.24)

Finally letting $a(0) = h[(N-1)/2]$, and $a(n) = 2h[(N-1)/2 - n]$, $n = 1, 2, \ldots, (N-1)/2$, Eq. (3.25) can be written as

$$H(e^{j\omega}) = e^{-j\omega(N-1)/2} \left[\sum_{n=0}^{(N-1)/2} a(n) \cos(\omega n) \right]$$

(3.25)

which is the desired result. Thus for Case 1

$$H^*(e^{j\omega}) = \sum_{n=0}^{(N-1)/2} a(n) \cos(\omega n)$$

(3.26)

Case 2   Symmetrical Impulse Response, N Even
For this case $H(e^{j\omega})$ assumes the form

$$H(e^{j\omega}) = e^{-j\omega(N-1)/2} \left\{ \sum_{n=0}^{N/2-1} 2h(n) \cos\left[\omega\left(\frac{N}{2} - n - \frac{1}{2}\right)\right] \right\}$$

(3.27)

Letting

$$b(n) = 2h\left(\frac{N}{2} - n\right) \qquad n = 1, 2, \ldots, \frac{N}{2}$$

Eq. (3.27) becomes

$$H(e^{j\omega}) = e^{-j\omega(N-1)/2} \left\{ \sum_{n=1}^{N/2} b(n) \cos[\omega(n - \tfrac{1}{2})] \right\}$$

(3.28)

Thus for Case 2

$$H^*(e^{j\omega}) = \sum_{n=1}^{N/2} b(n) \cos[\omega(n - \tfrac{1}{2})]$$

(3.29)

It should be noted that at $\omega = \pi$, $H^*(e^{j\omega}) = 0$, independent of $b(n)$ [or $h(n)$]. This implies that filters with a frequency response that is nonzero at $\omega = \pi$ (e.g., a highpass filter) cannot be satisfactorily approximated with this type of filter.

Case 3   Anti-symmetrical Impulse Response, N Odd
The derivation for $H^*(e^{j\omega})$ for Case 3 is almost identical to that of Case 1 except, because of the anti-symmetry in $\{h(n)\}$, the cosine summations are

replaced by sine summations multiplied by j; i.e., Eq. (3.24) is replaced by

$$H(e^{j\omega}) = e^{-j\omega(N-1)/2}e^{j\pi/2}\left[\sum_{m=0}^{(N-3)/2} 2h\left(\frac{N-1}{2} - m\right)\cos(\omega m)\right] \quad (3.30)$$

where $h[(N-1)/2] = 0$ as explained previously. Making the substitution $c(n) = 2h[(N-1)/2 - n]$, $n = 1, 2, \ldots, (N-1)/2$, Eq. (3.30) becomes

$$H(e^{j\omega}) = e^{-j\omega(N-1)/2}e^{j\pi/2}\left[\sum_{n=1}^{(N-1)/2} c(n)\sin(\omega n)\right] \quad (3.31)$$

Thus for Case 3

$$H^*(e^{j\omega}) = \sum_{n=1}^{(N-1)/2} c(n)\sin(\omega n) \quad (3.32)$$

At the frequencies $\omega = 0$ and $\omega = \pi$, Eq. (3.32) shows $H^*(e^{j\omega}) = 0$, independent of $c(n)$ or equivalently $h(n)$. Furthermore, the factor $e^{j\pi/2} = j$ in Eq. (3.31) shows the frequency response to be imaginary to within a linear phase factor. Thus this case of filters is most suitable for such filters as Hilbert transformers and differentiators.

Case 4   Anti-symmetrical Impulse Response, N Even

This case is again similar to Case 2 except the cosine summations become sine summations multiplied by j. Thus Eq. (3.27) is replaced by

$$H(e^{j\omega}) = e^{-j\omega(N-1)/2}e^{j\pi/2}\left\{\sum_{n=0}^{(N/2)-1} 2h(n)\sin\left[\omega\left(\frac{N}{2} - n - \frac{1}{2}\right)\right]\right\} \quad (3.33)$$

Letting

$$d(n) = 2h\left(\frac{N}{2} - n\right) \qquad n = 1, 2, \ldots, \frac{N}{2}$$

Eq. (3.33) becomes

$$H(e^{j\omega}) = e^{-j\omega(N-1)/2}e^{j\pi/2}\left\{\sum_{n=1}^{N/2} d(n)\sin\left[\omega(n - \tfrac{1}{2})\right]\right\} \quad (3.34)$$

Thus for Case 4

$$H^*(e^{j\omega}) = \sum_{n=1}^{N/2} d(n)\sin\left[\omega(n - \tfrac{1}{2})\right] \quad (3.35)$$

For Case 4, at $\omega = 0$, $H^*(e^{j\omega}) = 0$. Thus this class of filters is most suitable for approximating such filters as differentiators and Hilbert transformers.

Figure 3.4 presents a comprehensive summary of the results of this section. Shown in this figure are typical impulse response sequences $h(n)$, the resulting shifted sequence [$a(n)$ through $d(n)$, depending on the case], and typical frequency response functions $H^*(e^{j\omega})$ for each of the four cases of linear phase FIR filter.

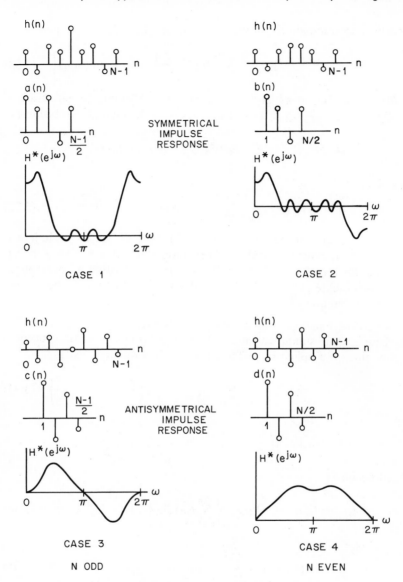

Fig. 3.4. Summary of cases of linear phase filters.

3.6 Positions of the Zeros of Linear Phase FIR Filters

The positions of the zeros of linear phase FIR filters are highly constrained by the symmetry conditions on the impulse response. It is easy to see where the zeros of such filters can lie by examining their z transforms, which can be put in the form

$$H(z) = \sum_{n=0}^{N-1} h(n)z^{-n} = h(0) + h(1)z^{-1} + h(2)z^{-2} + \cdots$$

$$\pm h(2)z^{-(N-3)} \pm h(1)z^{-(N-2)} \pm h(0)z^{-(N-1)} \quad (3.36)$$

where the $+$ sign corresponds to symmetry in the impulse response and the $-$ sign corresponds to anti-symmetry in the impulse response. Equation (3.36) may be conveniently put into the form

$$H(z) = z^{-(N-1)/2}\{h(0)[z^{(N-1)/2} \pm z^{-(N-1)/2}] + h(1)[z^{(N-3)/2} \pm z^{-(N-3)/2}]$$

$$+ h(2)[z^{(N-5)/2} \pm z^{-(N-5)/2}] + \cdots\} \quad (3.37)$$

If z is replaced by z^{-1} in Eq. (3.37), the result is

$$H(z^{-1}) = z^{(N-1)/2}\{h(0)[z^{-(N-1)/2} \pm z^{(N-1)/2}] + h(1)[z^{-(N-3)/2} \pm z^{(N-3)/2}]$$

$$+ h(2)[z^{-(N-5)/2} \pm z^{(N-5)/2}] + \cdots\} \quad (3.38)$$

or relating Eq. (3.37) to Eq. (3.38) shows

$$H(z^{-1}) = \pm z^{(N-1)}H(z) \quad (3.39)$$

Equation (3.39) shows that $H(z)$ and $H(z^{-1})$ are identical to within a delay of $(N-1)$ samples and a multiplier of ± 1. Thus the zeros of $H(z^{-1})$ are identical to the zeros of $H(z)$. Thus, assume $H(z)$ has a complex zero at $z_i = r_i e^{j\theta_i}$, with $r_i \neq 1$, $\theta_i \neq 0$, π. Equation (3.39) says that $H(z)$ must also have a mirror-image zero at $z_i^{-1} = (1/r_i)e^{-j\theta_i}$. Since the impulse response of the filter is real, every complex zero of $H(z)$ has a complex conjugate partner. Thus from the argument above for every complex zero of $H(z)$, off the unit circle ($r_i \neq 1$) and off the real axis ($\theta_i \neq 0$, π), $H(z)$ consists of at least the elemental factor $H_i(z)$ of the form or

$$H_i(z) = (1 - z^{-1}r_i e^{j\theta_i})(1 - z^{-1}r_i e^{-j\theta_i})\left(1 - z^{-1}\frac{1}{r_i}e^{j\theta_i}\right)$$

$$\times \left(1 - z^{-1}\frac{1}{r_i}e^{-j\theta_i}\right) \quad (3.40)$$

or

$$H_i(z) = 1 - 2\left(\frac{r_i^2 + 1}{r_i}\right)\cos\theta_i z^{-1}$$

$$+ \left(r_i^2 + \frac{1}{r_i^2} + 4\cos^2\theta_i\right)z^{-2} - 2\left(\frac{r_i^2 + 1}{r_i}\right)\cos\theta_i z^{-3} + z^{-4} \quad (3.41)$$

The zero configuration for the elemental section of Eq. (3.41) is shown in

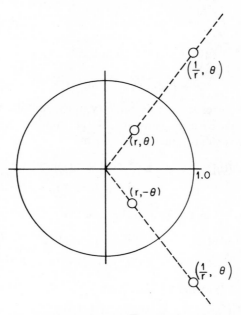

Fig. 3.5. Zero positions for linear phase filters.

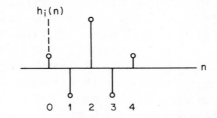

Fig. 3.6. Impulse response for elemental section of Fig. 3.5.

Fig. 3.5 and the equivalent elemental impulse response is shown in Fig. 3.6.

Equation (3.41) was obtained assuming $r_i \neq 1$ and $\theta_i \neq 0, \pi$. If $r_i = 1 (\theta_i \neq 0, \pi)$, the zero lies on the unit circle and its complex conjugate is its own mirror image. Thus zeros of $H(z)$ which are on the unit circle are also zeros of $H(z^{-1})$ which are on the unit circle. For such zeros the elemental factor is modified to the form

$$H_i(z) = (1 - z^{-1}e^{j\theta_i})(1 - z^{-1}e^{-j\theta_i}) = 1 - 2\cos\theta_i z^{-1} + z^{-2} \quad (3.42)$$

If $r_i \neq 1$ but $\theta_i = 0$ or π, the zeros are real. Hence there are no complex conjugate partners to include so for this case the elemental factor is

$$H_i(z) = (1 \pm r_i z^{-1})\left(1 \pm \frac{1}{r_i}z^{-1}\right) = 1 \pm \left(r_i + \frac{1}{r_i}\right)z^{-1} + z^{-2} \quad (3.43)$$

where the $+$ sign corresponds to $\theta_i = \pi$ and the $-$ sign corresponds to $\theta_i = 0$.

Finally if $r_i = 1$ and $\theta_i = 0$ or π, the zeros lie at either $z = +1$ or $z = -1$. For these cases the zero is its own complex conjugate as well as mirror-image partner. In this case the elemental factor is

$$H_i(z) = (1 \pm z^{-1}) \tag{3.44}$$

for such zeros. The factors of Eq. (3.44) are important because they represent networks with half-sample delays; hence an odd number of them must

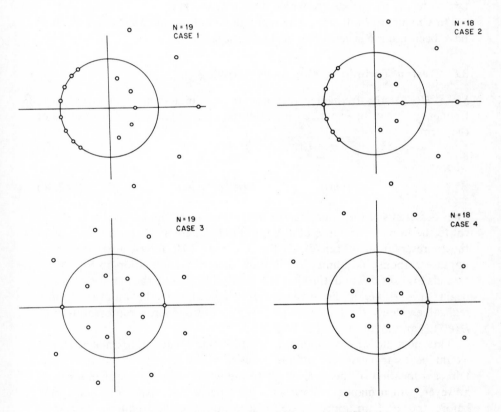

Fig. 3.7. Zero positions of typical linear phase filters.

be included when the filter impulse response duration N is even and an even number (possibly 0) must be included when N is odd.

Figure 3.7 shows a plots of typical zero positions for each of the four cases of linear phase FIR filters. The positions of the zeros are as predicted by the results of this section.

3.7   Design Techniques for Linear Phase FIR Filters

There are essentially three well-known classes of design techniques for linear phase FIR filters—namely, the window method, the frequency sampling, and optimal (in the Chebyshev sense) filter design methods. Since there are many factors that determine which class of filter will be used in a particular application, it is difficult, if not impossible, to recommend one class of design techniques to the exclusion of all others. Instead, the known weaknesses and strengths of each of these design techniques will be presented, and it will be left up to the user to decide which filter best meets his own specifications.

In Chapters 12 and 13, the reader will be able to see what types of filters have been used in several practical examples.

3.8   Design Technique No. 1—Windowing

Since $H(e^{j\omega})$, the frequency response of any digital filter, is periodic in frequency, it can be expanded in a Fourier series. The resultant series is of the form

$$H(e^{j\omega}) = \sum_{n=-\infty}^{\infty} h(n)e^{-j\omega n} \tag{3.45}$$

where

$$h(n) = \frac{1}{2\pi} \int_0^{2\pi} H(e^{j\omega})e^{j\omega n} \, d\omega \tag{3.46}$$

The coefficients of the Fourier series $h(n)$ are easily recognized as being identical to the impulse response of a digital filter. There are two difficulties with the representation of Eq. (3.45) for designing FIR filters. First, the filter impulse response is infinite in duration since the summation in Eq. (3.45) extends to $\pm\infty$. Second, the filter is unrealizable because the impulse response begins at $-\infty$; i.e., no finite amount of delay can make the impulse response realizable. Hence the filter resulting from a Fourier series representation of $H(e^{j\omega})$ is an unrealizable II R filter.

One possible way of obtaining an FIR filter that approximates $H(e^{j\omega})$ would be to truncate the infinite Fourier series (Eq. (3.45) at $n = \pm M$. Direct truncation of the series leads to the well-known Gibbs phenomenon, however, which manifests itself as a fixed percentage overshoot and ripple before and after an approximated discontinuity in the frequency response. Thus, for example, in the approximation of such standard filters as the ideal lowpass or bandpass filter, the largest ripple in the frequency response is about 9% of the size of the discontinuity and its amplitude does *not* decrease with increasing impulse response duration—i.e., including more and more terms in the Fourier series does not decrease the amplitude of the largest ripple. Instead, the overshoot is confined to a smaller and smaller frequency range as N is increased. Since any reasonable design technique must be capable of designing good approximations to ideal lowpass filters, direct truncation of Eq. (3.45) is not a reasonable way of obtaining an FIR filter.

A more successful way of obtaining an FIR filter is to use a finite weighting sequence $w(n)$, called a window, to modify the Fourier coefficients $h(n)$ in Eq. (3.45) to control the convergence of the Fourier series. The technique of windowing is illustrated in Fig. 3.8. At the top of this figure is shown the desired periodic frequency response $H(e^{j\omega})$ and its Fourier series coefficients $\{h(n)\}$. The next row shows a finite duration weighting sequence $w(n)$ with Fourier transform $W(e^{j\omega})$. $W(e^{j\omega})$, for most reasonable windows, consists of a central lobe which contains most of the energy of the window and side lobes which generally decay rapidly. To produce an FIR approximation to

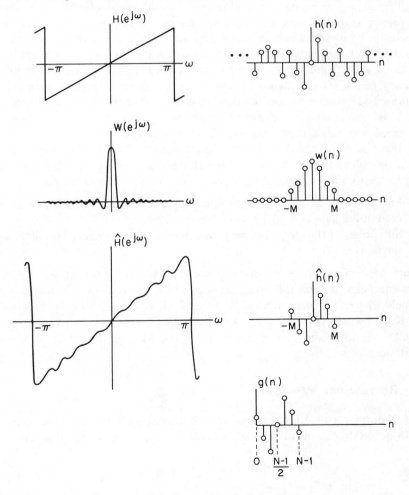

Fig. 3.8. Illustration of windowing.

$H(e^{j\omega})$, the sequence $\hat{h}(n) = h(n) \cdot w(n)$ is formed. Outside the interval $-M \le n \le M$, $\hat{h}(n)$ is zero exactly. The third row of Fig. 3.8 shows $\hat{h}(n)$ and its Fourier transform $\hat{H}(e^{j\omega})$, which is readily seen to be the circular convolution of $H(e^{j\omega})$ and $W(e^{j\omega})$, since $\hat{h}(n)$ is the product of the sequences $h(n)$ and $w(n)$. Finally the last row of Fig. 3.8 shows the realizable sequence $g(n)$, which is a shifted version of $\hat{h}(n)$ and may be used as the desired filter impulse response.

As seen in the simple example of Fig. 3.8, there are several noteworthy effects of windowing the Fourier coefficients of the filter on the resulting frequency response. A major effect is that discontinuities in $H(e^{j\omega})$ become transition bands between values on either side of the discontinuity. Since the final frequency response of the filter is the circular convolution of the ideal frequency response with the window's frequency response, it is clear that the width of these transition bands depends on the width of the main lobe of $W(e^{j\omega})$. A secondary effect of windowing is that ripple from the side lobes of $W(e^{j\omega})$ produces approximation errors (ripple in the resulting frequency response) for all ω. Finally since the filter frequency response is obtained via a convolution relation, it is clear that the resulting filters are *never optimal* in any sense, even though the windows from which they are obtained may satisfy some reasonable optimality criterion.

The discussion above leads to the questions of what are desirable window characteristics and how closely are they attained in practice. The first question can be answered relatively simply. Desirable window characteristics are

1. Small width of main lobe of the frequency response of the window containing as much of the total energy as possible.
2. Side lobes of the frequency response that decrease in energy rapidly as ω tends to π.

There have been many windows proposed that approximate the desired characteristics. In the following sections three windows will be discussed, namely, the rectangular window, the "generalized" Hamming window, and the Kaiser window. These three designs cover the entire range of issues with specific windows and give a reasonably good idea of the advantages and limitations of windows.

3.9 Rectangular Window

The N-point rectangular window, which corresponds to direct truncation (with no modification) of the Fourier series, has the weighting function

$$w_R(n) = \begin{cases} 1.0 & -\left(\dfrac{N-1}{2}\right) \le n \le \dfrac{N-1}{2} \\ 0.0 & \text{elsewhere} \end{cases} \tag{3.47}$$

(In this and following sections on windows, N is assumed odd. With simple modifications, similar results can be derived for the case when N is even. It is also assumed that the window sequence has zero delay.) The frequency response of the rectangular window is

$$W_R(e^{j\omega}) = \sum_{n=-(N-1)/2}^{(N-1)/2} e^{-j\omega n}$$

$$= \frac{e^{j\omega[(N-1)/2]}(1 - e^{-j\omega N})}{(1 - e^{j\omega})} \tag{3.48}$$

$$= \frac{e^{j\omega(N/2)} - e^{-j\omega(N/2)}}{e^{j(\omega/2)} - e^{-j(\omega/2)}}$$

$$W_R(e^{j\omega}) = \frac{\sin(\omega N/2)}{\sin(\omega/2)} \tag{3.49}$$

A sketch of Eq. (3.49) is shown in Fig. 3.9 for the case $N = 25$.

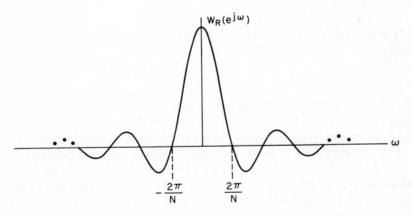

Fig. 3.9. Frequency response of a rectangular window.

3.10 "Generalized" Hamming Window

A second proposed window, called the generalized Hamming window, is of the form

$$w_H(n) = \begin{cases} \alpha + (1 - \alpha) \cos\left(\dfrac{2\pi n}{N}\right) & -\left(\dfrac{N-1}{2}\right) \le n \le \dfrac{N-1}{2} \\ 0.0 & \text{elsewhere} \end{cases} \tag{3.50}$$

where α is in the range $0 \leq \alpha \leq 1.0$. If $\alpha = 0.54$, the window is called a Hamming window; if $\alpha = 0.5$, it is called a hanning window.[2]

The frequency response of this window can readily be obtained by the observation that the window can be represented as the product of a rectangular window and an infinite duration window of the form of Eq. (3.50) but defined for all n; i.e.,

$$w_H(n) = w_R(n) \cdot \left[\alpha + (1 - \alpha) \cos \left(\frac{2\pi n}{N} \right) \right] \tag{3.51}$$

where $w_R(n)$ is a rectangular window as defined in Eq. (3.47). The frequency response of the generalized Hamming window is therefore the convolution (circular) of the frequency response of the rectangular window $W_R(e^{j\omega})$ with an impulse train, which can be written as

$$W_H(e^{j\omega})$$

$$= W_R(e^{j\omega}) * \left[\alpha u_0(\omega) + \frac{(1 - \alpha)}{2} u_0 \left(\omega - \frac{2\pi}{N} \right) + \frac{(1 - \alpha)}{2} u_0 \left(\omega + \frac{2\pi}{N} \right) \right] \tag{3.52}$$

or

$$W_H(e^{j\omega}) = \alpha W_R(e^{j\omega}) + \frac{(1 - \alpha)}{2} W_R[e^{j(\omega - 2\pi/N)}]$$

$$+ \frac{(1 - \alpha)}{2} W_R[e^{j(\omega + 2\pi/N)}] \tag{3.53}$$

Figure 3.10 shows a plot of the three components of $W_H(e^{j\omega})$ (at the top) and the resulting frequency response (at the bottom) for $\alpha = 0.54$, $N = 25$. Although the overall frequency response of the Hamming window appears to have no ripples beyond $\omega = 4\pi/N$, this is not the case. On the linear amplitude scale of Fig. 3.10, however, the ripples are not visible. A comparison of Figs. 3.9 and 3.10 shows that the main lobe of the frequency response of the Hamming window is twice the width of the main lobe of the frequency response of the rectangular window. The side lobe amplitudes of the Hamming window frequency response are considerably smaller, however, than the side lobe amplitudes of the rectangular window frequency response. For $\alpha = 0.54$, i.e., the conventional Hamming window, 99.96% of the spectral energy is in the main lobe and the peak side lobe ripple is down about 40 dB from the main lobe peak. In contrast, for the rectangular window, the spectral side lobes are down only about 14 dB from the main lobe peak.

Figure 3.10 gives a good idea of how the Hamming window trades off transition width for ripple cancellation. It can be seen that the side lobe

[2] The reader should note that the terms *Hamming* and *hanning* are often confused in the digital signal processing literature.

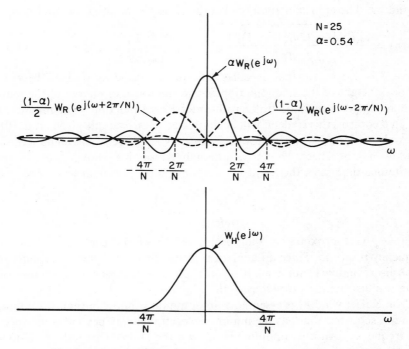

Fig. 3.10. Frequency response of a Hamming window for $\alpha = 0.54$.

ripples from the functions $W_R[e^{j(\omega \pm 2\pi/N)}]$ tend to cancel the side lobe ripples from $W_R(e^{j\omega})$, thereby reducing considerably the overall side lobe level. At the same time the width of the main lobe of the frequency response is proportionally bigger. As will be seen, for a lowpass filter increased main lobe width corresponds to increased transition bandwidth between the passband and the stopband, whereas decreased side lobe levels correspond to smaller ripples (more rejection) in the filter stopband and passband.

3.11 Kaiser Window

The problem of designing good windows essentially reduces to the mathematical problem of finding a time-limited function whose Fourier transform best approximates a bandlimited function—i.e., a time-limited function that has the minimum energy outside some selected frequency interval. For continuous-time functions this problem has been solved in closed form to give a class of functions called *prolate spheroidal wavefunctions*. The form of these functions is generally quite complicated. Kaiser has proposed a relatively simple approximation to these functions as an ideal candidate for a window. The approximation, called the Kaiser window, is of the form

$$w_K(n) = \frac{I_0(\beta\sqrt{1 - [2n/(N-1)]^2})}{I_0(\beta)}, \qquad -\left(\frac{N-1}{2}\right) \le n \le \frac{N-1}{2} \qquad (3.54)$$

where β is a constant that specifies a frequency response tradeoff between the peak height of the side lobe ripples and the width or energy of the main lobe and $I_0(x)$ is the modified zeroth-order Bessel function. The manner in which β controls this tradeoff will be illustrated in an example given on p. 101.

A closed form expression for the frequency response of the digital Kaiser window has not been obtained, although Kaiser has shown that, in the continuous-time case, the frequency response is proportional to

$$\frac{\sin[\beta\sqrt{(\omega/\omega_\beta)^2 - 1}]}{\sqrt{(\omega/\omega_\beta)^2 - 1}}$$

where ω_β is approximately the spectral width of the central lobe of the frequency response. Since an analytical result for the frequency response of the Kaiser window is not available, actual plots of its frequency response will serve to illustrate the properties of this window.

The Kaiser window is essentially an optimum window in that it is a finite duration sequence that has the minimum spectral energy beyond some specified frequency. Another optimum window is the Dolph–Chebyshev window, which has the minimum width of the main lobe of its frequency response for a fixed peak value of side lobe ripple. This window has the property that all its spectral side lobes are of equal amplitude. As mentioned previously, however, neither of these windows can be used to design optimum (minimax) approximations to any ideal frequency response function since they are convolved with the desired ideal response to give the actual filter response. Thus although there are optimum windows, there are no optimum window designed filters.

3.12   Examples of a Windowed Lowpass Filter

This section demonstrates how to use windows in a practical example—the design of an ideal lowpass filter. The use of three windows will be considered— the rectangular, Hamming, and Kaiser windows. Figures 3.11 to 3.16 show plots of the time and frequency responses of these three windows for $N = 257$. (The parameter β for the Kaiser window is 7.865.) Figures 3.11, 3.13, and 3.15 show the impulse responses and Figures 3.12, 3.14, and 3.16 show the frequency responses.[3] As discussed earlier, Fig. 3.12 shows the peak of the

---

[3] The apparent modulation of the minima of the log magnitude function (as in Fig. 3.12) is an anomaly due to slight inaccuracies in evaluating the zeros of the function and should not be regarded as significant.

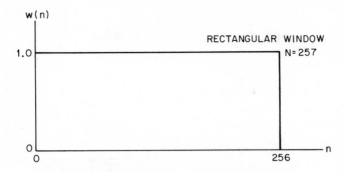

Fig. 3.11. A 257 point rectangular window.

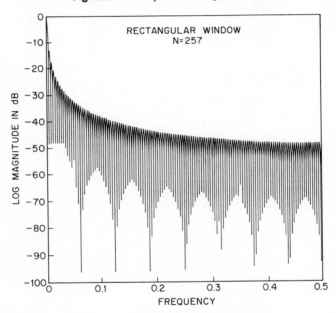

Fig. 3.12. Frequency response of a 257 point rectangular window.

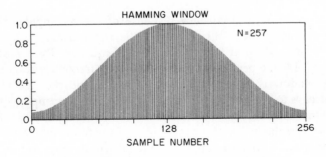

Fig. 3.13. A 257 point Hamming window.

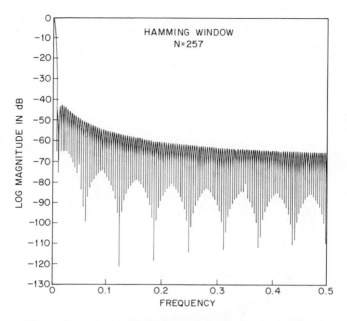

Fig. 3.14. Frequency response of a 257 point Hamming window.

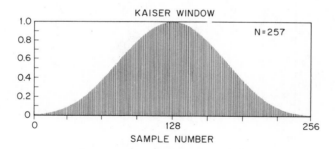

Fig. 3.15. A 257 point Kaiser window.

side lobe ripple for the frequency response of the rectangular window is 0.217 or, on a log scale, −13.27 dB and the ripple peaks decay slowly to about 0.004 or −48 dB at half the sampling frequency. By comparison, as seen in Fig. 3.14, the peak side lobe ripple of the frequency response of the Hamming window is about 0.0074 or −42.7 dB and the side lobe ripple envelope decays to about 0.000059 or −65 dB at half the sampling frequency. However, the width of the main lobe of the Hamming window frequency response is twice the width of the main lobe of the frequency response of the rectangular window. Thus in approximating discontinuities in the frequency response of an ideal filter (as shown below for a lowpass filter) the width of the

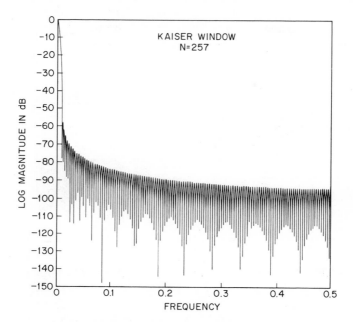

Fig. 3.16. Frequency response of Kaiser window.

transition band around the discontinuity using a Hamming window is at least twice the width when using a rectangular window. For the Kaiser window (Fig. 3.16), the peak ripple of the frequency response is 0.00133 or -57 dB and the side lobe ripple envelope decays to a value of about 0.00002 or -94 dB at half the sampling frequency. To compensate for such excellent ripple characteristics, however, the width of the main lobe of the frequency response is almost three times the width of the main lobe of the rectangular window's frequency response.

The next set of figures shows a lowpass filter designed using each of the three windows just described. Figures 3.17, 3.19, and 3.21 show the resulting filter's impulse and step response, whereas Fig. 3.18, 3.20, and 3.22 show the final frequency response of the filter. The ideal lowpass filter being designed has the Fourier series coefficients (impulse response)

$$h(n) = \frac{\sin (2\pi F_c n)}{\pi n} \qquad -\infty < n < \infty \qquad (3.55)$$

where $F_c = 0.1245$ in these examples. Figure 3.17 shows the product of $h(n)$ with a rectangular window. The basic $(\sin n/n)$ pattern of the impulse response may be observed in this figure. A comparison with Figs. 3.19 and 3.21 shows the attenuation of the higher values of $h(n)$ by the Hamming and Kaiser windows. Figure 3.18 shows the rectangular window lowpass

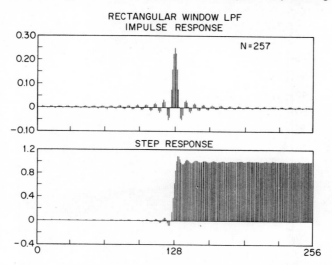

Fig. 3.17. Impulse and step responses of rectangular window lowpass filter.

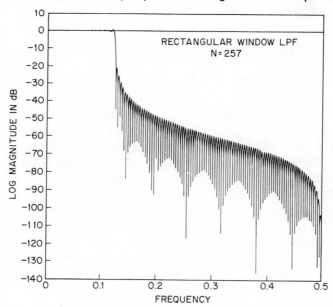

Fig. 3.18. Frequency response of rectangular window lowpass filter.

filter frequency response. The characteristic Gibbs phenomenon is clearly seen in this figure where the amplitude of the first ripple is approximately 0.09 or −21 dB on a log scale. The width of the transition band is 0.9375N ($N = 257$), indicating a very narrow transition band for this filter. Because of the Gibbs phenomenon ripple, however, this filter is impractical for

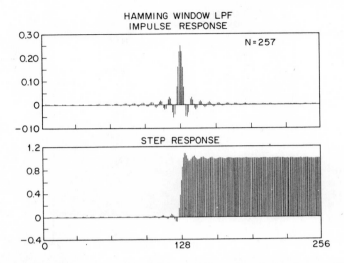

Fig. 3.19. Impulse and step responses of Hamming window lowpass filter.

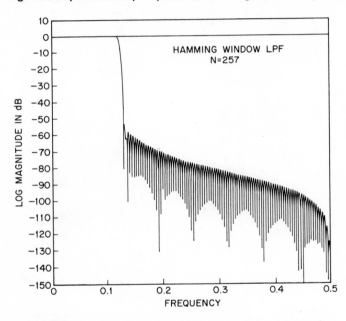

Fig. 3.20. Frequency response of Hamming window lowpass filter.

many applications. Figure 3.20 shows the Hamming window lowpass filter frequency response. For this filter the peak passband ripple is about 0.0018 and the peak stopband ripple is 0.002 or -53.6 dB. The width of the transition band is $3.3125N$, which is more than three times the width of the rectangular window. Finally, for the Kaiser window lowpass filter (Fig.

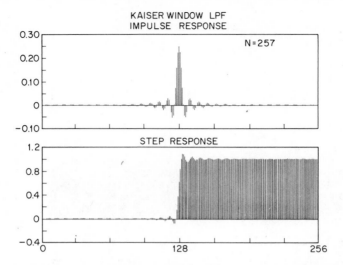

Fig. 3.21. Impulse and step responses of Kaiser window lowpass filter.

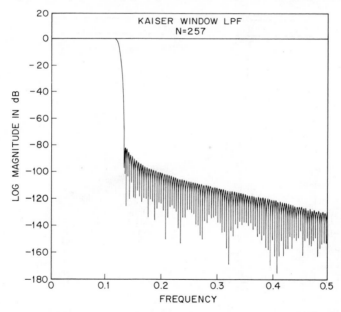

Fig. 3.22. Frequency response of Kaiser window lowpass filter.

3.22), the peak passband and stopband ripple in the frequency response is 0.0001 (for both bands) or −80 dB and the width of the transition band is 5.06 N.

As the examples above have shown, in order to achieve better approximations to the ideal lowpass filter, the width of the transition band is increased in order to decrease peak error of approximation (ripple). As mentioned earlier, the parameter β for the Kaiser window allows the designer the flexibility to trade off width of transition band $(D \cdot N)$ for peak ripple. Table 3.1 presents several values of β and the resulting values of D, passband

Table 3.1

β	D	Passband ripple (dB)	Stopband ripple (dB)
2.120	1.50	± 0.27	-30
3.384	2.23	± 0.0864	-40
4.538	2.93	± 0.0274	-50
5.658	3.62	± 0.00868	-60
6.764	4.32	± 0.00275	-70
7.865	5.0	± 0.000868	-80
8.960	5.7	± 0.000275	-90
10.056	6.4	± 0.000087	-100

(Courtesy of J. Kaiser, Bell Laboratories.)

ripple (in dB), and stopband ripple (in dB). The data are based on ideal integrations of the continuous Kaiser window and are reasonably good approximations for large values of N.

3.13   Issues with Windowing

Although windowing appears to be a most attractive technique for designing FIR filters, there are several issues that often impede their use in many cases. One problem is that one needs a closed form expression for the Fourier series coefficients $h(n)$, where $h(n)$ is defined as

$$h(n) = \frac{1}{2\pi} \int_0^{2\pi} H(e^{j\omega}) e^{j\omega n} \, d\omega \tag{3.56}$$

When $H(e^{j\omega})$ is complicated or cannot easily be put into a closed form mathematical expression (and sometimes even when it can), Eq. (3.56) is often cumbersome or difficult to evaluate. Without an expression for the unwindowed coefficients, it is difficult to contemplate the use of windows.

Another issue with windows is that windowing offers little design flexibility. For example, in the design of a lowpass filter, the passband edge frequency generally cannot be specified exactly since the window "smears" the discontinuity in frequency. Thus the ideal lowpass filter frequency response $\hat{H}(e^{j\omega})$ of Fig. 3.23 with cutoff frequency F_c is smeared by the window to give a frequency response with passband cutoff frequency F_1 and stopband cutoff frequency F_2, as shown in the figure. Although in many

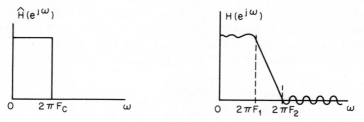

Fig. 3.23. Smearing due to windows.

designs such as the ideal lowpass filter this effect can be compensated, in more complicated filters this is no trivial matter.

3.14 Some Practical Techniques with Windows

The difficulties with windows discussed in the Sec. 3.13 do not preclude their use in many practical situations. There generally are partial solutions to all the problems encountered but one must be aware that these problems do exist and understand the nature of the approximations made in solving them.

In order to determine the unwindowed Fourier series coefficients, when analytic evaluation of Eq. (3.56) is either cumbersome or difficult, a summation approximation to the integral can be used. Instead of evaluating $h(n)$ as an integral of $H(e^{j\omega}) \cdot e^{j\omega n}$ over one period, the approximate sequence $\underset{\sim}{h}(n)$ is computed as

$$\underset{\sim}{h}(n) = \frac{1}{M} \sum_{k=0}^{M-1} H[e^{j(2\pi/M)k}]e^{j(2\pi/M)kn} \tag{3.57}$$

where $H(e^{j\omega}) \cdot e^{j\omega n}$ is evaluated at the M points $\omega_k = 2\pi k/M$. Clearly Eq. (3.57) may be evaluated efficiently as the M-point inverse DFT of the sequence $\underset{\sim}{H}(k) = H[e^{j(2\pi/M)k}]$. Since Eq. (3.57) is a sampled version of Eq. (3.56), it is easy to show that $\underset{\sim}{h}(n)$ is related to $h(n)$ by the aliasing relation

$$\underset{\sim}{h}(n) = \sum_{k=-\infty}^{\infty} h(n - kM) \tag{3.58}$$

Equation (3.55) says that as M gets larger, the difference between $\underset{\sim}{h}(n)$ and $h(n)$ becomes smaller, especially near $n = 0$. Since the window looks only at N points of $\underset{\sim}{h}(n)$, a necessary restriction is that $M \gg N$.

For generating the function $I_0(x)$ there exists a simple, powerful algorithm that merits mentioning. The function $I_0(x)$ has the power series expansion

$$I_0(x) = 1 + \sum_{k=1}^{\infty} \left[\frac{(x/2)^k}{k!} \right]^2$$

This expression for $I_0(x)$ can be evaluated (to within specified accuracy) by the FORTRAN subroutine (due to Kaiser):

```
          SUBROUTINE INO (X, N)
  C       X BETWEEN O. AND 20.
          Y = X/2.
          T = 1. E-08
          E = 1.
          DE = 1.
          DO 1 I = 1,25
          DE = DE*Y/FLOAT (I)
          SDE = DE**2
          E = E + SDE
          IF (E*T-SDE) 1,1,2
  1       CONTINUE
  2       X = E
          N = I
          RETURN
          END
```

This subroutine requires 5 terms in the sum for $x = 0.5$, and 25 terms for $x \geq 19.0$ [N in the calling statement is the number of terms needed to evaluate $I_0(x)$ to the specified accuracy].

3.15 Additional Examples of Window Designed Filters

This section presents examples to illustrate some typical window designed FIR filters. Figures 3.24 to 3.26 show the log magnitude responses of a highpass filter, a bandpass filter, and another lowpass filter. The highpass filter of Fig. 3.24 was designed using a hanning window [$\alpha = 0.5$ in Eq. (3.50)] with an impulse response duration of 45 samples and an ideal cutoff frequency of 0.35. The peak ripple in the passband is 0.00635 and the peak stopband ripple is 0.00635 or -43.94 dB. The transition bandwidth is 0.07233 or 3.25 N. The bandpass filter of Fig. 3.25 was designed using a Kaiser window with $\beta = 3.38$. The impulse response duration is ($N = 46$) samples and the ideal band edges were 0.15 and 0.27. The peak ripple in the passband is 0.0078 and the peak stopband ripples are 0.00792 (or -42 dB) in the lower stopband and 0.00909 (or -40.8 dB) in the upper stopband. The final example of Fig. 3.26 is a lowpass filter designed using a Dolph–Chebyshev window. The impulse response duration of this filter is ($N = 46$) samples and the ideal cutoff frequency is 0.25. The peak passband ripple is 0.00688 and the peak stopband ripple is 0.0054 or -45.3 dB. The transition bandwidth of the filter is 0.065 or $2.97 \cdot N$.

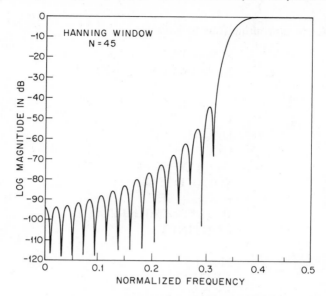

Fig. 3.24. Hanning window highpass filter frequency response.

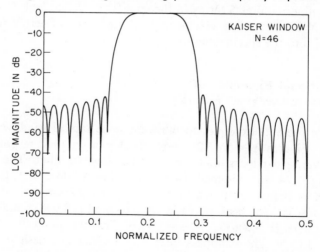

Fig. 3.25. Frequency response of Kaiser window bandpass filter.

These examples were obtained by analytically obtaining the infinite impulse response of the ideal filters and then weighting the impulse responses by the appropriate window. One of the main advantages of windowing is that it is reasonably straightforward to obtain the filter impulse response with minimal computational effort. The main disadvantage of the technique is that the resulting FIR filters satisfy no known optimality criterion; hence their performance can be considerably improved in most cases.

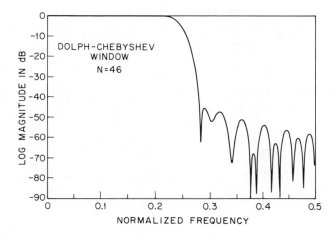

Fig. 3.26. Frequency response of Dolph-Chebyshev window lowpass filter.

3.16   Summary of Windows

The general technique of window design of FIR filters has been shown to be a reasonable design method that has certain advantages and disadvantages. The windows discussed here included the rectangular, Hamming, and Kaiser windows. Other well-known, and often widely used, windows include the Dolph–Chebyshev window, the triangular window, and the Blackman window. This last window is a generalized Hamming window with five terms instead of three. No great attempt has been made at cataloging all these windows because the design techniques to be discussed in the following sections tend to supersede the window technique for most applications. The major reasons for the relative success of windows is their simplicity and ease of use and the fact that closed form expressions are often available for the window coefficients.

3.17   Design Technique No. 2—Frequency Sampling

An FIR filter can be uniquely specified by giving either the impulse response coefficients $\{h(n)\}$ or, equivalently, the DFT coefficients $\{H(k)\}$. The reader will recall that these sequences are related by the DFT relations

$$H(k) = \sum_{n=0}^{N-1} h(n)e^{-j(2\pi/N)nk} \qquad \text{DFT} \qquad (3.59)$$

$$h(n) = \frac{1}{N}\sum_{k=0}^{N-1} H(k)e^{j(2\pi/N)nk} \qquad \text{IDFT} \qquad (3.60)$$

It has also been shown that the DFT samples $H(k)$ for an FIR sequence can be regarded as samples of the filter's z transform, evaluated at N points equally spaced around the unit circle; i.e.,

$$H(k) = H(z)\big|_{z=e^{j(2\pi/N)k}} \tag{3.61}$$

Thus the z transform of an FIR filter can easily be expressed in terms of its DFT coefficients by substituting Eq. (3.60) into the z transform to give

$$H(z) = \sum_{n=0}^{N-1} h(n)z^{-n} = \sum_{n=0}^{N-1}\left[\frac{1}{N}\sum_{k=0}^{N-1} H(k)e^{j(2\pi/N)nk}\right]z^{-n} \tag{3.62}$$

Interchanging orders of summation and summing over the n index gives

$$H(z) = \sum_{k=0}^{N-1}\frac{H(k)}{N}\sum_{n=0}^{N-1}[e^{j(2\pi/N)k}z^{-1}]^n$$

$$= \sum_{k=0}^{N-1}\frac{H(k)}{N}\frac{(1 - e^{j2\pi k}z^{-N})}{[1 - e^{j(2\pi/N)k}z^{-1}]} \tag{3.63}$$

Since $e^{j2\pi k} = 1$, Eq. (3.63) reduces to

$$H(z) = \frac{(1 - z^{-N})}{N}\sum_{k=0}^{N-1}\frac{H(k)}{[1 - z^{-1}e^{j(2\pi/N)k}]} \tag{3.64}$$

which is the desired result.

The interpretation of Eq. (3.64) is that, to approximate any continuous frequency response, one could *sample in frequency* at N equispaced points around the unit circle (the frequency samples) and evaluate the continuous frequency response as an interpolation of the sampled frequency response. The approximation error would then be exactly zero at the sampling frequencies and be finite between them. The smoother the frequency response being approximated, the smaller the error of interpolation between the sample points. An example of this process is shown in Fig. 3.27. Part (a) shows a desired frequency response (the solid curve) and a set of frequency samples (the heavy dots). Part (b) shows a continuous interpolation of the frequency samples.

Although the procedure can be used directly as stated to design an FIR filter, in order to improve the quality of the approximation, i.e., to make the approximation error smaller, a number of the frequency samples can be made unconstrained variables. The values of these unconstrained variables are generally optimized by computer to minimize some simple function of the approximation error—e.g., the peak error of approximation. For example, one might choose as unconstrained variables the frequency samples that lie in a transition band between two frequency bands in which the frequency response is specified, i.e., in the band between the passband and stopband of a lowpass filter.

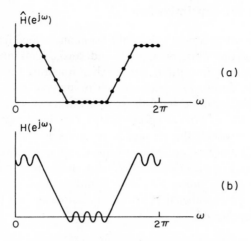

Fig. 3.27. Explanation of frequency sampling.

To see why this technique of optimizing only a few of the N frequency samples can often drastically reduce the approximation error, Eq. (3.64) must be evaluated on the unit circle to obtain a continuous frequency interpolation formula for the filter—i.e., the frequency response of the filter. The resulting formula is

$$H(e^{j\omega}) = \frac{e^{-j\omega(N-1)/2}}{N} \sum_{k=0}^{N-1} \frac{H(k)e^{-j(\pi k/N)} \sin(\omega N/2)}{\sin(\omega/2 - \pi k/N)} \tag{3.65}$$

Equation (3.65) shows the frequency response of the filter to be a linear combination of the frequency samples $H(k)$ with frequency interpolating functions that are of the form

$$S(\omega, k) = e^{-j(\pi k/N)} \frac{\sin(\omega N/2)}{\sin(\omega/2 - \pi k/N)}$$

$$= \pm e^{-j(\pi k/N)} \frac{\sin[N(\omega/2 - \pi k/N)]}{\sin(\omega/2 - \pi k/N)} \tag{3.66}$$

Thus every frequency sample with value $H(k)$ produces a frequency response proportional to a constant times $\sin(N\omega/2)/\sin(\omega/2)$ displaced by $\pi k/N$ in frequency. The interpolation functions [i.e., $\sin(N\omega/2)/\sin(\omega/2)$] from frequency samples in transition bands have been found to provide good ripple cancellation in adjacent frequency bands. Thus by allowing only those frequency samples that lie in the preselected transition bands to be unconstrained, it has been found that very good filters can be designed by optimizing the values of these frequency samples.

3.18 Solution for Optimization

In order to solve for optimum values of the unconstrained frequency samples, a series of equations must be written and solved that mathematically expresses the desired optimization. Rather than illustrate this procedure in the general case, the necessary equations will be derived for a simple example and then the results will be generalized to arbitrary cases.

Figure 3.28 shows typical specifications for the design of a frequency sampling filter in which the frequency response is specified in two bands (regions 1 and 2) and unspecified in a transition band between these regions. The solid curve of Fig. 3.28 represents $\hat{H}(e^{j\omega})$, the desired frequency response, and the dots show the preset frequency samples. The frequency samples in the transition band are labeled T_1 and T_2 for convenience. These samples are the ones to be optimized.

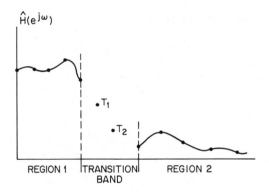

Fig. 3.28. Specifications for a frequency sampling filter.

Figure 3.28 only shows half of the frequency samples since in order for the filter impulse response to be real the frequency samples $H(k)$ form a sequence that has complex conjugate symmetry around the midpoint of the sequence. Further restrictions are placed on $\{H(k)\}$, in order for the filter to have an exactly linear phase response. A discussion of exactly what these restrictions are will be deferred to Sec. 3.21 and 3.22. For the time being, it is assumed that $H(e^{j\omega})$ can be expressed in the form

$$H(e^{j\omega}) = e^{-j\omega(N-1)/2} \sum_{k=0}^{KM} H(k)S(\omega, k) \tag{3.67}$$

$$= e^{-j\omega(N-1)/2} H^*(e^{j\omega}) \tag{3.68}$$

where $S(\omega, k)$ is the resulting frequency interpolation function and $KM + 1$

is the number of frequency samples that need to be specified. Equations (3.67) and (3.68) imply

$$H^*(e^{j\omega}) = \sum_{k=0}^{KM} H(k)S(\omega, k) \qquad (3.69)$$

The real function $H^*(e^{j\omega})$ will be used in the design equations since the linear phase term in Eq. (3.68) can clearly be ignored for design purposes.

For the example of Fig. 3.28, $H^*(e^{j\omega})$ can be simplified to the form

$$H^*(e^{j\omega}) = B(\omega) + T_1 A_1(\omega) + T_2 A_2(\omega) \qquad (3.70)$$

where $B(\omega)$ represents the contribution to $H^*(e^{j\omega})$ of all the fixed frequency samples and $A_1(\omega)$ and $A_2(\omega)$ represent the contributions of the unconstrained frequency samples with magnitudes T_1 and T_2.

To solve for the unknown frequency samples in the example of Fig. 3.28, a set of constraint equations must be written for frequencies in regions 1 and 2. A typical set of such constraints might be

1. $|H^*(e^{j\omega}) - \hat{H}(e^{j\omega})| \leq \varepsilon$ ω in region 1
2. minimize   maximum $|H^*(e^{j\omega}) - \hat{H}(e^{j\omega})|$
    $\{T_1, T_2\}$ $\{\omega$ in region 2$\}$

where ε is a fixed tolerance. A second possible set of constraints might be

$$\text{minimize} \quad \text{maximum} \, |W(e^{j\omega})[H^*(e^{j\omega}) - \hat{H}(e^{j\omega})]|$$

$$\{T_1, T_2\} \qquad \{\omega \text{ in both regions 1 and 2}\}$$

where $W(e^{j\omega})$ is a known weighting function on the frequency response approximation error.

The set of constraints above may be restated in a more formal way mathematically by using Eq. (3.70) and by evaluating each of the constraint equations at a dense set of frequencies in the given regions. Thus, for example, the first set of constraints given above lead to the following set of inequalities:

$$\left. \begin{aligned} T_1 A_1(\omega_m) + T_2 A_2(\omega_m) &\leq \varepsilon - B(\omega_m) + \hat{H}(e^{j\omega_m}) \\ -T_1 A_1(\omega_m) - T_2 A_2(\omega_m) &\leq \varepsilon + B(\omega_m) - \hat{H}(e^{j\omega_m}) \end{aligned} \right\} \quad \omega_m \text{ in region 1}$$

$$\left. \begin{aligned} T_1 A_1(\omega_m) + T_2 A_2(\omega_m) - T_3 &\leq -B(\omega_m) + \hat{H}(e^{j\omega_m}) \\ -T_1 A_1(\omega_m) - T_2 A_2(\omega_m) - T_3 &\leq B(\omega_m) - \hat{H}(e^{j\omega_m}) \end{aligned} \right\} \quad \omega_m \text{ in region 2}$$

where T_3 represents the maximum approximation error in region 2. The set of inequalities above is in a form suitable for solution by linear programming techniques. In a similar manner, the second set of constraints could also be written as a set of linear inequalities in the variables T_1, T_2, and T_3.

In fact the general case in which the frequency response is specified in several regions separated by transition bands in which the frequency samples are unconstrained can be described by a set of linear inequalities in the unconstrained frequency samples, which can be solved using linear programming techniques. Since linear programming will be relied on at several places in this chapter, the next section presents a slight digression to discuss briefly this powerful mathematical technique for solving systems of linear inequalities.

To summarize, frequency sampling filters can be designed by specifying the DFT coefficients (frequency samples) of the filter which lie in frequency bands of interest to be samples of the desired frequency response and by leaving the DFT coefficients which fall in transition bands unspecified. A set of linear inequalities in the unconstrained frequency samples is obtained that describes the desired constraints on the frequency response. This set of linear inequalities is solved using linear programming techniques, yielding values for the unconstrained frequency samples.

3.19   Linear Programming

The general linear programming problem can be mathematically stated in the form

find $\{x_j\}, j = 1, 2, \ldots, N$, subject to the constraints

$$x_j \geq 0 \qquad j = 1, 2, \ldots, N \tag{3.71}$$

$$\sum_{j=1}^{N} c_{ij} x_j = b_i \qquad i = 1, 2, \ldots, M \quad (M < N) \tag{3.72}$$

such that

$$\sum_{j=1}^{N} a_j x_j \text{ is minimized} \tag{3.73}$$

where c_{ij}, b_i, and a_j are constants.

The problem above is referred to as the *primal problem* and, by a duality principle, can be shown to be mathematically equivalent to the "dual problem"

find $\{y_i\}, i = 1, 2, \ldots, M$ subject to the constraints

$$\sum_{i=1}^{M} c_{ij} y_i \leq a_j \qquad j = 1, 2, \ldots, N \tag{3.74}$$

such that

$$\sum_{i=1}^{M} b_i y_i \text{ is maximized} \tag{3.75}$$

One characteristic of linear programs is that given there is a solution, it is guaranteed to be a unique solution and there are several well-defined procedures for arriving at this solution within $(M + N)$ iterations. There are also straightforward techniques for determining if the solution is unconstrained or poorly constrained.

A simple graphical interpretation of the linear program in two dimensions is shown in Fig. 3.29. Each constraint equation in this figure, C_1 to C_5, is a linear inequality in x_1 and x_2. Therefore a straight line can be drawn representing the equivalent linear equality and half of the solution space (shown by shaded lines) is eliminated as a possible solution. When all the constraint lines have been drawn, only a small region of the x_1, x_2 plane (generally a convex polyhedron) is still admissible for a solution to the minimization phase. Assuming some chosen value for the minimum the linear equation in (x_1, x_2) satisfying the minimum is then plotted in the (x_1, x_2) plane. By

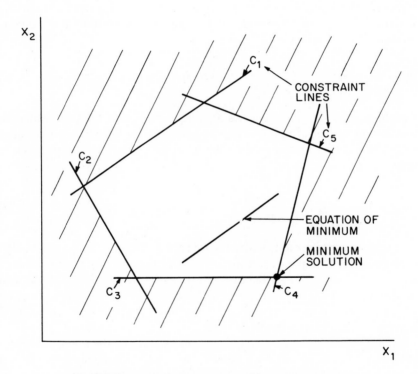

Fig. 3.29. Graphical interpretation of linear programming.

successively reducing the value assumed for the minimum, this straight line eventually intersects one of the vertices of the polyhedron (in this ease the point where constraint lines C_3 and C_4 intersect) and this point becomes the solution—i.e., the absolute minimum consistent with the constraints. It should be noted that this is *not* the method generally used to find the minimum in linear programming algorithms.

An important general property of the linear programming problem is that since the constraint equations form a convex polyhedron, the minimum generally occurs at a vertex of the polyhedron—i.e., where M of the inequalities attain equality. Thus the approach used to find the minimum is to examine only the vertices of the polyhedron in an efficient manner. An algorithm for performing this procedure is called the *simplex* method and is described in the References on linear programming.

3.20   Types 1 and 2

Frequency sampling filters are based on specification of a set of samples of the desired frequency response at N uniformly spaced points around the unit circle. The set of frequencies that have been used until this point is determined by the relation

$$f_k = \frac{k}{N} \qquad k = 0, 1, \ldots, N-1 \qquad (3.76)$$

corresponding to the N frequencies at which an N-point DFT is evaluated. There is a second set of uniformly spaced frequencies for which a frequency sampling structure can conveniently be obtained. This set of frequencies is determined by the relation

$$f_k = \frac{(k + 1/2)}{N} \qquad k = 0, 1, \ldots, N-1 \qquad (3.77)$$

Figure 3.30 illustrates exactly where the frequency sampling points are located for the two sets of frequencies [called type 1 for those of Eq. (3.76) and type 2 for those of Eq. (3.77)] for the case where N is both even and odd. As seen from this figure, if $\theta = 1/N$ is the angular spacing between frequency sampling points, the type 1 designs have the initial point at $f = 0$, whereas the type 2 designs have the initial point at $f = \theta/2$.

The importance of type 2 frequency samples lies in the additional flexibility it gives the design method to specify the desired frequency response at a second possible set of frequencies. Thus a given band edge frequency may be much closer to a type 2 frequency sampling point than to a type 1 point—in which case a type 2 design would be used in the optimization procedure. Since the optimization procedure is insensitive to whether the initial design is type 1 or type 2, either design can be used to obtain the filter coefficients, once the real function $H^*(e^{j\omega})$ (Eq. (3.69) has been obtained for each type filter. Thus, in the next sections, expressions are derived for $H^*(e^{j\omega})$ for types 1 and 2 designs for N even and odd filters, in the case of linear phase filters.

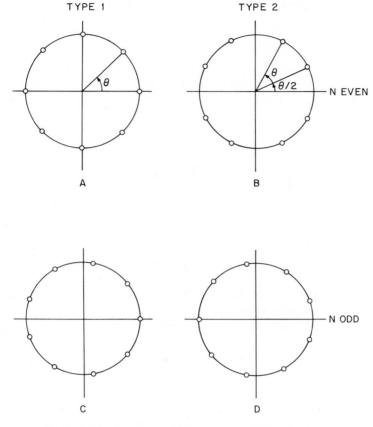

Fig. 3.30. The four cases of frequency sampling grid points.

3.21 Type 1 Designs—Linear Phase Constraints

For type 1 designs the frequency response was previously shown to be of the form

$$H(e^{j\omega}) = \frac{e^{-j\omega[(N-1)/2]}}{N} \sum_{k=0}^{N-1} \frac{H(k)e^{-j(\pi k/N)} \sin (\omega N/2)}{\sin (\omega/2 - \pi k/N)} \tag{3.78}$$

In the case of a linear phase filter [with delay $(N - 1)/2$ samples], $H(K)$ can be expressed in the form

$$H(k) = |H(k)| \, e^{j\theta(k)} \qquad k = 0, 1, \ldots, N - 1 \tag{3.79}$$

where

$$|H(k)| = |H(N - k)| \qquad k = 0, 1, \ldots, N - 1 \tag{3.80}$$

and, when N is even,

$$
\theta(k) = \begin{cases}
-\dfrac{2\pi}{N} k\left(\dfrac{N-1}{2}\right) & k = 0, 1, \ldots, \dfrac{N}{2} - 1 \\[3mm]
\dfrac{2\pi}{N} (N-k)\left(\dfrac{N-1}{1}\right) & k = \dfrac{N}{2} + 1, \ldots, N - 1 \\[3mm]
0 & k = \dfrac{N}{2}
\end{cases} \tag{3.81}
$$

$$
H\left(\frac{N}{2}\right) = 0 \tag{3.82}
$$

and, when N is odd,

$$
\theta(k) = \begin{cases}
-\dfrac{2\pi}{N} k\left(\dfrac{N-1}{2}\right) & k = 0, 1, \ldots, \dfrac{N-1}{2} \\[3mm]
\dfrac{2\pi}{N} (N-k)\left(\dfrac{N-1}{2}\right) & k = \dfrac{N+1}{2}, \ldots, N - 1
\end{cases} \tag{3.83}
$$

Equation (3.82) reflects the fact that $H(e^{j\omega}) = 0$ at $\omega = \pi$ for a linear phase filter with even impulse response duration, as discussed previously.

Using Eq. (3.81) and (3.82), Eq. (3.79) becomes (for N even)

$$
H(k) = \begin{cases}
|H(k)|\, e^{-j(2\pi/N)k[(N-1)/2]} & k = 0, 1, \ldots, \dfrac{N}{2} - 1 \\[3mm]
0 & k = \dfrac{N}{2} \\[3mm]
|H(k)|\, e^{j(2\pi/N)(N-k)[(N-1)/2]} & k = \dfrac{N}{2} + 1, \ldots, N - 1
\end{cases} \tag{3.84}
$$

Plugging Eq. (3.84) into Eq. (3.78) and reducing terms gives

$$
H(e^{j\omega}) = \frac{e^{-j\omega(N-1)/2}}{N} \sin\left(\frac{\omega N}{2}\right)
$$

$$
\times \left[\sum_{k=0}^{(N/2)-1} \frac{|H(k)|\,(-1)^k}{\sin(\omega/2 - \pi k/N)} - \sum_{k=N/2+1}^{N-1} \frac{|H(k)|\,(-1)^k}{\sin(\omega/2 - \pi k/N)} \right] \tag{3.85}
$$

Substituting $l = N - k$ in the second summation of Eq. (3.85) gives

$$
H(e^{j\omega}) = \frac{e^{-j\omega(N-1)/2}}{N} \sin\left(\frac{\omega N}{2}\right) \left\{ \sum_{k=0}^{(N/2)-1} \frac{|H(k)|\,(-1)^k}{\sin(\omega/2 - \pi k/N)} \right.
$$

$$
\left. - \sum_{l=1}^{(N/2)-1} \frac{|H(N-l)|\,(-1)^l}{\sin[\omega/2 - \pi(N-l)/N]} \right\} \tag{3.86}
$$

Using Eq. (3.80) and combining terms gives

$$H(e^{j\omega}) = \frac{e^{-j\omega(N-1)/2}}{N} \sin\left(\frac{\omega N}{2}\right)\left\{\frac{H(0)}{\sin(\omega/2)} + \sum_{k=1}^{(N/2)-1} |H(k)|\,(-1)^k\right.$$

$$\left.\times \left[\frac{1}{\sin(\omega/2 - \pi k/N)} + \frac{1}{\sin(\omega/2 + \pi k/N)}\right]\right\} \quad (3.87)$$

Finally, using trigonometric identities, Eq. (3.87) reduces to

$$H(e^{j\omega}) = e^{-j\omega(N-1)/2}\left(\frac{|H(0)|}{N}\frac{\sin(\omega N/2)}{\sin(\omega/2)} + \sum_{k=1}^{(N/2)-1}\frac{|H(k)|}{N}\right.$$

$$\left.\times \left\{\frac{\sin[N(\omega/2 - \pi k/N)]}{\sin(\omega/2 - \pi k/N)} + \frac{\sin[N(\omega/2 + \pi k/N)]}{\sin(\omega/2 + \pi k/N)}\right\}\right) \quad (3.88)$$

which is the desired result. The term inside the outermost parentheses of Eq. (3.88) is the desired real quantity $H^*(e^{j\omega})$ for N even, type 1, linear phase frequency sampling designs.

For type 1, N odd, linear phase designs, an almost identical derivation gives, for $H^*(e^{j\omega})$, the expression

$$H^*(e^{j\omega}) = \left(\frac{H(0)}{N}\frac{\sin(\omega N/2)}{\sin(\omega/2)} + \sum_{k=1}^{(N-1)/2}\frac{|H(k)|}{N}\right.$$

$$\left.\times \left\{\frac{\sin[N(\omega/2 - \pi k/N)]}{\sin(\omega/2 - \pi k/N)} + \frac{\sin[N(\omega/2 + \pi k/N)]}{\sin(\omega/2 + \pi k/N)}\right\}\right) \quad (3.89)$$

3.22 Type 2 Designs—Linear Phase Constraints

For type 2 designs the frequency samples $H(k)$ are obtained as

$$H(k) = H(z)\big|_{z=e^{j(2\pi/N)(k+1/2)}} \quad (3.90)$$

Equation (3.90) can be used to relate the set of frequency samples $H(k)$ to the filter's impulse response $h(n)$ giving

$$H(k) = \sum_{n=0}^{N-1} h(n)e^{-j(2\pi/N)n(k+1/2)}$$

$$= \sum_{n=0}^{N-1} \underbrace{h(n)e^{-j(\pi n/N)}}_{g(n)}e^{-j(2\pi/N)nk} \quad (3.91)$$

Equation (3.91) shows that $H(k)$ is the DFT of the sequence

$$g(n) = h(n)e^{-j(\pi n/N)} \quad (3.92)$$

Thus, $g(n)$ is the inverse DFT of the $H(k)$'s, or

$$g(n) = h(n)e^{-j(\pi n/N)} = \frac{1}{N}\sum_{k=0}^{N-1}H(k)e^{j(2\pi/N)nk} \tag{3.93}$$

or

$$h(n) = \frac{1}{N}\sum_{k=0}^{N-1}H(k)e^{j(2\pi/N)n(k+1/2)} \tag{3.94}$$

Equation (3.94) can be used in the z transform for the filter, giving

$$H(z) = \sum_{n=0}^{N-1}\left[\frac{1}{N}\sum_{k=0}^{N-1}H(k)e^{j(2\pi/N)(k+1/2)n}\right]z^{-n} \tag{3.95}$$

which, by interchanging the limits of summation and summing over n, can be put in the form

$$H(z) = \frac{(1 + z^{-N})}{N}\sum_{k=0}^{N-1}\frac{H(k)}{1 - z^{-1}e^{j(2\pi/N)(k+1/2)}} \tag{3.96}$$

Evaluation of Eq. (3.96) on the unit circle gives

$$H(e^{j\omega}) = \frac{e^{-j\omega[(N-1)/2]}}{N}\left\{\sum_{k=0}^{N-1}\frac{H(k)e^{-j(\pi/N)(k+1/2)}\cos(\omega N/2)}{j\sin[\omega/2 - (\pi/N)(k+1/2)]}\right\} \tag{3.97}$$

For type 2 designs, in the case of a linear phase filter [with a delay of $(N-1)/2$ samples] $H(k)$ can be expressed as

$$H(k) = \begin{cases} |H(k)|\,e^{-j(2\pi/N)[(N-1)/2](k+1/2)} & k = 0, 1,\ldots,\dfrac{N}{2} - 1 \\[2mm] |H(k)|\,e^{j(2\pi/N)[(N-1)/2](N-k-1/2)} & k = \dfrac{N}{2},\ldots, N - 1 \end{cases} \tag{3.98}$$

when N is even, or

$$H(k) = \begin{cases} |H(k)|\,e^{-j(2\pi/N)[(N-1)/2](k+1/2)} & k = 0, 1,\ldots,\dfrac{N-3}{2} \\[2mm] \left|H\!\left(\dfrac{N-1}{2}\right)\right| & k = \dfrac{N-1}{2} \\[2mm] |H(k)|\,e^{j(2\pi/N)[(N-1)/2](N-k-1/2)} & k = \dfrac{N+1}{2},\ldots, N - 1 \end{cases} \tag{3.99}$$

when N is odd, where

$$|H(k)| = |H(N - 1 - k)| \qquad k = 0, 1,\ldots, N - 1 \tag{3.100}$$

Equations (3.98) to (3.100) can be substituted into Eq. (3.97), as was done for the type 1 designs, leading to the following expressions for $H^*(e^{j\omega})$:

$$H^*(e^{j\omega}) = \left[\sum_{k=0}^{(N/2)-1} \frac{|H(k)|}{N} \left(\frac{\sin \{N[\omega/2 - (\pi/N)(k + 1/2)]\}}{\sin [\omega/2 - (\pi/N)(k + 1/2)]} \right.\right.$$

$$\left.\left. + \frac{\sin \{N[\omega/2 + (\pi/N)(k + 1/2)]\}}{\sin [\omega/2 + (\pi/N)(k + 1/2)]} \right)\right] \quad (3.101)$$

when N is even, and

$$H^*(e^{j\omega}) = \left\{ \frac{|H[(N - 1)/2]|}{N} \frac{\sin (\omega N/2)}{\sin (\omega/2)} \right.$$

$$+ \sum_{k=0}^{(N-3)/2} \left[\frac{|H(k)|}{N} \left(\frac{\sin \{N[\omega/2 - (\pi/N)(k + 1/2)]\}}{\sin [\omega/2 - (\pi/N)(k + 1/2)]} \right.\right.$$

$$\left.\left.\left. + \frac{\sin \{N[\omega/2 + (\pi/N)(k + 1/2)]\}}{\sin [\omega/2 + (\pi/N)(k + 1/2)]} \right)\right]\right\} \quad (3.102)$$

when N is odd.

Thus any of the four equations [Eq. (3.88), (3.89), (3.101), (3.102)] can be used in an optimization procedure to design linear phase FIR digital filters. The choice of N even or odd or type 1 or 2 design is up to the user and depends primarily on the filter being designed. Section 3.23 presents some results on actual filters designed using frequency sampling techniques.

3.23 Some General Results on the Design of Frequency Sampling Filters

1. Lowpass Filters

The technique of frequency sampling can be used to design a wide variety of filters. Figures 3.31 to 3.33 show some typical lowpass filters designed by this technique. For these cases the criterion for optimization was the minimization of the peak stopband ripple. Figure 3.31 shows an ($N = 256$) type 1 design with three variable frequency samples in the transition band. The peak stopband ripple is about 0.05. Figure 3.32 shows a narrowband lowpass filter ($N = 65$) where the cutoff frequency is 0.0306. Three transition samples are varied to give a peak stopband ripple of about 0.00002 or -93 dB. Figure 3.33 shows a wideband lowpass filter ($N = 64$) where the cutoff frequency is 0.4355. Three variable frequency samples are used, and the resulting peak stopband ripple is about 0.000002 or -115 dB.

For the general case of designing lowpass filters using one variable transition sample (and minimizing the peak stopband ripple) an out-of-band rejection of about 44 to 54 dB can be obtained. With two variable transition samples, 65 to 75 dB of rejection can be obtained; with three transition samples, 85 to 95 dB of rejection is possible. Figures 3.34 and 3.35 show the

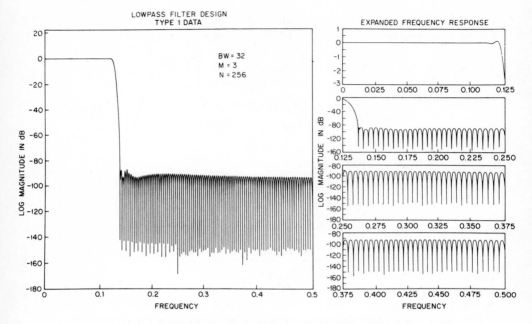

Fig. 3.31 Frequency response of a frequency sampling lowpass filter.

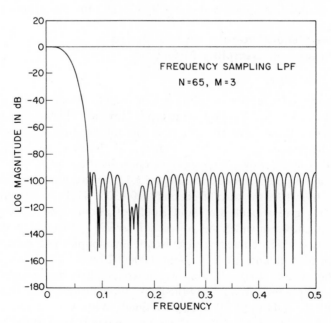

Fig. 3.32 Frequency response of a narrowband frequency sampling
lowpass filter.

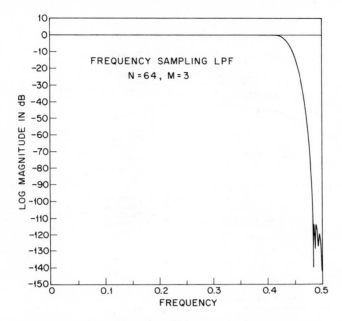

Fig. 3.33 Frequency response of a wideband frequency sampling lowpass filter.

values of the transition coefficients and the peak stopband ripple (the minimax solution) for the three-transition sample type 2 data for various values of N. These data are plotted as a function of the filter bandwidth. It is interesting to note that at both small and large values of filter bandwidth the results are generally *superior* (i.e., more out-of-band rejection) to midrange values. One possible explanation of this result is as follows. For small values of normalized bandwidth, there is very little ripple to cancel; hence the transition coefficients provide excellent cancellation of this ripple. For large values of normalized bandwidth, there is a very small region over which the ripple from the fixed frequency samples must be canceled; hence the transition coefficients again provide efficient designs here.

2. Bandpass Filters

Bandpass filters can be designed directly in much the same manner as lowpass filters. Figure 3.36 shows a type 1 design bandpass filter ($N = 128$) with three variable frequency samples on either side of the passband. The frequency samples on either side of the passband are symmetric. The peak stopband ripple is about 0.000025 or -91 dB and the ripple envelope falls to 10^{-6} or 120 dB at both 0 frequency and at half the sampling frequency.

3. Wideband Differentiators

In many systems the need arises for a wideband differentiator. The magnitude response, phase response, and the equivalent imaginary response of an

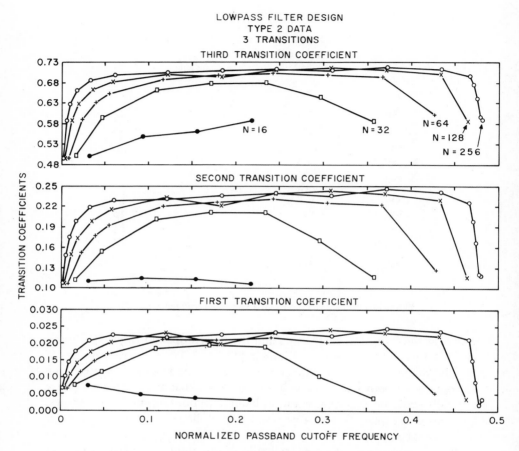

Fig. 3.34 Values of transition coefficients for a wide variety of frequency sampling lowpass filters.

ideal differentiator are shown in Fig. 3.37. At $\omega = \pi$ there is a discontinuity in the imaginary part of the frequency response. This discontinuity cannot occur in a band of zero bandwidth; hence a transition band is normally required around $\omega = \pi$. By cascading the ideal wideband differentiator with an ideal half-sample delay network, the frequency response discontinuity at $\omega = \pi$ can be eliminated, as shown in Fig. 3.37. The center curves of Fig. 3.37 show the magnitude and phase responses of an ideal half-sample delay network and the lower curves show the result of cascading an ideal wideband differentiator with an ideal half-sample delay network. At $\omega = \pi$ there is no longer any phase discontinuity. Thus a transition band is no longer needed here. At $\omega = 0$ there is a phase discontinuity of π radians but this is accounted for by the zero in the magnitude response at $\omega = 0$. Thus

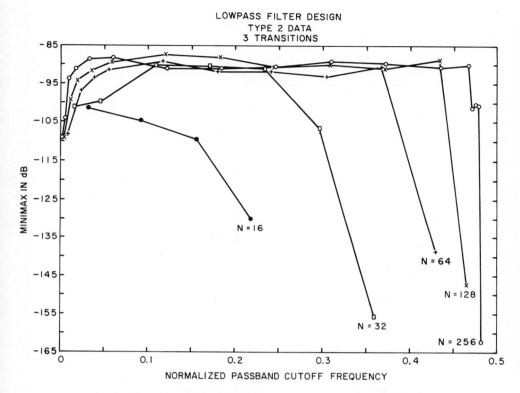

LOWPASS FILTER DESIGN
TYPE 2 DATA
3 TRANSITIONS

Fig. 3.35 Peak stopband attenuations for a wide variety of frequency sampling lowpass filters.

approximation of wide band differentiators is entirely straightforward using FIR filters where half-sample delays are easy to achieve with even length impulse response sequences.

Typical differentiators designed in this manner are shown in Fig. 3.38 and 3.39 for values of N of 16 and 256. For $N = 16$ the peak magnitude error is about 0.015 (the phase error is, of course, zero), whereas for $N = 256$ the peak magnitude error is about 0.0008.

3.24 Summary of Frequency Sampling Design

The main idea of the frequency sampling design method is that a desired frequency response can be approximated by sampling it at N evenly spaced points and then obtaining an interpolated frequency response that passes through the frequency samples. For filters with reasonably smooth frequency responses, the interpolation error is generally small. In the case of band select filters, where the desired frequency response changes radically across bands, the frequency samples which occur in transition bands are made to be

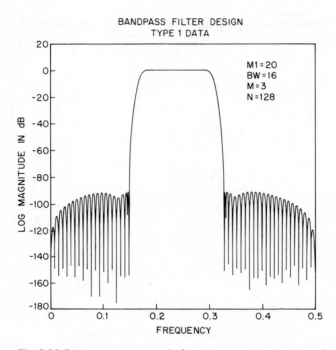

Fig. 3.36 Frequency response of a frequency sampling bandpass filter.

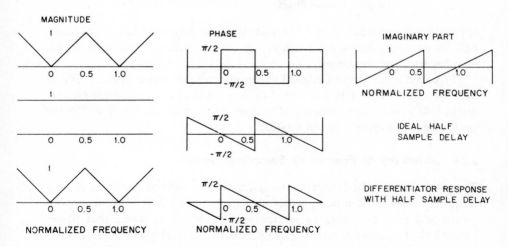

Fig. 3.37 Responses of an ideal half-sample delay differentiator realized as a cascade of an ideal differentiator and an ideal half-sample delay network.

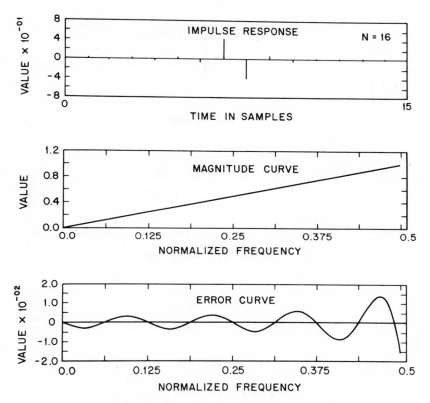

Fig. 3.38 Frequency responses of a 16-point differentiator.

unspecified variables whose values are chosen by an optimization algorithm which minimizes some function of the approximation error of the filter. Simple linear programming techniques may also be used to perform the necessary minimization. Finally, it was shown that there were two distinct types of frequency sampling filters, depending on where the initial frequency sample occurred. Expressions were derived for the frequency response of both types of linear phase frequency sampling filter for use by the optimization algorithm.

3.25   Design Technique No. 3—Optimal (Minimax Error) Filters

By considering the linear phase FIR filter design problem as a Chebyshev approximation problem, it is possible to derive a set of conditions for which it can be proved that the solution is optimal (in the sense that the peak approximation error over the entire interval of approximation is minimized) and unique. It is also easy to show how several standard optimization

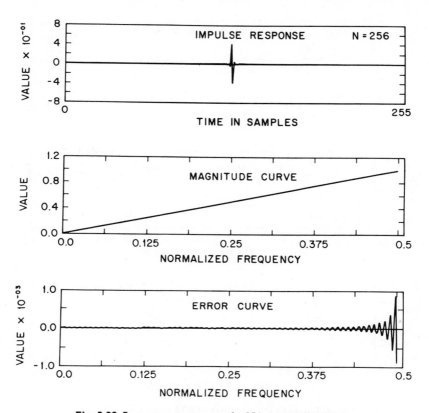

Fig. 3.39 Frequency responses of a 256-point differentiator.

procedures including linear programming can be used to solve for the filter coefficients of the optimal (minimax) solution. Here and in the following sections, the optimal FIR filter design problem is formulated and techniques for its solution are discussed. Since the resulting solutions are of such great practical importance, succeeding sections will discuss particular classes of optimal filters and their properties—e.g., lowpass filters, bandpass filters, differentiators, Hilbert transformers, and finally multiband and arbitrary specification filters.

3.26 Weighted Chebyshev Approximation[4]

In Sec. 3.5 it was shown that the frequency response of the four cases of linear phase filters could be written in the form

$$H(e^{j\omega}) = e^{-j\omega(N-1)/2}e^{j(\pi/2)L}H^*(e^{j\omega}) \tag{3.103}$$

---

[4] Many of the results in this section and several following ones are based on the work of T. Parks and J. McClellan.

Values for L and the form for $H^*(e^{j\omega})$ are given in Table 3.2 for each of the four cases of linear phase.

TABLE 3.2

	L	$H^*(e^{j\omega})$
Case 1—N odd Symmetrical impulse response	0	$\displaystyle\sum_{n=0}^{(N-1)/2} a(n) \cos{(\omega n)}$
Case 2—N even Symmetrical impulse response	0	$\displaystyle\sum_{n=1}^{N/2} b(n) \cos{\left[\omega\left(n - \frac{1}{2}\right)\right]}$
Case 3—N odd Anti-symmetrical impulse response	1	$\displaystyle\sum_{n=1}^{(N-1)/2} c(n) \sin{(\omega n)}$
Case 4—N even Anti-symmetrical impulse response	1	$\displaystyle\sum_{n=1}^{N/2} d(n) \sin{\left[\omega\left(n - \frac{1}{2}\right)\right]}$

Using simple trigonometric identities, each of the expressions for $H^*(e^{j\omega})$ in Table 3.2 can be written as a product of a fixed function of ω [call this $Q(e^{j\omega})$] and a term that is a sum of cosines [call this $P(e^{j\omega})$]. Thus the expressions for $H^*(e^{j\omega})$ in Table 3.2 become

Case 1 No change

Case 2

$$\sum_{n=1}^{N/2} b(n) \cos{[\omega(n - \tfrac{1}{2})]} = \cos\left(\frac{\omega}{2}\right) \sum_{n=0}^{(N/2)-1} \tilde{b}(n) \cos{(\omega n)} \qquad (3.104a)$$

Case 3

$$\sum_{n=1}^{(N-1)/2} c(n) \sin{(\omega n)} = \sin{(\omega)} \sum_{n=0}^{(N-3)/2} \tilde{c}(n) \cos{(\omega n)} \qquad (3.104b)$$

Case 4

$$\sum_{n=1}^{N/2} d(n) \sin{[\omega(n - \tfrac{1}{2})]} = \sin\left(\frac{\omega}{2}\right) \sum_{n=0}^{(N/2)-1} \tilde{d}(n) \cos{(\omega n)} \qquad (3.104c)$$

Table 3.3 shows a summary of the functions $Q(e^{j\omega})$ and $P(e^{j\omega})$ for each of the four cases of linear phase filters.

TABLE 3.3

	$Q(e^{j\omega})$	$P(e^{j\omega})$
Case 1	1	$\sum\limits_{n=0}^{(N-1)/2} \tilde{a}(n) \cos(\omega n)$
Case 2	$\cos\left(\dfrac{\omega}{2}\right)$	$\sum\limits_{n=0}^{(N/2)-1} \tilde{b}(n) \cos(\omega n)$
Case 3	$\sin(\omega)$	$\sum\limits_{n=0}^{(N-3)/2} \tilde{c}(n) \cos(\omega n)$
Case 4	$\sin\left(\dfrac{\omega}{2}\right)$	$\sum\limits_{n=0}^{(N/2)-1} \tilde{d}(n) \cos(\omega n)$

where $\tilde{a}(n) = a(n)$. For Cases 2 to 4, $Q(e^{j\omega})$ is constrained to be zero at either $\omega = 0$ or $\omega = \pi$ or both.

To show how the optimal linear phase FIR filter design problem can be formulated as a Chebyshev approximation problem, it is necessary to define $D(e^{j\omega})$, the desired (real) frequency response of the filter, and $W(e^{j\omega})$, a weighting function on the approximation error that enables the designer to choose the relative size of the error in different frequency bands. The weighted error of approximation $E(e^{j\omega})$ is, by definition,

$$E(e^{j\omega}) = W(e^{j\omega})[D(e^{j\omega}) - H^*(e^{j\omega})] \qquad (3.105)$$

By writing $H^*(e^{j\omega})$ as a product of $P(e^{j\omega})$ and $Q(e^{j\omega})$, $E(e^{j\omega})$ can be rewritten as

$$E(e^{j\omega}) = W(e^{j\omega})[D(e^{j\omega}) - P(e^{j\omega})Q(e^{j\omega})] \qquad (3.106)$$

Since $Q(e^{j\omega})$ is a fixed function of frequency, it can be factored out of Eq. (3.106), giving

$$E(e^{j\omega}) = W(e^{j\omega})Q(e^{j\omega})\left[\frac{D(e^{j\omega})}{Q(e^{j\omega})} - P(e^{j\omega})\right] \qquad (3.107)$$

Equation (3.107) is valid everywhere except possibly at $\omega = 0$ and/or $\omega = \pi$ Defining $\hat{W}(e^{j\omega})$ and $\hat{D}(e^{j\omega})$ as

$$\hat{W}(e^{j\omega}) = W(e^{j\omega})Q(e^{j\omega}) \qquad (3.108)$$

and

$$\hat{D}(e^{j\omega}) = \frac{D(e^{j\omega})}{Q(e^{j\omega})} \qquad (3.109)$$

the error function may be written as

$$E(e^{j\omega}) = \hat{W}(e^{j\omega})[\hat{D}(e^{j\omega}) - P(e^{j\omega})] \qquad (3.110)$$

The Chebyshev approximation problem may now be stated as finding the set of coefficients [$\tilde{a}(n)$, $\tilde{b}(n)$, $\tilde{c}(n)$, or $\tilde{d}(n)$] to minimize the maximum absolute value of $E(e^{j\omega})$ over the frequency bands in which the approximation is being performed. Using the notation $\|E(e^{j\omega})\|$ to denote this minimum value [i.e., the L_∞-norm of $E(e^{j\omega})$], the Chebyshev approximation problem may be stated mathematically as

$$\|E(e^{j\omega})\| = \min_{\{\text{coefficients}\}} \left[\max_{\omega \in A} |E(e^{j\omega})| \right] \tag{3.111}$$

where A represents the disjoint union of all the frequency bands of interest.

A well-known property of this class of Chebyshev approximation problems may be used to obtain a solution to Eq. (3.111). This is the so-called alternation theorem, which may be stated as follows:

THEOREM   If $P(e^{j\omega})$ is a linear combination of r cosine functions,

$$\text{i.e., } P(e^{j\omega}) = \sum_{n=0}^{r-1} \alpha(n) \cos(\omega n)$$

then a necessary and sufficient condition that $P(e^{j\omega})$ be the unique, best weighted Chebyshev approximation to a continuous function $\hat{D}(e^{j\omega})$ on A, a compact subset of $(0, \pi)$, is that the weighted error function $E(e^{j\omega})$ exhibit at least $(r + 1)$ extremal frequencies in A; i.e., there must exist $(r + 1)$ points ω_i in A such that $\omega_1 < \omega_2 < \cdots < \omega_r < \omega_{r+1}$ and such that $E(e^{j\omega_i}) = -E(e^{j\omega_{i+1}})$, $i = 1$, $2, \ldots, r$, and $|E(e^{j\omega_i})| = \max_{\omega \in A} [E(e^{j\omega})]$.

The alternation theorem above is extremely powerful in that it expresses a necessary and sufficient set of conditions for obtaining the optimal Chebyshev solution. A number of techniques have been devised for obtaining this optimal solution, depending on the interpretation of this theorem. In the following sections several of these techniques will be discussed, both for historical reasons and to aid in gaining an understanding of the nature of the optimal filter. Before discussing any specific algorithm for designing optimal filters, Sec. 3.27 presents an important result on the maximum number of extrema of a linear phase FIR filter.

3.27   Constraint on the Number of Extrema of the Frequency Response of a Linear Phase Filter

The alternation theorem of Sec. 3.26 states that for the optimal linear phase FIR filter the error function has at least $(r + 1)$ extrema where r is the number of cosine functions being used in the approximation. Since for most general cases of interest the extrema of $H^*(e^{j\omega})$ are also the extrema of $E(e^{j\omega})$ [i.e., both $dW(e^{j\omega})/d\omega$ and $dD(e^{j\omega})/d\omega$ will be zero when

$dH^*(e^{j\omega})/d\omega$ is zero], it is important to know the maximum number of extrema of $H^*(e^{j\omega})$. By adding to this number the number of extrema of $E(e^{j\omega})$ that are not extrema of $H^*(e^{j\omega})$, the total maximum number of extrema of $E(e^{j\omega})$ can be found.

Thus consider $H^*(e^{j\omega})$ for Case 1; i.e.,

$$H^*(e^{j\omega}) = \sum_{n=0}^{(N-1)/2} \tilde{a}(n) \cos{(\omega n)} \tag{3.112}$$

To find the maximum number of extrema of $H^*(e^{j\omega})$ in the interval $0 \leq \omega \leq \pi$, it is advantageous to convert Eq. (3.112) into an ordinary polynomial in $\cos \omega$. Thus each term $\cos{(\omega n)}$ may be expressed in the form

$$\cos{(\omega n)} = \sum_{m=0}^{n} \alpha_{mn}(\cos \omega)^m \tag{3.113}$$

where the α_{mn} are real coefficients that are readily obtained in standard books of tables such as the CRC standard mathematic tables. Using Eq. (3.113) in Eq. (3.112) gives

$$H^*(e^{j\omega}) = \sum_{n=0}^{(N-1)/2} \tilde{a}(n) \left[\sum_{m=0}^{n} \alpha_{mn}(\cos \omega)^m \right] = \sum_{n=0}^{(N-1)/2} \bar{a}(n)(\cos \omega)^n \tag{3.114}$$

where $\{\bar{a}(n)\}$ is obtained by collecting like power terms of $(\cos \omega)^n$. To find the extremal points, $H^*(e^{j\omega})$ is differentiated giving

$$\frac{d}{d\omega} [H^*(e^{j\omega})] = \sum_{n=0}^{(N-1)/2} n\bar{a}(n)(\cos \omega)^{n-1}(-\sin \omega)$$

$$= \sin \omega \sum_{m=0}^{(N-3)/2} \bar{b}(m)(\cos \omega)^m \tag{3.115}$$

where the coefficients $\bar{b}(m) = -(m+1)\bar{a}(m+1)$. To find the maximum number of extrema [i.e., values of ω where Eq. (3.115) is zero], it is useful to convert Eq. (3.115) to an ordinary polynomial in the variable x via the transformation $x = \cos \omega$. The resulting function $G(x)$ is of the form

$$G(x) = \frac{d}{d\omega} [H^*(e^{j\omega})]\Big|_{\omega=\cos^{-1} x} = \sqrt{1-x^2} \sum_{m=0}^{(N-3)/2} \bar{b}(m)x^m \tag{3.116}$$

which is of the form $G(x) = F_1(x) \cdot F_2(x)$ where

$$F_1(x) = \sqrt{1-x^2} \tag{3.117a}$$

$$F_2(x) = \sum_{m=0}^{(N-3)/2} \bar{b}(m)x^m \tag{3.117b}$$

Clearly $F_1(x)$ is zero only at $x = +1$ (the mapping of $\omega = 0$) and $x = -1$ (the mapping of $\omega = \pi$). $F_2(x)$ is a polynomial of degree $(N-3)/2$; hence it can have at most $[(N - 3)/2]$ zeros in the open interval $-1 < x < 1$. Therefore $G(x)$ can have at most $[(N + 1)/2]$ zeros on the closed interval $-1 \leq x \leq 1$. This means $H^*(e^{j\omega})$ can have at most $[(N + 1)/2]$ extrema in the closed interval $0 \leq \omega \leq \pi$. Thus, for Case 1 linear phase filters N_e, the number of extrema of $H^*(e^{j\omega})$ obeys the constraint

$$N_e \leq \frac{(N + 1)}{2} \qquad \text{Case 1} \qquad (3.118\text{a})$$

Rather than repeating similar derivations, the results for Cases 2 to 4 will be stated. Their verification is left to the reader. The number of extrema of $H^*(e^{j\omega})$ obeys the following constraints:

$$N_e \leq \frac{N}{2} \qquad \text{Case 2} \qquad (3.118\text{b})$$

$$N_e \leq \frac{(N - 1)}{2} \qquad \text{Case 3} \qquad (3.118\text{c})$$

$$N_e \leq \frac{N}{2} \qquad \text{Case 4} \qquad (3.118\text{d})$$

Equation (3.118) only constrains the number of extrema of $H^*(e^{j\omega})$. It is readily seen that if the approximation problem is being solved over a union of disjoint frequency bands, the error function can obtain an extremum at *each* band edge, whereas these points will generally *not* be extrema of $H^*(e^{j\omega})$. The exception to this rule is when the band edges are at either $\omega = 0$ or $\omega = \pi$ where $H^*(e^{j\omega})$ will often have an extremum. Thus, for example, the error function for a Case 1 lowpass filter (a two-band approximation problem) can have a maximum of $(N + 5)/2$ extrema, i.e., $[(N + 1)/2]$ extrema of $H^*(e^{j\omega})$ and two extra extrema for the passband and stopband edges. The error function for a Case 1 bandpass filter (a three-band approximation problem) can have a maximum of $[(N + 9)/2]$ extrema, i.e., $[(N + 1)/2)$ extrema of $H^*(e^{j\omega})$ and four extra extrema for the passband and stopband edges.

Foreknowledge of the maximum number of extrema of $E(e^{j\omega})$ is important because it relates to the exact ways in which design techniques have been devised to design optimal filters. For example, there are two well-known techniques that are only capable of designing optimal filters with the *maximum* possible number of extrema. These design techniques are clearly of limited utility in that the alternation theorem shows that, in general, filters with the maximum number of extrema in their error functions are special cases of the theorem and hence are only a subset of the larger class of optimal filters. In the following sections a discussion of the various optimal filter design algorithms is given. Both for historical reasons and for development purposes we first describe two algorithms that only are capable of designing a subclass of the set of optimal filters—i.e., those with the maximum possible number of extrema in their error functions. Then a discussion is given of a Remez-type algorithm and application of linear programming for designing any optimal, linear phase, FIR filter.

3.28 Nonlinear Equation Solution for Maximal Ripple FIR Filters

In Sec. 3.27, it was shown that the number of frequencies at which $H^*(e^{j\omega})$ could attain an extremum is strictly a function of the case of the linear phase filter under investigation. At each extremum, the value of $H^*(e^{j\omega})$ is predetermined by a combination of the weighting function $W(e^{j\omega})$, the desired frequency response $D(e^{j\omega})$, and the quantity δ that represents the peak error of approximation. By distributing the frequencies at which $H^*(e^{j\omega})$ attained an extremal value among the different frequency bands over which a desired response was being approximated and by requiring the resulting filter to have the maximum number of extremal frequencies, a unique optimal filter can be obtained. Since these filters have the maximum number of alternations, or ripples, in their error of approximation curve, they have been called maximal ripple filters. For the case of lowpass filters these maximal ripple filters have also been called extraripple filters because only a single extra ripple above the minimum number required for optimality is present.

The manner in which a set of nonlinear equation is obtained for describing the maximal ripple filter is as follows. (Herrmann and Schuessler originally proposed this method.) At each of the N_e unknown extremal frequencies, $E(e^{j\omega})$ attains the maximum value of either $\pm\delta$, and $E(e^{j\omega})$ or, equivalently, $H^*(e^{j\omega})$ has zero derivative. Thus two N_e equations of the form

$$H^*(e^{j\omega_i}) = \frac{\pm\delta}{W(e^{j\omega_i})} + D(e^{j\omega_i}) \qquad i = 1, 2, \ldots, N_e$$

$$\frac{d}{d\omega}[H^*(e^{j\omega})]\bigg|_{\omega=\omega i} = 0 \qquad i = 1, 2, \ldots, N_e$$

are obtained. These equations represent a set of two N_e nonlinear equations in two N_e unknowns [N_e impulse response coefficients and N_e frequencies at which $H^*(e^{j\omega})$ obtains the extremal value]. The set of two N_e equations may be solved iteratively using a nonlinear optimization procedure such as the well-known Fletcher–Powell algorithm.

Two facts should be noted about this procedure. First the quantity δ (i.e., the peak error) is a fixed quantity and is not minimized by the optimization scheme. Thus the shape of $H^*(e^{j\omega})$ is postulated a priori and only the frequencies at which $H^*(e^{j\omega})$ attains the extremal values are unknown. The second fact is that the design procedure has no way of specifying band edges for the different frequency bands of the filter. Thus the optimization algorithm does not work on given frequency bands but instead is free to select exactly where the bands will lie. This lack of control over frequency band edges diminishes the utility of this and the next algorithm to be discussed.

To illustrate a specific set of equations for optimization we consider the design of a Case 1 lowpass filter with $N = 15$, a peak ripple of $\delta = \delta_2$, a weighting function defined as

$$W(e^{j\omega}) = \begin{cases} \delta_2/\delta_1 & \omega \text{ in the passband} \\ 1 & \omega \text{ in the stopband} \end{cases}$$

and a desired response of

$$D(e^{j\omega}) = \begin{cases} 1 & \omega \text{ in the passband} \\ 0 & \omega \text{ in the stopband} \end{cases}$$

Figure 3.40 illustrates $H^*(e^{j\omega})$ for this example. The extremal frequencies are the set $\omega = 0; \omega_1, \omega_2, \omega_3, \omega_4, \omega_5, \omega_6,$ and $\omega_7 = \pi$. At $\omega = 0$ and $\omega = \pi$, $H^*(e^{j\omega})$ has zero derivative, independent of the impulse response coefficients for a Case 1 design. For this example, the $N_e = 8$ extremal frequencies are divided so that $N_p = 3$ occur in the passband and $N_s = 5$ occur in the stopband. Thus for this set of conditions the following equations are obtained:

Function Constraints	*Derivative Constraints*
$H^*(e^{j0}) = 1 + \delta_1$	
$H^*(e^{j\omega_1}) = 1 - \delta_1$	$\dfrac{d}{d\omega} H^*(e^{j\omega})\bigg|_{\omega=\omega_1} = 0$
$H^*(e^{j\omega_2}) = 1 + \delta_1$	$\dfrac{d}{d\omega} H^*(e^{j\omega})\bigg|_{\omega=\omega_2} = 0$
$H^*(e^{j\omega_3}) = -\delta_2$	$\dfrac{d}{d\omega} H^*(e^{j\omega})\bigg|_{\omega=\omega_3} = 0$
$H^*(e^{j\omega_4}) = +\delta_2$	$\dfrac{d}{d\omega} H^*(e^{j\omega})\bigg|_{\omega=\omega_4} = 0$
$H^*(e^{j\omega_5}) = -\delta_2$	$\dfrac{d}{d\omega} H^*(e^{j\omega})\bigg|_{\omega=\omega_5} = 0$
$H^*(e^{j\omega_6}) = +\delta_2$	$\dfrac{d}{d\omega} H^*(e^{j\omega})\bigg|_{\omega=\omega_6} = 0$
$H^*(e^{j\pi}) = -\delta_2$	

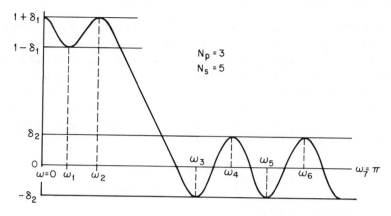

Fig. 3.40 Specifications for a maximal ripple lowpass filter.

Once the set of equations above has been solved for the unknown frequencies and the impulse response coefficients, the passband and stopband edges may be solved for by searching for the frequency beyond ω_2 where $H^*(e^{j\omega})$ exactly equals $1 - \delta_1$ (passband edge) and the frequency before ω_3 where $H^*(e^{j\omega})$ exactly equals $+\delta_2$ (stopband edge).

The optimization procedure above has been used to design lowpass and bandpass filters for values of N up to about 61. Section 3.29 discusses another technique for designing maximal ripple filters where much larger filters can be designed efficiently.

3.29 Polynomial Interpolation Solution for Maximal Ripple FIR Filters

A second more efficient method has been proposed by Hofstetter, Oppenheim, and Siegel for designing maximal ripple filters. This algorithm is basically an iterative technique for producing a polynomial $[H^*(e^{j\omega})]$ that has extrema of desired values. The algorithm begins by making an initial estimate of the frequencies at which the extrema in $H^*(e^{j\omega})$ will occur and then uses the well-known Lagrange interpolation formula to obtain a polynomial that alternatingly goes through the maximum allowable ripple values at these frequencies. It has been experimentally found that the initial guess of extremal frequencies does not affect the ultimate convergence of the algorithm but instead affects the number of iterations required to achieve the desired result.

It is instructive to consider the design of a Case 1 lowpass filter as an example of how the algorithm works. Figure 3.41 shows the frequency response of a lowpass filter with $N = 11$, peak ripple $\delta = \delta_2$, weighting function $W(e^{j\omega})$ and desired frequency response $D(e^{j\omega})$, as defined in Sec. 3.28. The number of extremal frequencies N_e is six for this example. They

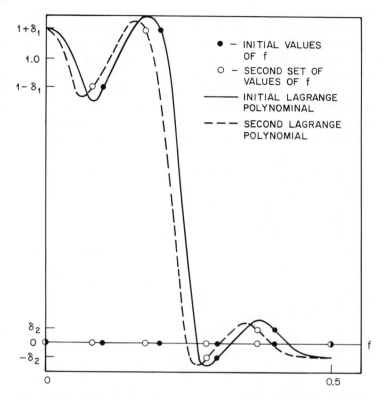

Fig. 3.41 Iterative solution for a maximal ripple lowpass filter. (After Hofstetter, Oppenheim and Siegel.)

are divided into $N_p = 3$ passband extrema and $N_s = 3$ stopband extrema. The filled dots along the frequency axis of Fig. 3.41 show the initial guess as to the extremal frequencies of $H^*(e^{j\omega})$. The solid line shows the initial Lagrange polynomial obtained by choosing polynomial coefficients so that the values of the polynomial at the guessed set of frequencies are identical to the assigned extreme values. As seen in Fig. 3.41, the polynomial associated with the initial guess does not have extrema that achieve the maximum allowable ripples but rather it has extrema that exceed these values. The next stage of the algorithm is to locate the frequencies at which the extrema of the first Lagrange interpolation occur. These frequencies are used as an updated, improved guess of the frequencies for which the extrema of the filter response will achieve the desired ripple values. This second set of frequencies is indicated by the open dots in Fig. 3.41. The algorithm uses these new frequencies to construct another Lagrange polynomial (shown by the dotted line in Fig. 3.41) that achieves the desired values at these frequencies. At this point the iterative nature of the algorithm has emerged. By locating the extrema of the new polynomial, another iteration of the algorithm is begun. This algorithm is quite similar to the well-known Remez multiple exchange algorithm of the Chebyshev approximation theory.

Some typical maximal ripple filters obtained by Hofstetter *et al.* using this algorithm are shown in Figs. 3.42 to 3.44. Figure 3.42 shows the log magnitude response of a Case 1 bandpass filter with $N = 41$ (i.e., $N_e = 21$) with 6 extrema of $H^*(e^{j\omega})$ in each stopband and 9 extrema in the passband. The peak ripple in the stopbands is $\delta_2 = 0.00001$ (or -100 dB), whereas the peak ripple in the passband is 0.005. Figure 3.43 shows the log magnitude

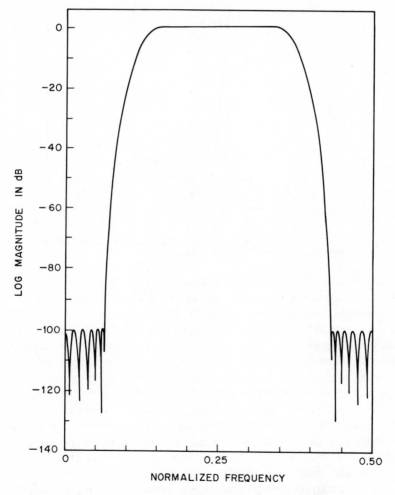

Fig. 3.42 Maximal ripple bandpass filter frequency response. (After Hofstetter, Oppenheim and Siegel.)

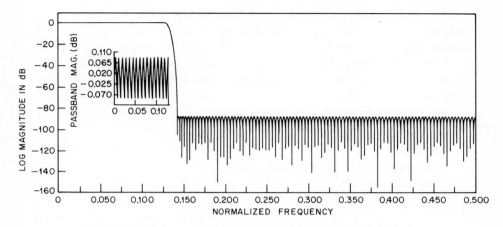

Fig. 3.43 Frequency response of maximal ripple lowpass filter. (After Hofstetter, Oppenheim and Siegel.)

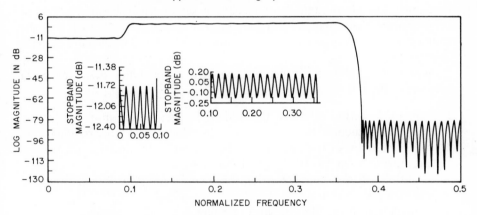

Fig. 3.44 Frequency response of maximal ripple multiband filter. (After Hofstetter, Oppenheim and Siegel.)

response of a Case 1 lowpass filter with $N = 251$ and with 33 extrema of $H^*(e^{j\omega})$ in the passband and 94 extrema in the stopband. The peak ripple in the passband is $\delta_1 = 0.01$ and the peak ripple in the stopband is $\delta_2 = 0.00004$ (or -88 dB). Finally, Fig. 3.44 shows the log magnitude response of a Case 1 modified lowpass filter with two different passbands and one stopband. For this filter, $N = 121$ and the extrema are distributed as follows: 12 in the first passband, 31 in the second passband, and 18 in the stopband. The desired response in the first passband is 0.25, whereas the desired response in the second passband is 1.0, as usual. The peak passband errors are $\delta_{11} = 0.01$ for the first passband and $\delta_{12} = 0.02$ for the second passband. The peak stopband error is $\delta_2 = 0.0001$.

Although this improved algorithm has essentially eliminated the difficulty of designing filters with large values of N, the inherent difficulty still remains that the filter band edge frequencies cannot be specified a priori; they must be calculated from the final solution. Furthermore both this and the preceding technique are only capable of designing maximal ripple filters, which (as discussed earlier) are a subclass of the class of optimal filters. In the next sections, design techniques are presented that are capable of designing any optimal (minimax) filter.

3.30   Remez Exchange Algorithm Design of Optimal FIR Filters

As shown earlier, the optimal linear phase FIR filter design problem can be formulated as a Chebyshev approximation problem where the approximating function [$P(e^{j\omega})$ in Eq. (3.110)] is a sum of r independent cosine functions. The alternation theorem gave a set of necessary and sufficient conditions on the weighted error function $E(e^{j\omega})$ [Eq. (3.110)] such that the solution was the unique best approximation to the desired frequency response $\hat{D}(e^{j\omega})$.

Based on this alternation theorem, a general optimal filter design program consists of the following steps:

1.   An input section in which the desired frequency response $D(e^{j\omega})$, the weighting function $W(e^{j\omega})$, and the filter duration N are specified.
2.   A formulation of the appropriate equivalent approximation problem—i.e., forming $\hat{D}(e^{j\omega})$, $\hat{W}(e^{j\omega})$, and $P(e^{j\omega})$.
3.   Solution of the approximation problem using the Remez multiple exchange algorithm.
4.   Calculation of the filter impulse response.

Step 1 in the design program is the interface between the designer and the filter design algorithm. The user communicates the type of filter to be designed and the particular specifications that must be met. The next step, that of formulating an appropriate equivalent design problem, has already been discussed in Sec. 3.26.

Figure 3.45 shows a block diagram representation of how the Remez exchange algorithm can be used to obtain the solution to the approximation problem—i.e., step 3 in the procedure. A dense grid of frequency points is used to find the set of $(r + 1)$ extremal frequencies required by the alternation theorem. An initial guess of these $(r + 1)$ extremal frequencies is made for which the error function is forced to have magnitude δ with alternating signs. From the initial specification of the problem, this requirement says that for the given set of extremal frequencies $\{\omega_k\}$, $k = 0, 1, \ldots, r$, the

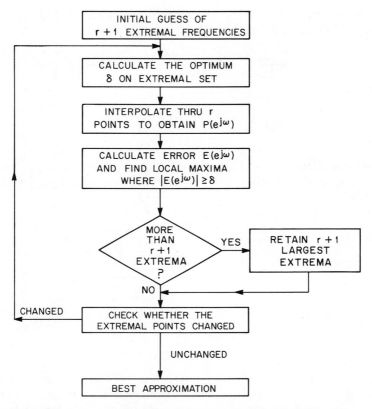

Fig. 3.45 Flowchart of Remez algorithm. (After Parks and McClellan.)

following set of equations must be solved:

$$\hat{W}(e^{j\omega_k})[\hat{D}(e^{j\omega_k}) - P(e^{j\omega_k})] = (-1)^k\delta \qquad k = 0, 1, \ldots, r \qquad (3.119)$$

or, in matrix form $\left(\text{assuming } P(e^{j\omega}) = \sum_{n=0}^{r-1} \alpha(n) \cos \omega n\right)$,

$$
\begin{bmatrix}
1 & \cos \omega_0 & \cos 2\omega_0 & \ldots & \cos[(r-1)\omega_0] & \dfrac{1}{\hat{W}(e^{j\omega_0})} \\
1 & \cos \omega_1 & & & & \\
\cdot & & & & & \\
\cdot & & & & & \\
\cdot & & & & & \\
& & & & & \dfrac{(-1)^r}{\hat{W}(e^{j\omega_r})} \\
1 & \cos \omega_r & & & &
\end{bmatrix}
\begin{bmatrix}
\alpha(0) \\
\alpha(1) \\
\cdot \\
\cdot \\
\cdot \\
\alpha(r-1) \\
\delta
\end{bmatrix}
=
\begin{bmatrix}
\hat{D}(e^{j\omega_0}) \\
\hat{D}(e^{j\omega_1}) \\
\cdot \\
\cdot \\
\cdot \\
\hat{D}(e^{j\omega_r})
\end{bmatrix}
$$

$$(3.120)$$

Since direct solution of Eq. (3.120) is both difficult and slow, it is more efficient to calculate δ analytically as

$$\delta = \frac{a_0 \hat{D}(e^{j\omega_0}) + a_1 \hat{D}(e^{j\omega_1}) + \ldots + a_r \hat{D}(e^{j\omega_r})}{a_0/\hat{W}(e^{j\omega_0}) - a_1/\hat{W}(e^{j\omega_1}) + \ldots + (-1)^r a_r/\hat{W}(e^{j\omega_r})} \qquad (3.121)$$

where

$$a_k = \prod_{\substack{i=0 \\ i \neq k}}^{r} \frac{1}{(x_k - x_i)} \qquad (3.122)$$

and

$$x_i = \cos \omega_i \qquad (3.123)$$

After calculating δ, the Lagrange interpolation formula in the barycentric form is used to interpolate $P(e^{j\omega})$ on the r points $\omega_0, \omega_1, \ldots, \omega_{r-1}$ to the values

$$C_k = \hat{D}(e^{j\omega_k}) - (-1)^k \frac{\delta}{\hat{W}(e^{j\omega_k})} \qquad k = 0, 1, \ldots, r - 1 \qquad (3.124)$$

and

$$P(e^{j\omega}) = \frac{\displaystyle\sum_{k=0}^{r-1} \left(\frac{\beta_k}{x - x_k}\right) C_k}{\displaystyle\sum_{k=0}^{r-1} \left(\frac{\beta_k}{x - x_k}\right)} \qquad (3.125)$$

where

$$\beta_k = \prod_{\substack{i=0 \\ i \neq k}}^{r-1} \frac{1}{(x_k - x_i)} \qquad (3.126)$$

and $x = \cos \omega$.

Note that $P(e^{j\omega})$ will also interpolate to $\hat{D}(e^{j\omega_r}) - [(-1)^r \delta/\hat{W}(e^{j\omega_r})]$ since it satisfies Eq. (3.119). The next step in the process is to evaluate $E(e^{j\omega})$ on the dense set of frequencies. If $|E(e^{j\omega})| \leq \delta$ for all frequencies in the dense set, then the optimal approximation has been found. If $|E(e^{j\omega})| > \delta$ for some frequencies in the dense set, then a new set of $(r + 1)$ frequencies must be chosen as candidates for the extremal frequencies. The new points are chosen as the peaks of the resulting error curve, thereby forcing δ to increase and ultimately converge to its upper bound, which corresponds to the solution to the problem. In the event that these are more than $(r + 1)$ extrema in $E(e^{j\omega})$ at any iteration, the $(r + 1)$ frequencies at which $|E(e^{j\omega})|$ is largest are retained as the guessed set of extremal frequencies for the next iteration.

Step 4 in the procedure, obtaining the filter impulse response, is implemented by evaluating $P(e^{j\omega})$ at 2^M equally spaced frequencies (where $2^M \geq N$) and using the DFT to obtain the sequence $\{\alpha(n)\}$, from which the impulse response coefficients may be derived. Depending on which case linear phase filter is derived, a unique formula can be written for obtaining $h(n)$ from $\alpha(n)$.

Based on the discussion above, a general-purpose linear phase FIR design procedure has been coded and is included in an appendix on p. 194.

Figure 3.46 shows a block diagram of how this program works. This program has a built-in interface (step 1 in the process) for designing multiple bandpass/bandstop filters (including lowpass, highpass, bandpass, and bandstop filters), differentiators, and Hilbert transformers. The capabilities and limitations of the program given in the appendix are

1. Filter length (called NFILT) satisfies the constraint, $3 \leq \text{NFILT} \leq 256$.
2. Filter type (called JTYPE):
    (a) Multiple bandpass/bandstop (JTYPE = 1)
    (b) Differentiator (JTYPE = 2)
    (c) Hilbert transformer (JTYPE = 3).

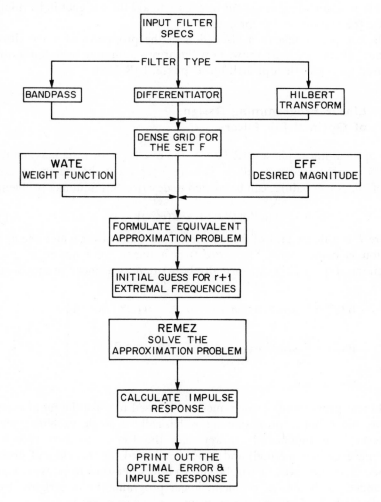

Fig. 3.46 Flowchart of design algorithm.

3.   Frequency regions (bands) specified by upper and lower cutoff frequencies with up to a maximum of 10 bands.
4.   Desired frequency response in each band is independently specified by the user.
5.   Weighting function in each band is independently specified by the user.

The set F in Fig. 3.46 consists of the dense set of frequencies in each band for which an approximation is being made. The subroutines WATE and EFF provide automatic weighting and magnitude functions for the pre-interfaced designs mentioned above. The subroutine REMEZ works exactly as discussed previously in this section. Several filter examples designed by this program are given in the appendix to aid the designer in learning how to use the features of this program.

Before proceeding to a discussion of the properties of some classes of optimal filters, Sec. 3.31 shows how linear programming techniques can also be used to obtain the optimal, linear phase, FIR filter.

3.31   Linear Programming Design of Optimal FIR Filters

The optimal linear phase FIR filter is the one for which the maximum error $E(e^{j\omega})$ is minimized over all ω. Letting δ represent the maximum error, a set of linear inequalities can be written to describe this minimax problem; i.e.,

$$-\delta \leq \hat{W}(e^{j\omega_i})[\hat{D}(e^{j\omega_i}) - P(e^{j\omega_i})] \leq \delta \qquad \omega_i \in F \qquad (3.127)$$

where F is a dense grid of frequencies in the bands over which the approximation is being made. Since $P(e^{j\omega})$ is a linear combination of r cosine functions, Eq. (3.127) can formally be written as a linear program—i.e.,

$$\left.\begin{aligned} -\hat{W}(e^{j\omega_i}) \sum_{m=0}^{r-1} \alpha(m) \cos(m\omega_i) - \delta &\leq -\hat{W}(e^{j\omega_i})\hat{D}(e^{j\omega_i}) \\ \hat{W}(e^{j\omega_i}) \sum_{m=0}^{r-1} \alpha(m) \cos(m\omega_i) - \delta &\leq \hat{W}(e^{j\omega_i})\hat{D}(e^{j\omega_i}) \end{aligned}\right\} \qquad \omega_i \in F$$

minimize δ.

Linear programming techniques can be used to solve the set of equations above. Since linear programming is basically a single exchange method, however, it is significantly slower than the Remez method (which is a multiple exchange method) and hence is avoided for this class of problems. In Sec. 3.39, however, it will be shown how when time response constraints are added to the design problem, linear programming is perhaps the only simple method of solving the problem.

3.32   Characteristics of Optimal Case 1 Lowpass Filters

For a lowpass filter the optimal design problem consists of specifying N, the passband cutoff frequency F_p, the stopband cutoff frequency F_s, and the ripple ratio $K = \delta_1/\delta_2$, which describes the desired weighting function $W(e^{j\omega})$ as

$$W(e^{j\omega}) = \begin{cases} \dfrac{1}{K} = \dfrac{\delta_2}{\delta_1} & 0 \le \omega \le 2\pi F_p \\ 1 & 2\pi F_s \le \omega \le \pi \end{cases} \tag{3.128}$$

where δ_1 is the passband ripple and δ_2 is the stopband ripple. Figure 3.47

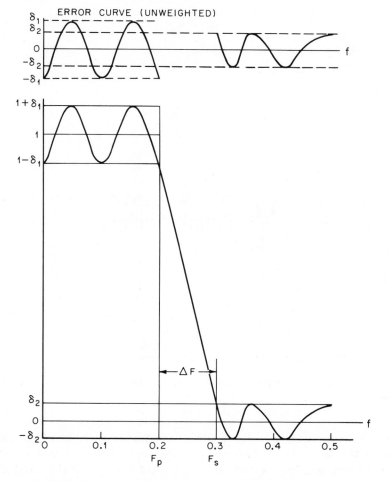

Fig. 3.47 Frequency response of optimal (minimax error) lowpass filter.

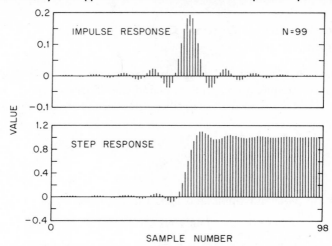

Fig. 3.48 Impulse and step responses of minimax error lowpass filter.

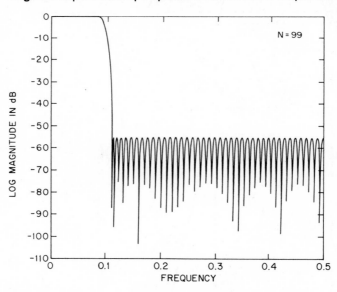

Fig. 3.49 Frequency response of optimal (minimax error) lowpass filter.

shows the frequency response of a Case 1 lowpass filter. The auxiliary parameter ΔF is defined as

$$\Delta F = F_s - F_p \tag{3.129}$$

and serves as a measure of the width of the transition band of the filter. Figures 3.48 and 3.49 show the time and frequency responses of an ($N = 99$), Case 1, lowpass filter with specifications $F_p = 0.0808$, $F_s = 0.1111$, and $K = 1.0$. The resulting values of δ_1 and δ_2 are 0.001724 for both ripples.

It was shown earlier that the error curve for the optimal lowpass filter could have either $(r + 1)$ or $(r + 2)$ extrema where $r = (N + 1)/2$ for Case 1 and $r = N/2$ for Case 2. It is important to understand the nature of the optimum lowpass filter to see under what conditions the number of ripple extrema attains the maximum value. It has been found experimentally that a reasonably straightforward and informative way of summarizing the behavior of the optimum filter is to plot the transition width of the filter (ΔF) versus passband cutoff frequency F_p for fixed values of N, δ_1, and δ_2. Figure 3.50 shows such a plot for Case 1 data for $N = 11$, $\delta_1 = \delta_2 = 0.1$. As seen in this figure, the curve of ΔF versus F_p has an oscillatory behavior, alternating between sharp minima and flat-topped maxima. The local minima of the curves (labeled ER1 to ER5) have been found to be the maximal ripple (extraripple) filters for the particular choice of N, δ_1, and δ_2. {Recall these extraripple filters have $[(N + 5)/2]$ equal amplitude extrema in their error curves.} There are exactly $(N - 1)/2$ of these extraripple filters for fixed values of δ_1 and δ_2. In between the extraripple solutions, it has been found that there are two types of optimum filters—scaled extraripple filters and filters with exactly $[(N + 3)/2]$ equal amplitude extrema in their error curves.

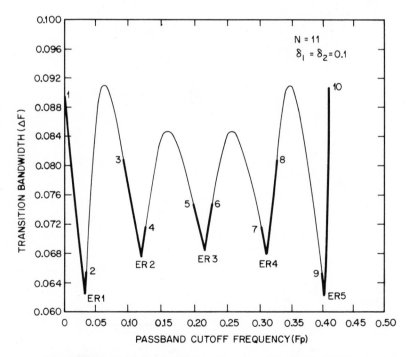

Fig. 3.50 Transition bandwidth as a function of passband cutoff frequency for optimal lowpass filters.

The scaled extraripple filters (shown in Fig. 3.50 by the heavy lines on the curve) have error curves with $[(N + 3)/2]$ equal amplitude extrema, as well as one smaller amplitude extremum that is at either $f = 0$ or $f = 0.5$. These filters can be derived from their neighboring extraripple filter by a simple scaling procedure as illustrated in Fig. 3.51. The upper left-hand corner of Fig. 3.51 shows the frequency response of an extraripple filter with $[(N + 5)/2]$ extrema in the error curve. The frequency response is in the form of a trigonometric polynomial in f. By applying a transformation of the form $x = \cos(2\pi f)$, the resulting frequency response is an ordinary polynomial in x, as shown in the upper right-hand corner of Fig. 3.51. Now a simple linear scaling of x of the form $x' = \alpha x + \beta$ can be used, where $x = -1$ is mapped to $x' = -1$ and $x = X_H$ is mapped to $x' = 1$. The value X_H is chosen to lie between $+1$ and the preceding extremum value. In this manner the resulting polynomial in x' (lower right-hand corner of Fig. 3.51) has $[(N + 3)/2]$ equal amplitude extrema, whereas at $x' = 1$ the value of the error is smaller than at the other extrema. By applying the inverse transformation $f = \cos^{-1}(x')/(2\pi)$, the resulting frequency response (lower

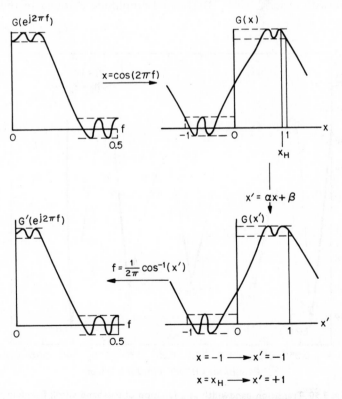

Fig. 3.51 Scaling procedure for optimal lowpass filters.

left-hand corner of Fig. 3.51) is that of a scaled extraripple solution with $[(N + 3)/2]$ equal amplitude extrema and a smaller extremum at $f = 0$. Since the optimum filter must have at least $[(N + 3)/2]$ equal amplitude extrema, the scaling can only be used until the next-to-last ripple extremum is scaled to lie at $x' = 1$. Scaling beyond this point leaves a nonoptimum filter.

The optimum filters between scaled extraripple solutions all have exactly $[(N + 3)/2]$ equal amplitude extrema in their error curves. No simple linear scaling procedure has been found to account for their presence; however, when viewed as an ordinary polynomial in x, one can account for the resulting behavior by adjusting the position and amplitude of the "invisible" ripple outside the range $-1 \leq x \leq +1$. As shown in Fig. 3.52, as F_p increases, the position of the invisible ripple tends to $-\infty$ and its amplitude tends to $+\infty$. Eventually the position of the invisible ripple is at $\pm\infty$, at which point the N-point filter is equivalent to an extraripple $(N - 2)$-point filter since all

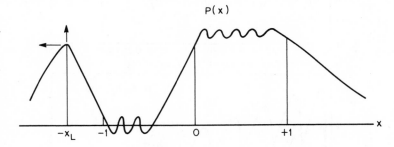

Fig. 3.52 The "invisible" ripple.

the ripples are in the region $-1 \leq x \leq +1$. This effect is illustrated in Fig. 3.53, which shows the curves of transition width versus passband cutoff frequency for both $N = 9$ and $N = 11$ with $\delta_1 = \delta_2 = 0.1$. Each extraripple solution for $N = 9$ is equivalent to a nonextraripple solution for $N = 11$, as predicted above.

As F_p is increased further beyond the point at which the invisible ripple is at $-\infty$, the invisible ripple comes back from $+\infty$ and starts heading toward $x = +1$. Eventually the invisible ripple becomes visible (at $x = +1$) and becomes a scaled extraripple solution. The description of the mechanics of the different types of optimal filters above, although somewhat qualitative, has been experimentally verified and shown to be exceedingly useful in understanding the nature of the optimal filter.

Figure 3.54 presents a summary of the types of optimal filters that may be obtained by varying the filter cutoff frequencies. The first filter shown is an extraripple solution with $N = 25$, $\delta_1 = \delta_2 = 0.05$. Below it are two different scaled solutions where the frequency response is 0 at $f = 0.5$ for the first

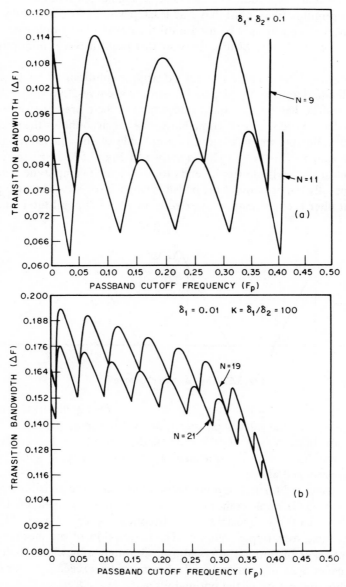

Fig. 3.53 Transition bandwidth as a function of passband cutoff frequency for several optimal lowpass filters.

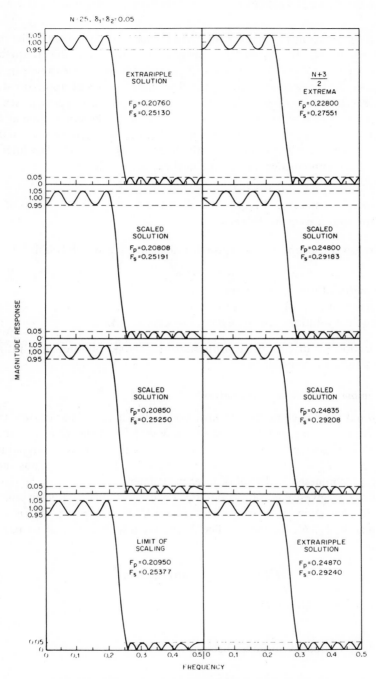

Fig. 3.54 Summary of the various types of optimal lowpass filters.

filter and then 0.03 at $f = 0.5$ for the second case. The last filter in the first column represents the maximum possible scaling—i.e., the frequency response is 0.05 at $f = 0.5$ with $[(N + 3)/2]$ equal amplitude extrema in the error curve. In the second column, the first filter is a point approximately midway between extraripple solutions. The next two filters represent scaled extraripple solutions where the error at $f = 0$ is not of the same value as the other error extrema. For the first of these filters the error is about -0.005 at $f = 0$, whereas for the second it is about 0.015 at $f = 0$. The last filter in the second column corresponds to the next extraripple solution.

3.33   Some Additional Properties of Case 1 Optimal Lowpass Filters

There are several properties of optimal filters that are worth noting. These include

1. Symmetry of filter parameters.
2. Behavior of transition width for large K.
3. Analytical solutions for the Chebyshev case.

In this section we discuss these properties and show how they lend insight into the filter design problem.

1. Symmetry of Filter Parameters

The optimal filter is completely characterized by the set of parameters that include N, F_p, F_s, δ_1, and δ_2. We now show that for each optimal lowpass filter with the specifications above, another optimal filter exists with parameters $N' = N$, $F_p' = 0.5 - F_s$, $F_s' = 0.5 - F_p$, $\delta_1' = \delta_2$, $\delta_2' = \delta_1$. Thus there is a kind of symmetry in the set of filter parameters.

Figure 3.55 gives a graphical description of how this symmetry can be shown. It also shows the frequency response of an optimal filter with parameters N, F_p, F_s, δ_1, and δ_2. The frequency response may be written in the form

$$H^*(e^{j\omega}) = \sum_{n=0}^{(N-1)/2} a(n) \cos(\omega n) \tag{3.130}$$

By making the substitution

$$\omega \to \pi - \omega \tag{3.131}$$

the resulting frequency response becomes

$$H^*[e^{j(\pi-\omega)}] = \sum_{n=0}^{(N-1)/2} a(n) \underbrace{\cos[(\pi - \omega)n]}_{(-1)^n \cos(\omega n)} \tag{3.132}$$

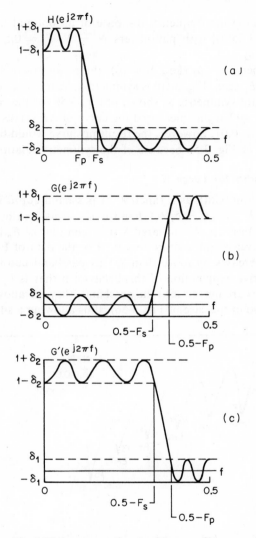

Fig. 3.55 Symmetry of optimal lowpass filter parameters.

or setting $G(e^{j\omega}) = H^*[e^{j(\pi-\omega)}]$ we find

$$G(e^{j\omega}) = \sum_{n=0}^{(N-1)/2} g(n) \cos(\omega n) \qquad (3.133)$$

which is the frequency response of an optimal filter (highpass), as shown in Fig. 3.55(b). Finally letting

$$G'(e^{j\omega}) = 1 - G(e^{j\omega}) = 1 - \sum_{n=0}^{(N-1)/2} g(n) \cos(\omega n)$$

$$= \sum_{n=0}^{(N-1)/2} g'(n) \cos(\omega n) \qquad (3.134)$$

results in $G'(e^{j\omega})$ being the frequency response of an optimal lowpass filter [as shown in Fig. 3.55(c)] with parameters $N' = N$, $F_p' = 0.5 - F_s$, $F_s' = 0.5 - F_p$, $\delta_1' = \delta_2$, $\delta_2' = \delta_1$.

For the case when $\delta_1 = \delta_2$ (i.e., $\delta_1' = \delta_2'$), the result above says that the curve of ΔF versus F_p (e.g., Fig. 3.50) is symmetric. When $\delta_1 \neq \delta_2$, the curve of ΔF versus F_p is not symmetric as shown in Fig. 3.56 for the case $N = 21$, $\delta_1 = 0.01$, $\delta_2 = 0.0001$. If one measured the curve of ΔF versus F_p for $N = 21$, $\delta_1 = 0.0001$, $\delta_2 = 0.01$, however, the resulting curve would be the mirror image of the curve of Fig. 3.56 because of the symmetry relations above.

2. Transition Width for Large K

The curve of transition width as a function of passband cutoff frequency is greatly different for $K > 1$ ($K = \delta_1/\delta_2$) than for $K = 1$ as seen by comparing Fig. 3.50 and 3.56. For the case of large K it is seen that as F_p is increased, transition width decreases quite dramatically. For the data of Fig. 3.56, the transition width decreases by more than 2:1 as passband cutoff frequency increases. An intuitive explanation of this behavior is that as F_p is increased and stopband ripples are moved into the passband, they are allowed to grow by a factor of K (100 in the case of Fig. 3.56), thus allowing a smaller ΔF.

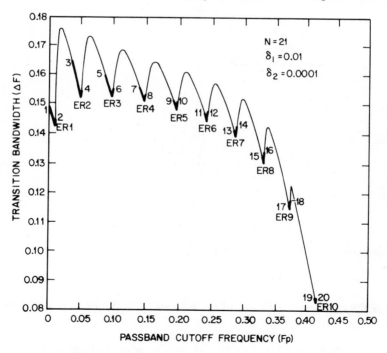

Fig. 3.56 Transition bandwidth versus passband cutoff frequencies for optimal lowpass filters.

This effect is not quite so dramatic as N becomes large. Figure 3.57 shows curves of the transition width versus passband cutoff frequency for $N = 101$ and $K = 1$, 10, and 100 with $\delta_1 = 0.1$, 0.01, 0.001, and 0.0001. (Only the data for extraripple filters are plotted here.) It is seen from this figure that this dramatic variation in transition width occurs, for the most part, near $F_s = 0.5$—i.e., only for very wide bandwidth filters.

3. Chebyshev Solutions

Although analytical solutions to the optimal filter design problem are not easily attainable, in the special case when there is either one passband or one stopband ripple, such an analytical solution exists. These solutions are the well-known Chebyshev polynomials.

Consider the Mth-degree Chebyshev polynomial $T_M(x)$ defined as

$$T_M(x) = \cos(M \cos^{-1} x) \qquad |x| \leq 1$$
$$= \cosh(M \cosh^{-1} x) \qquad |x| > 1 \qquad (3.135)$$

or, equivalently, defined as the ordinary polynomial

$$T_M(x) = \sum_{n=0}^{M} b(n)x^n \qquad (3.136)$$

Figure 3.58 shows a plot of $T_M(x)$ for $M = 4$. If we let $M = (N - 1)/2$, then it can easily be shown that the polynomial $\delta_2 T_{(N-1)/2}(x)$ is of the correct form for an optimal Case 1 filter with one passband ripple. Since the Chebyshev polynomial is defined in the x domain, the transformation required to map it into the f domain is of the form

$$x = \left(\frac{X_0 + 1}{2}\right) \cos(2\pi f) + \left(\frac{X_0 - 1}{2}\right) \qquad (3.137)$$

which maps the interval $-1 \leq x \leq X_0$ into the interval $0 \leq f \leq 0.5$ and converts the ordinary polynomial in x into a trigonometric polynomial in f. The value X_0 in Eq. (3.137) is the point where $T_{(N-1)/2}(x) = (1 + \delta_1)/\delta_2$.

For the case of the Chebyshev polynomial solution, the passband and stopband cutoff frequencies are the mappings of the points $x = X_p$ [where $T_{(N-1)/2}(x) = (1 - \delta_1)/\delta_2$] and $x = +1$. Thus one can analytically solve for the transition width ΔF, which, in the case of large N, is of the form

$$\Delta F = F_s - F_p \approx \frac{1}{\pi(N - 1)}$$

$$\times \left(\cosh^{-1}\left(\frac{1 + \delta_1}{\delta_2}\right) - \left\{\left[\cosh^{-1}\left(\frac{1 + \delta_1}{\delta_2}\right)\right]^2 - \left[\cosh^{-1}\left(\frac{1 - \delta_1}{\delta_2}\right)\right]^2\right\}^{1/2}\right)$$

$$(3.138)$$

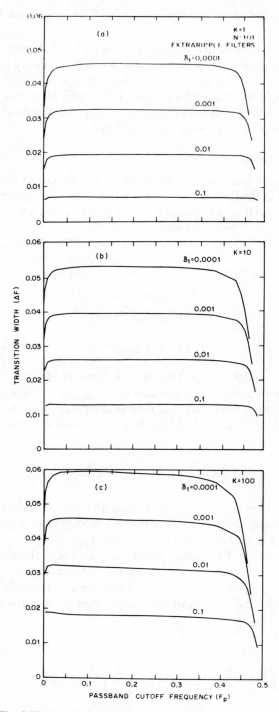

Fig. 3.57 Curves of transition width for optimal 101-point filters.

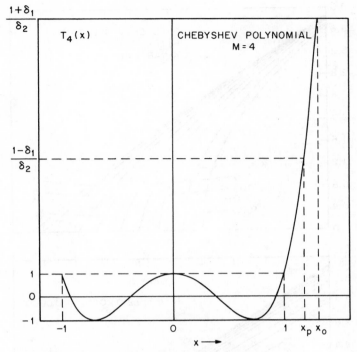

Fig. 3.58 A Chebyshev polynomial.

or, for $\delta_1 \ll 1$, Eq. (3.138) becomes

$$\Delta F \approx \frac{1}{\pi(N-1)}(\ln 2 - \ln \delta_2) \tag{3.139}$$

Thus the quantity D, defined as

$$D = (N-1)\,\Delta F \tag{3.140}$$

is independent of ΔF and N under the assumptions above. Figure 3.59 shows plots of D versus $\log_{10}\delta_2$ for $N = 127$ (i.e., large N) and various values of K and δ_1. This figure verifies that D is independent of δ_1 for small δ_1. Figure 3.60 shows plots of D versus $\log_{10}\delta_2$ for $K = 1$, 10, and 100 and various values of N from 3 to 127. This figure shows that D is essentially independent of N for $N \geq 51$, as shown above.

3.34 Relations Between Optimal Lowpass Filter Parameters

Except for the case of the Chebyshev solutions, no known analytical solution exists to the optimal filter design problem. As seen from earlier figures, the

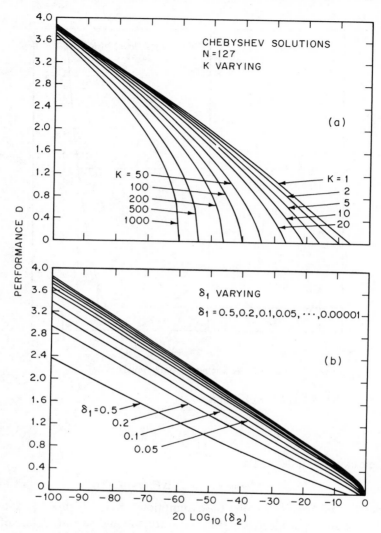

Fig. 3.59 Properties of optimal Chebyshev lowpass filters.

transition width for Chebyshev solutions is generally significantly smaller than the transition width for other optimal filters with the same values of N, δ_1, and δ_2 but different F_p and F_s. Based on experimental measurements on an extensive set of optimal filters, a set of approximate design relations between filter parameters has been obtained, from which a designer can choose any four of the five parameters N, F_p, F_s, δ_1, and δ_2 and can estimate the value of the missing parameter. To actually design the filter, the user would vary the unspecified parameter until specifications on all the remaining parameters were met or exceeded.

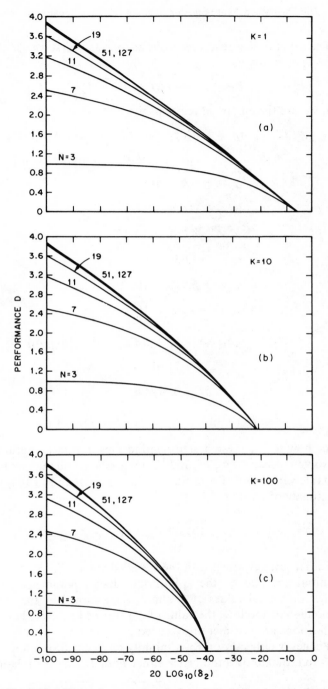

Fig. 3.60 Additional properties of optimal Chebyshev lowpass filters.

The set of design relationships is as follows. First we use the measure D, defined previously as

$$D = (N - 1)\,\Delta F = (N - 1)(F_s - F_p) \tag{3.141}$$

Then we use the empirical relationships

$$D_\infty(\delta_1, \delta_2) = [a_1(\log_{10} \delta_1)^2 + a_2 \log_{10} \delta_1 + a_3] \cdot \log_{10} \delta_2$$
$$+ [a_4(\log_{10} \delta_1)^2 + a_5 \log_{10} \delta_1 + a_6] \tag{3.142}$$

$$a_1 = 5.309 \times 10^{-3}$$

$$a_2 = 7.114 \times 10^{-2}$$

$$a_3 = -4.761 \times 10^{-1}$$

$$a_4 = -2.66 \times 10^{-3}$$

$$a_5 = -5.941 \times 10^{-1}$$

$$a_6 = -4.278 \times 10^{-1}$$

and

$$f(\delta_1, \delta_2) = b_1 + b_2(\log_{10} \delta_1 - \log_{10} \delta_2) \tag{3.143}$$

$$b_1 = 11.01217$$

$$b_2 = 0.51244$$

and

$$D_\infty(\delta_1, \delta_2) = (N - 1)\,\Delta F + f(\delta_1, \delta_2)(\Delta F)^2 \tag{3.144}$$

To see how to use these design relationships in a meaningful way, consider the problem where δ_1, δ_2, F_p, and F_s are specified, and the user must determine a suitable value of N. From Eq. (3.141), (3.142), and (3.144) we find for $\hat{N}$, the estimate of N, that

$$\hat{N} = \frac{D_\infty(\delta_1, \delta_2)}{\Delta F} - f(\delta_1, \delta_2)(\Delta F) + 1 \tag{3.145}$$

The initial value $\hat{N}$ along with the specified values F_p, F_s, and $K = \delta_1/\delta_2$ are the initial guess as to the appropriate design parameters. Depending on whether the actual values of δ_1 and δ_2 (as returned by the design program) are too high or too low, the value of N is raised or lowered accordingly until specifications are just met or exceeded.

Whenever either F_s or F_p is initially unspecified, the appropriate design rule is to estimate ΔF (and hence either F_s or F_p) via the relation

$$\widehat{\Delta F} = \frac{(N - 1)}{2f(\delta_1, \delta_2)}\left[\sqrt{1 + \frac{4f(\delta_1, \delta_2)D_\infty(\delta_1, \delta_2)}{(N - 1)^2}} - 1\right] \tag{3.146}$$

In this case the unspecified parameter is varied from the initial guess until specifications are met.

When δ_2 is the unspecified parameter, it may be estimated via the relation

$$\log_{10} \hat{\delta}_2 = \frac{(N-1)\Delta F + d_1(\Delta F)^2 - c_2}{c_1 + b_2(\Delta F)^2} \tag{3.147}$$

where

$$c_1 = a_1(\log_{10} \delta_1)^2 + a_2 \log_{10} \delta_1 + a_3$$

$$c_2 = a_4(\log_{10} \delta_1)^2 + a_5 \log_{10} \delta_1 + a_6$$

$$d_1 = b_1 + b_2 \log_{10} \delta_1$$

For this case the initial value of $\hat{\delta}_2$ enables the user to guess a value of $K = \delta_1/\delta_2$, which is then varied until δ_1 is within some specified tolerance.

The final possibility is that δ_1 is the unspecified parameter, in which case it may be estimated via the relation

$$\log_{10} \hat{\delta}_1 = -\frac{g_2}{2} + \sqrt{\frac{g_2^2}{4} - g_3} \tag{3.148}$$

where

$$g_1 = b_1 - b_2 \log_{10} \delta_2$$

$$g_2 = \frac{[e_2 - b_2(\Delta F)^2]}{e_1}$$

$$g_3 = \frac{e_3 - g_1(\Delta F)^2 - (N-1)\Delta F}{e_1}$$

$$e_1 = a_1 \log_{10} \delta_2 + a_4$$

$$e_2 = a_2 \log_{10} \delta_2 + a_5$$

$$e_3 = a_3 \log_{10} \delta_2 + a_6$$

For this case the initial value of $\hat{\delta}_1$ enables the user to guess a value of $K = \hat{\delta}_1/\delta_2$, which is then varied until δ_2 is within some specified tolerance.

The design relations above have been found to yield reasonably good estimates of the unspecified parameters over a wide range of design examples.

3.35 Properties of Case 2 Optimal Lowpass Filters

Most of the observed behavior for Case 1 optimal lowpass filters also occurs for Case 2 lowpass filters. Thus, for example, there are three basic types of optimal lowpass filters—the extraripple designs with $(N/2 + 2)$ equal amplitude extrema in their error curves, scaled extraripple designs with

$(N/2 + 1)$ equal amplitude extrema and one smaller extremum, and finally equiripple designs with exactly $(N/2 + 1)$ equal amplitude extrema. The major differences between Cases 1 and 2 is that for Case 2 filters the frequency response is constrained to be 0 at $\omega = \pi$. This causes some small differences in the properties of Case 2 lowpass filters that were not present for Case 1 lowpass filters. In this section these differences will be discussed and comparisons between Case 1 and Case 2 lowpass filters will be made.

Figure 3.61 shows a plot of the magnitude response of a typical optimal Case 2 lowpass filter. The parameters F_p, F_s, δ_1, and δ_2 are as defined for Case 1 lowpass filters. At $f = 0.5$, the magnitude response is exactly 0 because of the zero of $H^*(e^{j\omega})$ at $z = -1$. Figure 3.62(a) shows the magnitude responses of a Case 2 extraripple filter and Fig. 3.62(b), a Case 2 equiripple filter. For these examples, $N = 10$; hence the extraripple filter has $[(10/2 + 2) = 7]$ extrema in the error function, whereas the equiripple filter has only 6 extrema in the error function.

It is relatively easy to show that if we restrict ourselves to either odd or even values of N, an optimal filter with impulse response duration of $(N - 2)$ samples *cannot* achieve better specifications (i.e., smaller peak error) than an

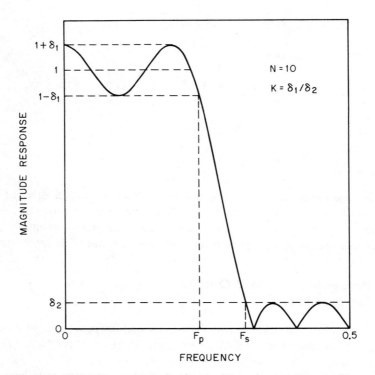

Fig. 3.61 Magnitude response of a 10-point (N even) optimal lowpass filter.

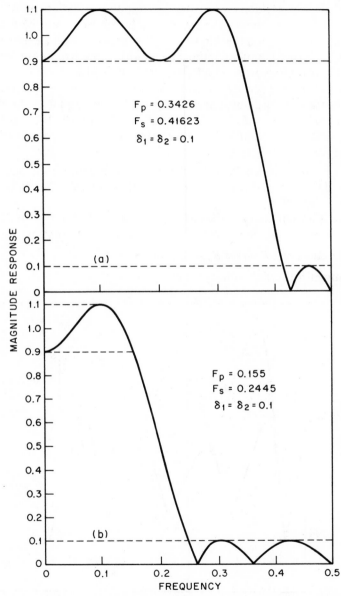

Fig. 3.62 Magnitude responses of an extraripple and an equiripple 10-point optimal lowpass filter.

optimal filter with impulse response duration of N samples. This is clear since the set of filters with impulse response duration of $(N-2)$ samples is a subset of the set of filters with impulse response duration of N samples. Thus an optimal member of the subset of a larger set cannot be better than the optimal member of the larger set. The argument above is not valid, however, when one compares optimal filters with impulse response duration of N samples with filters with impulse response durations of $(N-1)$ samples. A priori, one *cannot predict* which filter can achieve better specifications.

To illustrate the argument above, Fig. 3.63 shows a plot of the curves of transition width ($\Delta F = F_s - F_p$) versus F_p for lengths 9, 10, and 11 filters where $\delta_1 = \delta_2 = 0.1$ in all cases. From this figure several observations can be made:

1. The transition width for ($N = 10$) filters is sometimes smaller than for ($N = 11$) filters and is sometimes larger than for ($N = 9$) filters with the same values of F_p.

2. There is no symmetry in the curve of ΔF versus F_p for $N = 10$, although the curves for $N = 9$ and $N = 11$ are symmetrical in that for every point on the curve of ΔF versus F_p with coordinates $(\tilde{F}_p, \widetilde{\Delta F} = \tilde{F}_s - \tilde{F}_p)$ there is a symmetrical point with coordinates $(0.5 - \tilde{F}_s, \widetilde{\Delta F})$.

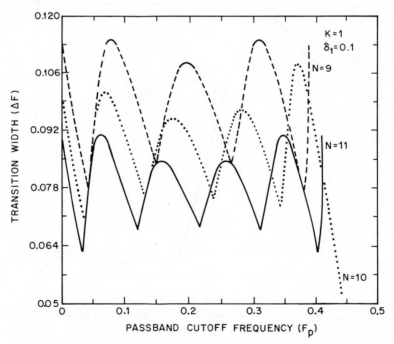

Fig. 3.63 Comparison of the transition widths of even and odd optimal lowpass filters.

3. The curve of ΔF versus F_p for $N = 10$ ends at an extraripple solution.

The explanation for observations 2 and 3 is related to the behavior of the optimal solutions at $f = 0.5$ for Case 2 filters. Since at this point $H^*(e^{j\pi}) = 0$, i.e., the error function does *not* have a peak as for Case 1 filters—the simple transformation of variables used to explain Case 1 symmetry is not applicable. Therefore, no simple symmetry in the curve of ΔF versus F_p is maintained. The explanation for observation 3 is given in Fig. 3.64, which shows the magnitude response of the last extraripple filter. Because of the constraint of a zero at $f = 0.5$, one cannot obtain filters with values of F_s arbitrarily close to 0.5, as for N odd.

The importance of observation 1 should not be underestimated. It is an unexpected and *surprising* result that a filter with $N = 10$ (i.e., approximation using five functions) can achieve given ripple specifications with a smaller transition width than a filter with $N = 11$ (i.e., approximation using six functions). Stated in a slightly different way, given fixed values of F_p, F_s, and K, a length 10 filter can achieve smaller ripple than a length 11 filter. For example, for the case $F_p = 0.3426$, $F_s = 0.41623$, $K = 1$, the length 11 filter achieves $\delta_1 = \delta_2 = 0.128215$, whereas the length 10 filter achieves $\delta_1 = \delta_2 = 0.1$. Expressed as a logarithm, the stopband attenuation of the length 10 filter is approximately 2.2 dB better than for the equivalent length 11 filter.

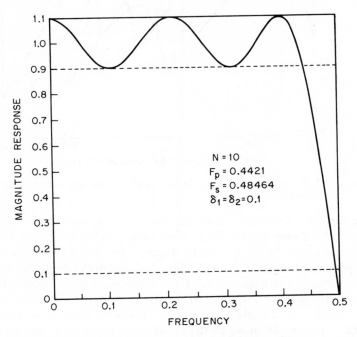

Fig. 3.64 The last N even extraripple lowpass filter.

Figure 3.65 shows a plot of ΔF versus F_p for $N = 9$, 10, and 11 with $K = 100$ ($\delta_1 = 0.1$, $\delta_2 = 0.001$). This figure shows F_p in the range $0.15 \leq F_p \leq 0.35$. The behavior of the curve of transition width versus F_p is similar to that of Fig. 3.62 in that the ($N = 10$) solutions sometimes have smaller transition widths than the ($N = 11$) solutions and sometimes have larger transition widths than the ($N = 9$) solutions.

Two of the interesting properties of Case 1 lowpass filters carry over only partially to Case 2 designs. These are the scaling procedures and the existence of Chebyshev solutions to the optimal design problem. Case 2 extraripple filters may be scaled in the vicinity of $f = 0$ directly as was

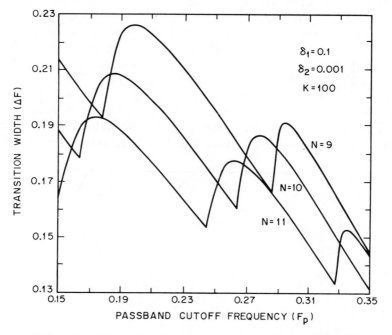

Fig. 3.65 Comparison of the transition widths of even and odd optimal lowpass filters.

possible for Case 1 designs; however, since the frequency response at the point $f = 0.5$ is constrained to be zero, no simple scaling procedure has been found in this region for Case 2 filters. Figure 3.66, which shows the magnitude response of five of the filters of Fig. 3.63, illustrates that *both* types of scaled extraripple filters do exist for Case 2 designs. Figure 3.66(a) shows a scaled extraripple filter response where the error at $f = 0$ is about 0.02, whereas the error at all other peaks is 0.1. Figure 3.66(b) shows the extraripple filter response from which the response of Fig. 3.66(a) was obtained. Figure 3.66(c) shows an optimal filter response where the error in the last

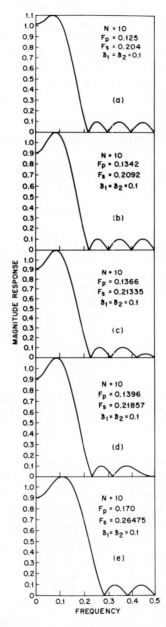

Fig. 3.66 Summary of the various possibilities for *N* even optimal lowpass filters.

extremum is much smaller than the other error extrema. A scaling procedure to account for how this type of filter response is obtained is not yet known. Figure 3.66(d) shows a filter response where the error curve at $f = 0.5$ has a triple zero due to the unexplained behavior noted above. Figure 3.66(e) shows the magnitude response of a filter with a larger value of F_p than the one of Fig. 3.66(d). This filter is an equiripple filter with $(N/2 + 1)$ peaks in the error function.

3.36   Characteristics of Optimal Differentiators

Optimal FIR differentiators are characterized by the desired frequency response

$$D(e^{j\omega}) = \begin{cases} \dfrac{j\omega}{\pi} & 0 \leq \omega \leq 2\pi F_p \\[2ex] \dfrac{j(2\pi - \omega)}{\pi} & 2\pi(1 - F_p) \leq \omega \leq 2\pi \end{cases} \tag{3.149}$$

where F_p represents the highest frequency for which differentiator action is required. Either Case 3 or Case 4 designs are used to approximate the response of Eq. (3.149) since $D(e^{j\omega})$ is purely imaginary. To minimize the peak *relative* error of the resulting response, the weighting function

$$W(e^{j\omega}) = \frac{1}{\omega} \qquad 0 < \omega \leq 2\pi F_p \tag{3.150}$$

is used. The differentiator cutoff frequency F_p can become as large as 0.5 for Case 4 designs, whereas for Case 3 designs it must be less than 0.5 or else the peak relative error will become almost 1.0 since at $f = 0.5$ the response is always 0.

Figures 3.67 to 3.70 show plots of the responses of several wideband differentiators. In Fig. 3.67 the magnitude, error, and relative error responses of an $(N = 16)$, differentiator $(F_p = 0.5)$ are shown. The peak relative error is 0.0136 and the relative error curve is equiripple in amplitude. Figure 3.68 shows these same curves for $N = 32$. In this case the peak relative error is reduced to 0.0062. Figure 3.69 shows the magnitude response of an $(N = 31)$, differentiator. Since the response is constrained to be zero at $f = 0.5$, the peak error occurs at $f = 0.5$ and is 1.0—i.e., an extremely poor differentiator is obtained. If the cutoff frequency is reduced to $F_p = 0.4$, for example, the resulting differentiator response for $N = 31$ becomes quite acceptable, as seen in Fig. 3.70(a). Figure 3.70(b) shows the magnitude response plotted from $f = 0$ to $f = 0.4$, the cutoff frequency, and Fig. 3.70(c) shows the error function over this range. The peak relative error is approximately 0.000028 over the interval $0 \leq f \leq 0.4$.

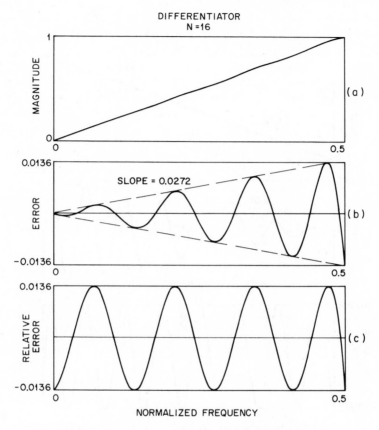

Fig. 3.67 Frequency responses of a 16-point optimal differentiator.

The basic differentiator parameters are N, F_p, and δ, the peak relative error of the differentiator. A large set of measurements of δ as a function of F_p and N are shown in Fig. 3.71 to 3.73. Figure 3.71 shows a plot of $20 \log_{10} \delta$ versus N for values of F_p of 0.5, 0.45, and 0.4 and for even and odd values of N in the range $3 \leq N \leq 128$. The curve for N odd, $F_p = 0.5$ is not included since $\delta = 1.0$ independent of N. From Fig. 3.71 it is seen that for the same value of F_p the values of δ for even-valued differentiators are approximately one to two orders of magnitude (20 to 40 dB) smaller than the values of δ for odd-valued differentiators. Another observation from these figures is that the smaller the bandwidth (F_p) of the differentiator, the faster the peak relative error decreases with increasing N. Thus for $F_p = 0.5$ the value of $20 \log_{10} \delta$ decreases by only about 35 dB as N varies from 4 to 128, whereas for $F_p = 0.45$ the value of $20 \log_{10} \delta$ decreases by about 98 dB as N varies from 4 to 52.

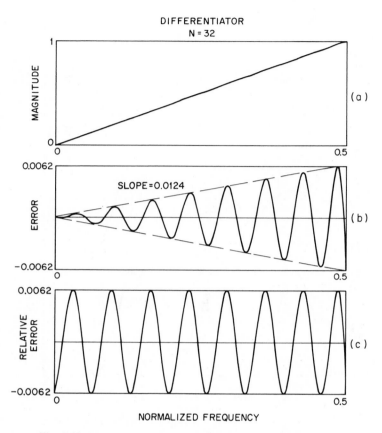

Fig. 3.68 Frequency responses of a 32-point optimal differentiator.

Figures 3.72 and 3.73 show plots of $20 \log_{10} \delta$ versus F_p for even values of N ($N = 4, 8, 16, 32, 64$) and odd values of N ($N = 5, 9, 17, 33, 65$). The data for even and odd values of N are presented on different figures because of the different nature of the solution for even and odd N. As seen in Fig. 3.73, as F_p tends to 0.5, the value of $20 \log_{10} \delta$ becomes 0, independent of N—i.e., the curves come together for odd values of N at $F_p = 0.5$. For even values of N, the curves are spaced apart for all values of F_p. The main observation from these figures is that the larger the value of N, the faster the peak relative error decreases with decreasing differentiator bandwidth.

Based on the observations above, in order to design the most efficient FIR differentiator (i.e., using the smallest possible value of N to obtain a desired value of peak relative error), the user should choose as small a bandwidth as possible and select an even value for N whenever possible. As an example,

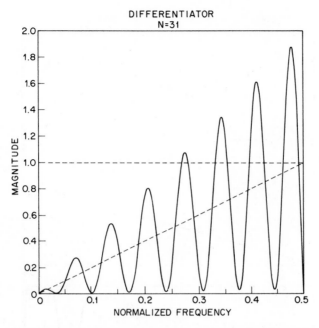

Fig. 3.69 Frequency response of a 31-point optimal differentiator.

to obtain a peak relative error less than 1 % requires the following values of N (as a function of F_p):

F_p	N (odd)	N (even)
0.5	impossible	22
0.45	27	10
0.40	15	6

whereas to obtain a peak relative error less than 0.1 % requires

F_p	N (odd)	N (even)
0.5	impossible	>128
0.45	41	18
0.40	21	12

The tables above indicate just how substantial the reductions in N can be as F_p varies and N goes from odd to even.

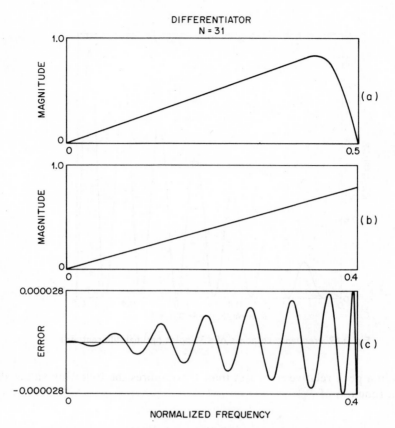

Fig. 3.70 Frequency responses of a 31-point optimal differentiator.

3.37 Characteristics of Optimal Hilbert Transformers

Optimal FIR Hilbert transformers are characterized by the desired frequency response

$$D(e^{j\omega}) = \begin{cases} -j & 2\pi F_L \leq \omega \leq 2\pi F_H \\ j & 2\pi(1 - F_H) \leq \omega \leq 2\pi(1 - F_L) \end{cases} \tag{3.151}$$

where F_L represents the lower cutoff frequency and F_H represents the higher cutoff frequency of the band for which the filter approximates the ideal Hilbert transformer response. As in the case of differentiators, either Case 3 or Case 4 designs are used to approximate Eq. (3.151). To minimize the peak error of the filter response, the weighting function $W(e^{j\omega}) = 1$ is used. For

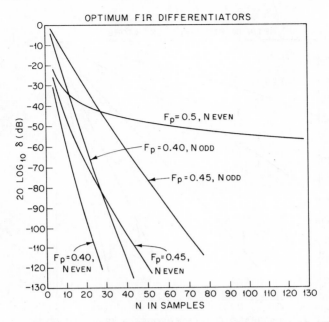

Fig. 3.71 Relative approximation error versus *N* for optimal differentiators.

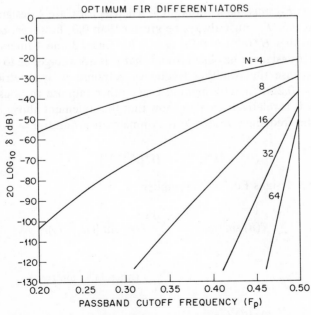

Fig. 3.72 Relative approximation error versus F_p for optimal differentiators.

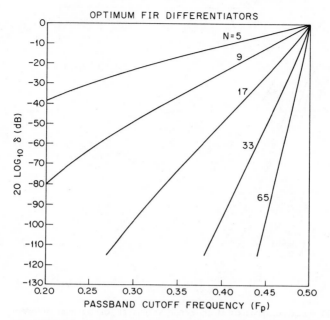

Fig. 3.73 Relative approximation error versus F_p for optimal differentiators.

Case 3 designs F_H must be less than 0.5, whereas for Case 4 designs F_H can be as large as 0.5. F_L must always be greater than 0.0, however, because of the constraint that $H^*(e^{j\omega}) = 0$ for $\omega = 0$ for Cases 3 and 4 filters.

For Case 3 designs it has been found that it is advantageous to set $F_L = 0.5 - F_H$ because the resulting frequency response is symmetric around $\omega = \pi/2$ ($f = 0.25$), thereby making every other impulse response sample exactly zero. It is relatively easy to show that the statement above is valid as follows. If the frequency response is symmetrical around $\omega = \pi/2$, then

$$H^*(e^{j\omega}) = H^*[e^{j(\pi-\omega)}] \tag{3.152}$$

For Case 3 designs Eq. (3.152) implies

$$\sum_{n=1}^{(N-1)/2} \hat{c}(n) \sin(\omega n) = \sum_{n=1}^{(N-1)/2} \hat{c}(n) \sin[(\pi - \omega)n]$$

$$= \sum_{n=1}^{(N-1)/2} \hat{c}(n)(-1)^{n+1} \sin(\omega n)$$

or

$$\sum_{n=1}^{(N-1)/2} \hat{c}(n) \sin(\omega n)[1 - (-1)^{n+1}] = 0$$

which implies

$$\hat{c}(n) = \begin{cases} 0 & n \text{ even} \\ \text{unconstrained} & n \text{ odd} \end{cases} \qquad (3.153)$$

which is the desired result. Equation (3.153) says that the direct form realization of a Case 3 Hilbert transformer with equal lower and upper transition widths requires only $[(N + 1)/4]$ multiplications per sample, whereas a Case 4 realization requires $N/2$ multiplications per sample even if the upper and lower transition widths are equal. This is because Case 4 filters can never have a symmetrical frequency response (e.g., there is no zero at $\omega = \pi$, whereas there is a zero at $\omega = 0$). Thus, in some sense, Case 3 FIR Hilbert transformers have a 2:1 speed advantage over their Case 4 counterparts. It remains to be shown that the Case 3 filters can use this advantage to design filters with smaller peak errors of approximation than Case 4 filters. This aspect of the design problem will be discussed later in this section.

Figures 3.74 to 3.77 show plots of the responses of several FIR Hilbert transformers. Figure 3.74 shows the impulse response, magnitude response, and error function of an ($N = 31$) Hilbert transformer where $F_L = 0.04$ and $F_H = 0.46$. Since the upper and lower transition widths are equal, every other impulse response coefficient is exactly zero. The peak approximation error for this case is 0.008094 and the error curve is seen to be equiripple, as stated earlier. Figure 3.75 shows the same responses as Fig. 3.74 for an ($N = 32$) Hilbert transformer where $F_L = 0.04$ and $F_H = 0.46$. As previously noted, even though the upper and lower transition widths are equal, all the impulse response coefficients are nonzero because of the lack of symmetry in the frequency response for Case 4 designs. As seen in Fig. 3.75(b), the magnitude response is unconstrained in the region $0.46 < f \le 0.5$; hence its value is completely unpredictable. Figure 3.75(c) shows the error curve to be again equiripple over the band of approximation. The peak approximation error for this case is 0.007175.

Figures 3.76 and 3.77 show the effects of making the upper and lower transition bandwidths unequal. Figure 3.76 shows three sets of conditions for an ($N = 15$) filter—i.e., $F_L = 0.02$, $F_H = 0.48$ in Fig. 3.76(a); $F_L = 0.10$, $F_H = 0.48$ in Fig. 3.76(b); and $F_L = 0.02$, $F_H = 0.40$ in Fig. 3.76(c). The peak approximation errors are 0.266551 for Fig. 3.76(a), 0.260817 for Fig. 3.76(b), and 0.260737 for Fig. 3.76(c). Thus even though one of the transition widths changes by a factor of 5:1 (i.e., from 0.02 to 0.10), the change in peak error is on the order of 2%. Furthermore, as seen in Fig. 3.76(b) and (c), the magnitude response of the filter peaks up significantly in the wide transition band—a generally undesirable result. Also when the transition bandwidths are unequal, the symmetry property of the frequency

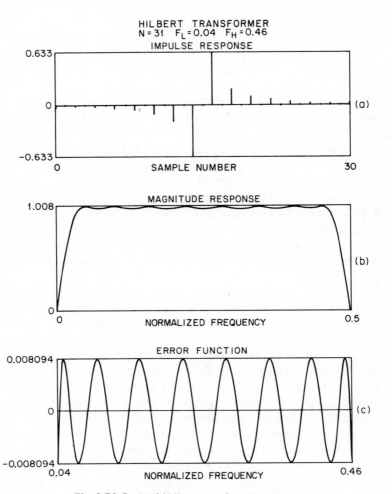

Fig. 3.74 Optimal Hilbert transformer responses.

response no longer holds and all the impulse response coefficients are non-zero. In conclusion the negligible decrease in peak error obtained by using unequal transition bandwidths is more than offset by the undesirable effects in the magnitude and impulse responses. Furthermore, based on this and other similar examples, it is seen that the peak error of approximation is determined primarily by the smaller of the two transition widths when the filter impulse response duration is odd.

Figure 3.77 shows similar examples of unequal transition widths for even length ($N = 16$) Hilbert transformers. The lower and upper cutoff frequencies for these examples are $-F_L = 0.02$, $F_H = 0.48$ for Fig. 3.77(a); $F_L = 0.02$, $F_H = 0.40$ for Fig. 3.77(b); and $F_L = 0.02$, $F_H = 0.50$ for Fig.

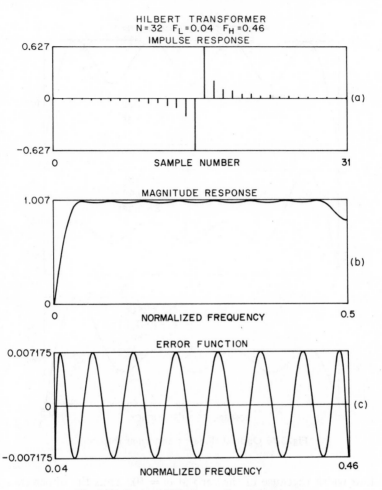

Fig. 3.75 Optimal Hilbert transformer responses.

3.57(c). The resulting peak approximation errors are 0.247920 for Fig. 3.77(a), 0.232594 for Fig. 3.77(b), and 0.248561 for Fig. 3.77(c). For the data of Fig. 3.77(b) (where the upper transition width is five times the lower transition width) the magnitude response again peaks up tremendously in the upper transition band. For this case the peak approximation error is about 6% smaller than the peak error for equal transition widths. Thus the slight decrease in peak error does not compensate the undesirable peaking in the frequency response. On the other hand, Fig. 3.77(c) shows that setting the upper transition width to 0, i.e., letting $F_H = 0.5$, produces a negligible change in the peak error. Based on these and other examples, it has been found that the peak approximation error depends almost entirely on the lower

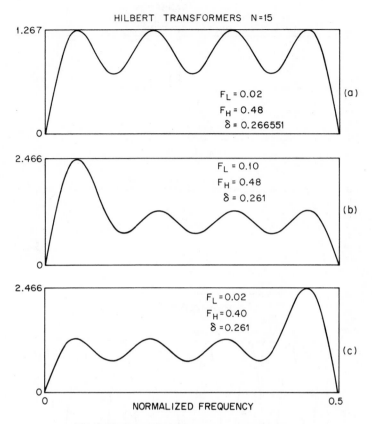

Fig. 3.76 Optimal Hilbert transformer responses.

transition width (because of the zero at $\omega = 0$). Thus the upper transition width should be made less than or equal to the lower transition width to minimize the peaking in the upper transition band.

The basic Hilbert transformer parameters are N, F_L, F_H, and δ, the peak approximation error (or the ripple) of the filter. It is assumed that $F_H = 0.5 - F_L$; i.e., the upper and lower transition widths are equal for the data to be presented in this section. Thus there are only three parameters—N, δ, and $\Delta F = F_L = 0.5 - F_H$. A large set of measurements of δ as a function of ΔF and N are shown in Fig. 3.78 and 3.79. Figure 3.78 shows a plot of 20 $\log_{10} \delta$ versus N for values of ΔF of 0.01, 0.02, 0.05, and 0.10 and for even and odd values of N in the range $3 \leq N \leq 128$. The curves for even and odd values of N, for fixed transition widths, are almost indistinguishable on the scales of Fig. 3.78; hence a single curve is given for both. Based on these curves it is seen that the larger the transition bandwidth of the Hilbert

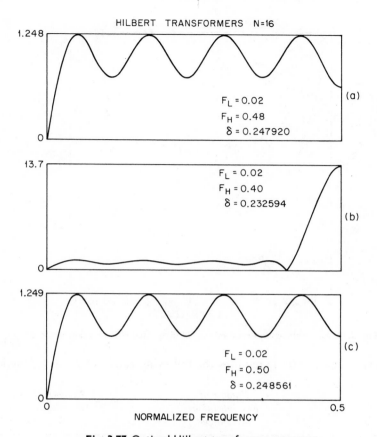

Fig. 3.77 Optimal Hilbert transformer responses.

transformer, the faster the peak error decreases with increasing N. Thus for $\Delta F = 0.01$ the value of $20 \log_{10} \delta$ decreases by only about 42 dB as N varies from 3 to 128, whereas for $\Delta F = 0.05$ the value of $20 \log_{10} \delta$ decreases by about 112 dB as N varies from 3 to 76.

Figure 3.79 shows plots of $20 \log_{10} \delta$ versus ΔF for even and odd values of N. The actual values used were $N = 3, 4, 7, 8, 15, 16, 31, 32, 63,$ and 64. As seen in this figure, as ΔF tends to 0, $20 \log_{10} \delta$ tends to 0 dB or a peak error of 1.0, independent of N. It is also seen from Fig. 3.79 that the larger the value of N, the faster the peak error decreases with increasing transition width of the Hilbert transformer.

Based on the discussion above, in order to design the most efficient FIR Hilbert transformer (i.e., using the smallest number of multiplications per sample to obtain a desired value of peak approximation error), the user should choose as large a transition bandwidth as possible and select an odd value of impulse response duration. As an example, to obtain a peak error

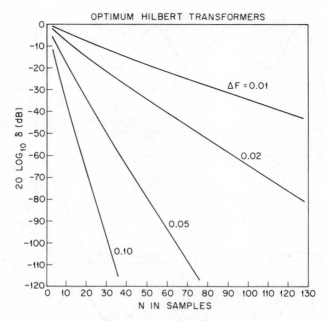

Fig. 3.78 Approximation error versus N for optimal Hilbert transformers.

of less than 1% ($\delta \leq 0.01$) requires the following values of N (as a function of ΔF):

ΔF	N (odd)	Number of Multiplications per Sample	N (even)	Number of Multiplications per Sample
0.01	119	30	118	59
0.02	59	15	60	30
0.05	27	7	24	12
0.10	11	3	12	6

whereas to obtain a peak error of less than 0.1% ($\delta \leq 0.001$) requires

ΔF	N (odd)	Number of Multiplications per Sample	N (even)	Number of Multiplications per Sample
0.01	>127	—	>127	—
0.02	95	24	94	47
0.05	39	10	38	19
0.10	19	5	18	9

The tables above indicate the substantial processing advantages of odd length Hilbert transformers with symmetrical frequency responses.

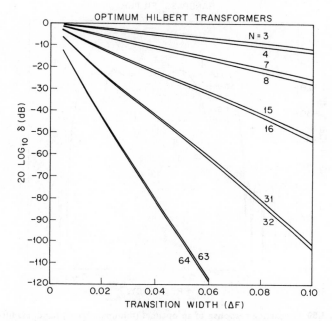

Fig. 3.79 Approximation error versus F_p for optimal Hilbert transformers.

3.38 Multiple Band Optimal FIR Filters

As mentioned earlier, the optimal design program given in the appendix on p. 194 is capable of designing multiple band filters including bandpass and bandstop designs. Figures 3.80 to 3.83 show the log magnitude responses of three typical multiple band filters. Figure 3.80 shows the response of an ($N = 32$) bandpass filter with band edge frequencies of $(0, 0.1)$ for the lower stopband $(0.2, 0.35)$ for the passband, and $(0.425, 0.5)$ for the upper stopband. The weighting function makes the stopband errors 10 times smaller than the passband error; thus the peak error is 0.00151 (56.4 dB) in the stopbands and 0.0151 in the passband. Figure 3.81 shows the response of an ($N = 31$) bandstop filter with band edge frequencies of $(0, 0.1)$ for the lower passband, $(0.15, 0.36)$ for the stopband, and $(0.41, 0.50)$ for the upper passband. The weighting function makes the stopband error 50 times smaller than the passband errors; the peak passband error is 0.144, whereas the peak stopband error is 0.00288 (50.8 dB).

Figure 3.82 shows the frequency response of a four-band filter with one passband and three stopbands. The band edge frequencies of the filter are $(0, 0.01786)$ for the passband, $(0.125, 0.1607)$ for the first stopband, $(0.2679, 0.3036)$ for the second stopband, and $(0.411, 0.4464)$ for the third stopband. The filter specifications were a peak error of less than 0.0144 in the passband

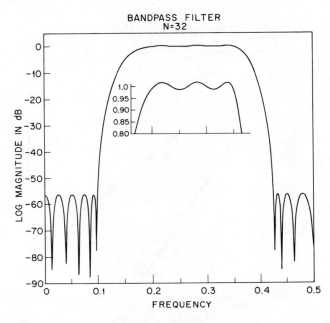

Fig. 3.80 Frequency response of an optimal (minimax error) bandpass filter.

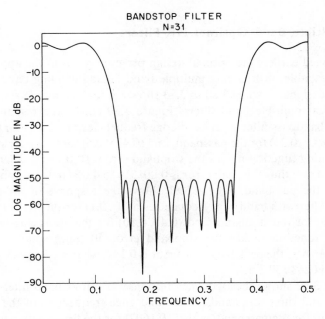

Fig. 3.81 Frequency response of an optimal (minimax error) bandstop filter.

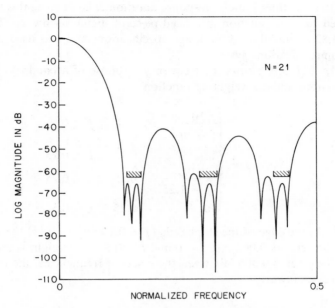

Fig. 3.82 Frequency response of an optimal multiband filter.

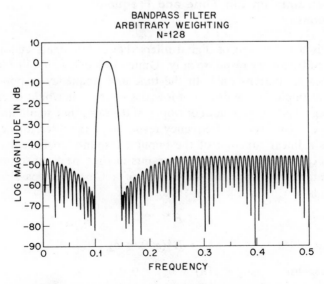

Fig. 3.83 Frequency response of an optimal bandpass filter with arbitrary stopband weighting.

179

and a peak error of less than 0.000708 in each of the stopbands. It was also required that the filter impulse response duration N be 21 since this filter was being used for interpolation in a fixed piece of digital hardware. The filter of Fig. 3.82 essentially met all design specifications and was used in a telephone communications system.

Finally, Fig. 3.83 shows the frequency response of a bandpass filter for which the nonstandard weighting function

$$W(e^{j2\pi f}) = \begin{cases} \dfrac{10}{1 - 9f} & 0 \leq f \leq 0.1 \\ 1 & 0.12 \leq f \leq 0.13 \\ \dfrac{10}{9f - 1.25} & 0.15 \leq f \leq 0.25 \\ 10 & 0.25 \leq f \leq 0.5 \end{cases}$$

was used. At the edge of the stopbands ($f = 0.1$ and $f = 0.15$) the weight is 100 and the error is 0.005. At $f = 0$ and $f = 0.25$, the weight is reduced to 10; thus the error is 0.005. Between these sets of frequencies, the maximum error increases linearly.

3.39 Design of Filters with Simultaneous Constraints on the Time and Frequency Response

We have discussed design of digital filters that approximate characteristics of a specified frequency response only. Quite often one would like to impose simultaneous restrictions on both the time and frequency response of the filter. For example, in the design of lowpass filters, one would often like to limit the step response overshoot or ripple, at the same time maintaining some reasonable control over the frequency response of the filter. Since the step response is a linear function of the impulse response coefficients, a linear program is capable of setting up constraints of the type discussed above. By way of example, we consider the design of a Case 1 lowpass filter with specifications

$$\textit{Passband} \qquad 1 - \delta_1 \leq H^*(e^{j\omega}) \leq 1 + \delta_1 \tag{3.154}$$

$$\textit{Stopband} \qquad -\delta_2 \leq H^*(e^{j\omega}) \leq \delta_2 \tag{3.155}$$

$$\textit{Step Response} \qquad -\delta_3 \leq g(n) \leq \delta_3 \qquad 0 \leq n \leq N_1 \tag{3.156}$$

where $g(n)$ is the step response of the filter and is defined as

$$g(n) = \sum_{m=0}^{n} h(m) \tag{3.157}$$

and N_1 is the region in which the step response oscillates around zero. Clearly $g(n)$ is a linear combination of the impulse response coefficients; hence Eq. (3.154) to (3.156) can be solved using linear programming techniques. For example, one could fix any one or two of the parameters δ_1, δ_2, or δ_3 and minimize the other(s). Alternatively one could set $\delta_1 = \alpha_1\delta$, $\delta_2 = \alpha_2\delta$, and $\delta_3 = \alpha_3\delta$, where α_1, α_2, and α_3 are constants, and simultaneously minimize all three parameters by minimizing δ.

Figure 3.84 and 3.85 demonstrate an example of the use of this technique. Figure 3.84 shows the step response and log magnitude response of a Case 1

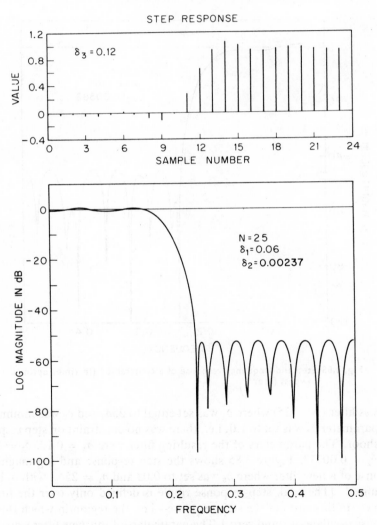

Fig. 3.84 Step and frequency response of an unconstrained optimal lowpass filter.

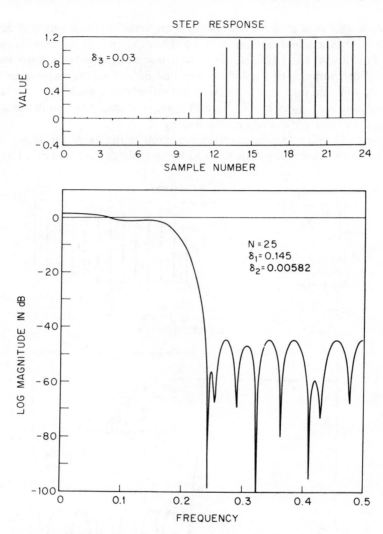

Fig. 3.85 Step and frequency response of a constrained (in time) optimal lowpass filter

lowpass filter ($N = 25$) where δ_1 was set equal to $25\delta_2$ and δ_2 was minimized. The parameter δ_3 was set to 1.0; i.e., there was no constraint on step response overshoot. The parameters of the resulting filter were $\delta_3 = 0.12$, $\delta_1 = 0.06$, and $\delta_2 = 0.00237$. Figure 3.85 shows the step response and log magnitude response of a new filter where δ_3 was set to 0.03 and $\delta_1 = 25\delta_2$, with δ_2 being minimized. [The peak step response ripple is defined only over the first 10 samples (in this case) of the step response—i.e., the region in which the step response oscillates around zero.] The parameters of this new filter were $\delta_1 = 0.145$, $\delta_2 = 0.00582$, and $\delta_3 = 0.03$. As noted in this figure, the equiripple character of the frequency response has been sacrificed in order to constrain the peak step response ripple. Using linear programming techniques, tradeoffs can be made between time and frequency response specifications to obtain a filter that is best suited to the particular application.

3.40   Direct Comparison Between FIR Lowpass Filters

In this chapter we have discussed three classes of FIR digital filters—i.e., window designs, frequency sampling designs, and optimal designs. Although there are a number of ways of comparing the resulting filters, perhaps the most objective (and most useful) is in terms of transition width for lowpass filters required to achieve certain specifications on passband ripple δ_1 and stopband ripple δ_2. By definition, such a comparison must favor optimal filters as they have the minimum transition width of all linear phase FIR filters that meet equivalent design specification. It is informative, however, to see how good (or how bad) the other classes of filters are by comparison.

Figure 3.86 shows such a comparison for the case of the Kaiser window, frequency sampling filters, and extraripple designs. In this figure the measure $D = (N - 1) \Delta F$ is plotted versus δ_2 with δ_1 as a parameter of the curves. The measure D is used because it has been shown to be approximately independent of N (for large N) and ΔF; thus it serves as a useful basis for comparison. This figure shows that, as expected, the extraripple designs have the smallest transition widths (D) for fixed values of δ_1 and δ_2.

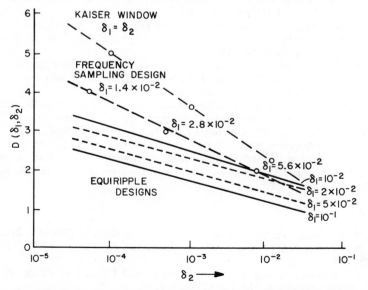

Fig. 3.86 Comparisons among window, frequency sampling, and optimal (minimax error) lowpass filters.

REFERENCES

Windows

1. J. F. KAISER, "Design Methods for Sampled Data Filters," *Proc. First Allerton Conf. on Circuit and System Theory*, 221–236, Nov., 1963.
2. J. F. KAISER, "Digital Filters," Chapter 7, in *System Analysis by Digital Computer*, (F. F. Kuo and J. F. Kaiser, ed.) John Wiley & Sons, Inc., New York, 1966.
3. R. B. BLACKMAN AND J. W. TUKEY, *The Measurement of Power Spectra*, Dover Publications, New York, 1958.
4. G. E. HEYLIGER, "The Scanning Function Approach to the Design of Numerical Filters," *Report R-63-2*, Martin Co., Denver, Colo., Apr., 1963.
5. G. E. HEYLIGER, "Design of Numerical Filters: Scanning Functions and Equal Ripple Approximation," *Proc. Fourth Allerton Conf. on Circuit and System Theory*, 175–185, 1966.
6. G. E. HEYLIGER AND C. A. HALIJAK, "Topics in the Design of Moving Average Numerical Filters," *Proc. Second Asilomar Conf. on Circuits and Systems*, 214–220, 1968.
7. H. D. HELMS, "Nonrecursive Digital Filters: Design Methods for Achieving Specifications on Frequency Response," *IEEE Trans. Audio and Electroacoustics*, **16**, No. 3, 336–342, Sept., 1968.

Frequency Sampling Filters

1. B. GOLD AND C. M. RADER, *Digital Processing of Signals*, McGraw-Hill Book Co., New York, 1969.
2. B. GOLD AND K. JORDAN, "A Note on Digital Filter Synthesis," *Proc. IEEE*, **56**, No. 10, 1717–1718, Oct., 1968.
3. B. GOLD AND K. JORDAN, "A Direct Search Procedure for Designing Finite Duration Impulse Response Filters," *IEEE Trans. on Audio and Electroacoustics*, **17**, No. 1, 33–36, Mar., 1969.
4. L. R. RABINER, B. GOLD, AND C. A. McGONEGAL, "An Approach to the Approximation Problem for Nonrecursive Digital Filters," *IEEE Trans. on Audio and Electroacoustics*, **18**, No. 2, 83–106, June, 1970.
5. L. R. RABINER AND K. STEIGLITZ, "The Design of Wide-Band Recursive and Nonrecursive Digital Differentiators," *IEEE Trans. on Audio and Electroacoustics*, **18**, No. 2, 204–209, June, 1970.
6. L. R. RABINER AND R. W. SCHAFER, "Recursive and Nonrecursive Realizations of Digital Filters Designed by Frequency Sampling Techniques," *IEEE Trans. on Audio and Electroacoustics*, **19**, No. 3, 200–207, Sept., 1971.
7. L. R. RABINER AND R. W. SCHAFER, "Correction to Recursive and Nonrecursive Realizations of Digital Filters Designed by Frequency Sampling Techniques," *IEEE Trans. on Audio and Electroacoustics*, **AU-20**, No. 1, 104–105, Mar., 1972.

Optimal Filters

1. L. R. RABINER, "Techniques for Designing Finite-Duration Impulse Response Digital Filters," *IEEE Trans. on Communication Technology*, **19**, No. 2, 188–195, 1971.
2. O. HERRMANN, "Design of Nonrecursive Digital Filters with Linear Phase," *Electronics Letters*, **6**, No. 11, 328–329, 1970.

3. O. HERRMANN AND H. W. SCHUESSLER, "Design of Nonrecursive Digital Filters with Minimum Phase," *Electronics Letters*, **6,** No. 11, 329–330, 1970.

4. H. D. HELMS, "Digital Filters with Equiripple or Minimax Responses," *IEEE Trans. on Audio and Electroacoustics*, **19,** No. 1, 87–94, 1971.

5. E. HOFSTETTER, A. OPPENHEIM, AND J. SIEGEL, "A New Technique for the Design of Nonrecursive Digital Filters," *Proc. Fifth Annual Princeton Conf. on Information Sciences and Systems*, 64–72, 1971.

6. E. HOFSTETTER, A. OPPENHEIM, AND J. SIEGEL, "On Optimum Nonrecursive Digital Filters," *Proc. Ninth Allerton Conf. on Circuit and System Theory*, 789–798, Oct., 1971.

7. T. W. PARKS AND J. H. MCCLELLAN, "Chebyshev Approximation for Nonrecursive Digital Filters with Linear Phase," *IEEE Trans. Circuit Theory*, **CT-19,** 189–194, Mar., 1972.

8. L. R. RABINER, "The Design of Finite Impulse Response Digital Filters Using Linear Programming Techniques," *Bell System Tech. J.*, **51,** No. 6, 1177–1198, July–Aug., 1972.

9. L. R. RABINER, "Linear Program Design of Finite Impulse Response (FIR) Digital Filters," *IEEE Trans. on Audio and Electroacoustics*, **AU-20,** No. 4, 280–288, Oct., 1972.

10. T. W. PARKS AND J. H. MCCLELLAN, "A Program for the Design of Linear Phase Finite Impulse Response Digital Filters," *IEEE Trans. on Audio and Electroacoustics*, **AU-20,** No. 3, 195–199, Aug., 1972.

11. T. W. PARKS, L. R. RABINER, AND J. H. MCCLELLAN, "On the Transition Width of Finite Impulse Response Digital Filters, *IEEE Trans. on Audio and Electroacoustics*, **AU-21,** No. 1, 1–4, Feb., 1973.

12. L. R. RABINER AND O. HERRMANN, "The Predictability of Certain Optimum Finite Impulse Response Digital Filters," *IEEE Trans. on Circuit Theory*, **CT-20,** No. 4, 401–408, July, 1973.

13. O. HERRMANN, L. R. RABINER, AND D. S. K. CHAN, "Practical Design Rules for Optimum Finite Impulse Response Lowpass Digital Filters," *Bell System Tech. J.*, **52,** No. 6, 769–799, July–Aug., 1973.

14. L. R. RABINER AND O. HERMANN, "On the Design of Optimum FIR Low-Pass Filters with Even Impulse Response Duration," *IEEE Trans. on Audio and Electroacoustics*, **AU-21,** No. 4, 329–336, Aug., 1973.

15. L. R. RABINER, "Approximate Design Relationships for Low-Pass FIR Digital Filters," *IEEE Trans. on Audio and Electroacoustics*, **AU-21,** No. 5, 456–460, Oct., 1973.

16. L. R. RABINER AND R. W. SCHAFER, "On the Behavior of Minimax Relative Error FIR Digital Differentiators," *Bell System Tech. J.*, Vol. 53, No. 2, 333–361, Feb., 1974.

17. L. R. RABINER AND R. W. SCHAFER, "On the Behavior of Minimax FIR Digital Hilbert Transformers," *Bell System Tech. J.*, Vol. 53, No. 2, 363–390, Feb., 1974.

18. J. H. MCCLELLAN, T. W. PARKS, AND L. R. RABINER, "A Computer Program for Designing Optimum FIR Linear Phase Digital Filters," *IEEE Trans. on Audio and Electroacoustics*, **AU-21,** No. 6, 506–526, Dec., 1973.

Linear Programming

1. G. DANTZIG, *Linear Programming and Extensions*, Princeton University Press, Princeton, N.J., 1963.

2. G. HADLEY, *Linear Programming*, Addison-Wesley Publishing Co., Reading, Mass., 1963.
3. T. C. HU, *Integer Programming and Newtork Flows*, Addison-Wesley Publishing Co., Reading, Mass., 1969.
4. W. A. SPIVEY AND R. M. THRALL, *Linear Optimization*, Holt, Rinehart and Winston, Inc., New York, 1970.
5. S. I. GASS, *Linear Programming*, McGraw-Hill Book Co., New York, 1969.
6. M. SIMONNARD, *Linear Programming*, Prentice-Hall, Inc., Englewood Cliffs, N.J., 1966.

Optimization Methods

1. E. W. CHENEY, *Introduction to Approximation Theory*, McGraw-Hill Book Co., New York, 1966.
2. J. R. RICE, *The Approximation of Functions*, Addison-Wesley Publishing Co., Reading, Mass., 1964.
3. M. AOKI, *Introduction to Optimization Techniques*, The Macmillan Co., New York, 1971.
4. G. MEINARDUS, *Approximation of Functions: Theory and Numerical Methods*, Springer-Verlag, New York, 1967.
5. E. YA REMEZ, "General Computational Methods of Chebyshev Approximation", Atomic Energy Translation 4491, Kiev, 1957.

Appendix

This appendix gives a FORTRAN program that is capable of designing a wide variety of optimal (minimax) FIR filters including lowpass, highpass, bandpass, and bandstop filters, as well as differentiators and Hilbert transformers.

To illustrate the use of this program, four complete examples are given. The program listing appears after the examples.

EXAMPLE 1. Design a 24-point lowpass filter with passband cutoff frequency of 0.08 and stopband cutoff frequency of 0.16 and ripple ratio of 1.0.

The input parameters for this example are $N = 24$, $F_p = 0.08$, $F_s = 0.16$, and $K = \delta_1/\delta_2 = 1.0$. The specific card input to the program is therefore

Card 1. 24, 1, 2, 0, 16
Card 2. 0, 0.08, 0.16, 0.5
Card 3. 1, 0
Card 4. 1, 1

where card 1 specifies (in order) N, filter type, number of bands, punch option, and grid interpolation; card 2 specifies the band edges for each of the bands; card 3 specifies the desired value in each band; and card 4 specifies the desired ripple weights in each band. Figure A-3.1 shows the computer output listing (including the run time on a Honeywell 6000 computer) and Fig. A-3.2 shows the frequency response of the resulting lowpass filter. As seen in Fig. A-3.1 the resulting values of deltas are $\delta_1 = \delta_2 = 0.0015$ (-56.4 dB).

****

```
                        FINITE IMPULSE RESPONSE (FIR)
                        LINEAR PHASE DIGITAL FILTER DESIGN
                        REMEZ EXCHANGE ALGORITHM
                        BANDPASS FILTER
                FILTER LENGTH =  24
                ***** IMPULSE RESPONSE *****
                        H(  1) =  0.33740917E-02 = H(  24)
                        H(  2) =  0.14938299E-01 = H(  23)
                        H(  3) =  0.10569360E-01 = H(  22)
                        H(  4) =  0.25415067E-02 = H(  21)
                        H(  5) = -0.15929992E-01 = H(  20)
                        H(  6) = -0.34085343E-01 = H(  19)
                        H(  7) = -0.38112177E-01 = H(  18)
                        H(  8) = -0.14629169E-01 = H(  17)
                        H(  9) =  0.40089541E-01 = H(  16)
                        H( 10) =  0.11540713E 00 = H(  15)
                        H( 11) =  0.18850752E 00 = H(  14)
                        H( 12) =  0.23354606E 00 = H(  13)

                        BAND 1          BAND 2          BAND
        LOWER BAND EDGE  0.              0.16000000
        UPPER BAND EDGE  0.08000000      0.50000000
        DESIRED VALUE    1.00000000      0.
        WEIGHTING        1.00000000      1.00000000
        DEVIATION        0.01243364      0.01243364
        DEVIATION IN DB -38.10803413    -38.10803413

        EXTREMAL FREQUENCIES
        0.              0.0364583       0.0677083       0.0800000       0.1600000
        0.1730208       0.2068750       0.2459375       0.2876042       0.3318750
        0.3787500       0.4256251       0.4751043
```

****
```
TIME=      0.7651562 SECONDS
```

Fig. A-3.1 The output listing for the design of a 24-point optimal lowpass
filter.

EXAMPLE 2.   Design a 32-point bandpass filter with stopband cutoff frequencies
of 0.1 and 0.425 and passband cutoff frequencies of 0.2 and 0.35 and with
ripple weights of 10 in the stopbands and 1 in the passband.

The card input for this example is

Card 1.   32, 1, 3, 0, 16
Card 2.   0, 0.1, 0.2, 0.35, 0.425, 0.5
Card 3.   0, 1, 0
Card 4.   10, 1, 10

Figure A-3.3 shows the resulting computer output for this filter and Fig.
3.80 on p. 178 shows the resulting frequency response.

EXAMPLE 3.   Design a 32-point differentiator with a cutoff frequency of 0.5.
The card input for this example is

Card 1.   32, 2, 1, 0, 16
Card 2.   0, 0.5
Card 3.   1
Card 4.   1

Figure A-3.4 shows the computer output for this differentiator and Fig. 3.68
on p. 166 shows its frequency response.

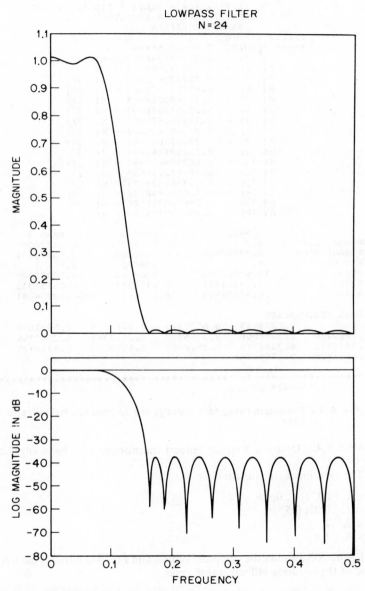

Fig. A-3.2 The frequency response on a linear and log scale for the optimal lowpass filter of Fig. A-3.1.

```
************************************************************************

                        FINITE IMPULSE RESPONSE (FIR)
                        LINEAR PHASE DIGITAL FILTER DESIGN
                        REMEZ EXCHANGE ALGORITHM
                        BANDPASS FILTER
           FILTER LENGTH =   32
           ***** IMPULSE RESPONSE *****
                 H (   1) =  -0.57534121E-02 = H (   32)
                 H (   2) =   0.99027198E-03 = H (   31)
                 H (   3) =   0.75733545E-02 = H (   30)
                 H (   4) =  -0.65141192E-02 = H (   29)
                 H (   5) =   0.13960525E-01 = H (   28)
                 H (   6) =   0.22951469E-02 = H (   27)
                 H (   7) =  -0.19994067E-01 = H (   26)
                 H (   8) =   0.71369560E-02 = H (   25)
                 H (   9) =  -0.39657363E-01 = H (   24)
                 H (  10) =   0.11260114E-01 = H (   23)
                 H (  11) =   0.66233643E-01 = H (   22)
                 H (  12) =  -0.10497223E-01 = H (   21)
                 H (  13) =   0.85136133E-01 = H (   20)
                 H (  14) =  -0.12024993E 00 = H (   19)
                 H (  15) =  -0.29678577E 00 = H (   18)
                 H (  16) =   0.30410917E 00 = H (   17)
```

	BAND 1	BAND 2	BAND 3	BAND
LOWER BAND EDGE	0.	0.20000000	0.42500000	
UPPER BAND EDGE	0.10000000	0.35000000	0.50000000	
DESIRED VALUE	0.	1.00000000	0.	
WEIGHTING	10.00000000	1.00000000	10.00000000	
DEVIATION	0.00151312	0.01513118	0.00151312	
DEVIATION IN DB	-56.40254641	-36.40254641	-56.40254641	

```
EXTREMAL FREQUENCIES
   0.            0.0273437    0.0527344    0.0761719    0.0937500
   0.1000000     0.2000000    0.2195312    0.2527344    0.2839844
   0.3132812     0.3386719    0.3500000    0.4250000    0.4328125
   0.4503906     0.4796875

************************************************************************
TIME=       0.8065938 SECONDS
```

Fig. A-3.3 The output listing for the design of a 32-point optimal bandpass filter.

EXAMPLE 4.   Design a 20-point Hilbert transformer with band edges of 0.05 and 0.5.

The card input for this example is

Card 1.   20, 3, 1, 0, 16
Card 2.   0.05, 0.5
Card 3.   1
Card 4.   1

Figure A-3.5 shows the computer output and Fig. A-3.6 shows the frequency response of the resulting Hilbert transformer.

```
************************************************************************
                    FINITE IMPULSE RESPONSE (FIR)
                    LINEAR PHASE DIGITAL FILTER DESIGN
                    REMEZ EXCHANGE ALGORITHM
                    DIFFERENTIATOR
            FILTER LENGTH =   32
            ***** IMPULSE RESPONSE *****
                 H(   1) =  -0.62713091E-03  =  -H(   32)
                 H(   2) =   0.85633433E-03  =  -H(   31)
                 H(   3) =  -0.42418549E-03  =  -H(   30)
                 H(   4) =   0.39901518E-03  =  -H(   29)
                 H(   5) =  -0.43437273E-03  =  -H(   28)
                 H(   6) =   0.49969450E-03  =  -H(   27)
                 H(   7) =  -0.59634961E-03  =  -H(   26)
                 H(   8) =   0.73277031E-03  =  -H(   25)
                 H(   9) =  -0.93002681E-03  =  -H(   24)
                 H(  10) =   0.12270042E-02  =  -H(   23)
                 H(  11) =  -0.17012820E-02  =  -H(   22)
                 H(  12) =   0.25272341E-02  =  -H(   21)
                 H(  13) =  -0.41601159E-02  =  -H(   20)
                 H(  14) =   0.81294555E-02  =  -H(   19)
                 H(  15) =  -0.22539097E-01  =  -H(   18)
                 H(  16) =   0.20266535E 00  =  -H(   17)

                       BAND  1              BAND
   LOWER BAND EDGE      0.
   UPPER BAND EDGE      0.50000000
   DESIRED SLOPE        1.00000000
   WEIGHTING            1.00000000
   DEVIATION            0.00620231

   EXTREMAL FREQUENCIES
       0.0019531    0.0332031    0.0664062    0.0996094    0.1328125
       0.1640625    0.1972656    0.2304687    0.2636719    0.2968750
       0.3300781    0.3632812    0.3945312    0.4277344    0.4589844
       0.4863281    0.5000000

************************************************************************
   TIME=      1.0845625 SECONDS
```

Fig. A-3.4 The output listing for the design of a 32-point optimal differentiator.

191

```
****************************************************************************
                        FINITE IMPULSE RESPONSE (FIR)
                        LINEAR PHASE DIGITAL FILTER DESIGN
                        REMEZ EXCHANGE ALGORITHM
                        HILBERT TRANSFORMER
                   FILTER LENGTH =  20
                   ***** IMPULSE RESPONSE *****
                        H(  1) =   0.16026190E-01 = -H(  20)
                        H(  2) =   0.14173287E-01 = -H(  19)
                        H(  3) =   0.20452437E-01 = -H(  18)
                        H(  4) =   0.28736882E-01 = -H(  17)
                        H(  5) =   0.39852581E-01 = -H(  16)
                        H(  6) =   0.55333299E-01 = -H(  15)
                        H(  7) =   0.78542752E-01 = -H(  14)
                        H(  8) =   0.11823755E 00 = -H(  13)
                        H(  9) =   0.20664125E 00 = -H(  12)
                        H( 10) =   0.63475619E 00 = -H(  11)

                        BAND 1              BAND
LOWER BAND EDGE         0.05000000
UPPER BAND EDGE         0.50000000
DESIRED VALUE           1.00000000
WEIGHTING               1.00000000
DEVIATION               0.02055604

EXTREMAL FREQUENCIES
   0.0500000    0.0656250    0.1031250    0.1468750    0.1937500
   0.2437500    0.2937500    0.3468750    0.3968751    0.4500001
   0.5000000

****************************************************************************
TIME=      0.4742500 SECONDS
```

Fig. A-3.5 The output listing for the design of a 20-point optimal Hilbert transformer.

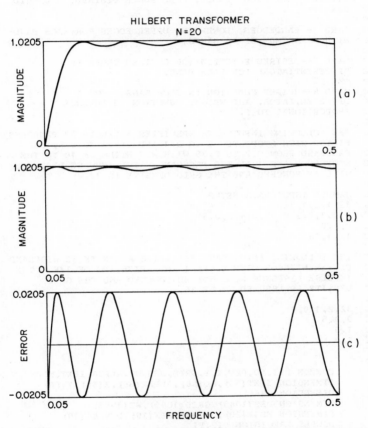

Fig. A-3.6 The frequency response and error response of the optimal Hilbert transformer of Fig. A-3.5.

FIR Design Program

```
C   PROGRAM FOR THE DESIGN OF LINEAR PHASE FINITE IMPULSE
C   RESPONSE (FIR) FILTERS USING THE REMEZ EXCHANGE ALGORITHM
C   JIM MCCLELLAN, RICE UNIVERSITY, APRIL 13, 1973
C   THREE TYPES OF FILTERS ARE INCLUDED--BANDPASS FILTERS
C   DIFFERENTIATORS, AND HILBERT TRANSFORM FILTERS
C
C   THE INPUT DATA CONSISTS OF 5 CARDS
C
C   CARD 1--FILTER LENGTH, TYPE OF FILTER.  1-MULTIPLE
C   PASSBAND/STOPBAND, 2-DIFFERENTIATOR, 3-HILBERT TRANSFORM
C   FILTER.  NUMBER OF BANDS, CARD PUNCH DESIRED, AND GRID
C   DENSITY.
C
C   CARD 2--BANDEDGES, LOWER AND UPPER EDGES FOR EACH BAND
C   WITH A MAXIMUM OF 10 BANDS.
C
C   CARD 3--DESIRED FUNCTION (OR DESIRED SLOPE IF A
C   DIFFERENTIATOR) FOR EACH BAND.
C
C   CARD 4--WEIGHT FUNCTION IN EACH BAND.  FOR A
C   DIFFERENTIATOR, THE WEIGHT FUNCTION IS INVERSELY
C   PROPORTIONAL TO F.
C
C   THE FOLLOWING INPUT DATA SPECIFIES A LENGTH 32 BANDPASS
C   FILTER WITH STOPBANDS 0 TO 0.1 AND 0.425 TO 0.5, AND
C   PASSBAND FROM 0.2 TO 0.35 WITH WEIGHTING OF 10 IN THE
C   STOPBANDS AND 1 IN THE PASSBAND.  THE IMPULSE RESPONSE
C   WILL BE PUNCHED AND THE GRID DENSITY IS 32.
C
C   SAMPLE INPUT DATA SETUP
C   32,1,3,1,32
C   0,0.1,0.2,0.35,0.425,0.5
C   0,1,0
C   10,1,10
C
C   THE FOLLOWING INPUT DATA SPECIFIES A LENGTH 32 WIDEBAND
C   DIFFERENTIATOR WITH SLOPE 1 AND WEIGHTING OF 1/F.  THE
C   IMPULSE RESPONSE WILL NOT BE PUNCHED AND THE GRID
C   DENSITY IS ASSUMED TO BE 16.
C
C   32,2,1,0,0
C   0,0.5
C   1.0
C   1.0
C
C
        COMMON PI2,AD,DEV,X,Y,GRID,DES,WT,ALPHA,IEXT,NFCNS,NGRID
        DIMENSION IEXT(66),AD(66),ALPHA(66),X(66),Y(66)
        DIMENSION H(66)
        DIMENSION DES(1045),GRID(1045),WT(1045)
        DIMENSION EDGE(20),FX(10),WTX(10),DEVIAT(10)
        DOUBLE PRECISION PI2,PI
        DOUBLE PRECISION AD,DEV,X,Y
        PI2=6.283185307179586
        PI=3.141592653589793
C
C   THE PROGRAM IS SET UP FOR A MAXIMUM LENGTH OF 128, BUT
C   THIS UPPER LIMIT CAN BE CHANGED BY REDIMENSIONING THE
C   ARRAYS IEXT, AD, ALPHA, X, Y, H TO BE NFMAX/2 + 2.
C   THE ARRAYS DES, GRID, AND WT MUST DIMENSIONED
C   16(NFMAX/2 + 2).
C
```

```
           NFMAX=128
      100  CONTINUE
           JTYPE=0
C
C    PROGRAM INPUT SECTION
C
           READ *,NFILT,JTYPE,NBANDS,JPUNCH,LGRID
           IF(NFILT.GT.NFMAX.OR.NFILT.LT.3) CALL ERROR
           IF(NBANDS.LE.0) NBANDS=1
C
C    GRID DENSITY IS ASSUMED TO BE 16 UNLESS SPECIFIED
C    OTHERWISE
C
           IF(LGRID.LE.0) LGRID=16
           JB=2*NBANDS
           READ *, (EDGE(J),J=1,JB)
           READ *, (FX(J),J=1,NBANDS)
           READ *, (WTX(J),J=1,NBANDS)
           IF(JTYPE.EQ.0) CALL ERROR
           NEG=1
           IF(JTYPE.EQ.1) NEG=0
           NODD=NFILT/2
           NODD=NFILT-2*NODD
           NFCNS=NFILT/2
           IF(NODD.EQ.1.AND.NEG.EQ.0) NFCNS=NFCNS+1
C
C    SET UP THE DENSE GRID.  THE NUMBER OF POINTS IN THE GRID
C    IS (FILTER LENGTH + 1)*GRID DENSITY/2
C
           GRID(1)=EDGE(1)
           DELF=LGRID*NFCNS
           DELF=0.5/DELF
           IF(NEG.EQ.0) GO TO 135
           IF(EDGE(1).LT.DELF) GRID(1)=DELF
      135  CONTINUE
           J=1
           L=1
           LBAND=1
      140  FUP=EDGE(L+1)
      145  TEMP=GRID(J)
C
C    CALCULATE THE DESIRED MAGNITUDE RESPONSE AND THE WEIGHT
C    FUNCTION ON THE GRID
C
           DES(J)=EFF(TEMP,FX,WTX,LBAND,JTYPE)
           WT(J)=WATE(TEMP,FX,WTX,LBAND,JTYPE)
           J=J+1
           GRID(J)=TEMP+DELF
           IF(GRID(J).GT.FUP) GO TO 150
           GO TO 145
      150  GRID(J-1)=FUP
           DES(J-1)=EFF(FUP,FX,WTX,LBAND,JTYPE)
           WT(J-1)=WATE(FUP,FX,WTX,LBAND,JTYPE)
           LBAND=LBAND+1
           L=L+2
           IF(LBAND.GT.NBANDS) GO TO 160
           GRID(J)=EDGE(L)
           GO TO 140
      160  NGRID=J-1
           IF(NEG.NE.NODD) GO TO 165
           IF(GRID(NGRID).GT.(0.5-DELF)) NGRID=NGRID-1
```

195

```
    165 CONTINUE
C
C   SET UP A NEW APPROXIMATION PROBLEM WHICH IS EQUIVALENT
C   TO THE ORIGINAL PROBLEM
C
        IF(NEG)  170,170,180
    170 IF(NODD.EQ.1) GO TO 200
        DO 175 J=1,NGRID
        CHANGE=DCOS(PI*GRID(J))
        DES(J)=DES(J)/CHANGE
    175 WT(J)=WT(J)*CHANGE
        GO TO 200
    180 IF(NODD.EQ.1) GO TO 190
        DO 185 J=1,NGRID
        CHANGE=DSIN(PI*GRID(J))
        DES(J)=DES(J)/CHANGE
    185 WT(J)=WT(J)*CHANGE
        GO TO 200
    190 DO 195 J=1,NGRID
        CHANGE=DSIN(PI2*GRID(J))
        DES(J)=DES(J)/CHANGE
    195 WT(J)=WT(J)*CHANGE
C
C   INITIAL GUESS FOR THE EXTREMAL FREQUENCIES--EQUALLY
C   SPACED ALONG THE GRID
C
    200 TEMP=FLOAT(NGRID-1)/FLOAT(NFCNS)
        DO 210 J=1,NFCNS
    210 IEXT(J)=(J-1)*TEMP+1
        IEXT(NFCNS+1)=NGRID
        NM1=NFCNS-1
        NZ=NFCNS+1
C
C   CALL THE REMEZ EXCHANGE ALGORITHM TO DO THE APPROXIMATION
C   PROBLEM
C
        CALL REMEZ(EDGE,NBANDS)
C
C   CALCULATE THE IMPULSE RESPONSE.
C
        IF(NEG)  300,300,320
    300 IF(NODD.EQ.0)  GO TO 310
        DO 305 J=1,NM1
    305 H(J)=0.5*ALPHA(NZ-J)
        H(NFCNS)=ALPHA(1)
        GO TO 350
    310 H(1)=0.25*ALPHA(NFCNS)
        DO 315 J=2,NM1
    315 H(J)=0.25*(ALPHA(NZ-J)+ALPHA(NFCNS+2-J))
        H(NFCNS)=0.5*ALPHA(1)+0.25*ALPHA(2)
        GO TO 350
    320 IF(NODD.EQ.0)  GO TO 330
        H(1)=0.25*ALPHA(NFCNS)
        H(2)=0.25*ALPHA(NM1)
        DO 325 J=3,NM1
    325 H(J)=0.25*(ALPHA(NZ-J)-ALPHA(NFCNS+3-J))
        H(NFCNS)=0.5*ALPHA(1)-0.25*ALPHA(3)
        H(NZ)=0.0
        GO TO 350
    330 H(1)=0.25*ALPHA(NFCNS)
        DO 335 J=2,NM1
```

```
 335 H(J)=0.25*(ALPHA(NZ-J)-ALPHA(NFCNS+2-J))
     H(NFCNS)=0.5*ALPHA(1)-0.25*ALPHA(2)
C
C  PROGRAM OUTPUT SECTION.
C
 350 PRINT 360
 360 FORMAT(1H1, 70(1H*)//25X,'FINITE IMPULSE RESPONSE (FIR)'/
    125X,'LINEAR PHASE DIGITAL FILTER DESIGN'/
    225X,'REMEZ EXCHANGE ALGORITHM'/)
     IF(JTYPE.EQ.1) PRINT 365
 365 FORMAT(25X,'BANDPASS FILTER'/)
     IF(JTYPE.EQ.2) PRINT 370
 370 FORMAT(25X,'DIFFERENTIATOR'/)
     IF(JTYPE.EQ.3) PRINT 375
 375 FORMAT(25X,'HILBERT TRANSFORMER'/)
     PRINT 378,NFILT
 378 FORMAT(15X,'FILTER LENGTH = ',I3/)
     PRINT 380
 380 FORMAT(15X,'***** IMPULSE RESPONSE *****')
     DO 381 J=1,NFCNS
     K=NFILT+1-J
     IF(NEG.EQ.0) PRINT 382,J,H(J),K
     IF(NEG.EQ.1) PRINT 383,J,H(J),K
 381 CONTINUE
 382 FORMAT(20X,'H(',I3,') = ',E15.8,' = H(',I4,')')
 383 FORMAT(20X,'H(',I3,') = ',E15.8,' = -H(',I4,')')
     IF(NEG.EQ.1.AND.NODD.EQ.1) PRINT 384,NZ
 384 FORMAT(20X,'H(',I3,') =  0.0')
     DO 450 K=1,NBANDS,4
     KUP=K+3
     IF(KUP.GT.NBANDS) KUP=NBANDS
     PRINT 385,(J,J=K,KUP)
 385 FORMAT(/24X,4('BAND',I3,8X))
     PRINT 390,(EDGE(2*J-1),J=K,KUP)
 390 FORMAT(2X,'LOWER BAND EDGE',5F15.9)
     PRINT 395,(EDGE(2*J),J=K,KUP)
 395 FORMAT(2X,'UPPER BAND EDGE',5F15.9)
     IF(JTYPE.NE.2) PRINT 400,(FX(J),J=K,KUP)
 400 FORMAT(2X,'DESIRED VALUE',2X,5F15.9)
     IF(JTYPE.EQ.2) PRINT 405,(FX(J),J=K,KUP)
 405 FORMAT(2X,'DESIRED SLOPE',2X,5F15.9)
     PRINT 410,(WTX(J),J=K,KUP)
 410 FORMAT(2X,'WEIGHTING',6X,5F15.9)
     DO 420 J=K,KUP
 420 DEVIAT(J)=DEV/WTX(J)
     PRINT 425,(DEVIAT(J),J=K,KUP)
 425 FORMAT(2X,'DEVIATION',6X,5F15.9)
     IF(JTYPE.NE.1) GO TO 450
     DO 430 J=K,KUP
 430 DEVIAT(J)=20.0*ALOG10(DEVIAT(J))
     PRINT 435,(DEVIAT(J),J=K,KUP)
 435 FORMAT(2X,'DEVIATION IN DB',5F15.9)
 450 CONTINUE
     PRINT 455,(GRID(IEXT(J)),J=1,NZ)
 455 FORMAT(/2X,'EXTREMAL FREQUENCIES'/(2X,5F12.7))
     PRINT 460
 460 FORMAT(/1X,70(1H*)/1H1)
     IF(JPUNCH.NE.0) PUNCH *,(H(J),J=1,NFCNS)
     IF(NFILT.NE.0) GO TO 100
     RETURN
     END
```

197

```
      FUNCTION EFF (TEMP,FX,WTX,LBAND,JTYPE)
C
C  FUNCTION TO CALCULATE THE DESIRED MAGNITUDE RESPONSE
C  AS A FUNCTION OF FREQUENCY.
C
      DIMENSION FX(5),WTX(5)
      IF (JTYPE.EQ.2) GO TO 1
      EFF=FX(LBAND)
      RETURN
    1 EFF=FX(LBAND)*TEMP
      RETURN
      END
```

```
      FUNCTION WATE (TEMP,FX,WTX,LBAND,JTYPE)
C
C  FUNCTION TO CALCULATE THE WEIGHT FUNCTION AS A FUNCTION
C  OF FREQUENCY.
C
      DIMENSION FX(5),WTX(5)
      IF (JTYPE.EQ.2) GO TO 1
      WATE=WTX(LBAND)
      RETURN
    1 IF (FX(LBAND).LT.0.0001) GO TO 2
      WATE=WTX(LBAND)/TEMP
      RETURN
    2 WATE=WTX(LBAND)
      RETURN
      END
```

```
      SUBROUTINE ERROR
      PRINT 1
    1 FORMAT(' *********** ERROR IN INPUT DATA **********')
      STOP
      END
```

```
      SUBROUTINE REMEZ(EDGE,NBANDS)
C
C  THIS SUBROUTINE IMPLEMENTS THE REMEZ EXCHANGE ALGORITHM
C  FOR THE WEIGHTED CHEBYCHEV APPROXIMATION OF A CONTINUOUS
C  FUNCTION WITH A SUM OF COSINES.  INPUTS TO THE SUBROUTINE
C  ARE A DENSE GRID WHICH REPLACES THE FREQUENCY AXIS, THE
C  DESIRED FUNCTION ON THIS GRID, THE WEIGHT FUNCTION ON THE
C  GRID, THE NUMBER OF COSINES, AND AN INITIAL GUESS OF THE
C  EXTREMAL FREQUENCIES.  THE PROGRAM MINIMIZES THE CHEBYCHEV
C  ERROR BY DETERMINING THE BEST LOCATION OF THE EXTREMAL
C  FREQUENCIES (POINTS OF MAXIMUM ERROR) AND THEN CALCULATES
C  THE COEFFICIENTS OF THE BEST APPROXIMATION.
C
      COMMON PI2,AD,DEV,X,Y,GRID,DES,WT,ALPHA,IEXT,NFCNS,NGRID
      DIMENSION EDGE(20)
```

```
      DIMENSION IEXT(66),AD(66),ALPHA(66),X(66),Y(66)
      DIMENSION DES(1045),GRID(1045),WT(1045)
      DIMENSION A(66),P(65),Q(65)
      DOUBLE PRECISION PI2,DNUM,DDEN,DTEMP,A,P,Q
      DOUBLE PRECISION AD,DEV,X,Y
C
C  THE PROGRAM ALLOWS A MAXIMUM NUMBER OF ITERATIONS OF  25
C
      ITRMAX=25
      DEVL=-1.0
      NZ=NFCNS+1
      NZZ=NFCNS+2
      NITER=0
  100 CONTINUE
      IEXT(NZZ)=NGRID+1
      NITER=NITER+1
      IF(NITER.GT.ITRMAX) GO TO 400
      DO 110 J=1,NZ
      DTEMP=GRID(IEXT(J))
      DTEMP=DCOS(DTEMP*PI2)
  110 X(J)=DTEMP
      JET=(NFCNS-1)/15+1
      DO 120 J=1,NZ
  120 AD(J)=D(J,NZ,JET)
      DNUM=0.0
      DDEN=0.0
      K=1
      DO 130 J=1,NZ
      L=IEXT(J)
      DTEMP=AD(J)*DES(L)
      DNUM=DNUM+DTEMP
      DTEMP=K*AD(J)/WT(L)
      DDEN=DDEN+DTEMP
  130 K=-K
      DEV=DNUM/DDEN
      NU=1
      IF(DEV.GT.0.0) NU=-1
      DEV=-NU*DEV
      K=NU
      DO 140 J=1,NZ
      L=IEXT(J)
      DTEMP=K*DEV/WT(L)
      Y(J)=DES(L)+DTEMP
  140 K=-K
      IF(DEV.GE.DEVL) GO TO 150
      CALL OUCH
      GO TO 400
  150 DEVL=DEV
      JCHNGE=0
      K1=IEXT(1)
      KNZ=IEXT(NZ)
      KLOW=0
      NUT=-NU
      J=1
C
C  SEARCH FOR THE EXTREMAL FREQUENCIES OF THE BEST
C  APPROXIMATION
C
  200 IF(J.EQ.NZZ) YNZ=COMP
      IF(J.GE.NZZ) GO TO 300
      KUP=IEXT(J+1)
```

199

```
        L=IEXT (J) +1
        NUT=-NUT
        IF (J.EQ.2)  Y1=COMP
        COMP=DEV
        IF (L.GE.KUP)  GO TO 220
        ERR=GEE (L,NZ)
        ERR= (ERR-DES (L) ) *WT (L)
        DTEMP=NUT*ERR-COMP
        IF (DTEMP.LE.0.0)  GO TO 220
        COMP=NUT*ERR
210     L=L+1
        IF (L.GE.KUP)  GO TO 215
        ERR=GEE (L,NZ)
        ERR= (ERR-DES (L) ) *WT (L)
        DTEMP=NUT*ERR-COMP
        IF (DTEMP.LE.0.0)  GO TO 215
        COMP=NUT*ERR
        GO TO 210
215     IEXT (J) =L-1
        J=J+1
        KLOW=L-1
        JCHNGE=JCHNGE+1
        GO TO 200
220     L=L-1
225     L=L-1
        IF (L.LE.KLOW)  GO TO 250
        ERR=GEE (L,NZ)
        ERR= (ERR-DES (L) ) *WT (L)
        DTEMP=NUT*ERR-COMP
        IF (DTEMP.GT.0.0)  GO TO 230
        IF (JCHNGE.LE.0)  GO TO 225
        GO TO 260
230     COMP=NUT*ERR
235     L=L-1
        IF (L.LE.KLOW)  GO TO 240
        ERR=GEE (L,NZ)
        ERR= (ERR-DES (L) ) *WT (L)
        DTEMP=NUT*ERR-COMP
        IF (DTEMP.LE.0.0)  GO TO 240
        COMP=NUT*ERR
        GO TO 235
240     KLOW=IEXT (J)
        IEXT (J) =L+1
        J=J+1
        JCHNGE=JCHNGE+1
        GO TO 200
250     L=IEXT (J) +1
        IF (JCHNGE.GT.0)  GO TO 215
255     L=L+1
        IF (L.GE.KUP)  GO TO 260
        ERR=GEE (L,NZ)
        ERR= (ERR-DES (L) ) *WT (L)
        DTEMP=NUT*ERR-COMP
        IF (DTEMP.LE.0.0)  GO TO 255
        COMP=NUT*ERR
        GO TO 210
260     KLOW=IEXT (J)
        J=J+1
        GO TO 200
300     IF (J.GT.NZZ)  GO TO 320
        IF (K1.GT.IEXT (1) )  K1=IEXT (1)
```

```
      IF(KNZ.LT.IEXT(NZ)) KNZ=IEXT(NZ)
      NUT1=NUT
      NUT=-NU
      L=0
      KUP=K1
      COMP=YNZ*(1.00001)
      LUCK=1
  310 L=L+1
      IF(L.GE.KUP) GO TO 315
      ERR=GEE(L,NZ)
      ERR=(ERR-DES(L))*WT(L)
      DTEMP=NUT*ERR-COMP
      IF(DTEMP.LE.0.0) GO TO 310
      COMP=NUT*ERR
      J=NZZ
      GO TO 210
  315 LUCK=6
      GO TO 325
  320 IF(LUCK.GT.9) GO TO 350
      IF(COMP.GT.Y1) Y1=COMP
      K1=IEXT(NZZ)
  325 L=NGRID+1
      KLOW=KNZ
      NUT=-NUT1
      COMP=Y1*(1.00001)
  330 L=L-1
      IF(L.LE.KLOW) GO TO 340
      ERR=GEE(L,NZ)
      ERR=(ERR-DES(L))*WT(L)
      DTEMP=NUT*ERR-COMP
      IF(DTEMP.LE.0.0) GO TO 330
      J=NZZ
      COMP=NUT*ERR
      LUCK=LUCK+10
      GO TO 235
  340 IF(LUCK.EQ.6) GO TO 370
      DO 345 J=1,NFCNS
  345 IEXT(NZZ-J)=IEXT(NZ-J)
      IEXT(1)=K1
      GO TO 100
  350 KN=IEXT(NZZ)
      DO 360 J=1,NFCNS
  360 IEXT(J)=IEXT(J+1)
      IEXT(NZ)=KN
      GO TO 100
  370 IF(JCHNGE.GT.0) GO TO 100


C  CALCULATION OF THE COEFFICIENTS OF THE BEST APPROXIMATION
C  USING THE INVERSE DISCRETE FOURIER TRANSFORM
C
  400 CONTINUE
      NM1=NFCNS-1
      FSH=1.0E-06
      GTEMP=GPID(1)
      X(NZZ)=-2.0
      CN=2*NFCNS-1
      DELF=1.0/CN
      L=1
      KKK=0
      IF(EDGE(1).EQ.0.0.AND.EDGE(2*NBANDS).EQ.0.5) KKK=1
```

```
      IF(NFCNS.LE.3) KKK=1
      IF(KKK.EQ.1) GO TO 405
      DTEMP=DCOS(PI2*GRID(1))
      DNUM=DCOS(PI2*GRID(NGRID))
      AA=2.0/(DTEMP-DNUM)
      BB=-(DTEMP+DNUM)/(DTEMP-DNUM)
  405 CONTINUE
      DO 430 J=1,NFCNS
      FT=(J-1)*DELF
      XT=DCOS(PI2*FT)
      IF(KKK.EQ.1) GO TO 410
      XT=(XT-BB)/AA
      FT=ARCOS(XT)/PI2
  410 XE=X(L)
      IF(XT.GT.XE) GO TO 420
      IF((XE-XT).LT.FSH) GO TO 415
      L=L+1
      GO TO 410
  415 A(J)=Y(L)
      GO TO 425
  420 IF((XT-XE).LT.FSH) GO TO 415
      GRID(1)=FT
      A(J)=GEE(1,NZ)
  425 CONTINUE
      IF(L.GT.1) L=L-1
  430 CONTINUE
      GRID(1)=GTEMP
      DDEN=PI2/CN
      DO 510 J=1,NFCNS
      DTEMP=0.0
      DNUM=(J-1)*DDEN
      IF(NM1.LT.1) GO TO 505
      DO 500 K=1,NM1
  500 DTEMP=DTEMP+A(K+1)*DCOS(DNUM*K)
  505 DTEMP=2.0*DTEMP+A(1)
  510 ALPHA(J)=DTEMP
      DO 550 J=2,NFCNS
  550 ALPHA(J)=2*ALPHA(J)/CN
      ALPHA(1)=ALPHA(1)/CN
      IF(KKK.EQ.1) GO TO 545
      P(1)=2.0*ALPHA(NFCNS)*BB+ALPHA(NM1)
      P(2)=2.0*AA*ALPHA(NFCNS)
      Q(1)=ALPHA(NFCNS-2)-ALPHA(NFCNS)
      DO 540 J=2,NM1
      IF(J.LT.NM1) GO TO 515
      AA=0.5*AA
      BB=0.5*BB
  515 CONTINUE
      P(J+1)=0.0
      DO 520 K=1,J
      A(K)=P(K)
  520 P(K)=2.0*BB*A(K)
      P(2)=P(2)+A(1)*2.0*AA
      JM1=J-1
      DO 525 K=1,JM1
  525 P(K)=P(K)+Q(K)+AA*A(K+1)
      JP1=J+1
      DO 530 K=3,JP1
  530 P(K)=P(K)+AA*A(K-1)
      IF(J.EQ.NM1) GO TO 540
      DO 535 K=1,J
```

202

```
  535 Q(K)=-A(K)
      Q(1)=Q(1)+ALPHA(NFCNS-1-J)
  540 CONTINUE
      DO 543 J=1,NFCNS
  543 ALPHA(J)=P(J)
  545 CONTINUE
      IF(NFCNS.GT.3) RETURN
      ALPHA(NFCNS+1)=0.0
      ALPHA(NFCNS+2)=0.0
      RETURN
      END


      DOUBLE PRECISION FUNCTION D(K,N,M)
C
C  FUNCTION TO CALCULATE THE LAGRANGE INTERPOLATION
C  COEFFICIENTS FOR USE IN THE FUNCTION GEE.
C
      COMMON PI2,AD,DEV,X,Y,GRID,DES,WT,ALPHA,IEXT,NFCNS,NGRID
      DIMENSION IEXT(66),AD(66),ALPHA(66),X(66),Y(66)
      DIMENSION DES(1045),GRID(1045),WT(1045)
      DOUBLE PRECISION AD,DEV,X,Y
      DOUBLE PRECISION Q
      DOUBLE PRECISION PI2
      D=1.0
      Q=X(K)
      DO 3 L=1,M
      DO 2 J=L,N,M
      IF(J-K) 1,2,1
    1 D=2.0*D*(Q-X(J))
    2 CONTINUE
    3 CONTINUE
      D=1.0/D
      RETURN
      END


      DOUBLE PRECISION FUNCTION GEE(K,N)
C
C  FUNCTION TO EVALUATE THE FREQUENCY RESPONSE USING THE
C  LAGRANGE INTERPOLATION FORMULA IN THE BARYCENTRIC FORM
C
      COMMON PI2,AD,DEV,X,Y,GRID,DES,WT,ALPHA,IEXT,NFCNS,NGRID
      DIMENSION IEXT(66),AD(66),ALPHA(66),X(66),Y(66)
      DIMENSION DES(1045),GRID(1045),WT(1045)
      DOUBLE PRECISION P,C,D,XF
      DOUBLE PRECISION PI2
      DOUBLE PRECISION AD,DEV,X,Y
      P=0.0
      XF=GRID(K)
      XF=DCOS(PI2*XF)
      D=0.0
      DO 1 J=1,N
      C=XF-X(J)
      C=AD(J)/C
      D=D+C
    1 P=P+C*Y(J)
      GEE=P/D
      RETURN
      END
```

```
      SUBROUTINE OUCH
      PRINT 1
1 FORMAT(' *********** FAILURE TO CONVERGE **********'/
 1' 0PROBABLE CAUSE IS MACHINE ROUNDING ERROR'/
 2' 0THE IMPULSE RESPONSE MAY BE CORRECT'/
 3' 0CHECK WITH A FREQUENCY RESPONSE')
      RETURN
      END
```

4

Theory and Approximation of Infinite Impulse Response Digital Filters

4.1 Introduction

In this chapter we shall discuss techniques for designing infinite impulse response (IIR) digital filters, with the restriction that the filters be realizable and, of course, stable. The impulse response sequence $h(n)$ for such a filter therefore must obey the restrictions

$$h(n) = 0 \qquad n < 0 \tag{4.1}$$

$$\sum_{n=0}^{\infty} |h(n)| < \infty \tag{4.2}$$

The most general form of the z transform of IIR filters can be written as

$$H(z) = \sum_{n=0}^{\infty} h(n)z^{-n} = \frac{\displaystyle\sum_{i=0}^{M} b_i z^{-i}}{1 + \displaystyle\sum_{i=1}^{N} a_i z^{-i}} \tag{4.3}$$

where at least one of the a_i is nonzero and all the roots of the denominator are not canceled exactly by the roots of the numerator. For example, a filter with z transform

$$H(z) = \frac{(1 - z^{-8})}{(1 - z^{-1})} \tag{4.4}$$

is of the form of Eq. (4.3). The denominator root at $z = 1$ is canceled by a numerator root at $z = 1$, however, making $H(z)$ a finite polynomial in z^{-1} or, equivalently, $h(n)$ is an FIR sequence.

The filter of Eq. (4.3) has, in general, M finite zeros and N finite poles. The zeros of $H(z)$ can be anywhere in the z plane but the poles must lie inside the unit circle for stability. In most cases, especially digital filters derived from analog designs, M will be less than or equal to N. Systems of this type are called Nth-order systems. When $M > N$, the order of the system is no longer unambiguous. In this case, $H(z)$ may be taken to be an Nth order system in cascade with an FIR filter of order $(M - N)$. All design techniques in this chapter start from the premise that $M \leq N$.

In contrast to FIR filters, except for the special case where all the poles of $H(z)$ lie on the unit circle, stable, realizable IIR filters cannot attain an exactly linear phase characteristic. As shown in Chapter 3, to obtain linear phase required

$$H(z) = H(z^{-1}) \tag{4.5}$$

(to within a linear phase component) which, in the case of IIR filters, implies that for every pole strictly inside the unit circle (pole magnitude <1.0) in $H(z)$ there had to be a mirror-image pole outside the unit circle—thereby making the filter unstable. Therefore the filter design problem for IIR filters always involves approximation of both magnitude and phase response specifications. There is one special case of an IIR filter in which the magnitude of the frequency response of the filter is constant and only the phase response changes as the pole and zero positions vary. This type of filter is called an all-pass network. The necessary condition for $H(z)$ to be an all-pass network is that for every pole at $z = re^{j\theta}$ there is a corresponding zero at $z = (1/r)e^{j\theta}$ and of course (for real $h(n)$) the complex conjugate pole and zero. A typical section of an all-pass filter is shown in Fig. 4.1. In this case $H(z)$ can be written as

$$H(z) = \frac{[z - (1/r)e^{j\theta}][z - (1/r)e^{-j\theta}]}{(z - re^{j\theta})(z - re^{-j\theta})} \tag{4.6}$$

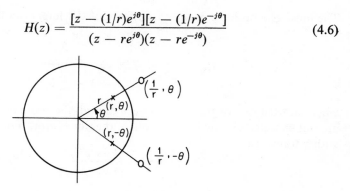

Fig. 4.1 Pole–zero pattern of a second-order all-pass filter.

which can be put in the form

$$H(z) = \frac{[z^2 - (2/r)(\cos\theta)z + (1/r^2)]}{(z^2 - 2r(\cos\theta)z + r^2)} = \frac{1}{r^2}\frac{(r^2z^2 - 2r(\cos\theta)z + 1)}{(z^2 - 2r(\cos\theta)z + r^2)} \quad (4.7)$$

Evaluation of $H(z)$ on the unit circle gives

$$|H(e^{j\omega})| = \text{constant}$$

and a typical phase function (for the case $r = 0.9$, $\theta = 36°$) is shown in Fig. 4.2. The importance of these all-pass filters lies in the fact that it can be shown that a cascade of all-pass sections can be used to equalize a given phase (or group delay) characteristic.

If the realizability restriction on the filter impulse response is removed, there are two distinct ways of achieving a linear phase characteristic for an IIR filter. These two methods are shown in Fig. 4.3. For both these methods, the filter $H(z)$ is a realizable IIR filter and the boxes labeled TIME REVERSAL

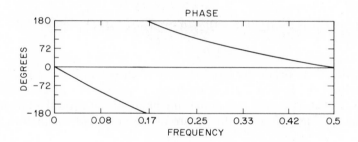

Fig. 4.2 Phase response of a first-order allpass filter.

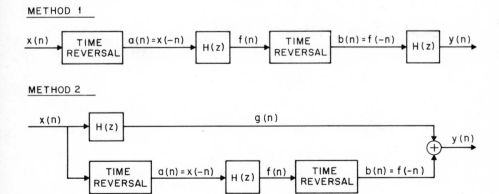

Fig. 4.3 Two theoretical methods for achieving a zero-phase IIR filter.

have input–output relations of the form

$$z(n) = w(-n) \tag{4.8}$$

where $w(n)$ is the input sequence and $z(n)$ is the output sequence. The limitations on the existence of such boxes will be discussed after it is shown that in both these methods the *equivalent filter* has linear (zero) phase. The definition of the equivalent filter is shown in Fig. 4.4. As seen from this figure, $H_{eq}(z)$ is defined as

$$H_{eq}(z) = \frac{Y(z)}{X(z)} \tag{4.9}$$

For method 1 we obtain the relations (referring to Fig. 4.3 and 4.4)

$$A(z) = X(z^{-1}) \tag{4.10a}$$

$$F(z) = H(z)A(z) = H(z)X(z^{-1}) \tag{4.10b}$$

$$B(z) = F(z^{-1}) = H(z^{-1})X(z) \tag{4.10c}$$

$$Y(z) = H(z)B(z) = X(z)H(z)H(z^{-1}) \tag{4.10d}$$

$$H_{eq}(z) = H(z)H(z^{-1}) \tag{4.10e}$$

$$H_{eq}(e^{j\omega}) = |H(e^{j\omega})|^2 \tag{4.10f}$$

In deriving these results we have used the relation that if a sequence $x(n)$ has z transform $X(z)$, then a (time reversed) sequence $x(-n)$ has z transform $X(z^{-1})$. Equation (4.10f) shows that the phase of the equivalent filter is zero and the filter magnitude is the square of the magnitude of the original IIR filter.

For method 2 we find the relations

$$A(z) = X(z^{-1}) \tag{4.11a}$$

$$F(z) = H(z)A(z) = H(z)X(z^{-1}) \tag{4.11b}$$

$$B(z) = F(z^{-1}) = H(z^{-1})X(z) \tag{4.11c}$$

$$G(z) = H(z)X(z) \tag{4.11d}$$

$$Y(z) = B(z) + G(z) = X(z)[H(z) + H(z^{-1})] \tag{4.11e}$$

$$H_{eq}(z) = H(z) + H(z^{-1}) \tag{4.11f}$$

$$H_{eq}(e^{j\omega}) = 2|H(e^{j\omega})| \cos[\varphi(\omega)] \tag{4.11g}$$

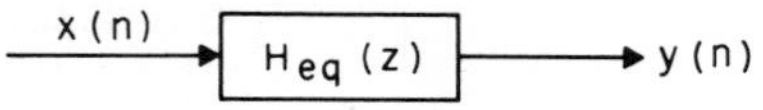

Fig. 4.4 The definition of an equivalent filter.

where

$$H(e^{j\omega}) = |H(e^{j\omega})| \, e^{j\varphi(\omega)} \tag{4.11h}$$

In this case the equivalent filter again has zero phase, but now the magnitude is twice the original filter magnitude times the cosine of the original phase function. This phase multiplication in the filter tends to make method 2 somewhat less desirable than method 1.

In practice, neither methods 1 nor 2 can be carried out exactly because of the need for reversing infinite sequences before they are available. Arbitrarily small approximation errors can be obtained, however, by truncating the infinite sequences after an appropriate number of terms. The details of such a realization are available in the thesis of Gibbs.

4.2 Some Elementary Properties of IIR Filters— Magnitude-Squared Response, Phase Response, and Group Delay

In the course of the discussion of design techniques for IIR filters, reference will be made at several points to some elementary properties of the filter transfer function. The three most important functions that will be used throughout this chapter are the magnitude-squared response, the phase response, and the group delay of the filter. The reason all these functions are of interest for IIR filters is that the general approximation problem involves a complex transfer function in ω. Thus both a magnitude and a phase response must be examined to determine the nature of the resulting filter approximation. Further, since the phase function is generally highly nonlinear, the filter group delay response is often used as a measure of how much dispersal a typical input will undergo when being processed by the filter. In this section each of the three responses is defined for reference later in the chapter.

1. Magnitude-Squared Response

When an IIR filter is determined strictly in terms of a magnitude approximation (i.e., the phase is completely disregarded), it is convenient to consider designs in terms of the magnitude-square function, which is defined as

$$|H(e^{j\omega})|^2 = |H(z)H(z^{-1})|\big|_{z=e^{j\omega}} \tag{4.12}$$

The poles and zeros of a magnitude-squared function are distributed with mirror-image symmetry with respect to the unit circle in the z plane. The poles of $H(z)$ are uniquely determined from the magnitude-squared function as those lying inside the unit circle. The zeros are not generally uniquely determined (except in the important case where all zeros are on the unit circle) as zeros may lie anywhere in the z plane. In most cases the zeros of $H(z)$ are also chosen to be the zeros of the magnitude-squared function lying inside or on the unit circle in the z plane. In this case the resulting filter is a minimum phase filter.

2. Phase Response

Since the transfer function of an IIR filter is generally a complex function of ω, it has both a magnitude and a phase response. The phase response is defined as

$$\beta(e^{j\omega}) = \tan^{-1}\left\{\frac{\text{Im }[H(z)]}{\text{Re }[H(z)]}\right\}_{z=e^{j\omega}} \tag{4.13}$$

which may be written in the alternate form

$$\beta(e^{j\omega}) = \frac{1}{2j}\ln\left\{\frac{H[z]}{H[z^{-1}]}\right\}_{z=e^{j\omega}} \tag{4.14}$$

as may be seen by writing $H(z)$ in the form

$$H(z) = |H(z)|\,e^{j\beta(z)} \tag{4.15}$$

and recognizing that

$$H(z^{-1}) = |H(z)|\,e^{-j\beta(z)} \tag{4.16}$$

thereby leading to Eq. (4.14).

3. Group Delay

The group delay of a filter is a measure of the average delay of the filter as a function of frequency and is defined as

$$\tau_g(e^{j\omega}) = -\frac{d\beta(e^{j\omega})}{d\omega} = -jz\frac{d\beta}{dz}\bigg|_{z=e^{j\omega}} \tag{4.17}$$

Using Eq. (4.14), $\tau_g(e^{j\omega})$ can be written as

$$\tau_g(e^{j\omega}) = -\text{Re}\left[z\frac{dH(z)/dz}{H(z)}\right]_{z=e^{j\omega}}$$

$$= -\text{Re}\left\{z\frac{d}{dz}\left[\ln H(z)\right]\right\}_{z=e^{j\omega}} \tag{4.18}$$

A desirable group delay characteristic of a filter is one that is approximately constant over the band(s) of frequencies that the filter passes.

4.3 Techniques for Determining IIR Filter Coefficients

The problem of designing filters is one of finding filter coefficients [e.g., the b_i's and a_i's of Eq. (4.3)] such that some aspect of the filter's response (e.g., time response, frequency response, group delay, etc.) approximates a desired behavior in a specified manner (e.g., minimum mean square or minimax error). As such, the "filter design problem" is basically a mathematical approximation problem and can be approached strictly from a mathematician's point of view. The domain in which the approximation problem is solved determines how and where the resulting filter can be used. Thus if the approximation problem is solved in the z plane, the resulting filter is a digital filter. If it is solved in the s plane, the resulting filter is an analog filter. Other possible filter classes that may result include optical filters and transmission line filters. The common ground among all these classes of filters is the mathematics of functions with filter-like properties.

Thus, contrary to some popular beliefs, a great deal of IIR digital filter design does not depend intrinsically on continuous-time filter designs but instead makes use of the wide body of knowledge that is available in the literature on the design of such filters. Instead of redeveloping the theory (i.e., restructuring the mathematics to the case of digital filters) for digital filters, simple mapping procedures can be used to transform filters in one domain to filters in the other domain. This technique of designing an appropriate continuous-time filter and digitizing the resulting design to give a digital filter is the most popular IIR design technique and is most useful for designing standard filters such as lowpass, bandpass, bandstop, and high pass filters where a considerable body of theory on continuous-time filters is available.

A second method for designing IIR digital filters is direct closed form design in the z plane. Beginning with the desired response of the filter, one can often decide where to place poles and zeros to approximate this response directly. A third way in which IIR filters are often designed is by using optimization procedures to place poles and zeros at appropriate positions in the z plane to approximate in some sense the desired response. This design procedure does not generally yield closed form expressions for pole and zero positions (i.e., filter coefficients) as a function of the desired response. Iterative techniques are generally used to arrive at the desired filter.

4.4   Digital Filter Design from Continuous-Time Filters

As discussed in Sec. 4.3, the most popular technique for designing IIR filters is to digitize an analog filter that satisfies the design specifications. There are many techniques for designing analog prototype filters when the specifications are of the form of a lowpass, bandpass, highpass, or band reject filter. Among the well-known analog filter classes are the Butterworth, Bessel, Chebyshev Types I and II, and Cauer or elliptic filters. We shall review the properties of these filters and present appropriate design relations for them in Sec. 4.9. First we shall discuss several procedures for transforming or digitizing an existing analog filter to an "equivalent" digital filter.

In the remainder of this section we assume that the coefficients of an analog filter with Laplace transform (i.e., transfer function)

$$H(s) = \frac{\sum\limits_{i=0}^{M} b_i s^i}{\sum\limits_{i=1}^{N} a_i s^i} = \frac{\prod\limits_{i=1}^{M} (s + c_i)}{\prod\limits_{i=1}^{N} (s + d_i)} \tag{4.19}$$

are known: i.e., the a_i's and b_i's (or, equivalently, c_i's and d_i's) have been determined. The differential equation of the filter is thus of the form

$$\sum_{i=0}^{N} a_i \frac{d^i y(t)}{dt^i} = \sum_{i=0}^{M} b_i \frac{d^i x(t)}{dt^i} \tag{4.20}$$

where $x(t)$ is the input to the filter and $y(t)$ is the output. The four most widely used procedures for digitizing the transfer function of Eq. (4.19) include

1. The method of mapping of differentials.
2. The impulse invariant transformation.
3. The bilinear transformation.
4. The matched z-transform technique.

The next sections discuss each of these techniques in detail.

4.5 Mapping of Differentials

One of the simplest ways of digitizing a continuous system is to replace differentials in the differential equation of the continuous system with finite differences in order to obtain a difference equation that approximates the given differential equation. The simplest replacement that can be made is to use a forward or backward difference to replace a first differential. Thus Eq. (4.20) can be digitized in this manner to give

$$\sum_{i=0}^{N} a_i \Delta_i[y(n)] = \sum_{i=0}^{M} b_i \Delta_i[x(n)] \tag{4.21}$$

where $x(n)$ is the input sequence to the digital filter, $y(n)$ is the output sequence, and $\Delta_i[w(n)]$ is the ith difference defined by

$$\Delta_{i+1}[w(n)] = \Delta_1\{\Delta_i[w(n)]\} \tag{4.22}$$
and

$$\Delta_1[w(n)] = \begin{cases} \dfrac{1}{T} [w(n) - w(n-1)] & \text{backward difference} \\[2mm] \dfrac{1}{T} [w(n+1) - w(n)] & \text{forward difference} \end{cases} \tag{4.23}$$

Thus for backward differences, $\Delta_2[w(n)]$ is defined as

$$\Delta_2[w(n)] = \frac{1}{T}\{\Delta_1[w(n)] - \Delta_1[w(n-1)]\}$$

$$= \frac{1}{T}\left\{\frac{1}{T}[w(n) - w(n-1)] - \frac{1}{T}[w(n-1) - w(n-2)]\right\}$$

$$= \frac{1}{T^2}[w(n) - 2w(n-1) + w(n-2)] \qquad (4.24)$$

Two desirable properties of any mapping from continuous to discrete space are

1. The $j\Omega$ axis in the s plane should be mapped to the unit circle in the z plane.
2. Points in the left-half s plane {Re $[s] < 0$} should be mapped inside the unit circle ($|z| < |$).

Property 1 preserves (through the uniformity of the mapping) the frequency selective properties of the continuous system, whereas property 2 ensures that stable continuous systems (filters) are mapped into stable discrete systems. We shall now examine how well the replacement of differentials by forward and backward differences satisfies the criteria above of a good mapping.

Backward Differences. When using backward differences we make the replacement

$$\frac{dy}{dt} \leftrightarrow \frac{y(n) - y(n-1)}{T} \qquad (4.25)$$

which, in terms of operators, corresponds to

$$s = \frac{1 - z^{-1}}{T} \qquad (4.26)$$

or

$$z = \frac{1}{1 - sT} \qquad (4.27)$$

When $s = j\Omega$, Eq. (4.27) shows

$$z = \frac{1}{1 - j\Omega T} = \frac{1}{2}\left(1 + \frac{1 + j\Omega T}{1 - j\Omega T}\right)$$

$$= \tfrac{1}{2}(1 + e^{2j\tan^{-1}\Omega T}) \qquad (4.28)$$

Taking real and imaginary parts of z gives

$$\text{Re}\,[z] = \frac{1}{2} + \frac{\cos(2\tan^{-1}\Omega T)}{2}$$

$$\text{Im}\,[z] = \frac{\sin(2\tan^{-1}\Omega T)}{2} \qquad (4.29)$$

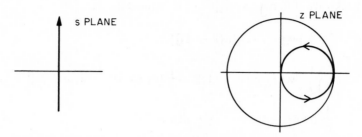

Fig. 4.5 s plane to z plane mapping of $j\Omega$ axis for method of backward differences.

Thus the mapping of the line $s = j\Omega$ (from $\Omega = -\infty$ to ∞) in the z plane is the circle described by

$$\{\text{Re } [z] - \tfrac{1}{2}\}^2 + \text{Im } [z]^2 = (\tfrac{1}{2})^2 \tag{4.30}$$

i.e., a circle with center at Re $[z] = \tfrac{1}{2}$ and radius $\tfrac{1}{2}$, as shown in Fig. 4.5. Except for extremely small values of ΩT, the image of the $j\Omega$ axis in the s plane is off the unit circle in the z plane. Therefore property 1 is not satisfied. To see whether property 2 of the mapping is satisfied we set

$$sT = \alpha + j\beta \tag{4.31}$$

where α and β are real and $\alpha < 0$. Equation (4.27) shows

$$z = \frac{1}{1 - \alpha - j\beta} \tag{4.32}$$

or

$$|z| = \frac{1}{\sqrt{(1-\alpha)^2 + \beta^2}} < 1 \tag{4.33}$$

Therefore stable analog filters map into stable digital filters using backward differences but their frequency selective properties are not maintained.

Forward Differences. When using forward differences we make the replacement

$$\frac{dy}{dt} \leftrightarrow \frac{y(n + 1) - y(n)}{T} \tag{4.34}$$

which corresponds to

$$s = \frac{z - 1}{T} \tag{4.35}$$

or

$$z = 1 + sT \tag{4.36}$$

When $s = j\Omega$, Eq. (4.36) shows

$$z = 1 + j\Omega T \tag{4.37}$$

Fig. 4.6 s plane to z plane mapping of $j\Omega$ axis for method of forward differences.

The contours in the s plane and z plane for this mapping are shown in Fig. 4.6. As seen there, property 1 of the mapping is not satisfied. Property 2 of the mapping is also not satisfied since when

$$sT = \alpha + j\beta \tag{4.38}$$

then

$$z = 1 + \alpha + j\beta \tag{4.39}$$

and

$$|z| > 1$$

when $\beta^2 > 1 - (1 + \alpha)^2$.

Generalized Differences. A more sophisticated procedure for digitizing analog designs by replacing differentials by differences is to use higher-order differences to replace lower-order differentials. For example, consider defining Δ_1 of Eq. (4.23) by the expression

$$\Delta_1[w(n)] = \frac{1}{T}\sum_{i=1}^{L}\alpha_i[w(n + i) - w(n - i)] \tag{4.40}$$

where L is the order of difference to be used. The mapping between s plane and z plane then becomes

$$s = \frac{1}{T}\sum_{i=1}^{L}\alpha_i(z^i - z^{-i}) \tag{4.41}$$

To show that the mapping of Eq. (4.41) satisfies property 1, we shall show that when $z = e^{j\omega T}$, then s is of the form $s = j\beta(\omega)$ (i.e., points on the unit circle in the z plane are the mappings of points on the $j\Omega$ axis in the s plane). By setting $z = e^{j\omega T}$ in Eq. (4.41) we obtain

$$s = \frac{1}{T}\sum_{i=1}^{L}\alpha_i(e^{j\omega iT} - e^{-j\omega iT}) \tag{4.42}$$

$$= \frac{1}{T}\sum_{i=1}^{L}2j\alpha_i \sin(\omega iT) = j\beta(\omega) \tag{4.43}$$

By proper choice of the coefficients α_i, the function $\beta(\omega)$ can approximate almost any desired odd function of ω, thereby mapping the $j\Omega$ axis in the s plane monotonically to the unit circle in the z plane. Furthermore the mapping of Eq. (4.41) can be shown to be conformal, thereby mapping the left-half s plane to the inside of the unit circle in the z plane. For such cases both desirable properties of the mapping are preserved. However, because of the difficulty in determining suitable coefficients for the mapping (i.e., the α_i's), as well as the availability of other techniques for digitizing filters, such higher-order differences have not been used in any practical examples.

General Comments on Replacing Differentials by Differences. The attractive feature of replacing differentials by *simple* differences is that rational transfer functions in s become rational transfer functions in z by using simple substitutions—e.g., Eq. (4.26) or (4.35). However, since neither simple forward nor simple backward differences adequately preserve the filter characteristics, alternate techniques are generally used to digitize analog filters.

4.6 Impulse Invariant Transformation

A second technique for digitizing an analog filter is called impulse invariant transformation. The characteristic property preserved by this transformation is that the impulse response of the resulting digital filter is a sampled version of the impulse response of the analog filter. In consequence of this result, the frequency response of the digital filter is an aliased version of the frequency response of the corresponding analog filter.

To demonstrate how an analog filter is digitized using the impulse invariant transformation, we rewrite Eq. (4.19), in its partial fraction expansion, as

$$H(s) = \sum_{i=1}^{N} \frac{c_i}{s + d_i} \tag{4.44}$$

where

$$c_i = H(s) \cdot (s + d_i)\big|_{s=-d_i} \tag{4.45}$$

and d_i is the location of the ith pole. In writing Eq. (4.44) in the given form, we have assumed that in Eq. (4.19) the order of the numerator M is less than the order of the denominator N and that all the poles of $H(s)$ are simple. The assumption that $M < N$ must be valid for the system to be digitized— otherwise the aliasing in the digital system would be intolerable. If the poles of $H(s)$ are not simple, the discussion in this section can be appropriately modified.

The impulse response of the analog filter $h(t)$ corresponding to Eq. (4.44) is of the form

$$h(t) = \sum_{i=1}^{N} c_i e^{-d_i t} u_{-1}(t) \tag{4.46}$$

The corresponding digital impulse response $h(nT)$ can be written as the sampled version of Eq. (4.46); i.e.,

$$h(nT) = \sum_{i=1}^{N} c_i e^{-d_i nT} u_{-1}(nT) \qquad (4.47)$$

where T is the sampling period. The z transform of Eq. (4.47) is

$$H(z) = \sum_{n=0}^{\infty} h(nT)z^{-n} = \sum_{n=0}^{\infty} \sum_{i=1}^{N} c_i e^{-d_i nT} z^{-n} \qquad (4.48)$$

Interchanging orders of summation and summing over n gives

$$H(z) = \sum_{i=1}^{N} c_i \sum_{n=0}^{\infty} (e^{-d_i T} z^{-1})^n$$

$$H(z) = \sum_{i=1}^{N} \frac{c_i}{1 - e^{-d_i T} z^{-1}} \qquad (4.49)$$

By comparing Eq. (4.49) and (4.44) it is seen that $H(z)$ is obtained from $H(s)$ by using the mapping relation

$$\frac{1}{s + d_i} \rightarrow \frac{1}{1 - z^{-1}e^{-d_i T}} \qquad (4.50)$$

for simple poles. When d_i is complex, then c_i, the residue in Eq. (4.44), is also complex. Since $h(t)$ is real, this implies that there is also a pole at d_i^* with residue c_i^*. By combining these terms in Eq. (4.44) we obtain a term of the form

$$\frac{c_i}{s + d_i} + \frac{c_i^*}{s + d_i^*} = \frac{(c_i + c_i^*)s + c_i d_i^* + c_i^* d_i}{s^2 + (d_i + d_i^*)s + d_i d_i^*} \qquad (4.51)$$

If we let $d_i = \sigma_i + j\Omega_i$ and $c_i = g_i + jh_i$, then Eq. (4.51) becomes

$$\frac{c_i}{s + d_i} + \frac{c_i^*}{s + d_i^*} = \frac{2g_i s + 2(\sigma_i g_i + \Omega_i h_i)}{s^2 + 2\sigma_i s + (\sigma_i^2 + \Omega_i^2)} \qquad (4.52)$$

By applying the substitution of Eq. (4.50) to each of the individual terms of Eq. (4.51), we find

$$\frac{c_i}{1 - z^{-1}e^{-d_i T}} + \frac{c_i^*}{1 - z^{-1}e^{-d_i^* T}}$$

$$= \frac{(c_i + c_i^*) - z^{-1}(c_i e^{-d_i T} + c_i^* e^{-d_i T})}{1 - z^{-1}(e^{-d_i T} + e^{-d_i^* T}) + z^{-2} e^{-(d_i + d_i^*)T}} \qquad (4.53)$$

$$= \frac{2g_i - z^{-1} e^{-\sigma_i T}[2g_i \cos(\Omega_i T) - 2h_i(\sin \Omega_i T)]}{1 - 2z^{-1}e^{-\sigma_i T} \cos(\Omega_i T) + z^{-2}e^{-2\sigma_i T}} \qquad (4.54)$$

Combining Eq. (4.52) and (4.54) (dropping the subscript i and dividing the numerators by $2g$) gives the substitution

$$\frac{s + \sigma + \Omega(h/g)}{s^2 + 2\sigma s + \sigma^2 + \Omega^2} \rightarrow \frac{1 - z^{-1}e^{-\sigma T}[\cos(\Omega T) - (h/g)\sin(\Omega T)]}{1 - 2z^{-1}e^{-\sigma T}\cos(\Omega T) + z^{-2}e^{-2\sigma T}} \quad (4.55)$$

Useful forms of Eq. (4.55) [for analog filters with impulse responses of the form $h_1(t) = e^{-\sigma t}\cos(\Omega t)u_{-1}(t)$ and $h_2(t) = e^{-\sigma t}\sin(\Omega t)u_{-1}(t)$] are

$$H_1(s) = \frac{s + \sigma}{s^2 + 2\sigma s + \sigma^2 + \Omega^2} \rightarrow \frac{1 - z^{-1}e^{-\sigma T}\cos\Omega T}{1 - 2z^{-1}e^{-\sigma T}\cos\Omega T + z^{-2}e^{-2\sigma T}} \quad (4.56)$$

$$H_2(s) = \frac{\Omega}{s^2 + 2\sigma s + \sigma^2 + \Omega^2} \rightarrow \frac{z^{-1}e^{-\sigma T}\sin\Omega T}{1 - 2z^{-1}e^{-\sigma T}\cos\Omega T + z^{-2}e^{-2\sigma T}} \quad (4.57)$$

As stated earlier, the frequency response of a digital system, realized from an impulse invariant transformation, is an aliased version of the frequency response of the continuous system from which it was derived. Thus we can write

$$H(e^{j\Omega T}) = \frac{1}{T}\sum_{l=-\infty}^{\infty} H(j\Omega + jl\Omega_s) \quad (4.58)$$

where $\Omega_s = 2\pi/T$ is the radian sampling frequency of the digital system. Figure 4.7 shows the mapping from the s plane to the z plane corresponding to impulse invariant transformation. Each horizontal strip of the s plane of width $2\pi/T$ is mapped into the entire z plane. Adjacent strips in the s plane are thus aliased or folded over into each other in the z plane. From this figure it is clear that for the frequency responses of an analog filter and the equivalent digital filter obtained by impulse invariant transformation to correspond, the analog filter must be bandlimited to the range $-\pi/T \leq \Omega \leq \pi/T$. This generally requires that a guard filter (i.e., a suitable lowpass filter) be used to guarantee that the analog filter be suitably bandlimited prior to transformation.

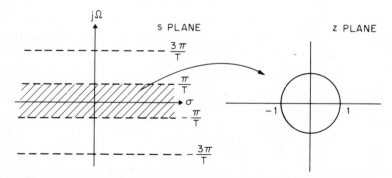

Fig. 4.7 The mapping from the s plane to the z plane corresponding to impulse invariant transformation.

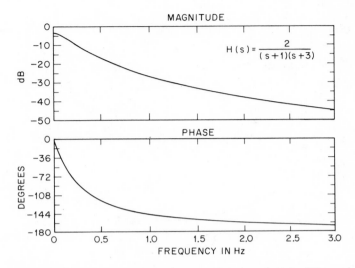

Fig. 4.8 Magnitude and phase responses of a continuous-time filter.

Example of Impulse Invariance. To illustrate this technique we shall digitize an analog filter with transfer function

$$H(s) = \frac{2}{(s+1)(s+3)} = \frac{1}{s+1} - \frac{1}{s+3}$$

Direct application of Eq. (4.50) gives

$$H(z) = \frac{1}{1 - z^{-1}e^{-T}} - \frac{1}{1 - z^{-1}e^{-3T}} = \frac{z^{-1}(e^{-T} - e^{-3T})}{1 - z^{-1}(e^{-T} + e^{-3T}) + e^{-4T}z^{-2}}$$

The frequency response of the analog system can be written as

$$H(j\Omega) = \frac{2}{(3 - \Omega^2) + 4j\Omega}$$

and its magnitude and phase are plotted in Fig. 4.8. The corresponding digital frequency responses are plotted in Fig. 4.9 for various values of $T = 1/F_s$, the sampling period. Clearly as T gets smaller, (i.e. F_s gets larger), the effects of aliasing become negligible and the digital and analog frequency responses become comparable.

4.7 Bilinear Transformation

The first technique we discussed for deriving a digital filter from an analog filter, i.e., the replacement of differentials by finite differences, had the advantage that the z transform of the digital filter was trivially derived from

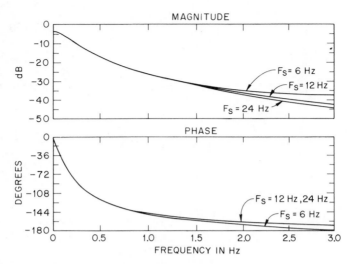

Fig. 4.9 Magnitude and phase responses of the digital filter(s) obtained via impulse invariant transformation of the filter in Fig. 4.8.

the Laplace transform of the analog filter by an algebraic substitution. The disadvantages of these mappings were that the $j\Omega$ axis in the s plane generally did not map into the unit circle in the z plane and (for the case of forward differences) stable analog filters didn't always map into stable digital filters.

A simple conformal mapping from the s plane to the z plane which eliminates the disadvantages mentioned above and which preserves the desired algebraic form is the bilinear transformation defined by

$$s \rightarrow \frac{2}{T} \frac{(1 - z^{-1})}{(1 + z^{-1})} \tag{4.59}$$

The nature of this mapping is best understood from Fig. 4.10, which shows how the s plane is mapped into the z plane. As seen in this figure, the entire $j\Omega$ axis in the s plane is mapped onto the unit circle; the left-half s plane is mapped inside the unit circle in the z plane and the right-half s plane is mapped outside the z plane unit circle. These properties are easily shown by solving for z in terms of s from Eq. (4.59), giving

$$z = \frac{(2/T) + s}{(2/T) - s} \tag{4.60}$$

When $s = j\Omega$, we find

$$z = \frac{(2/T) + j\Omega}{(2/T) - j\Omega} \tag{4.61}$$

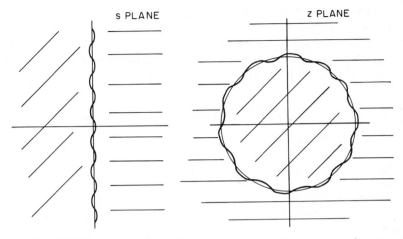

Fig. 4.10 The mapping from the s plane to the z plane corresponding to the bilinear transformation.

From Eq. (4.61) we find $|z| = 1$. When $\Omega = 0$, $z = 1$; when $\Omega = \infty$, $z = -1$; between these limits the angle of z varies monotonically from 0 to π. If we let $s = \sigma + j\Omega$ in Eq. (4.60), we obtain

$$z = \frac{(2/T) + \sigma + j\Omega}{(2/T) - \sigma - j\Omega} \qquad (4.62)$$

and when $\sigma < 0$ (left-half s plane), we find $|z| < 1$ (inside the unit circle).

The transfer function of the digital filter $H(z)$ is obtained from the bilinear transformation by making the algebraic substitution of Eq. (4.59); i.e.,

$$H(z) = H(s)\big|_{s=(2/T)[(1-z^{-1})/(1+z^{-1})]} \qquad (4.63)$$

It is readily shown that the orders of the denominators of $H(z)$ and $H(s)$ are the same; however, the orders of the numerators may differ. Thus, as an example, consider

$$H(s) = \frac{1}{s + a}$$

which has a numerator of zeroth degree and a first-degree denominator. Then $H(z)$ obtained from the bilinear transformation is

$$H(z) = \frac{1}{(2/T)[(1 - z^{-1})/(1 + z^{-1})] + a} = \frac{1 + z^{-1}}{2/T + a + z^{-1}[a - (2/T)]}$$

which has a first-degree numerator and denominator. The reason this occurs is that $H(s)$ has a zero at $s = \infty$; the bilinear transformation maps this zero to $z = -1$.

Since the *entire $j\Omega$* axis of the s plane is mapped onto the unit circle in the z plane, the aliasing errors inherent with impulse invariant transformations are eliminated. There is a highly nonlinear relationship, however, between analog frequency Ω and digital frequency ω. The nature of this nonlinearity is seen by evaluating Eq. (4.59) for $z = e^{j\omega T}$ and $s = j\Omega$, giving

$$j\Omega \rightarrow \frac{2}{T} \frac{(1 - e^{-j\omega T})}{(1 + e^{-j\omega T})} \tag{4.64}$$

which can be written as

$$j\Omega \rightarrow \frac{2}{T} \frac{(e^{j(\omega T/2)} - e^{-j(\omega T/2)}]}{[e^{j(\omega T/2)} + e^{-j(\omega T/2)}]}$$

or

$$j\Omega \rightarrow \frac{2}{T} j \tan\left(\frac{\omega T}{2}\right)$$

$$\Omega \rightarrow \frac{2}{T} \tan\left(\frac{\omega T}{2}\right) \tag{4.65}$$

The form of this nonlinearity is shown in Fig. 4.11 for the case $T = 2$. For small values of ω, the mapping is almost linear. For most of the frequency scale, however, the mapping is highly nonlinear. This imposes a strong restriction on when the bilinear transformation can be used. It implies that the amplitude response of the continuous system being transformed must be piece-wise constant. If this is not the case, Eq. (4.65) shows that the digital frequency response will be a warped version of the analog frequency response.

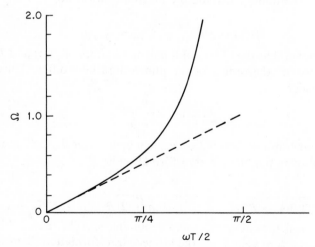

Fig. 4.11 The relation between analog and digital frequency scales for the bilinear transformation.

Thus, for example, the bilinear transform cannot be used to transform an analog differentiator to a digital filter. Fortunately there is a large class of filters for which one can compensate for the frequency warping of Eq. (4.65). This class includes such important filters as lowpass, bandpass, bandstop, and highpass filters. The compensation used is straightforward and is illustrated in Fig. 4.12. The desired critical set of digital filter cutoff frequencies is determined, as shown at the lower right-hand corner of Fig. 4.12. In this example we assume there are four such frequencies (ω_1, ω_2, ω_3, ω_4). Using the frequency warping relation [Eq. (4.65)] between the digital and analog frequency scales, the filter cutoff frequencies are converted to a new set of analog cutoff frequencies (Ω_1, Ω_2, Ω_2, Ω_4) as shown at the upper right of the figure. Finally an analog filter is designed with the appropriate warped cutoff frequencies as shown at the upper left of the figure. Applying the

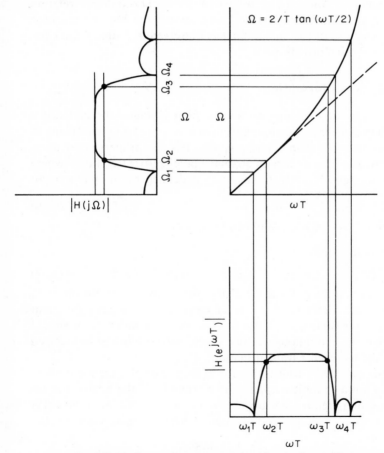

Fig. 4.12 Technique for compensating the nonlinear frequency warping of the bilinear transformation.

bilinear transformation to this analog filter gives a digital filter with the desired cutoff frequencies. Examples of the use of the bilinear transformation in the design of lowpass and bandpass filters will be given later in this chapter.

In summary, the bilinear transformation provides a simple mapping between continuous and digital filters which is algebraic in nature and which maps the entire $j\omega$ axis in the s plane to the unit circle in the z plane. It also has the property that realizable, stable continuous systems are mapped to realizable, stable digital systems. Furthermore, wideband sharp cutoff continuous filters can be mapped to wideband sharp cutoff digital filters without the aliasing in the frequency response that is inherent in filters obtained from impulse invariant transformation. One disadvantage of this technique is that the frequency response of the continuous system must be piece-wise constant to compensate for the effects of the nonlinear relation between analog and digital frequencies. Also neither the impulse response nor the phase response of the analog filter is preserved in a digital filter obtained by bilinear transformation.

4.8   Matched z Transformation

A fourth technique for digitizing an analog filter is called the matched z transformation and is a direct mapping from poles and zeros in the s plane to poles and zeros in the z plane. The mapping has the property that an s-plane pole (zero) at $s = -a$ maps to a z-plane pole (zero) at $z = e^{-aT}$ where T is the sampling period. Thus the mapping is defined by the replacement relationship

$$s + a \rightarrow 1 - z^{-1}e^{-aT} \tag{4.66}$$

For complex poles (zeros) Eq. (4.66) can be written as

$$(s + a - jb)(s + a + jb)$$
$$= (s + a)^2 + b^2 \rightarrow 1 - 2z^{-1}e^{-aT} \cos(bT) + e^{-2aT} \tag{4.67}$$

It should be noted that the poles of the digital filter derived from the matched z transformation are identical to the poles derived from impulse invariant transformation of the same filter; the zeros, however, are different. It should also be clear that the continuous transfer function $H(s)$ must be in factored form to apply the matched z transformation.

Although the matched z transformation is easy to apply, there are many cases when it is not a suitable mapping. For example, if the analog system has zeros with center frequencies greater than half the sampling frequency, their z-plane positions will be greatly aliased. Consider, as an example, the transfer function

$$H(s) = \frac{s^2 + 2s + 5626}{s^2 + 2s + 2}$$

which has zeros at $s = -1 \pm j75$ and poles at $s = -1 \pm j1$. In Fig. 4.13 we show the s-plane positions of the poles and zeros and the log magnitude response of the filter. If we set the sampling period T to $\frac{1}{12}$, using the matched z transformation we obtain the digital transfer function

$$H(z) = \frac{1 - 2e^{-1/12} \cos\left(\frac{7.5}{12}\right)z^{-1} + e^{-1/6}z^{-2}}{1 - 2e^{-1/12} \cos\left(\frac{1}{12}\right)z^{-1} + e^{-1/6}z^{-2}}$$

The z-plane positions of the poles in polar coordinates are $z = e^{-1/12}e^{\pm j(1/12)}$ and the positions of the zeros are $z = e^{-1/12}e^{\pm j75/12}$. The approximate positions of the poles and zeros in the z plane and the log magnitude response of the filter are shown in Fig. 4.14. As seen in this figure the position of the zero is aliased from a high frequency in the analog system to a low frequency in the digital system when using the matched z transformation.

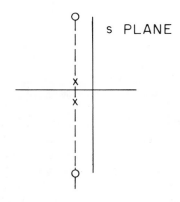

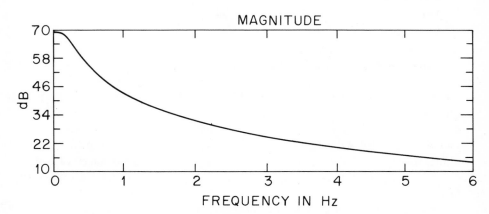

Fig. 4.13 Positions of the poles and zeros in the s plane and log magnitude response of an analog filter.

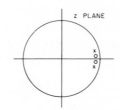

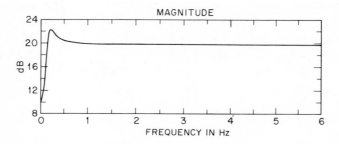

Fig. 4.14 Positions of the poles and zeros in the z plane and log magnitude response of the digital filter obtained by taking the matched z transformation of the poles and zeros of the filter of Fig. 4.13.

Another case where the matched z transformation is unsuitable is where the continuous transfer function is an all-pole system. Then the digital transfer function is an all-pole system that, in many cases, does not adequately represent the desired continuous system. The expediency of artificially adding zeros at $z = -1$ to the digital system has been suggested as a method of alleviating the difficulty—but this ad hoc technique is at best only a stopgap measure. In general, use of impulse invariant or bilinear transformation is to be preferred over the matched z transformation.

4.9 Review of Design Techniques for Analog Lowpass Filters

Since a great deal of the theory of design of IIR digital filters depends on understanding continuous-time filter design techniques, in this section we review pertinent design formulas for several standard types of analog filters including Butterworth, Bessel, Chebyshev I and II, and elliptic or Cauer filters. Complete descriptions of the advantages and disadvantages of each of these approximation techniques are available in several texts on analog filter design. Therefore in this section we shall only summarize the properties of each filter type and present the design equations that are necessary to be able to determine the analog filter coefficients.

It is assumed that a normalized lowpass filter is to be designed with a cutoff frequency of $\Omega = 1$ rad/sec. The approximation will generally be in terms of a magnitude-squared approximation (except in the case of Bessel filters). It is further assumed that the continuous filter transfer function is a rational function in s of the form

$$H(s) = \frac{\sum_{i=0}^{m} b_i s^i}{1 + \sum_{i=1}^{n} a_i s^i} \qquad (4.68)$$

1. Butterworth Filters

Butterworth lowpass filters are characterized by the property that the magnitude characteristic is maximally flat at the origin of the s plane. This means that as many derivatives of the magnitude response as possible are zero at the origin. The squared-magnitude response of a normalized (to 1 rad/sec cutoff frequency) Butterworth filter is

$$|H(\Omega)|^2 = \frac{1}{1 + (\Omega^2)^n} \qquad (4.69)$$

where n is the filter order. By analytic continuation, Eq. (4.69) can be written as

$$H(s)H(-s) = \frac{1}{1 + (-s^2)^n} \qquad (4.70)$$

The poles of Eq. (4.70) occur at equally spaced points on the unit circle in the s plane. By associating the pole in the left-half s plane with $H(s)$ we find

$$H(s) = \frac{k_0}{\prod_{k=1}^{n} (s - s_k)} \qquad (4.71)$$

where

$$s_k = e^{j\pi[1/2 + (2k-1)/2n]} \qquad k = 1, 2, \ldots, n \qquad (4.72)$$

and k_0 is a normalizing constant. From Eq. (4.69) and (4.72) several properties of Butterworth lowpass filters are apparent. These include

1. Butterworth filters are all-pole designs [i.e., the zeros of $H(s)$ are all at $s = \infty$].
2. Butterworth filters have magnitude $1/\sqrt{2}$ at $\Omega = 1$ (i.e., the magnitude response is down 3 dB at the cutoff frequency).
3. The filter order n completely specifies the filter.

In practice the order of a Butterworth filter is generally determined by the desired attenuation at a specified frequency $\Omega_t > 1$. The filter order required to obtain a magnitude of $1/A$ at $\Omega = \Omega_t$ can be obtained from the relation

$$n = \frac{\log_{10}(A^2 - 1)}{2 \log_{10} \Omega_t)} \qquad (4.73)$$

As an example, if the filter specifications require an attenuation of 0.01 ($A = 100$) at $\Omega_t = 2$ rad/sec, then

$$n = \frac{\log_{10}(9999)}{2 \log_{10} 2} \sim \frac{4}{2(0.301)} = 6.64$$

Rounding n to the next largest integer gives a value of 7 to meet the specifications above.

EXAMPLE 1. We want to design a Butterworth filter with at least 66-dB attenuation at $\Omega = 2000\pi$ rad/sec and 3-dB attenuation at $\Omega = 1000\pi$ rad/sec.

Solution: The design specifications are $1/A = 0.0005$ (66-dB attenuation) and $\Omega_t = 2$, which gives a value of 10.97 for n which becomes $n = 11$ when rounded to the nearest integer. The s-plane pole positions of this filter are shown in Fig. 4.15. The log magnitude, phase, and group delay response for this filter are shown in Fig. 4.16.

2. Bessel Filters

Bessel lowpass filters are characterized by the property that the group delay is maximally flat at the origin of the s plane. The step response of Bessel filters exhibits extremely low overshoot, typically less than 1%, and both the impulse response and magnitude response tend toward Gaussian as the filter order is increased. It can be shown that the maximally flat group delay property of continuous Bessel filters is not generally preserved by the methods of digitization discussed in this chapter. For further explanation on this point the interested reader is referred to the work of Thiran.

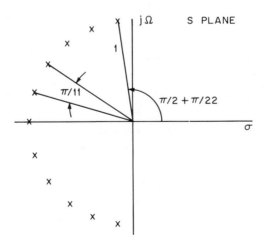

Fig. 4.15 Pole positions of an analog Butterworth lowpass filter.

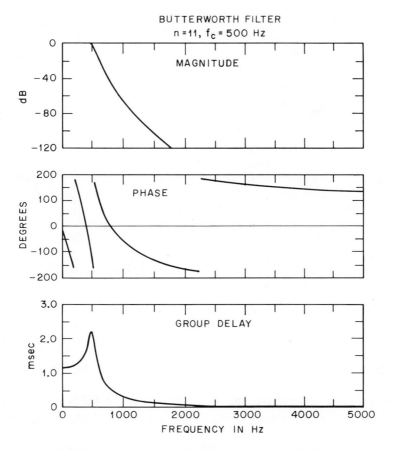

Fig. 4.16 Frequency responses (magnitude, phase, and group delay) of an analog Butterworth lowpass filter

Bessel filters have the transfer function of the form

$$H(s) = \frac{d_0}{B_n(s)} \qquad (4.74)$$

where $B_n(s)$ is the nth-order Bessel polynomial and d_0 is a normalizing constant of the form

$$d_0 = \frac{(2n)!}{2^n n!} \qquad (4.75)$$

The Bessel polynomials result from truncating a continued fraction expansion representation of the unit delay e^{-s} and satisfy the recursion relation

$$B_n(s) = (2n - 1)B_{n-1}(s) + s^2 B_{n-2}(s) \qquad (4.76)$$

with initial conditions $B_0(s) = 1$, $B_1(s) = s + 1$. $B_n(s)$ can be put in the form

$$B_n(s) = \sum_{k=0}^{n} d_k s^k \tag{4.77}$$

where

$$d_k = \frac{(2n - k)!}{2^{n-k} k! (n - k)!} \qquad k = 0, 1, \ldots, n \tag{4.78}$$

It can be shown that Bessel filters are all-pole filters with poles lying approximately on a circle centered on the positive real axis in the s plane.

One difficulty with Bessel filters is that unlike Butterworth filters the cutoff frequency Ω_c varies with the filter order. The cutoff frequency can be determined by examining the high-frequency behavior of an nth-order Bessel filter. From Eq. (4.75) and (4.78) we obtain

$$\lim_{\Omega \to \infty} |H(j\Omega)| \to \frac{d_0}{d_n \Omega^n} = \frac{d_0}{2\Omega^n} \tag{4.79}$$

To find the asymptotic cutoff frequency we find the frequency Ω_c at which $|H(j\Omega)|$ is $\frac{1}{2}$. Equation (4.79) gives

$$|H(j\Omega_c)| \to \frac{1}{2} = \frac{d_0}{2\Omega_c^n} \tag{4.80}$$

or

$$\Omega_c = d_0^{1/n} \tag{4.81}$$

To normalize Ω_c to 1 rad/sec, the roots of the filter are divided by $d_0^{1/n}$. The delay of the filter becomes $d_0^{1/n}$ instead of 1 and the magnitude response at 1 rad/sec decreases with increasing n.

Bessel filters are generally designed by specifying the order n and the cutoff frequency and by looking up the roots in a table.

By way of example, the log magnitude, phase, and group delay response of a tenth-order Bessel lowpass filter are shown in Fig. 4.17. The asymptotic cutoff frequency is 1000π rad/sec or 500 Hz.

3. Chebyshev Filters

Chebyshev filters are characterized by the property that over a prescribed band of frequencies the peak magnitude of the approximation error is minimized. The magnitude error is, in fact, equiripple over the band of frequencies—i.e., the error oscillates between maxima and minima of equal amplitude. Depending on whether the band of frequencies over which the error is minimized is the passband or the stopband, the filter designs are called Type I or Type II.

Chebyshev Type I filters are all-pole designs that exhibit equiripple passband behavior and monotonic stopband response. The squared-magnitude

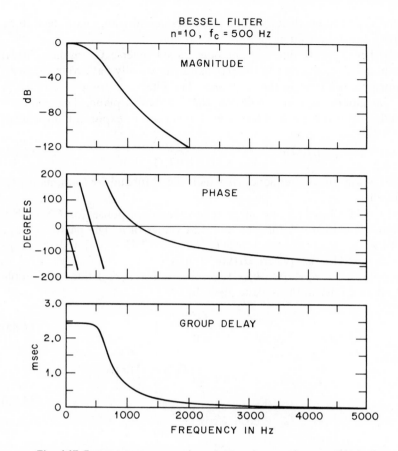

Fig. 4.17 Frequency responses (magnitude, phase, and group delay) of an analog Bessel lowpass filter.

response of an nth-order Type I filter can be expressed as

$$|H(\Omega)|^2 = \frac{1}{1 + \varepsilon^2 T_n^2(\Omega)} \tag{4.82}$$

where $T_n(\Omega)$ is the nth-order Chebyshev polynomial defined as

$$T_n(\Omega) = \begin{cases} \cos\left(n \cos^{-1} \Omega\right) & |\Omega| \le 1 \\ \cosh\left(n \cosh^{-1} \Omega\right) & |\Omega| > 1 \end{cases} \tag{4.83}$$

and ε is a parameter that is related to the passband ripple.

The optimality property that nth-order Type I Chebyshev filters satisfy is that there is no better nth-order *all-pole* filter with equal or better performance in both the passband and stopband. In other words, if a different nth-order all-pole filter has better characteristics in the passband than the

nth-order Type I filter, then its stopband characteristics are guaranteed to be worse than the Type I filter.

Chebyshev Type II filters (sometimes called inverse Chebyshev filters) exhibit monotonic behavior in the passband (maximally flat around $\Omega = 0$) and equiripple behavior in the stopband. The filters are comprised of zeros on the imaginary axis and poles in the left-half s plane. The squared-magnitude response for nth-order Type II filters can be expressed in the form

$$|H(\Omega)|^2 = \frac{1}{1 + \varepsilon^2[T_n(\Omega_r)/T_n(\Omega_r/\Omega)]^2} \tag{4.84}$$

where Ω_r is the lowest frequency at which the stopband loss attains a pre-scribed value.

Figure 4.18 illustrates the magnitude-squared response for Chebyshev Types I and II filters for both even and odd values of n. The passband edge for all these designs is at $\Omega = 1$ [where $|H(1)|^2 = 1/(1 + \varepsilon^2)$] and the stop-band edge is at $\Omega = \Omega_r$ [where $|H(\Omega)|^2 = 1/A^2$].

The poles ($s_k = \sigma_k + j\Omega_k, k = 1, 2, \ldots, n$) of a Type I filter are simple and lie on an ellipse in the s plane given by

$$\frac{\sigma_k^2}{\sinh^2 \varphi} + \frac{\Omega_k^2}{\cosh^2 \varphi} = 1 \tag{4.85}$$

where

$$\sigma_k = -\sinh \varphi \sin \left[\frac{(2k - 1)\pi}{2n} \right]$$
$$\Omega_k = \cosh \varphi \cos \left[\frac{(2k - 1)\pi}{2n} \right] \tag{4.86}$$

$$\sinh \varphi = \frac{\gamma - \gamma^{-1}}{2}$$
$$\cosh \varphi = \frac{\gamma + \gamma^{-1}}{2} \tag{4.87}$$

and

$$\gamma = \left(\frac{1 + \sqrt{1 + \varepsilon^2}}{\varepsilon} \right)^{1/n} \tag{4.88}$$

Type II filters have both poles and zeros. The zeros are imaginary and are located at

$$s_k = j \frac{\Omega_r}{\cos \{[(2k - 1)/2n]\pi\}} \qquad k = 1, 2, \ldots, n \tag{4.89}$$

(*Note:* If n is odd, then for $k = (n + 1)/2$ the zero lies at infinity.) The poles of Type II filters can be found by solving for the singularities of the denomi-nator of Eq. (4.84).

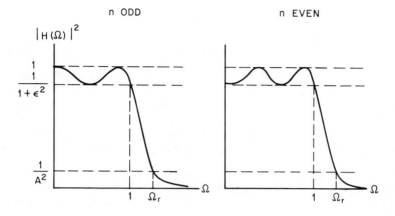

CHEBYSHEV TYPE I

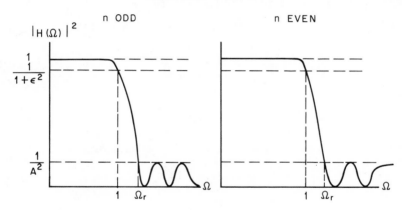

CHEBYSHEV TYPE II

Fig. 4.18 Magnitude-squared responses of analog Chebyshev filters.

With some simple manipulations, the poles $s_k = \sigma_k + j\Omega_k$, $k = 1, 2, \ldots$, can be shown to be at

$$\sigma_k = \frac{\Omega_r \alpha_k}{\alpha_k^2 + \beta_k^2}$$

$$\Omega_k = \frac{-\Omega_r \beta_k}{\alpha_k^2 + \beta_k^2}$$

(4.90)

where

$$\alpha_k = -\sinh \varphi \sin \left[\frac{(2k-1)\pi}{2n} \right]$$

$$\beta_k = \cosh \varphi \cos \left[\frac{(2k-1)\pi}{2n} \right]$$

(4.91)

and

$$\sinh \varphi = \frac{\gamma - \gamma^{-1}}{2}$$

$$\cosh \varphi = \frac{\gamma + \gamma^{-1}}{2}$$

(4.92)

and

$$\gamma = (A + \sqrt{A^2 - 1})^{1/n}$$

(4.93)

Chebyshev filters (both Types I and II) are completely specified by selecting values for any three of the following four parameters.

1. n, the filter order.
2. ε, the parameter related to passband ripple (see Fig. 4.18).
3. Ω_r, the lowest frequency at which the stopband loss attains the prescribed attenuation (see Fig. 4.18).
4. A, the parameter related to stopband loss (see Fig. 4.18).

The Chebyshev filter degree n required to achieve given values of ε, A, and Ω_r is given by

$$n = \frac{\log_{10}(g + \sqrt{g^2 - 1})}{\log_{10}(\Omega_r + \sqrt{\Omega_r^2 - 1})}$$

(4.94)

where

$$g = \sqrt{\frac{A^2 - 1}{\varepsilon^2}}$$

(4.95)

EXAMPLE 2.   Find the minimum-order Chebyshev filter required to meet the following specifications

$$\text{Passband ripple} = 2 \text{ dB}$$

$$\text{Transition ratio} = \frac{1}{\Omega_r} = 0.781$$

$$\text{Stopband loss} = 40 \text{ dB}$$

Solution: With reference to Fig. 4.18 the parameters ε, A, and ω_r can be obtained from the filter specifications as

$$-2 \text{ (dB)} = 20 \log_{10}\left(\frac{1}{\sqrt{1 + \varepsilon^2}}\right) \Rightarrow \varepsilon = 0.764$$

$$-30 \text{ (dB)} = 20 \log_{10}\left(\frac{1}{A}\right) \Rightarrow A = 31.62$$

$$\Omega_r = \frac{1}{0.781} = 1.28$$

From Eq. (4.95) the value of 41.33 is obtained for g and from Eq. (4.94) the required value $n = 6.03$ is obtained.

Figures 4.19 and 4.20 show the responses (log magnitude, phase, and group delay) of Types I and II Chebyshev filters that essentially meet the specifications of the example above. The filter cutoff frequency is $\Omega_c = 1000\pi$ or

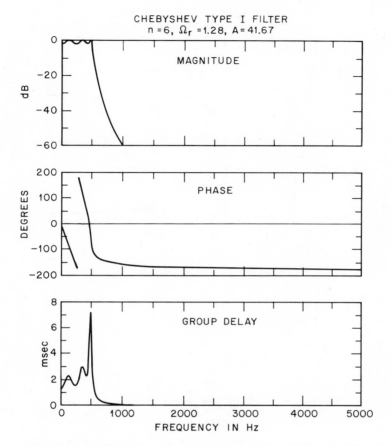

Fig. 4.19 Frequency responses (magnitude, phase, and group delay) of an analog Chebyshev Type I lowpass filter.

$f_c = 500$ Hz for both these examples. Comparison of these figures shows that group delay characteristics of Type II Chebyshev filters are generally much better than Type I filters for the in-band frequencies. This is due to the zeros of the Type II transfer function that are on the $j\Omega$-axis in the s plane, whereas the zeros of the Type I designs are all at $s = \infty$.

4. Elliptic Filters

Elliptic filters are characterized by a magnitude response that is equiripple in both the passband and the stopband. It can be shown that elliptic filters are optimum in the sense that for a given order and for given ripple specifications no other filter achieves a faster transition between the passband and stopband—i.e., has a narrower transition bandwidth. The frequency response

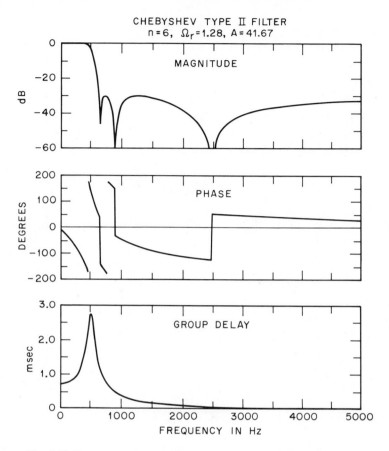

Fig. 4.20 Frequency responses (magnitude, phase, and group delay) of an analog Chebyshev Type II lowpass filter.

of a lowpass elliptic filter can be expressed as

$$|H(j\Omega)|^2 = \frac{1}{1 + \varepsilon^2 R_n^2(\Omega, L)} \tag{4.96}$$

where $R_n(\Omega, L)$ is called a Chebyshev rational function and L is a parameter describing the ripple properties of $R_n(\Omega, L)$. Figure 4.21 illustrates typical behavior of $R_5^2(\Omega, L)$ and helps to explain the behavior of elliptic filters. From this figure we see that $R_5^2(\Omega, L)$ oscillates between 0 and 1 in the passband ($-1 \le \Omega \le 1$) and beyond Ω_L it oscillates between L^2 and ∞. As the parameter L is varied, the frequency Ω_L also varies. This feature of the Chebyshev rational functions is the key to designing filters with arbitrary attenuations in both passband and stopband. In fact any three of the four parameters, filter order, in-band loss, out-of-band loss, and transition ratio

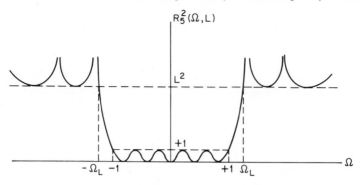

Fig. 4.21 Chebyshev rational function.

(or cutoff frequency Ω_L), can be chosen and the fourth parameter is uniquely defined.

The theory behind the determination of the function $R_n(\Omega, L)$ involves an understanding of the Jacobi elliptic function. To try to cover this highly mathematical subject in sufficient detail for the reader to truly understand the mathematics behind $R_n(\Omega, L)$ would take us too far afield. Instead, the interested reader is referred to the excellent treatment of this topic by Daniels. Before proceeding to an example of a typical elliptic filter, we first present a design relation that allows the filter designer to select the value of elliptic filter order required to meet given ripple and transition ratio specifications. Figure 4.22 shows magnitude-squared responses of typical elliptic filters with an odd and an even value of n. The ripple parameters ε and A are defined in this figure for the elliptic filter identically to the way they were

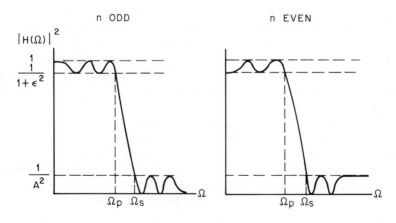

ELLIPTIC

Fig. 4.22 Magnitude-squared responses of analog elliptic filters.

defined for Chebyshev filters. The transition ratio k is defined as

$$k = \frac{\Omega_p}{\Omega_s} \qquad (4.97)$$

where Ω_p is the passband edge frequency and Ω_s is the stopband edge frequency. If the parameter k_1 is defined as

$$k_1 = \frac{\varepsilon}{\sqrt{A^2 - 1}} \qquad (4.98)$$

then the required elliptic filter order to meet given values of ε, A, Ω_p, and Ω_s is

$$n = \frac{K(k)K(\sqrt{1 - k_1^2})}{K(k_1)K(\sqrt{1 - k^2})} \qquad (4.99)$$

where $K(\cdot)$ is the complete elliptic integral of the first kind. In Sec. 4.10 a graphical procedure is given for evaluating filter orders to meet specifications for elliptic, Chebyshev, and Butterworth filters.

An example of an elliptic lowpass filter is shown in Fig. 4.23. The filter parameters are $n = 6$, $\Omega_c = 1000\pi$ ($f_c = 500$ H,z), transition ratio $= 0.781$, $A = 31.62$, and $\varepsilon = 0.01$. Figure 4.23 shows the log magnitude, phase, and group delay response for this elliptic filter.

4.10   Design Charts for Lowpass Filters

Although a set of equations has been given that completely describes the design relations for the lowpass filters being discussed, it is generally quite helpful to see the relationships among filter parameters displayed in a meaningful way. Since in general there are five filter parameters (i.e., two ripple parameters, two band edge frequencies, and the filter order), there is no simple way of presenting all parameter relationships on a single plot. There is, however, a simple and straightforward way of including all design relations for *both* analog and digital filters (obtained by the bilinear transformation) using a sequence of three charts. In this section the procedure for generating and using these design charts is presented.

Figure 4.24 shows three different sets of ripple parameters for representing passband and stopband ripple for a lowpass filter. For the magnitude response of Fig. 4.24(a), the filter magnitude oscillates between $1 + \delta_1$ and $1 - \delta_1$ in the passband ($0 \leq f \leq F_p$) and between 0 and δ_2 in the stopband ($F_s \leq f \leq 0.5$). In Fig. 4.24(b) the filter magnitude oscillates between 1 and $1 - \hat{\delta}_1$ in the passband and 0 and $\hat{\delta}_2$ in the stopband. The magnitude response of Fig. 4.24(a) corresponds to that of FIR filters, as discussed in Chapter 3, whereas the magnitude response of Fig. 4.24(b) [and Fig. 4.24(c), as will be discussed] is more typical of IIR filters obtained by the bilinear transformation.

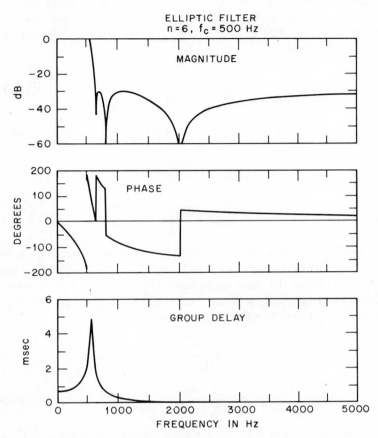

Fig. 4.23 Frequency responses (magnitude, phase, and group delay) of an analog elliptic lowpass filter.

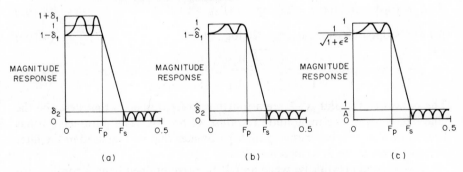

Fig. 4.24 Equivalent specifications for the magnitude response of a digital filter.

It is straightforward to relate δ_1, δ_2, $\hat{\delta}_1$, and $\hat{\delta}_2$ so the resulting magnitude characteristics of Fig. 4.24(a) and (b) are equivalent. If the magnitude characteristic of Fig. 4.24(a) is scaled by $1/(1 + \delta_1)$, then the following relationships are obtained.

$$\hat{\delta}_1 = \frac{2\delta_1}{1 + \delta_1} \tag{4.100}$$

$$\hat{\delta}_2 = \frac{\delta_2}{1 + \delta_1} \tag{4.101}$$

$$\delta_1 = \frac{\hat{\delta}_1}{2 - \hat{\delta}_1} \tag{4.102}$$

$$\delta_2 = \frac{2\hat{\delta}_2}{2 - \hat{\delta}_1} \tag{4.103}$$

Although the notation of Fig. 4.24(b) is acceptable for the magnitude response of an IIR filter, it is not the most widely used form for these filters. Figure 4.24(c) shows the same magnitude response described in terms of passband parameter ε and stopband parameter A. Comparing Fig. 4.24(b) and (c), it is easy to relate ε and A to $\hat{\delta}_1$ and $\hat{\delta}_2$ as

$$\varepsilon = \frac{\sqrt{2 - \hat{\delta}_1} \sqrt{\hat{\delta}_1}}{(1 - \hat{\delta}_1)} \tag{4.104a}$$

$$A = \frac{1}{\hat{\delta}_2} \tag{4.104b}$$

At this point it is convenient to define the additional filter terms E, ATT, and η as

$$E = \text{(in-band) ripple} = 20 \log_{10} \sqrt{(1 + \varepsilon^2)} \tag{4.105}$$

$$ATT = \text{stopband attenuation} = 20 \log_{10} A \tag{4.106}$$

$$\eta = \frac{\varepsilon}{\sqrt{A^2 - 1}} = \frac{\sqrt{\hat{\delta}_1} \sqrt{2 - \hat{\delta}_1}\, \hat{\delta}_2}{\sqrt{1 - \hat{\delta}_2^2}\,(1 - \hat{\delta}_1)} = \frac{2\sqrt{\delta_1}\, \delta_2}{(1 - \delta_1)\sqrt{(1 + \delta_1)^2 - \delta_2^2}} \tag{4.107}$$

The parameters E and ATT are a fourth set of parameters that describe the characteristics of the magnitude response of an IIR. The parameter η has been shown to be a basic analog filter parameter that will be used in the filter design curves to be described next.

As discussed previously, when an IIR filter is obtained from a continuous-time prototype filter via the bilinear transformation, the respective frequency scales are related by a simple frequency warping. Thus the quantity k, the lowpass filter transition ratio, may be defined as

$$k = \frac{\Omega_p}{\Omega_s} \qquad \text{continuous filters} \qquad (4.108)$$

$$k = \frac{\tan(\omega_p/2)}{\tan(\omega_s/2)} \qquad \text{digital filters} \qquad (4.109)$$

where Eq. (4.109) expresses the warped (in frequency) transition ratio.

In order to be able to relate required filter order n to ripple parameters (δ_1, δ_2) or $(\hat{\delta}_1, \hat{\delta}_2)$ or (ε, A) and band edge frequencies (Ω_p, Ω_s) or (ω_p, ω_s), the relation for n for each filter type must be used directly. To review these relations, we have already shown that the appropriate equations are

$$n = \frac{K(k)K(\sqrt{1 - k_1^2})}{K(k_1)K(\sqrt{1 - k^2})} \qquad \text{elliptic filters} \qquad (4.110)$$

$$n = \frac{\cosh^{-1}(1/k_1)}{\ln\left(\dfrac{1 + \sqrt{1 - k^2}}{k}\right)} \qquad \text{Chebyshev filters} \qquad (4.111)$$

and

$$n = \frac{\ln k_1}{\ln k} \qquad \text{Butterworth filters} \qquad (4.112)$$

where $k_1 = \eta$ of Eq. (4.107). We now give a simple and straightforward way of graphically portraying the design relations for *both* digital and analog filters for each of the filter classes above, using a sequence of three charts.

The first chart(s) relates the filter design parameter η to the passband and stopband ripple specifications δ_1 and δ_2 or their equivalents. The second chart(s) graphs the filter design equation relating filter order n, design parameter η, and transition ratio k. The third chart(s) relates transition ratio k to passband cutoff frequency F_p and transition bandwidth ν.

Figures 4.25(a) to (d) show four possibilities for chart 1. The graphs of Fig. 4.25(a) and (b) correspond to filters with δ_1 as a parameter [Fig. 4.25(a)] or $20 \log_{10}(1 + \delta_1)$ (dB) as a parameter [Fig. 4.25(b)]. The graphs of Fig. 4.25(c) and (d) correspond to filters with absolute ripple $\hat{\delta}_1$ as a parameter [Fig. 4.25(c)] or total ripple $20 \log_{10}[1/(1 + \hat{\delta}_1)]$ (dB) as a parameter [Fig. 4.25(d)].

Chart 2 in Fig. 4.26(a) to (c) represents the design relations particular to the prototype filters—i e., Eq. (4.110) for elliptic filters, Eq. (4.111) for Chebyshev filters, and Eq. (4.112) for Butterworth filters. For these plots the parameter η is plotted versus transition ratio k with filter order n as the parameter. Figures 4.26(a) to (c) show the resulting plots for elliptic filters, Chebyshev filters, and Butterworth filters. The horizontal scale on each of

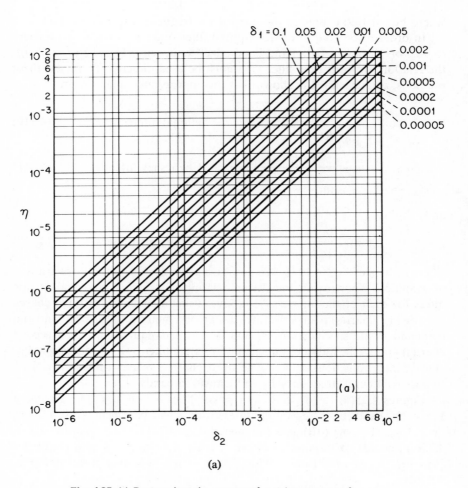

(a)

Fig. 4.25 (a) Design chart 1—η versus δ_2 with parameter δ_1.

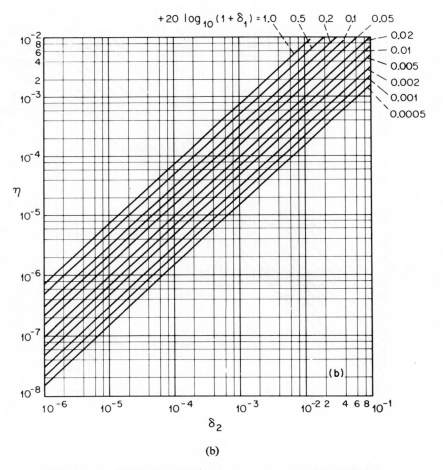

Fig. 4.25 (cont.) (b) Design chart 1—η versus δ_2 with parameter $20 \log_{10}(1 + \delta_1)$.

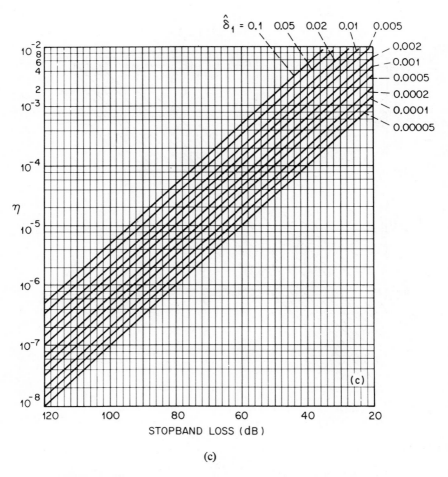

Fig. 4.25 (cont.) (c) Design chart 1—η versus stopband loss (dB) with parameter $\hat{\delta}_1$.

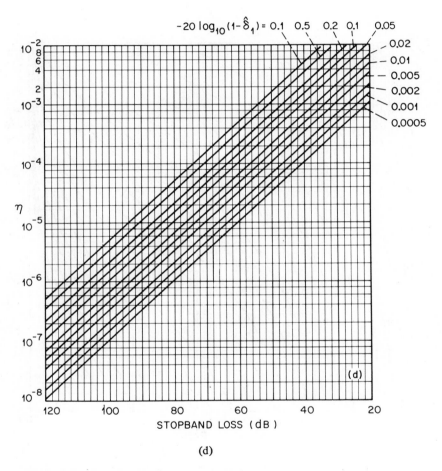

Fig. 4.25 (cont.) (d) Design chart $1-\eta$ versus stopband loss (dB) with parameter $20 \log_{10}(1 - \mathring{\delta}_1)$.

245

ELLIPTIC FILTERS

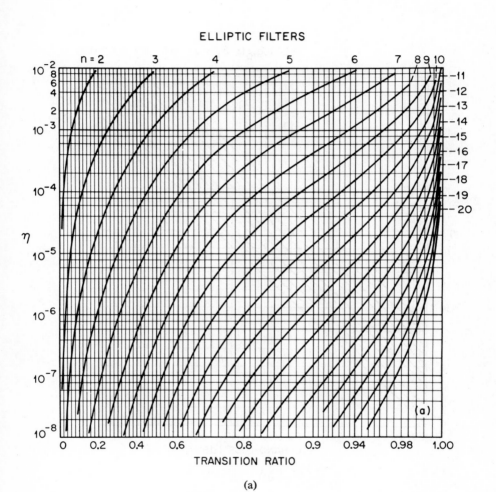

(a)

Fig. 4.26 (a) Design chart 2—η versus transition ratio for elliptic filters with parameter n (filter order).

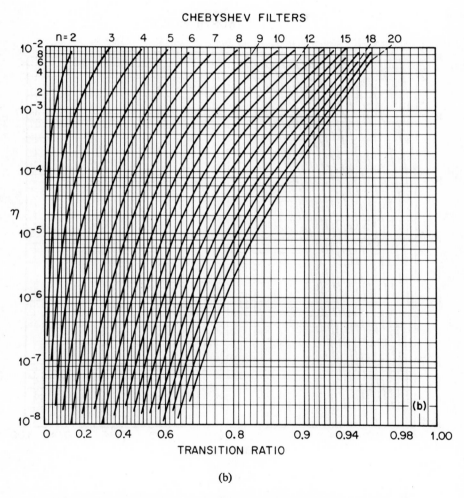

CHEBYSHEV FILTERS

(b)

Fig. 4.26 (cont.) (b) Design chart 2—η versus transition ratio for Chebyshev filters with parameter n (filter order).

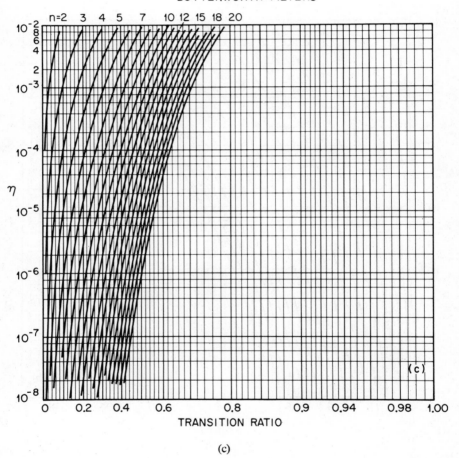

BUTTERWORTH FILTERS

Fig. 4.26 (cont.) (c) Design chart 2—η versus transition ratio for Butterworth filters with parameter *n* (filter order).

these plots is a nonuniform scale that was used to provide a reasonably good spacing of the curves for the various values of n. The actual nonlinear scale used is represented by the equation

$$x = \frac{k + k^8}{2} \tag{4.113}$$

where x is the x-axis coordinate $(0 \le x \le 1)$ and k is the transition width. Thus, the scale is linear for small values of k and highly nonlinear near $k = 1.0$.

Chart 3 in Fig. 4.27(a) and (b) represents the relation between the transition ratio and the filter cutoff frequencies [Eq. (4.108) and (4.109)]. For these graphs the passband cutoff frequency F_p is plotted versus transition ratio k

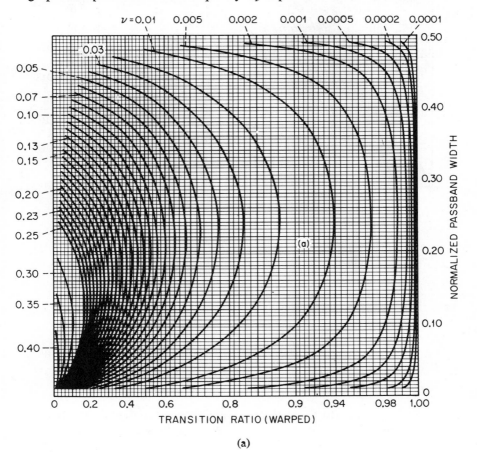

(a)

Fig. 4.27 (a) Design chart 3—normalized passband width versus transition ratio for digital filters.

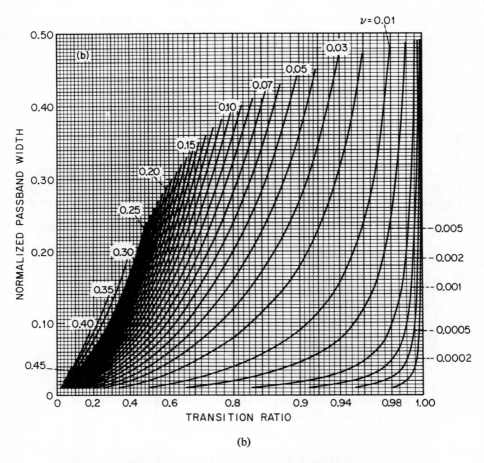

(b)

Fig. 4.27 (cont.) (b) Design chart 3—normalized passband width versus transition ratio for continuous filters.

for various values of normalized transition width v, defined as

$$v = F_s - F_p = \frac{\omega_s - \omega_p}{2\pi} \qquad \text{digital} \qquad (4.114)$$

or

$$v = \Omega_s - \Omega_p \qquad \text{continuous}$$

Figures 4.27(a) and (b) show the resulting graphs for digital and analog filters. The scale for transition ratio is identical to the scale used for chart 2 in Fig. 4.26.

Use of Charts. To illustrate how to use the set of charts of Figs. 4.25 to 4.27, consider the determination of digital elliptic filter order n required to meet the following specifications.

$$\delta_1 = 0.01 \ (\approx \pm 0.086\text{-dB passband ripple})$$

$$\delta_2 = 0.0001 \ (80\text{-dB stopband loss})$$

Passband cutoff frequency = 480 Hz
Stopband edge frequency = 520 Hz
Sampling frequency = 8000 Hz

Normalizing the band edge frequencies gives

$$F_p = \tfrac{480}{8000} = 0.06$$

$$F_s = \tfrac{520}{8000} = 0.065$$

For the determination of filter order n for a digital filter of the elliptic type, the charts of Fig. 4.25(a), 4.26(a), and 4.27(a) are used [Fig. 4.26(a) specializes the design to the elliptic type]. To obtain the value of η on Fig. 4.25(a) we use the curve $\delta_1 = 0.01$ and find its intersection with the line $\delta_2 = 0.0001$, which yields a value of η approximately equal to 2×10^{-5}. To obtain the transition ratio we use Fig. 4.27(a) by finding the intersection of the curve $v = F_s - F_p = 0.005$ with line $F_p = 0.06$; this yields a value of 0.923 for the transition ratio (this agrees well with $F_p/F_s = 0.06/0.065 = 0.923$, an alternate way of arriving at the same result). Finally the filter order n can now be determined from Fig. 4.26(a) by finding the intersection of the lines $\eta = 2 \times 10^{-5}$ and transition ratio = 0.923; thus the required theoretical elliptic filter order is approximately equal to 11.5. In order to meet specifications on all four parameters, a twelfth-order filter must be used.

There are several tradeoffs possible, however, for the final filter specifications. For example, if η is held fixed at 2×10^{-5} and transition ratio is changed to approximately 0.94 to lie on the ($n = 12$) curve, then either F_s or F_p can be varied to match this new value of transition ratio. The tradeoffs here are obtained from Fig. 4.27(a). If the transition ratio is held fixed, then for $n = 12$ we find $\eta \approx 1.0 \times 10^{-5}$; from chart 1 [Fig. 4.25(a)] we can observe the tradeoff as δ_1 and δ_2 are varied for this new value of η. Finally both transition ratio and η can be varied, e.g., to 0.93 for transition ratio and 1.5×10^{-5} for η, so as to make their intersection remain on the ($n = 12$) curve; now all four filter parameters can be varied to match the new values of η and transition ratio.

It is interesting to note that if a Chebyshev or Butterworth filter type would have been specified in place of the elliptic, the designer need only substitute Fig. 4.26(b) or (c) for (a) as chart 2 and proceed as before. In both cases for the example given the required filter order considerably exceeds the maximum limit of 20 of the curves; thus the "efficiency" of the elliptic design is definitely seen.

Clearly this design procedure presents a tremendous amount of flexibility to the designer—more so than is generally available in most programs for filter design. *Furthermore, the insight into the design problem afforded by this graphical technique allows the designer to get a feeling for the way in which small changes in filter specification affect the required filter order.* Quite often the designer is willing to change his ideas on "required specifications, especially if it reduces the filter order necessary to meet his specifications.

4.11   Comparisons Between Impulse Invariant and Bilinear Transformation for Elliptic Filters

The next set of figures illustrate several of the issues involved in using either impulse invariant transformation or bilinear transformation to digitize continuous filters. Figure 4.28 shows the log magnitude response of a continuous elliptic lowpass filter with parameters $n = 6$, $k =$ transition ratio $= 0.9$, $A = 1000$ (60-dB stopband attenuation) and $\varepsilon \cong 4.29$. The lowpass cutoff frequency is 1000 Hz. The second plot in this figure shows the log magnitude response of a digital filter (sampling frequency is 10,000 Hz in all examples) obtained by impulse invariant transformation of the continuous filter. The effects of aliasing in the region above 1500 Hz are evident. The equiripple character of the stopband response has been destroyed by the aliasing; further, the peak stopband attenuation has decreased from 60 to about 55 dB. The third plot shows the log magnitude response of a digital filter obtained by first prewarping the critical frequencies of the analog design and then applying the bilinear transformation. Although the exact locations of the maxima and minima have moved slightly with respect to the analog filter locations, the log magnitude response is still equiripple.

The next group of figures show similar effects for an elliptic bandstop filter and an elliptic highpass filter. Figure 4.29 shows the log magnitude responses of an analog elliptic bandstop filter and the two digital filters obtained by impulse invariant transformation and bilinear transformation of the analog filter. Since the original analog filter was *not* bandlimited, the aliasing in the digital filter obtained by impulse invariant transformation renders it entirely useless as a digital bandstop filter; whereas the bilinearly transformed filter is essentially identical to the original analog filter. Figure 4.30 shows a similar set of results for a highpass filter. Again the digital filter obtained by impulse invariant transformation is unacceptable since the original analog filter was not bandlimited.

The last group of plots of Fig. 4.31 shows an interesting consequence of using the bilinear transformation in digitizing filters such as Butterworth filters. The top plot shows the log magnitude response of a lowpass Butterworth filter $(n = 6, f_c = 3500$ Hz$)$. At 5000 Hz the filter response has

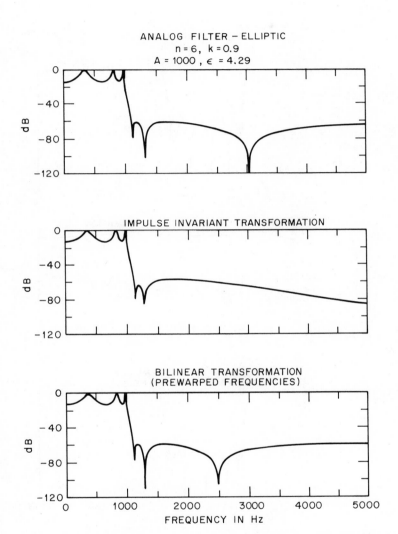

Fig. 4.28 Comparison of impulse invariant and bilinear transformation for an elliptic lowpass filter.

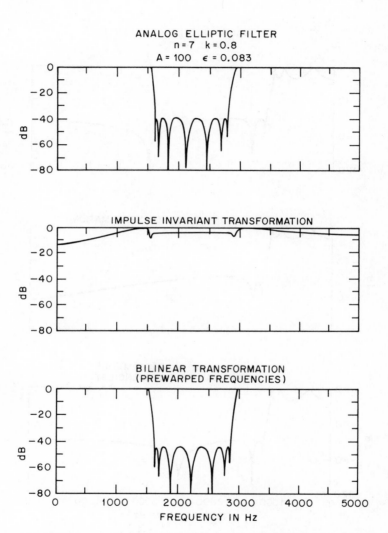

Fig. 4.29 Comparison of impulse invariant transformation and bilinear transformation for an elliptic bandstop filter.

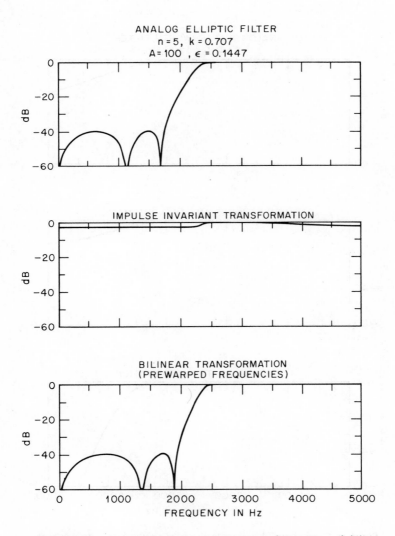

Fig. 4.30 Comparison of impulse invariant transformation and bilinear transformation for an elliptic highpass filter.

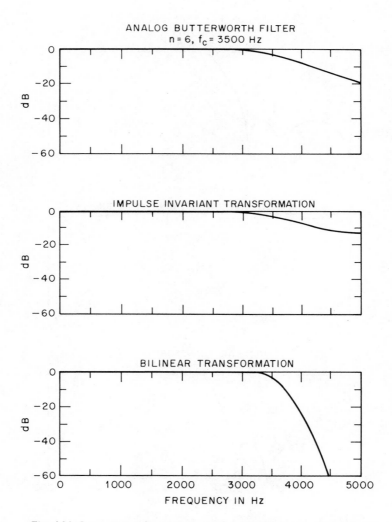

Fig. 4.31 Comparison of impulse invariant and bilinear transformation for a wideband Butterworth lowpass filter.

fallen to about -20 dB. The digital filter obtained by impulse invariant transformation (the second plot) shows the effects of aliasing in that its response at 5000 Hz falls only to about -12 dB. The digital filter obtained by bilinear transformation, however, has a response that has fallen to -60 dB by 4500 Hz. In other words, the bilinearly transformed Butterworth filter is far superior to its analog counterpart. This is of course due to the nonlinear warping of the frequency scale, which, in this case, has greatly enhanced the resulting design.

It should be noted that the preceding set of examples concentrated only on comparing the log magnitude responses of the original analog filter with digital filters obtained by impulse invariant and bilinear transformation. If the filter designer is interested in the resulting group delay response (or time response) of the digital filter, then additional considerations must be weighed. It can be shown that the bilinear transformation will generally not preserve the time response or group delay response of an analog filter. Thus, for example, the bilinear transformation of an analog Bessel filter does *not* preserve the maximally flat group delay property of these filters. Furthermore since the amplitude response of Bessel filters is *not* essentially constant in the passband, the bilinear transform will distort the amplitude as well as the group delay response. In most cases, when the designer is primarily interested in preserving the temporal characteristics of the filter, the impulse invariant transformation is used. In almost all other cases, the bilinear transformation is used.

4.12 Frequency Transformations

The preceding sections have discussed techniques for designing continuous-time lowpass filters and methods of digitizing these designs. There are two distinct approaches to the design of digital filters of the bandpass, highpass, and bandstop types. These approaches are summarized in Fig. 4.32. These

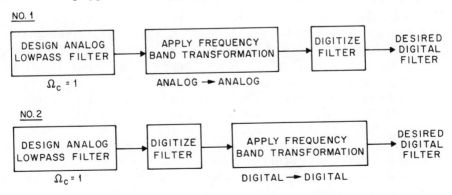

Fig. 4.32 Frequency transformations for lowpass filters.

two approaches differ in that technique 1 transforms a normalized analog lowpass filter to another analog filter that can be digitized to give the desired digital filter, whereas technique 2 digitizes the normalized lowpass filter immediately and then applies a digital frequency band transformation to give the desired digital filter. Since we have already discussed both how to design the normalized lowpass filter and how to digitize it, in this section we shall consider appropriate frequency band transformations for the continuous and discrete cases.

1. Continuous Frequency Band Transformations

There are a variety of techniques for transforming a 1-rad/sec cutoff lowpass filter to another lowpass filter (with different cutoff frequency) or a bandpass, highpass, or bandstop filter. A particularly simple set of transformations is listed below.

$$s \to \frac{s}{\Omega_u} \qquad \text{lowpass} \to \text{lowpass} \qquad (4.115)$$

$$s \to \frac{\Omega_u}{s} \qquad \text{lowpass} \to \text{highpass} \qquad (4.116)$$

$$s \to \frac{s^2 + \Omega_l \Omega_u}{s(\Omega_u - \Omega_l)} \qquad \text{lowpass} \to \text{bandpass} \qquad (4.117)$$

$$s \to \frac{s(\Omega_u - \Omega_l)}{s^2 + \Omega_u \Omega_l} \qquad \text{lowpass} \to \text{bandstop} \qquad (4.118)$$

where

$$\Omega_l = \text{lower cutoff frequency}$$

$$\Omega_u = \text{upper cutoff frequency}$$

The examples in Fig. 4.33 show how a normalized lowpass filter (an elliptic filter) is converted to a lowpass, highpass, bandpass, and bandstop filter using the transformations of Eq. (4.115 to (4.118). It is seen from these equations that the band transformations are highly nonlinear. For transforming the standard lowpass filters discussed earlier (except the Bessel filter), however, this nonlinearity provides no difficulty as the frequency response of the filters being transformed is an approximation to a piecewise constant characteristic in the frequency bands of interest. Thus, for example, the nonlinearity of the mapping affects the spacing of the ripple peaks and valleys for elliptic designs but has no affect on the amplitudes of these ripples. Therefore the filters designed by this transformation preserve the equiripple character of the prototype filter.

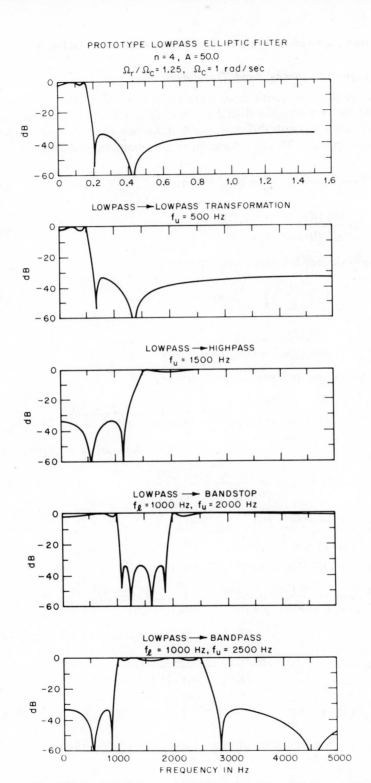

Fig. 4.33 Continuous frequency band transformations of a lowpass filter.

2. Digital Frequency Band Transformations

Similar to the continuous case, there exists a set of simple transformations suitable for converting a digital lowpass filter (of cutoff frequency ω_c) to another lowpass (with different cutoff frequency ω_u), highpass, bandpass, or bandstop filter. The transformations are summarized below.

1. $z^{-1} \rightarrow \dfrac{z^{-1} - \alpha}{1 - \alpha z^{-1}}$ lowpass $\rightarrow$ lowpass   (4.119)

$$\alpha = \frac{\sin\{[(\omega_c - \omega_u)/2]T\}}{\sin\{[(\omega_c + \omega_u)/2]T\}} \tag{4.120}$$

$\omega_u =$ desired lowpass cutoff frequency

2. $z^{-1} \rightarrow -\left(\dfrac{z^{-1} + \alpha}{1 + \alpha z^{-1}}\right)$ lowpass $\rightarrow$ highpass   (4.121)

$$\alpha = -\frac{\cos\{[(\omega_c - \omega_u)/2]T\}}{\cos\{[(\omega_c + \omega_u)/2]T\}} \tag{4.122}$$

$\omega_u =$ desired highpass cutoff frequency

3. $z^{-1} \rightarrow -\dfrac{\{z^{-2} - [2\alpha k/(k+1)]z^{-1} + (k-1)/(k+1)\}}{\{(k-1)/(k+1)z^{-2} - [2\alpha k/(k+1)]z^{-1} + 1\}}$

$$\text{lowpass} \rightarrow \text{bandpass} \tag{4.123}$$

$$\alpha = \cos(\omega_0 T) = \frac{\cos\{[(\omega_u + \omega_l)/2]T\}}{\cos\{[(\omega_u - \omega_l)/2]T\}} \tag{4.124}$$

$$k = \cot\left[\left(\frac{\omega_u - \omega_l}{2}\right)T\right]\tan\left(\frac{\omega_c T}{2}\right) \tag{4.125}$$

$\omega_0 =$ center frequency of bandpass filter

4. $z^{-1} \rightarrow \dfrac{\{z^{-2} - [2\alpha/(1+k)]z^{-1} + (1-k)/(1+k)\}}{\{[(1-k)/(1+k)]z^{-2} - [2\alpha/(1+k)]z^{-1} + 1\}}$

$$\text{lowpass} \rightarrow \text{bandstop} \tag{4.126}$$

$$\alpha = \cos(\omega_0 T) = \frac{\cos\{[(\omega_u + \omega_l)/2]T\}}{\cos\{[(\omega_u - \omega_l)/2]T\}} \tag{4.127}$$

$$k = \tan\left(\frac{\omega_u - \omega_l}{2}T\right)\tan\left(\frac{\omega_c T}{2}\right) \tag{4.128}$$

$\omega_0 =$ center frequency of reject band of bandstop filter

The transformations of Eq. (4.119) to (4.128) are readily seen to be of the all-pass type—i.e., the unit circle is mapped into itself one or more times. Thus although the frequency scale is warped by the mapping, the magnitude response of the original lowpass filter is preserved. Thus, for example, an elliptic lowpass filter is transformed into new elliptic filters.

Several of the properties of the transformations are illustrated in the set of examples shown in Fig. 4.34. In this case we begin with a Chebyshev low-pass prototype filter (cutoff frequency ω_c radians per second) and transform it to a new lowpass filter (cutoff frequency ω_u), a highpass filter, a bandpass filter, and a bandstop filter. Certain observations can be made from this figure. First, each one of the transformed filters is essentially obtained by first transforming the prototype lowpass filter to a suitable lowpass design and then applying a simple first- or second-order substitution for z^{-1} (i.e., $z^{-1} \rightarrow -z^{-1}$ for the highpass filter case). Second, the nature of the mapping can be studied in terms of what frequencies the critical frequencies of the low-pass prototype are mapped to. These transformed frequencies are given in Table 4.1.

Table 4.1
TRANSFORMED FREQUENCIES FOR DIGITAL TRANSFORMATIONS

		Frequency of Prototype Filter			
		0	$-\omega_c$	$+\omega_c$	$\omega_s/2$
	Lowpass	0	$-\omega_u$	$+\omega_u$	$\omega_s/2$
Frequency of Transformed Filter	*Highpass*	$\omega_s/2$	ω_u	$-\omega_u$	0
	Bandpass	$\pm\omega_0$	$\pm\omega_l$	$\pm\omega_u$	0 $\omega_s/2$
	Bandstop	0 $\omega_s/2$	$\pm\omega_u$	$\pm\omega_l$	$\pm\omega_0$

These digital frequency band transformations are easy to use since they map a rational transfer function into a new rational transfer function. Since they introduce no distortion errors in the amplitude scale, they provide an alternative to the continuous transformations as a means of obtaining trans-formed filters from a lowpass prototype. In fact it has been found that the bilinear transformation of a normalized analog lowpass filter followed by a digital all-pass transformation is equivalent to a continuous frequency transformation followed by the bilinear transformation. Figure 4.35 illus-trates this result in the case of the design of a lowpass filter. Figure 4.35(a) shows a continuous normalized lowpass filter which is bilinearly transformed to give a "normalized" digital lowpass filter (i.e., the passband cutoff frequency is $\hat{\omega}_p = \pi/2$) which is then transformed using a digital lowpass-to-lowpass

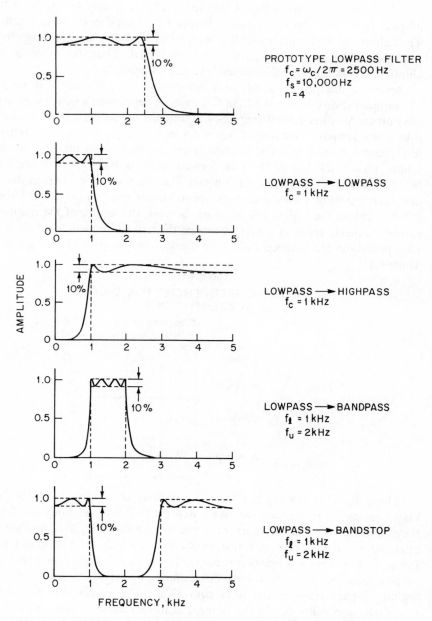

Fig. 4.34 Digital frequency transformations of a lowpass filter. (After Constantinides.)

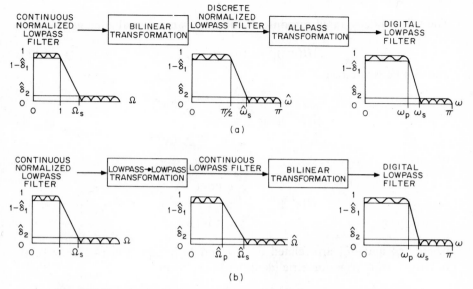

Fig. 4.35 Two methods of obtaining a digital filter from an analog filter.

transformation to give the desired digital filter. Figure 4.35(b) shows an equivalent procedure in which the normalized lowpass filter is first frequency transformed to give an unnormalized lowpass filter that is then bilinearly transformed to give the desired digital filter. There appears to be little numerical advantage of either of these techniques over the other.

4.13 Direct Design of Digital Filters

In the preceding sections we have discussed techniques for digitizing continuous filters. A second possible method for designing infinite impulse response digital filters is direct digital design in either the frequency or the time domains. Among the techniques that fall into the category of direct digital design are magnitude-squared function design and time domain methods. In the following we shall briefly describe these techniques and discuss their applicability.

1. Magnitude-Squared Function Design

Let $H(z)$ be the z transform of an arbitrary realizable IIR filter. $H(z)$ is of the form

$$H(z) = \frac{\sum\limits_{i=0}^{m-1} b_i z^{-i}}{\sum\limits_{i=0}^{n-1} a_i z^{-i}} \tag{4.129}$$

The magnitude-squared response of the filter is easily evaluated as

$$|H(e^{j\omega})|^2 = H(z)H(z^{-1})\big|_{z=e^{j\omega}} \tag{4.130}$$

and can be written as

$$|H(e^{j\omega})|^2 = \frac{\displaystyle\sum_{i=0}^{m-1} b_i e^{-j\omega i} \sum_{k=0}^{m-1} b_k e^{j\omega k}}{\displaystyle\sum_{i=0}^{n-1} a_i e^{-j\omega i} \sum_{k=0}^{n-1} a_k e^{j\omega k}} \tag{4.131}$$

or

$$|H(e^{j\omega})|^2 = \frac{\displaystyle\sum_{i=0}^{m-1} c_i \cos(\omega i)}{\displaystyle\sum_{i=0}^{n-1} d_i \cos(\omega i)} \tag{4.132}$$

where $\{c_i\}$ and $\{d_i\}$ are related to $\{b_i\}$ and $\{a_i\}$. Equation (4.132) is often rewritten using trigonometric identities in the form

$$|H(e^{j\omega})|^2 = \frac{\displaystyle\sum_{i=0}^{m-1} e_i \cos^2(\omega i/2)}{\displaystyle\sum_{i=0}^{n-1} f_i \cos^2(\omega i/2)} \tag{4.133}$$

showing that the magnitude-squared function can always be expressed as the ratio of two trigonometric functions of ω.

Equation (4.133) forms the basis of many methods for the synthesis of digital filters with prescribed magnitude-squared characteristics. Equation (4.133) also relates to analog filters whose magnitude-squared function is a ratio of polynomials in Ω^2. Using the substitution (for example)

$$\Omega = \cos\left(\frac{\omega}{2}\right) \tag{4.134}$$

Eq. (4.133) assumes a form appropriate to analog filter transfer functions.

Equation (4.133) can be simplified to the form

$$|H(e^{j\omega})|^2 = \frac{1}{1 + A_n^2(\omega)} \tag{4.135}$$

where $A_n^2(\omega)$ is an nth-order rational trigonometric polynomial. By suitable choice of the function $A_n^2(\omega)$, various types of digital filters can be designed to match prescribed magnitude characteristics. For the Butterworth case, the form of $A_n^2(\omega)$ is

$$A_n^2(\omega) = \frac{\tan^{2n}(\omega/2)}{\tan^{2n}(\omega_c/2)} \tag{4.136}$$

where ω_c is the lowpass cutoff frequency. For the Chebyshev case the form for $A_n^2(\omega)$ is

$$A_n^2(\omega) = \varepsilon^2 T_n^2 \left[\frac{\tan (\omega/2)}{\tan (\omega_c/2)} \right] \qquad (4.137)$$

where $T_n(x)$ is the Chebyshev polynomial of nth order and ε is a ripple parameter. It can be shown that the digital Butterworth and Chebyshev filters obtained by the magnitude-squared function are simply related to the bilinear transformation of continuous Butterworth and Chebyshev filters. Thus no further discussion of these filter classes will be given here.

IIR filter design by specification of the magnitude-squared function is readily extendable to several other classes of filters and need not be restricted to lowpass filters. The difficulties with this technique are twofold. First, a suitable rational trigonometric polynomial $A_n^2(\omega)$ must be found to provide the desired filtering. Second, the magnitude-squared function $|H(e^{j\omega})|^2|_{z=e^{j\omega T}}$ must be factored to find the poles and zeros. This factorization is generally nontrivial and therefore makes this an undesirable filter design method.

2. Time Domain Design of IIR Filters

Just as it is possible to design filters to approximate an arbitrary frequency response, it is also possible to design an IIR filter whose impulse response approximates a desired impulse response. Assume the z transform of the filter is of the form

$$H(z) = \frac{\sum\limits_{i=0}^{m-1} b_i z^{-i}}{\sum\limits_{i=0}^{n-1} a_i z^{-i}} = \sum\limits_{k=0}^{\infty} h(k) z^{-k} \qquad (4.138)$$

and it is desired that the filter impulse response $h(k)$ approximate a desired response $g(k)$, over the range $0 \leq k \leq P - 1$. Under a wide variety of conditions, Burrus and Parks, and Brophy and Salazar, among others, have shown that it is possible to find a set of a_i, b_i such that

$$\langle \varepsilon \rangle = \sum\limits_{k=0}^{P-1} [g(k) - h(k)]^2 w(k) \qquad (4.139)$$

is minimized over all possible choices of a_i, b_i [where $w(k)$ is a positive weighting function on the error sequence]. Since $h(k)$ is a nonlinear function of the filter parameters ($\{a_i\}$, $\{b_i\}$), generally the minimization of ε can only be achieved using iterative techniques. In the case when $P = n + m - 1$, the filter parameters to minimize ε can be obtained by solving $(n + m)$ linear equations. To see how this is done we write the filter response directly (assuming $a_0 = b_0 = 1$) as

$$g(k) = -a_1 h(k-1) - a_2 h(k-2) - \ldots - a_n h(k-n) + b_k \qquad 1 < k \leq m$$

$$\text{(4.140)}$$

$$g(k) = -a_1 h(k-1) - a_2 h(k-2) - \ldots - a_n h(k-n) \qquad k > m \quad \text{(4.141)}$$

Letting $g(k) = h(k)$ for $k = 1, 2, \ldots, m$, Eq. (4.141) can be solved for the set of a's that give $g(k) = h(k)$ for $k = m+1, m+2, \ldots, m+n$. Given the a's, Eq. (4.140) can be solved for the b's that give $g(k) = h(k)$, $k = 1, 2, \ldots, m$. This procedure amounts to equating the first $(n+m+1)$ terms in a series expansion of $H(z)$ of Eq. (4.138) to the truncated z transform of the desired impulse response $g(k)$ after $(m+n)$ terms. This method of approximation of a power series by a rational function is often referred to as the Pade approximant procedure. By approximating the desired digital filter response by reproducing the first $(m+n+1)$ samples, it is *hoped* that the overall time or frequency response from the approximating filter will not deviate too much from the desired one. In practice there is no simple way of even approximately bounding the deviation on either of these responses. Rather than discuss these issues further we now give some practical examples, due to Brophy and Salazar, of IIR filters designed using this technique.

Figures 4.36 and 4.37 show two filters which have application to data transmission systems which were designed using the Pade approximant procedure. Figure 4.36 shows the magnitude response of a desired bandpass shape (curve A). The sampling frequency for this and the following examples is 7200 Hz. The design specifications for this filter included 3-dB loss at 200 and 3200 Hz, less than 0.25-dB peak-to-peak passband ripple, linear phase in the passband, and greater than 12-dB/octave slopes in the stopbands. Curve B in this figure shows the magnitude response of a twenty-fourth-order filter obtained using a Pade approximation. The largest absolute error in the samples of the filter impulse response was 0.0018. The resulting filter phase response is shown at the bottom of Fig. 4.36.

Figure 4.37 shows similar comparisons for a tenth-order filter designed to approximate a bandpass function with a specified cosine roll-off in frequency (curve A). Curve B shows the resulting magnitude approximation; the resulting phase response is shown at the bottom of the figure.

It should be noted that since the Pade approximation method only involves time domain considerations, the resulting frequency domain approximation to filter stopband responses where as little as 40-dB stopband loss is desired is generally unacceptable. The filter coefficients obtained by this technique, however, may often be suitable as an initial guess to more sophisticated optimization algorithms for designing IIR filters in terms of frequency response specifications. We discuss such techniques in the following sections.

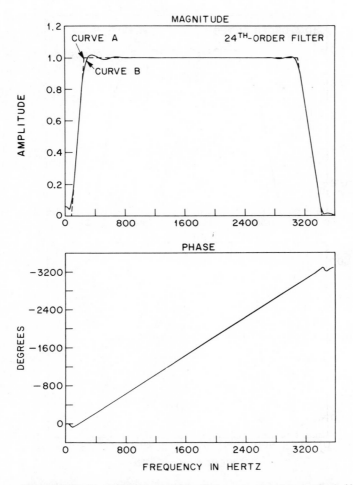

Fig. 4.36 Bandpass filter designed by Pade approximation method. (After Brophy and Salazar.)

4.14 Optimization Methods for Designing IIR Filters

The final class of design methods for IIR filters to be studied in this chapter can be classified as optimization methods. The salient characteristic of these methods is that the set of design equations cannot be solved explicitly to give the filter coefficients. Instead a mathematical optimization procedure is used to determine the filter coefficients that minimize some error criterion, subject to the appropriate design equations. Using this iterative procedure, either

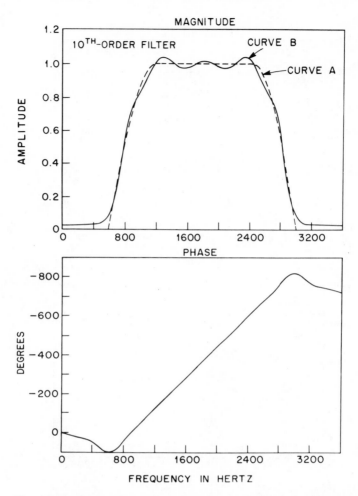

Fig. 4.37 Bandpass filter designed by Pade approximation method. (After Brophy and Salazar.)

the error eventually reaches a minimum value or a specified maximum number of iterations is performed and the procedure terminates. In this section several such optimization design procedures will be discussed.

1. Minimum Mean Squared Error Design

Assume the IIR filter's z transform is of the form

$$H(z) = A \prod_{k=1}^{K} \frac{1 + a_k z^{-1} + b_k z^{-2}}{1 + c_k z^{-1} + d_k z^{-2}} \tag{4.142}$$

i.e., the filter is realized as a cascade of second-order sections. Let the desired magnitude response of the filter be denoted as $H_d(e^{j\omega})$. Let $\{\omega_i, i = 1, 2, \ldots, M\}$ be the discrete set of frequencies (not necessarily uniformly spaced) at which the error between the actual response and desired response is evaluated. Then, following the treatment of Steiglitz, the squared error in frequency (as a function of the filter parameters) can be expressed as

$$Q(\underline{\theta}) = \sum_{i=1}^{M} [|H(e^{j\omega_i})| - |H_d(e^{j\omega_i})|]^2 \qquad (4.143)$$

where $\underline{\theta}$ is the vector $[(4K + 1)$ components$]$ of unknown coefficients

$$\underline{\theta} = (a_1, b_1, c_1, d_1, \ldots, a_K, b_K, c_K, d_K, A) \qquad (4.144)$$

To minimize the squared error [Eq. (4.143)] implies finding the optimum value of $\underline{\theta}$, say $\underline{\theta}^*$, such that

$$Q(\underline{\theta}^*) \le Q(\underline{\theta}) \qquad \text{for } \underline{\theta} \ne \underline{\theta}^* \qquad (4.145)$$

To find $\underline{\theta}^*$, i.e., to minimize Eq. (4.143), requires use of a nonlinear optimization procedure such as the Fletcher–Powell algorithm. To use this algorithm, the gradient of the function being minimized must be known. Before showing how to calculate the gradient it is expedient to eliminate the gain coefficient A from the calculations, as it may be analytically determined. Letting

$$\underline{\varphi} = (a_1, b_1, c_1, d_1, \ldots, a_K, b_K, c_K, d_K) \qquad (4.146)$$

and defining

$$G(z, \underline{\varphi}) = AH(z, \underline{\varphi}) \qquad (4.147)$$

then

$$Q(A, \underline{\varphi}) = \sum_{i=1}^{M} [|AH(e^{j\omega_i}, \underline{\varphi})| - |H_d(e^{j\omega_i})|]^2 \qquad (4.148)$$

To find A^*, the optimum value of A, Eq. (4.148) is differentiated with respect to A and the result is set to zero, giving

$$|A^*| = \frac{\displaystyle\sum_{i=1}^{M} |H(e^{j\omega_i}, \underline{\varphi})| \, |H_d(e^{j\omega_i})|}{\displaystyle\sum_{i=1}^{M} |H(e^{j\omega_i}, \underline{\varphi})|^2} \qquad (4.149)$$

The error function

$$\hat{Q}(\underline{\varphi}) = Q(A^*, \underline{\varphi}) \qquad (4.150)$$

is now minimized. The gradient of $\hat{Q}$ with respect to $\underline{\varphi}$ is

$$\frac{\partial \hat{Q}}{\partial \varphi_n} = \frac{\partial \hat{Q}(A^*, \underline{\varphi})}{\partial \varphi_n} + \frac{\partial Q(A^*, \underline{\varphi})}{\partial A^*} \frac{\partial A^*}{\partial \varphi_n} \qquad n = 1, 2, \ldots, 4K \quad (4.151)$$

The second term in Eq. (4.151) is zero since A^* minimizes Q. Thus Eq. (4.151) can be written as

$$\frac{\partial \hat{Q}}{\partial \varphi_n} = 2A^* \sum_{i=1}^{M} [A^* |H(e^{j\omega_i}, \varphi)| - |H_d(e^{j\omega_i})|] \cdot \frac{\partial |H(e^{j\omega_i}, \underline{\varphi})|}{\partial \varphi_n} \quad (4.152)$$

Since

$$|H(e^{j\omega_i}, \varphi)| = [H(e^{j\omega_i}, \underline{\varphi})H^*(e^{j\omega_i}, \underline{\varphi})]^{1/2} \quad (4.153)$$

then

$$\frac{\partial |H(e^{j\omega_i}, \varphi)|}{\partial \underline{\varphi}_n} = \frac{1}{|H(e^{j\omega_i}, \varphi)|} \cdot \text{Re}\left[H^*(e^{j\omega_i}, \underline{\varphi}) \frac{\partial H(e^{j\omega_i}, \underline{\varphi})}{\partial \underline{\varphi}_n}\right] \quad (4.154)$$

which is readily computed. Thus all the computations necessary for optimization algorithms such as the Fletcher–Powell algorithm are readily carried out for the filter design problem.

Since the optimization procedure deals strictly with the filter magnitude functions, it is possible that at convergence some poles and/or zeros may be outside the unit circle. At first thought one might strictly replace a pole outside the unit circle with polar coordinates (ρ, θ) by a pole inside the unit circle with polar coordinates $(1/\rho, \theta)$. The magnitude function of the filter is left unaltered (to within a constant) by this procedure, as one is replacing a pole by its mirror-image pole. It is possible, however, that having brought all poles inside the unit circle the squared error can be further lowered by additional searching. Since this is often the case, the optimization procedure should be run in a two-step sequence.

1.   Use the optimization program to minimize $\hat{Q}(\underline{\varphi})$ without constraining pole or zero locations.
2.   At convergence, invert all poles or zeros outside the unit circle. Continue running the optimization program until a new minimum is reached.

An example of a wideband differentiator designed by Steiglitz using this technique is shown in Fig. 4.38. For this example, $K = 3$ and

$$\omega_i = \pi(i - 1)(0.05) \qquad i = 1, 2, \ldots, 21$$

$$H_d(e^{j\omega}) = \frac{\omega}{\pi}$$

Figure 4.38 shows the error in approximation in both the magnitude and phase for the differentiator.

2. Minimum L_p Error Design

The minimum squared error criterion design just discussed can be extended to higher-order error criteria directly as shown by Deczky. Furthermore the approximation error can be defined for the group delay response of the filter as well as the magnitude response.

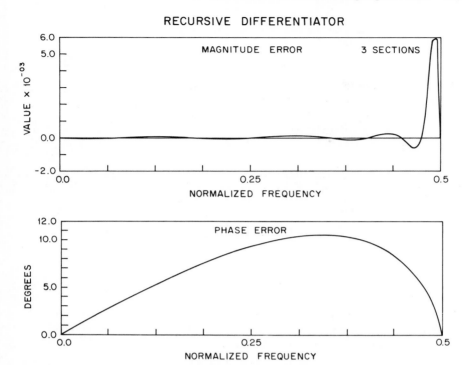

Fig. 4.38 Magnitude and phase error of a differentiator designed using minimum mean squared error criterion. (After Steiglitz.)

Expressing the z transform of the filter as a cascade of K second-order sections and using polar coordinates to express the pole and zero positions, the z transform of the filter can be written as

$$H(z) = A \prod_{k=1}^{K} \frac{z^2 - 2r_{0k}z \cos(\theta_{0k}) + r_{0k}^2}{z^2 - 2r_{pk}z \cos(\theta_{pk}) + r_{pk}^2} \quad (4.155)$$

Letting $\underline{\varphi}$ be the unknown parameter vector, defined as

$$\underline{\varphi} = (r_{01}, \theta_{01}, r_{02}, \theta_{02}, \ldots, r_{0k}, \theta_{0k}, r_{p1}, \theta_{p1}, r_{p2}, \theta_{p2}, \ldots, r_{pk}, \theta_{pk}, A) \quad (4.156)$$

then the magnitude response of the filter can be written as

$$|H(e^{j\omega})| = a(\underline{\varphi}, \omega)$$

$$= A \prod_{k=1}^{K} \frac{\begin{aligned}(1 - 2r_{0k}\cos(\omega - \theta_{0k}) + r_{0k}^2)^{1/2} \\ \times (1 - 2r_{0k}\cos(\omega + \theta_{0k}) + r_{0k}^2)^{1/2}\end{aligned}}{(1 - 2r_{pk}\cos(\omega - \theta_{pk}) + r_{pk}^2)^{1/2}} \quad (4.157)$$

$$\times (1 - 2r_{pk}\cos(\omega + \theta_{pk}) + r_{pk}^2)^{1/2}$$

and the group delay of the filter assumes the form

$$
\tau(\underline{\varphi}, \omega) = \sum_{k=1}^{K} \left[\frac{1 - r_{pk} \cos(\omega - \theta_{pk})}{1 - 2r_{pk} \cos(\omega - \theta_{pk}) + r_{pk}^2} + \frac{1 - r_{pk} \cos(\omega + \theta_{pk})}{1 - 2r_{pk} \cos(\omega + \theta_{pk}) + r_{pk}^2} \right.
$$

$$
\left. - \frac{1 - r_{0k} \cos(\omega - \theta_{0k})}{1 - 2r_{0k} \cos(\omega - \theta_{0k}) + r_{0k}^2} - \frac{1 - r_{0k} \cos(\omega + \theta_{0k})}{1 - 2r_{0k} \cos(\omega + \theta_{0k}) + r_{0k}^2} \right]
$$

$$(4.158)$$

The problem of designing an IIR filter that approximates a desired magnitude response $a_d(\omega)$ or group delay response $\tau_d(\omega)$ can be formulated as a minimum L_p approximation problem by defining the L_p errors as

$$
L_{2p}^a(\underline{\varphi}) = \sum_{j=1}^{J} w_a(\omega_j)[a(\underline{\varphi}, \omega_j) - a_d(\omega_j)]^{2p} \tag{4.159}
$$

$$
L_{2p}^\tau(\underline{\varphi}) = \sum_{j=1}^{J} w_\tau(\omega_j)[\tau(\underline{\varphi}, \omega_j) - \tau_d(\omega_j)]^{2p} \tag{4.160}
$$

where Eq. (4.159) expresses the error in the magnitude response and Eq. (4.160) expresses the error in the group delay response as a function of the parameter vector $\underline{\varphi}$. When $p = 1$ and $w_a(\omega_j) = 1$, all j, the L_p approximation is identical to the squared error criterion of the preceding section. As $p \to \infty$, it can be shown that the approximation tends to the minimax or Chebyshev error criterion.

The basic L_p design problem is thus the minimization, with respect to $\underline{\varphi}$, of either $L_{2p}^a(\underline{\varphi})$ or $L_{2p}^\tau(\underline{\varphi})$. It can be proved that for $2p \geq 2$ and for $w_\tau(\omega)$ [or $\omega_\tau(\omega)$] a positive weighting function, that a local minimum of $L_{2p}^a(\underline{\varphi})$ or $L_{2p}^\tau(\underline{\varphi})$ exists. This result says that an unconstrained minimization (such as the Fletcher–Powell algorithm) can be used to solve for the parameter vector $\underline{\varphi}^*$ that minimizes the appropriate error measure.

Figures 4.39 to 4.41 show examples of filters designed by Deczky using a minimum L_p error criterion. Figure 4.39 shows the approximation error for a one-section ($K = 1$) wideband differentiator, designed with a value of $2p = 4$. In this case the magnitude approximation error was minimized. The peak approximation error is less than 1% over the entire frequency band. Figure 4.40 shows the magnitude response of a two-passband, three-stopband filter designed using a tenth-order filter ($K = 5$). The peak stopband error is about 0.1 (or 20-dB loss). Finally Fig. 4.41 shows the group delay characteristics of an elliptic filter and a group delay equalized elliptic filter. The equalizer network consisted of all-pass filter networks in cascade so as to leave the magnitude response of the elliptic filter unaltered. The order of the all-pass equalizer was 10 ($K = 5$) and the error criterion index was $2p = 10$. The

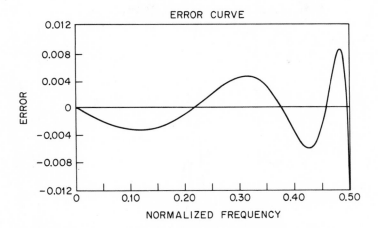

ERROR CURVE

Fig. 4.39 Approximation error for a differentiator designed with a minimum fourth-order error. (After Deczky.)

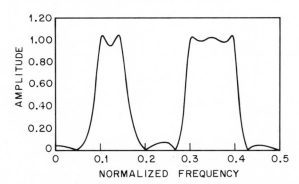

Fig. 4.40 Multiband filter designed via optimization techniques. (After Deczky.)

equalized group delay of the filter is essentially equiripple, as seen in Fig. 4.41.

3. Optimization Using All-Pass Sections in the w Plane

For designing IIR filters having an equiripple passband and an equiripple approximation to an arbitrary stopband (or vice versa), a fairly simple optimization procedure due to Deczky can be applied. To see how this technique works, consider writing the squared magnitude of the filter in the form

$$H(z)H(z^{-1})\big|_{z=e^{j\omega}} = \frac{1}{1 + \varepsilon^2 |T_n(z)|^2_{z=e^{j\omega}}} \qquad (4.161)$$

where $T_n(z)$ is a rational transfer function that is similar but not identical to the rational Chebyshev function used for designing elliptic filters. Rather than

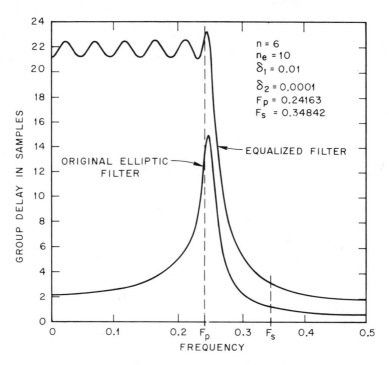

Fig. 4.41 Group delay equalization of a lowpass filter using an equalizer designed via optimization methods. (After Deczky.)

trying to design $T_n(z)$, it is convenient to map the desired z-plane approximation problem into a new plane (call it the w plane) such that the passband (or the stopband) of the filter is mapped to be the entire imaginary axis of the w plane. It is then fairly easy to construct an all-pass function in the w plane; from this all-pass function, create a function $|T_n(w)|^2$ that oscillates between 0 and 1 as w goes from 0 to ∞ along the imaginary axis in the w plane. Thus $|T_n(j\eta)|^2$ is an equiripple function in the w plane. Depending on the exact nature of the mapping, the equiripple behavior of $T_n(w)$ can be mapped to either the passband or the stopband of the filter in the z plane. The behavior in the other band is determined entirely by the values of the all-pass filter coefficients, which are as yet undetermined. Iterative procedures can be used to estimate the coefficients of the all-pass filter to provide equiripple approximations to any desired response in the unspecified band. We now show how $T_n(w)$ can be obtained.

To construct a function whose magnitude is constant along the entire imaginary axis, consider the all-pass function

$$F(w) = \prod_{i=1}^{n} \frac{w_i + w}{w_i - w} \tag{4.162}$$

where w_i either is real or occurs in complex conjugate pairs. Since $|F(j\eta)| = 1$, $F(j\eta)$ can be written as

$$F(j\eta) = e^{jf(\eta)} \tag{4.163}$$

where

$$f(\eta) = 2 \sum_{i=1}^{n} \tan^{-1}\left(\frac{\eta - \eta_i}{\xi_i}\right) \tag{4.164}$$

and

$$w_i = \xi_i + j\eta_i \tag{4.165}$$

The real function $|T_n(j\eta)|^2$ is formed as

$$|T_n(j\eta)|^2 = T_n(j\eta)T_n(-j\eta) = \tfrac{1}{4}[F(j\eta) + 1][F(-j\eta) + 1]$$

$$= \cos^2\left[\frac{f(\eta)}{2}\right] \tag{4.166}$$

Analytical continuation of Eq. (4.166) gives

$$T_n(w)T_n(-w) = \frac{\left[\displaystyle\prod_{i=1}^{n}(w_i + w) + \prod_{i=1}^{n}(w_i - w)\right]^2}{4\displaystyle\prod_{i=1}^{n}(w_i + w)(w_i - w)} \tag{4.167}$$

We now have constructed the desired function $|T_n(w)|^2$, which has equiripple magnitude along the $j\eta$ axis in the w plane independent of the all-pass filter coefficients w_i. We now examine the mappings from the z plane to the w plane that are appropriate for the case of equiripple passband, arbitrary stopband filters. Similar results can be obtained for the case of mappings of equiripple stopband, arbitrary passband filters but since these are generally of less importance than the previous case, they will not be discussed here.

For mapping the passband of a filter in the z plane to the entire imaginary axis in the w plane, the transformation

$$w^2 = \frac{z^2 - 2z\cos\omega_{pu} + 1}{z^2 - 2z\cos\omega_{pl} + 1} \tag{4.168}$$

can be used. The inverse transformation from the w plane to the z plane can be obtained from Eq. (4.168) as

$$z = p \pm \sqrt{p^2 - 1} \tag{4.169}$$

where

$$p = \frac{w^2\cos\omega_{pl} - \cos\omega_{pu}}{w^2 - 1} \tag{4.170}$$

The transformation of Eq. (4.168) maps the arc of the unit circle $\omega_{pl} \leq \omega \leq \omega_{pu}$ (i.e., the passband) in the z plane to the imaginary axis in the w plane. [Note that we are assuming there is a single passband and two (or more) stopbands.] The stopbands of the filter in the z plane are mapped as

$$0 \leq \omega \leq \omega_{sl} \longleftrightarrow \sqrt{\frac{1 - \cos \omega_{pu}}{1 - \cos \omega_{pl}}} \leq \xi \leq \sqrt{\frac{\cos \omega_{sl} - \cos \omega_{pu}}{\cos \omega_{sl} - \cos \omega_{pl}}}$$

$$\omega_{su} \leq \omega \leq \pi \longleftrightarrow \sqrt{\frac{\cos \omega_{su} - \cos \omega_{pu}}{\cos \omega_{su} - \cos \omega_{pl}}} \leq \xi \leq \sqrt{\frac{1 + \cos \omega_{pu}}{1 + \cos \omega_{pl}}} \quad (4.171)$$

where ξ is the real part of w. For a lowpass filter, $\omega_{sl} = \omega_{pl} = 0$ in the equations above.

Figure 4.42 illustrates the z-plane to w-plane transformation of Eq. (4.168). The passband (equiripple) and stopband regions are denoted in both the z plane and w plane in this figure.

All that remains now is to give a procedure for determining the w_i coefficients in Eq. (4.167) so as to approximate a desired stopband response and to give a procedure for obtaining $H(z)$ [or equivalently $H(w)$] from $T_n(w)$. We discuss this second problem first as it is fairly straightforward to determine $H(w)$ from $T_n(w)$.

In the case when the zeros of $H(z)$ are on the unit circle (as is generally true for equiripple approximations), the roots in the w plane will be real and occur in even multiplicity since (for zeros on the unit circle) both the zero and its complex conjugate partner map to the same real location in the w plane.

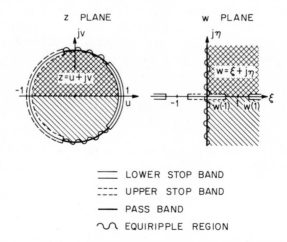

LOWER STOP BAND
UPPER STOP BAND
PASS BAND
EQUIRIPPLE REGION

Fig. 4.42 The mapping from the z plane to the w plane. (After Deczky.)

Thus Eq. (4.167) can be expressed as

$$T_n(w)T_n(-w) = \frac{\left[\prod_{i=1}^{n} (w_{0i} + w)^2 + \prod_{i=1}^{n} (w_{0i} - w)^2\right]}{4 \prod_{i=1}^{n} (w_{0i}^2 - w^2)^2} \tag{4.172}$$

where w_{0i} are the transformed zeros of the filter. Defining the auxiliary polynomials $A(w)$ and $B(w)$ as

$$A(w) = \frac{1}{2} \left[\prod_{i=1}^{n} (w_{0i} + w)^2 + \prod_{i=1}^{n} (w_{0i} - w)^2\right] \tag{4.173}$$

and

$$wB(w) = \frac{1}{2} \left[\prod_{i=1}^{n} (w_{0i} + w)^2 - \prod_{i=1}^{n} (w_{0i} - w)^2\right] \tag{4.174}$$

Eq. (4.161) can be written as

$$H(w)H(-w) = \frac{A^2(w) - w^2 B^2(w)}{(1 + \varepsilon^2)A^2(w) - w^2 B^2(w)} \tag{4.175}$$

which can be factored to give

$$H(w) = \frac{A(w) - wB(w)}{\sqrt{1 + \varepsilon^2} \, A(w) - wB(w)} \tag{4.176}$$

where the roots of $H(w)$ are in the right-half w plane, thereby guaranteeing that the resulting $H(z)$ is stable. [Furthermore there are numerical advantages to finding the w-plane roots of $H(w)$ and then transforming back to the z plane because the w-plane roots are generally more separated than the z-plane roots.]

The final step in the design is to give the algorithm for solving for the w_{0i} to give the desired stopband response. A simple approximate procedure is to use the log magnitude of the stopband response, defined as

$$\alpha = -20 \log_{10} |H(w)|_{w=\xi}$$

$$= 10 \log_{10} \left\{1 + \frac{\varepsilon^2}{4} \left[\prod_{i=1}^{n} \left(\frac{w_{0i} + w}{w_{0i} - w}\right) + \prod_{i=1}^{n} \left(\frac{w_{0i} - w}{w_{0i} + w}\right)\right]\right\}_{w=\xi} \tag{4.177}$$

or

$$\alpha \approx 20 \log_{10} \left(\frac{\varepsilon}{2}\right) + \sum_{i=1}^{n} 20 \log_{10} \left|\frac{w_{0i} + w}{w_{0i} - w}\right|_{w=\zeta} \tag{4.178}$$

A simple organized procedure, such as the well-known Remez algorithm, can be used to choose the w_{0i} such that the error in α in Eq. (4.178) is essentially minimax over the entire stopband.

To summarize the results of this section, we have shown that IIR filters with equiripple passband (or stopband) and arbitrary stopbands (passbands) can be readily designed by mapping the approximation problem from the z plane to the w plane such that the filter passband in the z plane maps to the entire imaginary axis in the w plane. An all-pass function is then synthesized whose magnitude is constant along the imaginary axis of the w plane. A simple substitution is used to create a w-plane transfer function from the all-pass function whose magnitude is equiripple along the imaginary axis of the w plane *independent* of the all-pass function coefficients. Finally a simple procedure is given for optimally choosing the all-pass coefficients to approximate the desired equivalent w-plane stopband response. We now illustrate the procedure with two examples due to Deczky.

Figures 4.43 and 4.44 show log magnitude responses of two equiripple passband filters with arbitrary stopband requirements. For the example in Fig. 4.43 the passband ripple requirement was 1-dB ripple and the required stopband ripple was

$$20 \log_{10}|H(e^{-j\omega})| \leq \begin{cases} -30\,\text{dB} & 0.35\pi \leq \omega \leq 0.45\pi \\ -40\,\text{dB} & 0.45\pi \leq \omega \leq 0.60\pi \\ -20\,\text{dB} & 0.60\pi \leq \omega \leq 0.65\pi \end{cases}$$

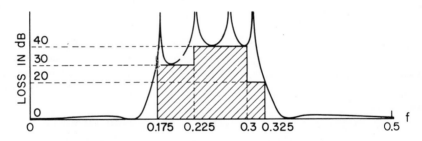

Fig. 4.43 Log magnitude response of a bandstop filter designed in the w plane. (After Deczky.)

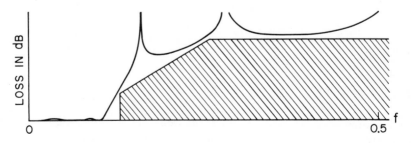

Fig. 4.44 Log magnitude response of a lowpass filter designed in the w plane. (After Deczky.)

The resulting filter was of eighth degree and met specifications to within 0.7 dB. The example of Fig. 4.44 is a lowpass filter with a sloping attenuation characteristic in the stopband.

4. Linear Program Design of IIR Filters

An equiripple approximation to an IIR filter with a prescribed magnitude characteristic can be obtained using linear programming techniques. If the transfer function of the digital filter is of the form

$$H(z) = \frac{N(z)}{D(z)} = \frac{\sum\limits_{i=0}^{m} b_i z^{-i}}{\sum\limits_{i=0}^{n} a_i z^{-i}} \tag{4.179}$$

then $H(z)H(z^{-1})$ can be written as

$$H(z)H(z^{-1}) = \frac{N(z)N(z^{-1})}{D(z)D(z^{-1})} = \frac{\left(\sum\limits_{i=0}^{m} b_i z^{-i}\right)\left(\sum\limits_{j=0}^{m} b_j z^{j}\right)}{\left(\sum\limits_{i=0}^{n} a_i z^{-i}\right)\left(\sum\limits_{j=0}^{n} a_j z^{j}\right)}$$

$$= \frac{\sum\limits_{i=-m}^{m} c_i z^{-i}}{\sum\limits_{i=-n}^{n} d_i z^{-i}} \tag{4.180}$$

where

$$c_i = c_{-i} \tag{4.181a}$$

$$d_i = d_{-i} \tag{4.181b}$$

The magnitude-squared function of the filter [i.e., Eq. (4.180) evaluated on the unit circle] is therefore a ratio of trigonometric polynomials; i.e.,

$$|H(e^{j\omega})|^2 = \frac{\hat{N}(\omega)}{\hat{D}(\omega)} = \frac{c_0 + \sum\limits_{i=1}^{m} 2c_i \cos(\omega i)}{d_0 + \sum\limits_{i=1}^{n} 2d_i \cos(\omega i)} \tag{4.182}$$

Both $\hat{N}(\omega)$ and $\hat{D}(\omega)$ are linear in the coefficients c_i and d_i. We now show how linear programming techniques can be used to determine the c_i's and d_i's such that $|H(e^{j\omega})|$ approximates a given magnitude-squared characteristic $F(\omega)$ and such that the peak approximation error is minimized—i.e., an equiripple approximation.

If we let $F(\omega)$ be the desired mag.iitude-squared characteristic, then the approximation problem consists of finding the filter coefficients such that

$$-\varepsilon(\omega) \le \frac{\hat{N}(\omega)}{\hat{D}(\omega)} - F(\omega) \le \varepsilon(\omega) \tag{4.183}$$

where $\varepsilon(\omega)$ is a tolerance function on the error that allows for unequal weighting of errors as a function of frequency. Since $F(\omega)$ and $\varepsilon(\omega)$ are generally known functions of frequency (or depend on some parameter, as illustrated below), Eq. (4.183) can be expressed as a set of linear inequalities in the c_i's and d_i's by writing it in the form

$$\hat{N}(\omega) - \hat{D}(\omega)F(\omega) \le \varepsilon(\omega)\hat{D}(\omega)$$
$$-\hat{N}(\omega) + \hat{D}(\omega)F(\omega) \le \varepsilon(\omega)\hat{D}(\omega) \tag{4.184}$$

or

$$\hat{N}(\omega) - \hat{D}(\omega)[F(\omega) + \varepsilon(\omega)] \le 0$$
$$-\hat{N}(\omega) + \hat{D}(\omega)[F(\omega) - \varepsilon(\omega)] \le 0 \tag{4.185}$$

The additional linear inequalities

$$-\hat{N}(\omega) \le 0 \tag{4.186}$$

$$-\hat{D}(\omega) \le 0 \tag{4.187}$$

completely define the approximation problem. To solve the system of linear inequalities of Eq. (4.184) to (4.187), an auxiliary variable w is subtracted from the left side of each constraint [Eq. (4.184) to (4.187)] and this variable is then minimized. If the resulting value of w is 0, then a solution exists to the approximation problem and the filter coefficients may be obtained directly as the output of the linear programming routine. If $w > 0$, then no solution to the approximation problem exists, and either $F(\omega)$, or $\varepsilon(\omega)$, or both must be modified in order to obtain a solution.

By way of example, consider the design of an equiripple lowpass filter with passband magnitude of 1 and desired stopband magnitude of 0. If we let δ be the peak approximation error in the stopband and $K\delta$ (K is a constant chosen by the user) be the peak approximation error in the passband, then the desired magnitude function for a lowpass filter is as shown in Fig. 4.45(a). The quantity δ is unknown and is to be minimized in the ultimate design program (of course, the resulting filter is an elliptic filter in this case but we are only using this an an example). The desired squared magnitude function of the filter is the square of the response in Fig. 4.45(a) and is shown in Fig. 4.45(b). This response can be viewed as an equiripple approximation of the function $F(\omega)$, shown in Fig. 4.45(c), with peak approximation $\varepsilon(\omega)$ as shown in Fig. 4.45(d). [The reader should verify that $F(\omega) + \varepsilon(\omega)$ is the upper bound on the squared magnitude response and $F(\omega) - \varepsilon(\omega)$ is the lower bound on

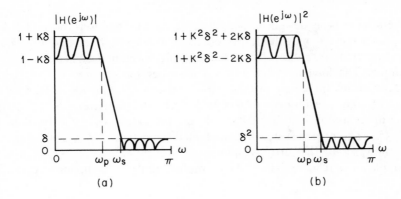

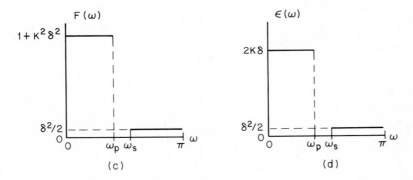

Fig. 4.45 Lowpass filter specifications for linear programming method.

this response.] For the given $F(\omega)$ and $\varepsilon(\omega)$, Eqs. (4.184) to (4.187) are iteratively solved for guessed values of δ. By definition, for a lowpass filter, δ is bounded by

$$0 < \delta < \frac{1}{K+1} \tag{4.188}$$

Thus an initial guess of δ is made, thereby specifying $F(\omega)$ and $\varepsilon(\omega)$ of Fig. 4.45. For this value of δ, the linear programming technique tells the user if the given set of inequalities has any solution. If not, the value of δ must be increased until a solution is obtained. If the equations have a solution, the input value of δ is an overbound on the minimum δ for which a solution exists. By iteratively raising the lower bound (for which no solution exists) and lowering the upper bound, the minimum value of δ can be obtained to within (theoretically) any desired accuracy.

Although this technique is subject to accuracy problems related to coefficient sensitivity in magnitude-squared function design, a large number of filters have been designed successfully using the method above. Figure 4.46 shows the log magnitude responses of a lowpass filter and Fig. 4.47 shows the error response of a wideband differentiator designed using this technique. The lowpass filter was a sixth-order design ($n = m = 6$) with passband cutoff frequency of 0.20, stopband cutoff frequency of 0.25, a K of 71.879, and a resulting δ of 0.0008252. The differentiator was a fourth-order differentiator with a peak ripple of 0.00000763 over the band $0 \leq f \leq 0.45$. (The response was unconstrained from $0.45 \leq f \leq 0.50$.) Figure 4.47 shows that the error response of the differentiator is essentially equiripple in a relative rather than absolute sense. The upper and lower

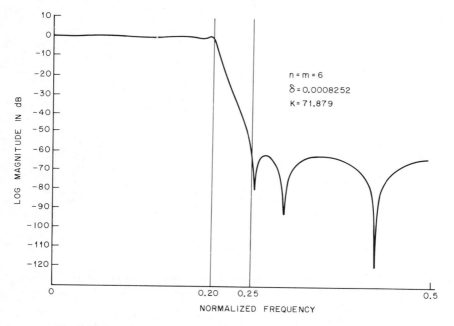

Fig. 4.46 Log magnitude response of a lowpass filter designed via linear programming.

straight lines demark the error slopes. The slight deviations from the error bounds are a result of the inaccuracies in the linear programming algorithm for such small values of δ.

4.15 Summary of IIR Filter Design Techniques

We have presented three classes of design techniques for IIR filters—namely, transformation of an analog filter to a digital filter, direct design of digital filters in the z plane, and optimization methods for design. As is

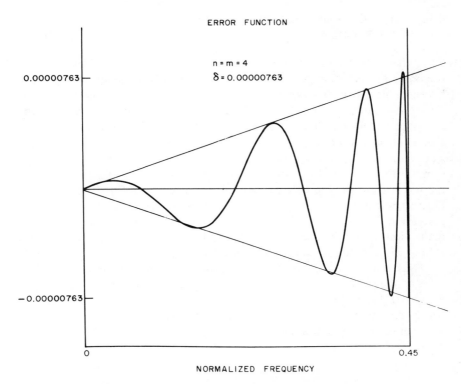

ERROR   FUNCTION

$n = m = 4$

$\delta = 0.00000763$

0.00000763

−0.00000763

0

0.45

NORMALIZED   FREQUENCY

Fig. 4.47 Error function for a differentiator designed via linear programming.

generally the case, it is impossible to say that one specific filter design procedure is recommended to the exclusion of all others. Depending on the specific application in mind, the availability of the various design procedures, and numerous other factors, almost any of the design techniques presented here may be most preferred. Thus we shall refrain from making value judgments of a controversial nature here. Instead we observe that many IIR digital filters are designed by bilinearly transforming a standard analog filter. This observation reflects the fact that in most cases the user is interested in a lowpass or bandpass filter of appropriate specifications, which is readily obtained using the bilinear transformation of an analog filter. In those cases when a nonstandard IIR filter is required, then another design technique must be applied. If the specifications are in terms of the time response of the filter, then impulse invariant transformation or the time domain technique discussed in Sec. 4.13 may be appropriate. In most other cases one of the available optimization algorithms will generally prove adequate for approximating unusual frequency domain specfications.

4.16 Some Comparisons Between FIR and IIR Filters

Since there are a large number of design techniques for both FIR and IIR filters, it is very difficult, if not impossible, to make objective comparisons between these two types of filters based on some performance measure. If one restricts the possibilities to optimum (minimax) FIR lowpass filters and elliptic IIR filters that meet equivalent frequency response specifications, then some quantitative comparisons can be made between these filter types. The main basis of comparison is the number of multiplications per input sample[1] required in the most standard realization of each filter type—i.e., the direct form for the FIR case and the cascade form for the elliptic case. Direct realization of an N-point (N odd) impulse response filter with linear phase requires $[(N+1)/2]$ multiplications per sample, whereas cascade realization of an nth-order elliptic filter (all zeros on the unit circle) requires $[(3n+3)/2]$ multiplications per sample[2] where $[\cdot]$ denotes *integer part of*.

Thus, one basis of comparison between equivalent filter designs [i.e., both meeting the same specifications on δ_1 (passband ripple), δ_2 (stopband ripple), F_p, and F_s] is in terms of the efficiency of the respective realizations—i.e., which structure requires fewer multiplications per sample. Equivalence between structures is attained when the condition

$$\left(\frac{3n+3}{2}\right) = \left(\frac{N+1}{2}\right) \tag{4.189}$$

or, equivalently,

$$\frac{N}{n} \approx 3 + \frac{1}{n} \tag{4.190}$$

Figure 4.48 shows two sets of curves indicating the ratio N/n as a function of n for two values of F_p and δ_1 and various values of δ_2. Figure 4.48(a) shows data for the case $F_p = 0.15$; $\delta_1 = 0.1$; $\delta_2 = 0.1, 0.01, 0.001, 0.0001$, whereas Fig. 4.48(b) shows data for the case $F_p = 0.35$, $\delta_1 = 0.00001$, and the same range of δ_2 as in Fig. 4.48(a). Also shown in these plots is the line $N/n = 3$, which indicates where the data lie with respect to the fixed portion of Eq. (4.190). As seen in Fig. 4.48(a), for certain values of F_p, δ_1, and δ_2, the ratio of N/n falls below the equivalence level of Eq. (4.190)—i.e., the FIR filter is

---

[1] The number of multiplications per input sample is a useful measure of the computational complexity of the filtering operations as it represents the number of multiply-add operations required for a software implementation of the algorithm as well as for a general hardware implementation.

[2] This number of multiplications per sample for the IIR filter assumes that any scaling between sections is an integer power of 2 and is performed entirely by shifts of the data. If finer scaling multipliers are included between each cascade section, the realization requires $[(4n+3)/2]$ multiplications per sample.

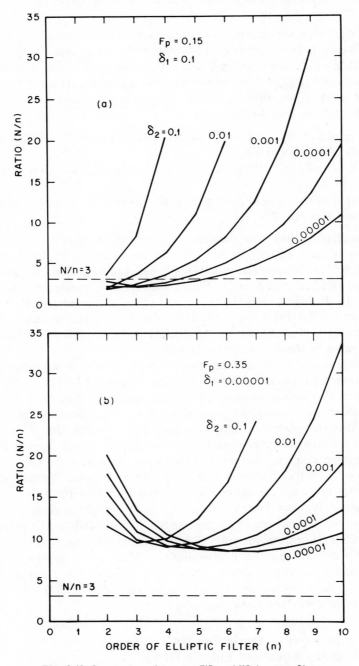

Fig. 4.48 Comparisons between FIR and IIR lowpass filters.

more efficient than the elliptic filter. In general, however, the elliptic filter is more efficient than the optimum FIR filter and, in the case of high-order elliptic designs, the ratio of N/n is often in the hundreds or thousands.

It has been found that the most favorable conditions for an FIR design are large values of δ_1, small values of δ_2, and large transition widths (i.e., small transition ratios). Furthermore,

1. For values of $F_p \geq 0.3$, the ratio N/n always exceeded $3 + 1/n$ for all values of δ_1, δ_2, and n.
2. For values of $n \geq 7$, the ratio N/n always exceeded $3 + 1/n$ for all values of δ_1, δ_2, and F_p.
3. The smaller the value of F_p, the larger the range of δ_1, δ_2, and n for which N/n was less than $3 + 1/n$.

Figure 4.49(a) shows the theoretical order n of an elliptic filter (n need not be an integer) required to match specifications on F_p, F_s, for $\delta_1 = 0.1$, $\delta_2 = 0.1$, 0.01, 0.001, 0.0001, and 0.00001, as a function of F_p for a set of optimum FIR filters with $N = 21$. Figure 4.49(b) shows similar measurements for $N = 41$. In Fig. 4.49(a) the theoretical point of equivalence is $n = 6.3$, whereas in Fig. 4.49(b) it is $n = 13$. From this figure it is seen that for these cases the elliptic filter is always more efficient than the equivalent FIR filter, as anticipated by the discussion in the preceding paragraph.

In summary, elliptic filters are generally more efficient in achieving given specifications on the magnitude response than optimum FIR filters. The FIR filters have the additional useful property, however, that their phase is exactly linear; i.e., there is no group delay distortion. For the elliptic filter, though, there is generally a large amount of group delay distortion (concentrated primarily near the band edge). A question of both theoretical and practical importance is whether, in cases when the additional requirement of a flat delay is specified, it is more desirable to equalize the delay of an elliptic filter or to use the equivalent optimum FIR filter (with its constant group delay). We now discuss various aspects of this question. It should be noted that the alternatives above are not the only possibilities for obtaining a digital filter that meets frequency domain specifications on both magnitude and group delay responses. For example, a filter can be designed, using modern optimization procedures, where the number of poles and zeros are unequal. In such cases, the comparisons between FIR and IIR filters are quite distinct from those to be discussed.

4.17 Comparisons of Optimum FIR Filters and Delay Equalized Elliptic Filters

As discussed in Sec. 4.14, it is possible to design an all-pass equalizer that can equalize the group delay of any digital filter to any desired accuracy over a

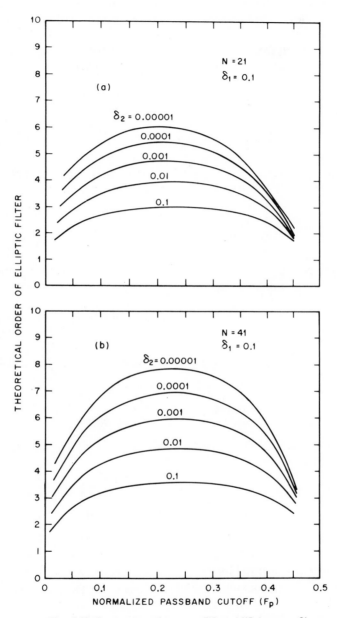

Fig. 4.49 Comparisons between FIR and IIR lowpass filters.

restricted band of frequencies. The difficulty with trying to equalize the group delay of a filter lies in the fact that the equalized filter must have a total delay greater than the largest delay in the unequalized filter. For the elliptic design, the peak delay always occurs near the passband cutoff frequency. It can be shown that an all-pass equalizer of degree n_e has the property

$$\frac{1}{2\pi} \int_0^\pi \tau_g(\omega) \, d\omega = 0.5 n_e \tag{4.191}$$

where $\tau_g(\omega)$ is the equalizer group delay and the integral is taken over half the sampling interval ($0 \le \omega \le \pi$). Since $\tau_g(\omega) \ge 0$; i.e., group delays add, to justify Eq. (4.191) it is sufficient to show that a first-degree all-pass equalizer has the required property. The z transform of a first-degree all-pass equalizer is

$$H(z) = \frac{(1 - z^{-1}/a)}{1 - az^{-1}} \tag{4.192}$$

where a is the pole position and $1/a$ is the zero position in the z plane. The group delay is commonly defined as

$$\tau_g(\omega) = - \frac{d[\measuredangle H(e^{j\omega})]}{d\omega} \tag{4.193}$$

where $\measuredangle H(e^{j\omega})$ is the phase of the transfer function. Using Eq. (4.191) and (4.192) we obtain

$$\tau_g(\omega) = \frac{1 - a^2}{1 + a^2 - 2a \cos \omega} \tag{4.194}$$

for the first-degree equalizer. Integrating Eq. (4.194) from 0 to π and normalizing by 2π gives

$$\frac{1}{2\pi} \int_0^\pi \frac{1 - a^2}{1 + a^2 - 2a \cos \omega} \, d\omega = \frac{1}{\pi} \tan^{-1} \left[\frac{(1 - a^2) \tan (\omega/2)}{(1 - a)^2} \right] \Big|_0^\pi = \frac{\pi}{2\pi} = 0.5$$

The significance of Eq. (4.191) is that one can estimate the minimum-order equalizer required to equalize a given group delay characteristic by determining the area between the line $\tau = \tau_{\max}$ and the curve $\tau_g(\omega)$ and dividing by π, where $\tau_{\max}$ is the maximum value of $\tau_g(\omega)$ in the passband. Of course, the required order of the equalizer must be greater than the estimate given above since this estimate assumes the delay of the equalizer exactly compensates the delay of the unequalized filter. As the degree of the equalizer is increased over the estimate, the peak error of approximation decreases monotonically.

The data for three sets of elliptic filters that were equalized using all-pass filters are given in Tables 4.2 to 4.4. Included in the table are the

TABLE 4.2

COMPARISONS OF OPTIMUM FIR AND EQUALIZED ELLIPTIC DIGITAL FILTERS

Set 1. $\delta_1 = 0.01$, $\delta_2 = 0.0001$

F_p	F_s	n	N	n_e	$\bar{\tau}_g$	e	$N_1{}^1$	$N_2{}^1$
0.0502	0.20273	5	21	2	28.7	12.1	11	11
				4	42.7	3.4		13
0.09846	0.25119	5	21	2	14.5	11.6	11	11
				4	22.2	4.1		13
				6	29.4	0.8		15
0.14722	0.29803	6	21	4	17.6	13.1	11	14
				6	23.0	6.3		16
				8	28.5	2.6		18
0.19507	0.34314	6	21	4	13.8	16.0	11	14
				6	17.8	8.7		16
				8	22.0	4.2		18
0.24163	0.38601	6	21	6	14.5	11.1	11	16
				8	18.3	7.0		18
				10	21.8	3.6		20
0.28664	0.42571	5	21	6	11.6	8.4	11	15
				8	14.5	3.8		17
				10	17.3	1.6		19
0.33014	0.46052	5	21	6	10.7	14.7	11	15
				8	13.1	8.3		17
				10	15.7	4.5		19
0.37254	0.48698	4	21	6	8.7	19.1	11	13
				8	11.1	6.5		15
				10	13.4	3.2		17
0.41665	0.49917	3	21	8	9.6	6.3	11	14
				10	11.8	3.2		16

[1] N_1 is the number of multiplications per sample for the optimum FIR filter; N_2 is the number of multiplications per sample for the equalized elliptic filter.

filter specifications (δ_1, δ_2, F_p, F_s), the required elliptic order n, the required FIR filter duration N, the equalizer order n_e, the average passband delay $\bar{\tau}_g$ (in samples) of the equalized filter, the percentage ripple e in the passband group delay of the equalized filter, and a comparison between the number of multiplications per sample required in both the optimum FIR filter and the equalized elliptic filters. The data in these tables indicate that to achieve equalization to within about a 3% error requires on the order of 30% more multiplications per sample for the equalized filter than for the optimum FIR

Table 4.3

COMPARISONS BETWEEN OPTIMUM FIR AND
EQUALIZED ELLIPTIC DIGITAL FILTERS

Set 2: $F_p = 0.25$, $\delta_1 = 0.02$, $\delta_2 = 0.001$

F_s	n	N	n_e	$\bar{\tau}_g$	e	$N_1{}^1$	$N_2{}^1$
0.4893	2	11	2	3.3	1.2	6	6
			4	5.6	0.1		8
0.44816	3	13	2	4.5	9.4	7	8
			4	7.3	1.0		10
0.39146	4	19	2	5.9	25.1	10	9
			4	8.8	8.0		11
			6	11.9	2.2		13
0.34153	5	29	2	8.4	37.4	15	11
			4	10.6	21.6		13
			6	13.7	11.6		15
			8	16.9	5.6		17
			10	20.3	2.4		19
0.30639	6	45	4	13.8	34.7	23	14
			6	16.0	25.0		16
			8	18.7	16.9		18
			10	22.0	11.7		20
			12	25.5	7.9		22
			14	29.4	5.2		24
			16	32.8	3.2		26
			18	36.3	1.8		28

[1] N_1 is the number of multiplications per sample for the optimum FIR filter; N_2 is the number of multiplications per sample for the equalized elliptic filter.

design, although in most cases the unequalized elliptic filter was more efficient than the optimum FIR designs. Thus it would appear that, at least for these restricted results, if constant group delay is required, in addition to the equiripple magnitude characteristics the optimum FIR filter is always more efficient than an equalized elliptic filter. It should also be noted that the delay of the optimum FIR filter $\{[(N-1)/2]$ samples$\}$ was *always* less than the delay of the equalized elliptic filter.

The examples of Tables 4.2 to 4.4 consider filters where the order of the unequalized elliptic filter was six or less. It can be argued that for higher-order elliptic designs the relative efficiency of the elliptic filter over the optimum FIR filter is far greater than for lower-order designs; hence in these cases perhaps the equalized filter may still be more efficient than the optimum FIR design. This conjecture turns out to be untestable because high-order elliptic filters have a peak passband delay $\tau_{\max}$ that is much larger than for

Table 4.4

COMPARISONS BETWEEN OPTIMUM FIR AND EQUALIZED ELLIPTIC DIGITAL FILTERS

Set 3: $F_p = 0.25$, $\delta_1 = 0.02$, $\delta_2 = 0.0001$

F_s	n	N	n_e	$\bar{\tau}_g$	e	N_1^1	N_2^1
0.49661	2	11	2	3.3	1.2	6	6
			4	5.6	0.1		8
0.47564	3	11	2	4.5	9.1	6	8
			4	7.3	1.0		10
0.43591	4	17	2	5.8	23.3	9	9
			4	8.8	7.0		11
			6	11.8	1.7		13
0.38983	5	21	2	8.0	33.4	11	11
			4	10.3	18.0		13
			6	13.5	8.7		15
			8	16.7	3.7		17
			10	20.0	1.4		19
0.34878	6	31	4	12.8	28.9	16	14
			6	15.5	19.2		16
			8	18.2	11.8		18
			10	22.0	7.7		20
			12	25.3	4.3		22
			14	28.8	2.2		24

[1] N_1 is the number of multiplications per sample for the optimum FIR filter; N_2 is the number of multiplications per sample for the equalized elliptic filter.

low-order filters; hence the order required for the equalizer becomes extremely large and thus is not even practical to consider if equalization over the entire passband is required. To illustrate this point, Fig. 4.50 shows the group delay of a tenth-order elliptic lowpass filter with $F_p = 0.25$. Using Eq. (4.191) to obtain an estimate of n_e we arrive at a value of $n_e = 45$. Since this value of n_e is only an underbound on the actual order of the equalizer, it is clear that it is not practical to try to obtain such a high-degree equalizer.

Another interesting question that arises when one considers the idea of equalizing an IIR filter is how does the combination of elliptic filter followed by an all-pass equalizer compare to the optimum IIR filter that best approximates both the desired magnitude and group delay characteristics. It is clear that the optimum IIR filter can be no worse than the cascade; the question remains as to how much better it can be. There is no clear-cut answer to this question. Based on experience with equalized elliptic filters, however, several observations can be made.

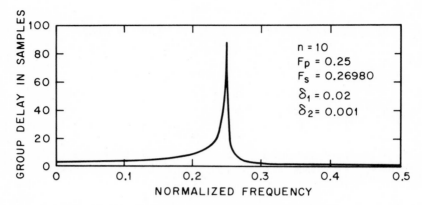

Fig. 4.50 Group delay of a tenth-order elliptic lowpass filter.

1. The zeros of the elliptic filter lie on the unit circle to give good stopband attenuation.
2. The zeros of the equalizer lie outside the unit circle to give positive delay.
3. The poles of the elliptic filter are constrained by the transition width requirements of the lowpass filter.
4. The poles of the equalizer lie approximately on a circle of fixed radius and are approximately equally spaced in the passband.

If the zeros of the optimum filter are not constrained to lie on the unit circle, then each second-order section will require four multiplications per sample rather than the three multiplications for each second-order section of the elliptic design and the two multiplications for each second-order section of the all-pass equalizer. Based on the observations above, it seems unlikely that there is much to gain using the optimum IIR filter over the equalized filter.

REFERENCES

General References

1. K. STEIGLITZ, "The Equivalence of Digital and Analog Signal Processing," *Information and Control*, **8**, No. 5, 455–467, Oct., 1965.
2. J. F. KAISER, "Design Methods for Sampled Data Filters," *Proc. First Allerton Conf. on Circuit and System Theory*, 221–236, Nov., 1963.
3. J. F. KAISER, "Digital Filters," Chapter 7 in *System Analysis by Digital Computer* (F. F. Kuo and J. F. Kaiser, ed.), John Wiley & Sons, Inc., New York, 1966.
4. A. J. GIBBS, "On the Frequency Domain Responses of Causal Digital Filters," Ph.D. Thesis, Univ. of Wisconsin, Madison, Wis., 1969.
5. A. J. GIBBS, "An Introduction to Digital Filters," *Australian Telecomm. Research*, **3**, No. 2, 3–14, Nov., 1969.

6. A. J. Gibbs, "The Design of Digital Filters," *Australian Telecomm. Research*, **4**, No. 1, 29–34, 1970.
7. C. M. Rader and B. Gold, "Digital Filter Design Techniques in the Frequency Domain," *Proc. IEEE*, **55**, No. 2, 149–171, Feb., 1967.
8. R. M. Golden and J. F. Kaiser, "Design of Wideband Sampled Data Filters," *Bell Sys. Tech. J.*, **43**, No. 4, 1533–1546, Part 2, July, 1964.
9. L. Weinberg, *Network Analysis and Synthesis*, McGraw-Hill Book Co., New York, 1962.
10. J. E. Storer, *Passive Network Synthesis*, McGraw-Hill Book Co., New York, 1957.
11. J. P. Thiran, "Recursive Digital Filters with Maximally Flat Group Delay," *IEEE Trans. Circuit Theory*, **CT-18**, 659–663, Nov., 1971.
12. J. P. Thiran, "Equal-Ripple Delay Recursive Digital Filters," *IEEE Trans. Circuit Theory*, **CT-18**, 664–677, Nov., 1971.
13. A. Fettweis, "A Simple Design of Maximally Flat Delay Digital Filters," *IEEE Trans. on Audio and Electroacoustics*, **AU-20**, No. 2, 112–114, June, 1972.
14. R. W. Daniels, *Approximation Methods for the Design of Passive, Active, and Digital Filters*, McGraw-Hill Book Co., 1974.

Frequency Transformations

1. L. Weinberg, *Network Analysis and Synthesis*, McGraw-Hill Book Co., New York, 1962.
2. A. G. Constantinides, "Spectral Transformation for Digital Filters," *Proc. IEE*, **117**, No. 8, 1585–1590, 1970.

Time Domain Design Techniques

1. C. S. Burrus and T. W. Parks, "Time Domain Design of Recursive Digital Filters," *IEEE Trans. Audio.*, **18**, 137–141, 1970.
2. J. L. Shanks, "Recursion Filters for Digital Processing," *Geophys*, **32**, 33–51, Feb., 1967.
3. F. Brophy and A. C. Salazar, "Considerations of the Padé Approximant Technique in the Synthesis of Recursive Digital Filters," *IEEE Trans. on Audio and Electroacoustics*, **AU-21**, No. 6, 500–505, Dec., 1973.
4. A. G. Evans and R. Fischl, "Optimal Least Squares Time-Domain Synthesis of Recursive Digital Filters," *IEEE Trans. on Audio and Electroacoustics*, **AU-21**, No. 1, 61–65, Feb., 1973.
5. F. Brophy and A. C. Salazar, "Recursive Digital Filter Synthesis in the Time Domain," *IEEE Trans. on Acoustics, Speech, and Signal Processing*, Vol. ASSP-22, No. 1, 45–55, Feb., 1974.

Optimization Techniques

1. K. Steiglitz, "Computer-Aided Design of Recursive Digital Filters," *IEEE Trans. on Audio and Electroacoustics*, **18**, 123–129, 1970.
2. A. G. Deczky, "Synthesis of Recursive Digital Filters using the Minimum P-Error Criterion," *IEEE Trans. on Audio and Electroacoustics*, **AU-20**, No. 4, 257–263, Oct., 1972.
3. H. D. Helms, "Digital Filters with Equiripple or Minimax Responses," *IEEE Trans. on Audio and Electroacoustics*, **19**, No. 1, 87–94, 1971.

4. A. DECZKY, "Computer Aided Synthesis of Digital Filters in the Frequency Domain," ScD. Thesis, Swiss Federal Institute of Technology, Zurich, Switzerland, 1973.

5. J. W. BANDLER AND B. J. BARDAKJIAN, "Least pth Optimization of Recursive Digital Filters," *IEEE Trans. on Audio and Electroacoustics*, **AU-21**, No. 5, 460–470, Oct., 1973.

6. P. THAJCHAYAPONG AND P. J. RAYNER, "Recursive Digital Filter Design by Linear Programming," *IEEE Trans. on Audio and Electroacoustics*, **AU-21**, No. 2, 107–112, Apr., 1973.

7. L. R. RABINER, N. Y. GRAHAM, AND H. D. HELMS, "Linear Programming Design of IIR Digital Filters with Arbitrary Magnitude Function," *IEEE Trans. on Acoustics, Speech, and Signal Processing*, Vol. ASSP-22, No. 2, 117–123, Apr., 1974.

8. R. FLETCHER AND M. J. D. POWELL, "A Rapidly Convergent Descent Method for Minimization," *Computer J.*, **6**, No. 2, 163–168, 1963.

5

Finite Word Length Effects in Digital Filters

5.1 Introduction

Up to this point we have been discussing digital filters as though both the coefficients and the variables of the filter were expressed to infinite precision. When implemented in special-purpose digital hardware or as a computer algorithm, attention must be paid to the effects of using finite register lengths to represent all relevant filter parameters. Included among these effects are

1. A/D conversion noise.
2. Uncorrelated roundoff noise.
3. Inaccuracies in filter response due to coefficient quantization.
4. Correlated roundoff noise or limit cycles.

Depending on the type of arithmetic used in the filter algorithm, the type of quantization used to reduce the word length to a desired size, and the exact filter structure used, one can generally estimate how filter performance is affected by each of the effects above in specific cases. In this chapter we present general discussions of each of the effects above and give detailed analyses of certain special cases.

There are several good reasons for choosing only certain special cases to study in detail. One of these is that the number of possible combinations of arithmetic type, method of quantization, and filter structure is quite large (i.e., 135 is not unreasonable for five types of arithmetic, three types of quantization, and nine filter structures). Another reason for restricting our attention to special cases is that many of the possibilities are only of academic interest—i.e., they would never be used in a practical system. Thus, for example, sign-magnitude arithmetic is generally avoided in digital systems because of the inherent problems it presents in performing simple arithmetic such as addition.

5.2 Analog-to-Digital Conversion

One of the most important ways of producing a sequence is by sampling a continuous-time or analog waveform. An analog-to-digital (A/D) converter is a device that operates on the analog waveform to produce a digital output consisting of a sequence of numbers each of which approximates a corresponding sample of the input waveform. Figure 5.1 shows a simple

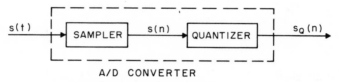

A/D CONVERTER

Fig. 5.1 Block diagram of A/D converter.

block diagram for the A/D converter that is conceptually represented as a two-stage process. In the first stage the sequence $s(n) = s(t)|_{t=nT}$ is created where $s(n)$ is expressed to infinite precision. In the second stage the numerical equivalent of each sample of $s(n)$ is expressed by a finite number of bits giving the sequence $s_Q(n)$. In actual A/D converters there are not two separate stages but instead a single stage with input $s(t)$ and output $s_Q(n)$. The difference signal $e(n) = s(n) - s_Q(n)$ is called quantizing noise or A/D conversion noise.

As discussed in Chapter 2, the analog signal $s(t)$ must be bandlimited in order for the quantized output $s_Q(n)$ to be a meaningful representation of $s(t)$. Thus generally a presampling analog lowpass filter precedes the A/D converter. The desirable characteristics of this analog filter are at least 40-dB loss for frequencies greater than half the sampling frequency and as little in-band ripple as possible in the signal baseband. Practical experience has shown that eighth-order (48-dB/octave slopes) lowpass filters are adequate for most speech processing applications.

Depending on the detailed way in which $s(n)$ is quantized, different distributions of quantization noise may be obtained. If the smallest quantization step used in the digital representation of $s_Q(n)$ is called Q, then the relation between $s_Q(n)$ and $s(n)$ for *rounding* is given in Fig. 5.2. Since there

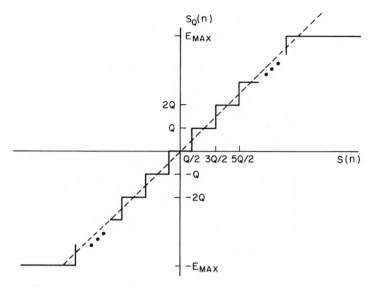

Fig. 5.2 Quantizer characteristic with rounding.

are only a finite number of quantization levels, all signals either exceeding the largest $(E_{\max})$ or falling below the smallest $(-E_{\max})$ quantization level are rounded to these numbers. Generally such overflows or underflows are avoided by judicious choice of the quantization step Q and by careful scaling of the analog input waveform. With these exceptions, it is clear from Fig. 5.2 that the error signal satisfies the relation

$$-\frac{Q}{2} \le e(n) \le \frac{Q}{2} \tag{5.1}$$

for all n. Under certain not overly restrictive assumptions it can be shown that the distribution of the error signal is uniform. Thus the probability distribution of the quantization error for rounding is as shown in Fig. 5.3.

To illustrate the process of rounding in an A/D converter, Fig. 5.4 shows an analog waveform $s(t)$ and the resulting digital samples $s_Q(n)$ that are obtained. At the bottom of Fig. 5.4, the quantization error (on a magnified

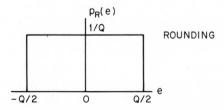

Fig. 5.3 Probability density function for roundoff error.

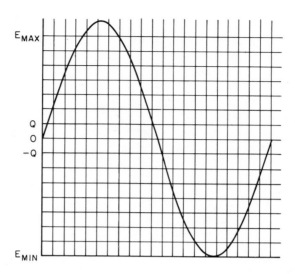

QUANTIZATION ERROR SIGNAL

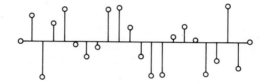

Fig. 5.4 Quantization error for a sine wave.

scale) is shown for each of the samples. This quantization error resembles a random sequence as predicted by the argument above.

Other types of quantization may be used to give $s_Q(n)$. For example, *truncation* is the rule where the signal is represented by the highest quantization level that is not greater than the signal. Figure 5.5 shows the relation between $s_Q(n)$ and $s(n)$ for truncation. Since truncation is equivalent to rounding less one-half a quantization step, the probability distribution of the error signal is as shown in Fig. 5.6.

From Fig. 5.3 and 5.6 it is clear that the quantization error has mean value of 0 for rounding and $Q/2$ for truncation, whereas the variance is $Q^2/12$ in both cases.

A third type of signal quantization is often used in digital processing. This type of quantization, called *sign-magnitude truncation*, is identical to truncation for positive signals; negative signals are approximated by the nearest quantization level that is greater than the signal. Thus, depending on whether $s_Q(n)$ is positive or negative, the distribution of Fig. 5.6, or its

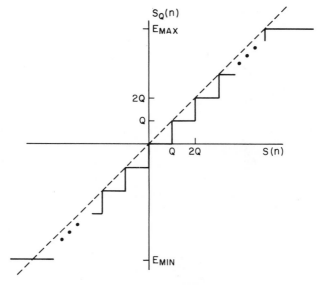

Fig. 5.5 Quantizer characteristic with truncation.

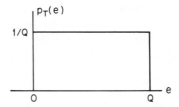

Fig. 5.6 Probability density function of truncation error.

mirror image, is used. The mean value of the quantization error is 0 for sign-magnitude truncation but the variance is $Q^2/3$, or four times that of either rounding or straight truncation. Because of the statistical considerations above, in most practical situations rounding is to be preferred to the other rules for quantizing a signal.

The importance of understanding A/D conversion and gaining insight into the distribution of the A/D conversion noise is that one can represent the quantized input to a digital system [with impulse response $h(n)$] as the infinite precision input [$s(n)$] plus the quantization error [$e(n)$] as shown in Fig. 5.7. Based on linearity, one can process the sequences $s(n)$ and $e(n)$ independently. Thus the output sequence (assuming infinite precision processing) can be represented as

$$y(n) = \underbrace{s(n) * h(n)}_{\text{signal}} + \underbrace{e(n) * h(n)}_{\text{noise}}$$

$$(5.2)$$

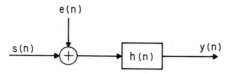

Fig. 5.7 Linear model of quantization noise in an LTI system.

In this manner one can define a signal and a noise output of the filter and thus obtain a measure of signal-to-noise ratio of the processed sequence. Of course the processing in the digital filter is performed with finite precision; thus the ideas above must be modified somewhat to account for this effect. However, the idea of representing a quantized signal as the sum of an unquantized signal plus a noise term is basic to most of the theory of finite word length effects.

Before proceeding to a discussion of the various types of arithmetic that are used in digital filter realizations, it is worthwhile emphasizing a small but important point about analog-to-digital conversion noise. All analog signals are corrupted by some form of noise—i.e., they have a finite signal-to-noise ratio. For example, a speech signal transmitted over an ordinary telephone line has a signal-to-noise ratio of about 36 dB. Thus the use of a large number of bits to represent such a signal does *not* give a digital signal with a higher signal-to-noise ratio than the analog signal. Hence even though the quantization step size Q may be very small relative to the peak signal, the lower-order bits of the sequence are merely providing a good representation of the noise inherent in the analog signal. This implies that, depending on the exact signal, increasing the number of bits in the A/D converter beyond a certain point merely increases the accuracy by which an analog noise is represented. Thus the word length of an A/D converter is dictated by the signals being converted.

5.3 Digital-to-Analog Conversion

A digital-to-analog (D/A) converter is a device that operates on a sequence $y(n)$ to give an analog signal $y(t)$, defined by

$$y(t) = \sum_n y(n)h(t - nT) \tag{5.3}$$

where $h(t)$ characterizes the particular D/A converter. One of the most commonly used waveforms for $h(t)$ is a square pulse of duration T seconds. In this case the device is generally called a zero-order hold D/A converter. Figure 5.8 shows a typical sequence $y(n)$ and the analog waveform $y(t)$ obtained from Eq. (5.3) using a zero-order hold D/A converter. It is clear from Fig. 5.8 that the analog waveform $y(t)$ contains a large amount of

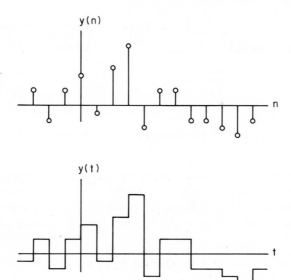

Fig. 5.8 Input and output waveforms for a D/A converter.

undesirable high-frequency energy. To eliminate this high-frequency energy, the D/A converter is usually followed by a linear, time-invariant analog lowpass filter, called a post filter, with cutoff frequency of at most $1/(2T)$ Hz (where $1/T$ is the sampling frequency). The combination of a D/A converter and a post filter is called a reconstruction device or reconstruction filter.

The process of D/A conversion does not inherently lead to any error in reconstructing an analog waveform that is, in some sense, equivalent to the digital sequence at its input. It is clear from Eq. (5.3), however, that $Y(\omega)$, the spectrum of the analog waveform $y(t)$, is not identical to $Y_D(e^{j\omega})$, the spectrum of the sequence $y(n)$ in the region $0 \leq f \leq 1/(2T)$, because $Y(\omega)$ and $Y_D(e^{j\omega})$ satisfy the relation

$$Y(\omega) = Y_D(e^{j\omega}) \cdot H(\omega) \tag{5.4}$$

where $H(\omega)$ is the spectrum of $h(t)$. In the case of the zero-order hold device,

$$|H(\omega)| = \frac{2 \sin (\omega T/2)}{\omega} \tag{5.5}$$

To compensate for this spectrum shaping by the D/A converter, the digital sequence is often preprocessed by a digital filter with a magnitude response approximating

$$|G(e^{j\omega})| = \frac{\omega}{2 \sin (\omega T/2)} = \frac{1}{|H(\omega)|} \tag{5.6}$$

Thus the cascade of $G(e^{j\omega})$ and $H(\omega)$ yields an overall flat magnitude response. Figure 5.9 shows the processing required to go from a digital sequence $y(n)$ to

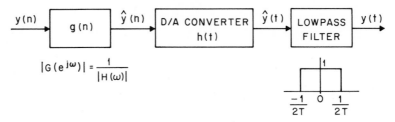

Fig. 5.9 Model of a compensating D/A converter.

an analog waveform $y(t)$ whose spectrum is equivalent to the digital spectrum in the interval $-1/(2T) \leq f \leq 1/(2T)$. It should be clear that the digital compensating network of Fig. 5.9 can be combined with any other digital processing in the system and need not be realized as a separate filter.

5.4 Types of Arithmetic in Digital Systems

As mentioned earlier there are a variety of types of arithmetic that are used in the implementation of digital systems. Among the most common are fixed point and floating point. Recently a hybrid between these arithmetic types was introduced called block floating point arithmetic. Although other forms of arithmetic can be used, we shall concentrate on these three types in our discussion of finite word length effects as they are the ones used most commonly. Many of the results presented in this chapter are straightforwardly extended to the new types of arithmetic as they are devised.

5.5 Fixed Point Arithmetic

It is assumed throughout this chapter that a word length of b bits is chosen to represent the numbers in a digital filter. Therefore 2^b different numbers may be represented exactly with a b-bit word. In fixed point representation it is assumed that the position of the binary point is fixed. The bits to the right represent the fractional part of the number and those to the left represent the integer part. Thus the binary number 01.01100 has the value $(0 \times 2^2) + (1 \times 2^1) + (0 \times 2^{-1}) + (1 \times 2^{-2}) + (1 \times 2^{-3}) + (0 \times 2^{-4}) + (0 \times 2^{-5})$ or 1.375 in base 10.

Depending on the way negative numbers are represented, there are three different forms of fixed point arithmetic. They are called *sign-magnitude*, *2's-complement*, and *1's-complement* representation. For sign-magnitude representation the leading binary digit is used to represent the sign (a 0 corresponds to $a +$ and 1 corresponds to $a -$), and the remaining $(b-1)$ bits are used to represent the magnitude. Thus, for example, with $b = 7$, the base 10 number -1.375 is represented as 11.01100 where the binary point is assumed to be located following the second bit. In sign-magnitude form the number 0 has two representations, i.e., 00.00000 or 10.00000 with $b = 7$. Thus with b bits, only $(2^b - 1)$ numbers are represented exactly.

For 2's-complement representation, a positive number is obtained identically as in the sign-magnitude system. To obtain the negative of a particular positive number, one simply complements all the bits of the positive number and adds one unit in the position of the least significant bit. For example, the negative of (01.01100) would be represented in 2's complement as

$$-(01.01100) = (10.10011) + (00.00001) = 10.10100$$

The largest positive number that can be represented exactly in 2's complement is 01.11111 (for $b = 7$), whereas the largest negative number that can be represented exactly is 10.00000, which is one unit larger (in magnitude) than the largest positive number. Since there is only one representation for the number 0, there are 2^b distinct numbers represented exactly with a b-bit word.

For the 1's-complement form, positive numbers are represented as in sign magnitude and 2's complement. The negative of a positive number is obtained by complementing all the bits of the positive number. For example, the negative of (01.01100) is $-(01.01100) = 10.10011$. Thus the number 0 has two representations, i.e., 00.00000 and 11.11111 in a 1's-complement system.

The choice among the three representations above is generally dictated by programming and/or hardware considerations. The operation of subtraction may be conveniently performed using an adder and a 2's complementer in 2's-complement arithmetic. Sign-magnitude representation is often appropriate for serial multipliers, which can multiply magnitudes and determine the sign of the resulting product by simple logic based on the signs of the two numbers being multiplied.

In most fixed point arithmetic realizations of digital filters, the position of the binary point is assumed to be just to the right of the first bit. Thus the range of numbers that can be represented is from -1.0 to $1.0 - 2^{-(b-1)}$ where b is the number of bits in the word. This results in no loss of generality as the specific signal level may be scaled to be in any desired range. In the case of digital filter coefficients, the position of the binary point is often moved further to the right to allow filter coefficients with magnitudes greater than 1.0.

5.6 Floating Point Arithmetic

In the floating point representation of a positive number, there are two fixed point numbers, the mantissa and the exponent. The floating point number f is obtained as the product of the mantissa m with the quantity that results when a given base (usually 2) is raised to the power denoted by the exponent a—i.e.,

$$f = m \cdot 2^a \qquad (5.7)$$

The mantissa m is generally normalized to be as large as possible but less than some number—e.g., 1.0. We shall assume in this chapter that m is normalized to the range

$$\tfrac{1}{2} \leq m < 1 \qquad (5.8)$$

Thus the decimal numbers 3.0, 1.5, and 0.75 have floating point representations (base 2) as $2^2 \times 0.75$, $2^1 \times 0.75$, and $2^0 \times 0.75$, respectively.

Negative floating point numbers are generally represented by considering the mantissa as a signed fixed point number. Thus the sign of the floating point number is obtained from the first bit of the mantissa. The exponent is also a signed fixed point number since negative exponents represent numbers with magnitudes less than 0.5.

The number of bits b in a floating point word must be broken up into the number of bits b_1 in the mantissa and the number of bits $b_2 = b - b_1$ in the exponent. For a fixed value of b, the more bits allocated to the exponent, the greater the dynamic range of numbers that can be represented but the smaller the accuracy in representing these numbers. In most practical situations where floating point arithmetic is used, e.g., computer simulations using higher level languages, b_1 is approximately $\tfrac{3}{4}b$. Thus, for example, on a computer with a 36-bit word length, 27 bits are used for the mantissa and 9 bits for the exponent. For the floating point number f, this gives a dynamic range of

$$0.5 \times 2^{-256} \leq |f| < 2^{256}$$

and a precision in the mantissa of about 2^{-27}.

In floating point arithmetic, multiplications are carried out as follows. Let f_1 and f_2 be two floating point numbers with representations

$$f_1 = m_1 \times 2^{a_1}$$

$$f_2 = m_2 \times 2^{a_2}$$

Then the product $f_3 = f_1 \times f_2$ is formed as

$$f_3 = (m_1 \times m_2)2^{(a_1+a_2)}$$

i.e., the mantissas are multiplied using fixed point arithmetic and the exponents are added. The product $|m_1 \times m_2|$ must be in the range of 0.25 to 1.0; hence it might not be properly normalized. To correct this problem, the exponent $a_1 + a_2$ must be altered. Thus the product of $(1.25)_{10} \times (1.25)_{10}$ is obtained as $(b_2 = 3, b_1 = 9)$

$(2^{001} \times 0.10100000) \times (2^{001} \times 0.10100000)$

$$= 2^{010} \times 0.01100100 = 2^{001} \times 0.11001000$$

Addition of two floating point numbers is somewhat more complicated than addition of fixed point numbers. In order to carry out an addition, the exponent of the smaller number must be changed to equal the exponent of the larger number. Therefore the mantissa of the smaller number must be unnormalized to correspond to the new exponent. The resulting mantissas are added using fixed point arithmetic and the sum is renormalized to be in the range of Eq. (5.8). If renormalization is used on the sum, the resulting exponent must be adjusted. As an example, consider the addition of the two floating point numbers given below. (The mantissas and exponents are represented using 2's-complement arithmetic.)

$$f_{1n} = 2^{010} \times 0.11000000 - f_1 \text{ normalized} = 3.0)_{10}$$

$+ \qquad f_{2n} = 2^{000} \times 0.10100000 - f_2 \text{ normalized} = 0.625)_{10}$

$$f_{2u} = 2^{010} \times 0.00101000 - f_2 \text{ unnormalized}$$

$f_{3u} = f_{1n} + f_{2u} = 2^{010} \times 0.11101000 - \text{sum unnormalized} = 3.625)_{10}$

$$f_{3n} = f_{3u} - \text{sum normalized}$$

In this case the number f_{2n} was unnormalized to give f_{2u}, which when added to f_{1n} gave the result f_{3u} which, in this example, turned out to be a normalized number.

It may be deduced from the discussion above that the floating point representation can lead to truncation or rounding errors for both addition and multiplication as opposed to the fixed point case where such errors occurred only for multiplication. The tradeoff occurs in that addition can lead to overflows in the fixed point case but this is highly unlikely in the floating point case because of the very large dynamic range.

5.7 Block Floating Point

The block floating point representation of numbers is a hybrid between the fixed and floating point systems. In this representation, instead of normalizing each number to be represented individually, as in the floating point case, a block or array of numbers has a fixed exponent associated with it. This fixed exponent is obtained by examining all the numbers in the block and representing the largest number as an ordinary floating point number with a normalized mantissa. The advantage of such a system is that the use of a single exponent for a large block of numbers saves memory. This representation of numbers is most suitable in implementations of such algorithms as the fast Fourier transform, although it can be used in implementing digital filters as well.

5.8 Types of Quantization in Digital Filters

As seen in previous chapters, the operations required in realizing digital filters include multiplications, additions, and shifts. Even though the input to a digital filter (e.g., from an A/D converter) is represented by a finite word length sequence, the results of processing generally lead to filter variables that require additional bits for accurate representation. For example, if a b-bit input word is multiplied by a b-bit filter coefficient, a register that is $2b$ bits long is required to store the product. If this product is not quantized back to b bits, then the number of bits required to store subsequent products will increase indefinitely for a recursive realization.

Consider, for example, the first-order recursive structure of Fig. 5.10.

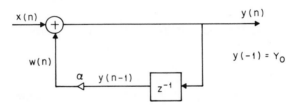

Fig. 5.10 A simple first-order recursive structure.

If the input sequence is quantized to b bits and the multiplier α is quantized to b bits, then after the first iteration $w(n)$ will require $2b$ bits, i.e., $\alpha(b$ bits$) \times y(-1)(b$ bits$) = w(0)(2b$ bits$)$. If $w(0)$ is not quantized, then $y(0) = w(0) + x(0)$ will require $2b$ bits to be represented. After the second iteration, $w(n)$ will require $3b$ bits—i.e., $\alpha(b$ bits$) \times y(0)(2b$ bits$) = w(1)(3b$ bits$)$. If this recursion is carried out indefinitely, the number of bits required to represent $w(n)$ [or $y(n)$] grows linearly to infinity. This is clearly an unacceptable situation. The common solution to this problem is to eliminate the low-order bits resulting from the multiplications (and, in some cases, the additions) in the digital filter. There are two standard methods for eliminating the low-order bits and these are truncation and rounding. We have already discussed these processes with respect to A/D conversion and therefore we shall only summarize their properties with respect to the different types of arithmetic used in realizing the digital filter.

5.9   Truncation

Truncation of a number is accomplished by discarding all bits less significant than the least significant bit that is retained. Thus for a 2's-complement representation, the error due to truncation of a positive number satisfies the inequality

$$0 \geq x_T - x > -2^{-b} \tag{5.9}$$

where b is the number of bits following the binary point that is retained and x_T is the truncated value of x and it is assumed that $|x| \leq 1.0$. For either a 1's-complement or sign-magnitude representation, the truncation error satisfies Eq. (5.9) only if $x > 0$. For $x < 0$, the truncation error satisfies the inequality

$$0 \leq x_T - x < 2^{-b} \qquad x < 0 \tag{5.10}$$

For truncation in floating point systems, the effect is seen only in the mantissa. Thus, consider the floating point word $x = 2^c \cdot m$ where m is the mantissa and c is the exponent. If the mantissa is truncated to b bits and the resulting truncation error $(x_T - x)$ is written as a quantity proportional to x, i.e.,

$$x_T - x = (1 + \varepsilon)x \tag{5.11}$$

then for a 2's-complement representation of the mantissa the truncation error satisfies the inequality

$$0 \geq \varepsilon x \geq -2^{-b} \cdot 2^c \tag{5.12}$$

If x is positive, then it satisfies the inequality

$$2^{c-1} \leq x < 2^c \qquad x > 0 \tag{5.13}$$

Since Eq. (5.12) shows εx is negative when x is positive, then ε must be negative. Thus Eq. (5.13) can be multiplied by ε, reversing the signs of the inequalities, giving

$$2^c \varepsilon < \varepsilon x \leq 2^{c-1} \varepsilon \qquad x > 0 \tag{5.14}$$

Combining the inequalities of Eq. (5.14) and (5.12) and solving for ranges of ε gives

$$0 \geq \varepsilon > -2^{-b} \cdot 2 \qquad x > 0 \tag{5.15}$$

Similarly, for $x < 0$, an inequality on ε can be obtained. The result is

$$0 \leq \varepsilon < 2^{-b} \cdot 2 \qquad x < 0 \tag{5.16}$$

In similar fashion, inequalities on ε may be obtained for the cases of 1's-complement or sign-magnitude representation of the mantissa. In these cases the resulting inequality is

$$0 \leq \varepsilon < -2 \cdot 2^{-b} \qquad \text{all } x \tag{5.17}$$

The most important aspect of the truncation error is that in all cases it lies between 0 and a number proportional to $\pm 2^{-b}$.

5.10 Rounding

Rounding of a number of b bits is accomplished by choosing the rounded result as the b-bit number closest to the original unrounded quantity. When the unrounded number lies midway between two adjacent b-bit numbers, a random choice ought to be made as to which of these numbers to round to. Thus 0.01010 rounded to two bits (following the binary point) is 0.01; whereas when it is rounded to three bits, it can be either 0.011 or 0.010 and the choice should be random. In most situations when a choice exists between which of two numbers to round to, an arbitrary decision to round up or down will have negligible effect on the accuracy of the computation.

For fixed point arithmetic, the error made by rounding a number to b bits following the binary point satisfies the inequality

$$-\frac{2^{-b}}{2} \leq x_T - x \leq \frac{2^{-b}}{2} \tag{5.18}$$

for all three types of number systems—i.e., two's complement, one's complement, and sign-magnitude.

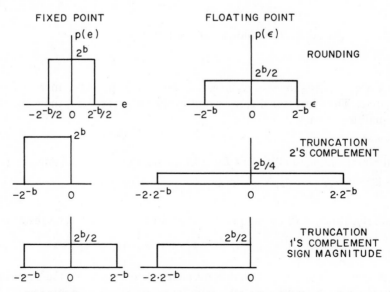

Fig. 5.11 Summary of quantization noise probability density functions.

For floating point arithmetic the error due to rounding satisfies the inequality

$$-2^c \cdot \frac{2^{-b}}{2} \le x_T - x \le 2^c \cdot \frac{2^{-b}}{2} \tag{5.19}$$

If the rounding error $(x_T - x)$ is again represented as being proportional to x, i.e., $x_T - x = \varepsilon x$, then the inequality on ε is of the form

$$-2^{-b} \le \varepsilon \le 2^{-b} \tag{5.20}$$

Equation (5.20) is valid for all three types of representation of the mantissa.

One convenient way of summarizing Eq. (5.9) to (5.20) on truncation and rounding errors is to show probability density functions for each of the types of error. Although the range of the errors is known, the probability of obtaining an error within the range is not known. It is quite reasonable to assume that all errors are equiprobable; i.e., the errors occur uniformly throughout the interval. Under these assumptions, Fig. 5.11 shows probability density functions for rounding and truncation for fixed and floating point realizations. For the fixed point case the variable is the error $e = x_T - x$, whereas for the floating point case the variable is $\varepsilon = (x_T - x)/x$.

5.11 Roundoff Noise in Recursive Structures—Fixed Point Arithmetic

The arithmetic operations involved in realizing a digital filter using fixed point arithmetic include multiplication by constants (the filter coefficients) and addition. As long as no overflow occurs, the addition of two or more fixed point numbers cannot lead to any inaccuracy in representing the sum. (Since overflows on addition can occur, this means that signal level dynamic range constraints must be taken into consideration in the realization of the filter. We shall return to this issue later in this chapter.) Multiplication, on the other hand, *cannot* cause overflow (assuming both numbers being multiplied are properly scaled); however, the results of a multiplication must be quantized. Unless otherwise stated, we shall assume that rounding is used to perform this quantization because of its desirable properties—i.e., the error signal is independent of the type of arithmetic, its mean is zero, unlike truncation, and no other method of quantization yields lower variance.

The model for fixed point roundoff noise following a multiplication is shown in Fig. 5.12. The multiplier is modeled as an infinite precision multiplier followed by an adder where roundoff noise is added to the product so that the overall result equals some quantization level. Each roundoff noise sample e is modeled as a random variable with uniform probability density as shown in Fig. 5.11 for the case of fixed point rounding. Thus each roundoff noise sample is a zero mean random variable with a variance of $2^{-2b}/12$

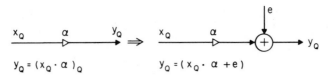

Fig. 5.12 Fixed point quantization noise model.

where $b + 1$ is the number of bits (including sign bit) used to represent the filter variables.

In order to model the effects of rounding due to multiplication in a digital filter, certain assumptions must be made concerning the statistical independence of the various noise sources that occur in realizing the filter. In particular, the following assumptions are generally made.

1. Any two different samples from the same noise source are uncorrelated.
2. Any two different noise sources (i.e., associated with different multipliers), regarded as random processes, are uncorrelated.
3. Each noise source is uncorrelated with the input sequence.

Thus each noise source is modeled as a discrete stationary white random process with a uniform power density spectrum of $2^{-2b}/12$.

One comment about the assumptions above is necessary. It is clear that the assumptions are not always valid. In particular, if the input is a constant, all three assumptions break down. In such cases the roundoff noise (error) is no longer uncorrelated with the input. We shall defer the questions associated with correlated roundoff noise or limit cycles to Sec. 5.31.

Based on the model above, Fig. 5.13 shows a block diagram for a fourth-order direct form digital filter where each finite precision multiplier has been replaced by an ideal multiplier and an additive roundoff noise. Since all the noise sources are added at the same point in the filter, there is equivalently a single noise source $e(n) = e_0(n) + e_1(n) + \ldots + e_8(n)$ with zero mean and variance $\sigma^2 = \sigma_0^2 + \sigma_1^2 + \ldots + \sigma_8^2 = (9 \times 2^{-2b})/12$ (by assumption 2) as shown at the bottom of Fig. 5.13.

If the fourth-order filter of Fig. 5.13 were realized as a cascade of two second-order systems, the noise model of one such structure would be as shown in Fig. 5.14. Again there are nine noise sources [$e_0(n)$ to $e_8(n)$] in this realization; however, they do not all add in at the same point as in Fig. 5.13. For this case, as for the case of Fig. 5.13, it is important to be able to determine the total noise variance at the *output* of the filter. Linear system theory can be used to determine the output variance due to any of the noise sources and, by assumption 2 above, the total variance can be obtained by adding the individual variances.

Consider the kth noise source $e_k(n)$. Let $h_k(n)$ be the impulse response from the noise source to the filter output. [Note that $h_k(n)$ may be obtained from the specific implementation being studied using the techniques of linear

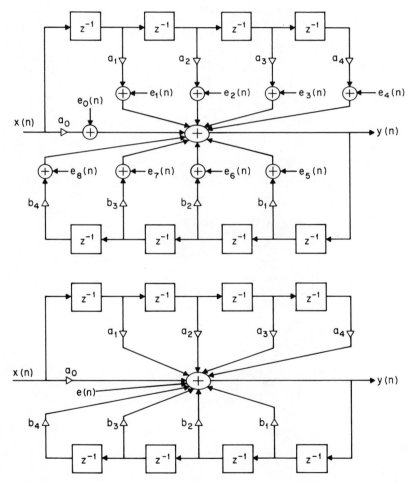

Fig. 5.13 Quantization noise model for a fourth-order recursive system.

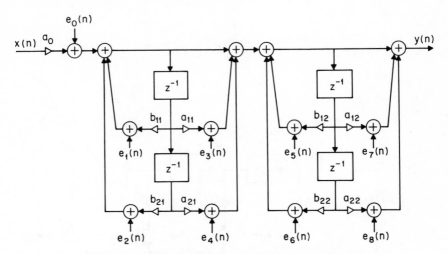

Fig. 5.14 Quantization noise in a cascade of recursive sections.

time-invariant systems.] The output noise component $\mathscr{E}_k(n)$, due solely to $e_k(n)$, may be obtained via convolution as

$$\mathscr{E}_k(n) = \sum_{m=0}^{n} h_k(m)e_k(n - m) \tag{5.21}$$

The variance of $\mathscr{E}_k(n)$ may be obtained as

$$\sigma_{0k}^2(n) = E\left[\sum_{m=0}^{n} h_k(m)e_k(n - m) \times \sum_{l=0}^{n} h_k(l)e_k(n - l)\right]$$

$$= \sum_{m=0}^{n} \sum_{l=0}^{n} h_k(m)h_k(l)E[e_k(n - m)e_k(n - l)]$$

$$= \sum_{m=0}^{n} \sum_{l=0}^{n} h_k(m)h_k(l) \, \delta(l - m)\sigma_e^2 \tag{5.22}$$

$$\sigma_{0k}^2(n) = \sigma_e^2 \sum_{m=0}^{n} h_k^2(m) \tag{5.23}$$

where assumptions 1 and 3 on page 310 have been used to derive Eq. (5.23) and $\sigma_e^2 = 2^{-2b}/12$.

In the limit, as n tends to infinity, the variance $\sigma_{0k}^2(n)$ tends to the steady-state limit

$$\sigma_{0k}^2 = \sigma_e^2 \sum_{m=0}^{\infty} h_k^2(m) \tag{5.24}$$

The total steady-state noise variance σ_0^2 is then

$$\sigma_0^2 = \sum_k \sigma_{0k}^2 \tag{5.25}$$

It is fairly straightforward to determine σ_0^2 in the case of first- and second-order sections. For the first-order filter, as shown in Fig. 5.15, there is only one noise source $e_0(n)$ with impulse response $h_0(n) = k^n u_{-1}(n)$. Thus σ_0^2 is

$$\sigma_0^2 = \sigma_{00}^2 = \frac{2^{-2b}}{12} \sum_{m=0}^{\infty} k^{2m} = \frac{2^{-2b}}{12} \frac{1}{1 - k^2}$$

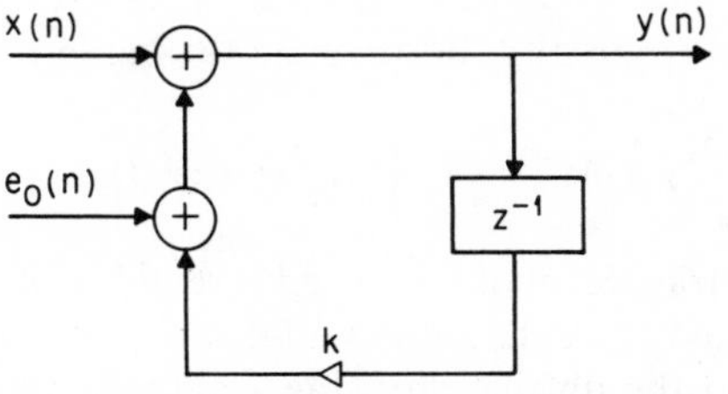

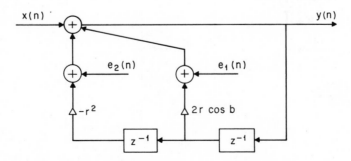

Fig. 5.15 Quantization noise model for a first-order system.

Figure 5.16 shows a realization of a second-order system with z transform

$$H(z) = \frac{1}{1 - 2r(\cos b)z^{-1} + r^2 z^{-2}}$$

which has two noise sources $e_1(n)$ and $e_2(n)$, each of which sees the impulse response

$$h_1(n) = h_2(n) = \{r^n \sin[(n+1)b]/\sin b\}u_{-1}(n).$$

In this case the steady-state output noise variance is

$$\sigma_0^2 = \sigma_{01}^2 + \sigma_{02}^2 = 2 \cdot \frac{2^{-2b}}{12} \sum_{m=0}^{\infty} r^{2m} \frac{\sin^2[(m+1)b]}{\sin^2 b}$$

Fig. 5.16 Quantization noise model for a second-order system.

which may be expressed in closed form as

$$\sigma_0^2 = \frac{2^{-2b}}{6} \left[\frac{(1 + r^2)}{(1 - r^2)} \frac{1}{(1 + r^4 - 2r^2 \cos{(2b)})} \right] \tag{5.25a}$$

The evaluation of the steady-state output noise variance [Eq. (5.24) or (5.25)] involves the infinite summation of $h_k^2(m)$. It is difficult to solve for $h_k(m)$ in a simple closed form and even more difficult to perform the desired summation. Sometimes, computations can be eased by making use of the Parseval relations of Chapter 2, however, to evaluate the infinite sum as

$$\sum_{m=0}^{\infty} h_k^2(m) = \frac{1}{2\pi j} \oint H_k(z) H_k(z^{-1}) z^{-1} \, dz \tag{5.26}$$

where $H_k(z)$ is the z transform of $h_k(m)$ and is readily obtained from examining the configuration used to realize the digital filter. The integral in Eq. (5.26) can be evaluated on the unit circle using Cauchy's residue theorem. Thus, for example, in the case of the second-order system of Fig. 5.16

$$H_1(z) = H_2(z) = \frac{1}{1 - 2r (\cos b) z^{-1} + r^2 z^{-2}}$$

which has poles at $z = re^{\pm jb}$ ($r < 1$ for stability). $H_1(z^{-1})$ and $H_2(z^{-1})$ have poles at $z = (1/r)e^{\pm jb}$ that are outside the unit circle. Thus to evaluate the integral of Eq. (5.26) one needs to find the residues of $H_1(z)H_1(z^{-1})/z$ at $z = re^{\pm jb}$. The result is identical to the expression given previously for the second-order system in Eq. (5.25a).

As another example of computing steady-state output noise variance, consider the cascade of first-order sections shown in Fig. 5.17. The noise source $e_2(n)$ sees a network with z transform

$$H_2(z) = \frac{1}{1 - k_2 z^{-1}}$$

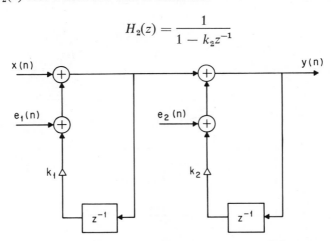

Fig. 5.17 Quantization noise in a cascade of two first-order sections.

whereas the noise source $e_1(n)$ sees a network with z transform

$$H_1(z) = \frac{1}{(1 - k_2 z^{-1})(1 - k_1 z^{-1})}$$

Thus using Eqs. (5.24) to (5.26) the total steady-state noise variance is

$$\sigma_0^2 = \frac{2^{-2b}}{12}\left[\frac{1}{2\pi j}\oint\left(\frac{1}{1 - k_2 z^{-1}}\right)\left(\frac{1}{1 - k_2 z}\right)z^{-1}\,dz\right.$$

$$\left. + \frac{1}{2\pi j}\oint\left(\frac{1}{1 - k_1 z^{-1}}\right)\left(\frac{1}{1 - k_1 z}\right)\left(\frac{1}{1 - k_2 z^{-1}}\right)\left(\frac{1}{1 - k_2 z}\right)z^{-1}\,dz\right]$$

The closed contour for evaluation of the integrals above is taken to be the unit circle in which case the first integral has only one pole (at $z = k_2$) inside the unit circle, whereas the second integral has two poles (at $z = k_1$ and $z = k_2$) inside the unit circle. Evaluating the residues gives

$$\sigma_0^2 = \frac{2^{-2b}}{12}\left[\frac{1}{1 - k_2^2} + \frac{k_1}{(1 - k_1^2)(k_1 - k_2)(1 - k_2 k_1)}\right.$$

$$\left. + \frac{k_2}{(1 - k_2^2)(k_2 - k_1)(1 - k_1 k_2)}\right]$$

$$\sigma_0^2 = \frac{2^{-2b}}{12}\left[\frac{1}{1 - k_2^2} + \frac{(1 + k_1 k_2)}{(1 - k_1 k_2)}\frac{1}{(1 - k_1^2)(1 - k_2^2)}\right]$$

To obtain this result without the help of Eq. (5.26) would have been far more tedious. Thus the ability to use spectral techniques to evaluate roundoff noise variance makes practical several of the scaling procedures to be discussed in Sec. 5.12.

5.12 Dynamic Range Constraints— Fixed Point Case

As mentioned earlier, additions in fixed point arithmetic cause no roundoff errors at all; however, they may cause far more disastrous overflows to occur in the digital filter. Although several reasonable ways of handling overflow problems after they are detected have been proposed, these techniques should not be relied on because of the nonlinearity of the processing following overflow. Instead the filter should be designed so that overflow is unlikely to occur under most normal operating conditions. To prevent overflows the signal levels at certain points in the digital filter must be scaled so that no overflow occurs following additions. This section presents a very

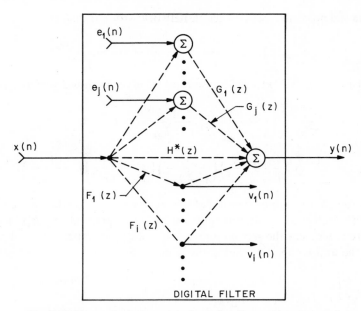

Fig. 5.18 Directed graph model of a digital filter. (After Jackson.)

general theory for choosing scale factors to prevent overflow and yet maintain the largest possible signal-to-roundoff noise ratio in the filter. The theoretical bases for these results are primarily due to Jackson and, although somewhat involved, they provide a reasonably straightforward method for both choosing the required filter word length to maintain a desired signal-to-roundoff noise ratio and deciding on the most appropriate configuration for realizing the filter. It is therefore hoped that the difficulties encountered in following the material of this section are more than compensated for by the insight gained into the filter realization problem. We shall generally follow the notation and method of exposition that Jackson has introduced.

Figure 5.18 shows a directed graph model of a digital filter where the multipliers and delays are represented by graph branches and the interconnection points, or nodes, are either summation nodes (corresponding to adders) or branch nodes (corresponding to wired interconnections).

The input to the filter is $x(n)$ and the output is $y(n)$. The output from the ith branch node is $v_i(n)$ and the roundoff error at the jth summation node is $e_j(n)$. The sequences $h(n)$ and $f_i(n)$ are the filter impulse response and the impulse response at the ith branch node, respectively. The sequence $g_j(n)$ is the impulse response due to the sequence $e_j(n) = u_0(n)$ with $x(n) = e_k(n) = 0$, $k \neq j$. The transforms $H(z)$, $F_i(z)$, and $G_j(z)$ are the z transforms of $h(n)$, $f_i(n)$, and $g_j(n)$, respectively.

To fix ideas, consider the direct form structure shown in Fig. 5.19. In this case there are two summation nodes and two branch nodes. The relevant z transforms of the model are

$$F_1(z) = 1$$

$$F_2(z) = \frac{1}{1 - b_1 z^{-1} - b_2 z^{-2} - b_3 z^{-3}}$$

$$H(z) = \frac{a_0 + a_1 z^{-1} + a_2 z^{-2} + a_3 z^{-3}}{1 - b_1 z^{-1} - b_2 z^{-2} - b_3 z^{-3}}$$

$$G_1(z) = H(z)$$

$$G_2(z) = 1$$

The goal of this modeling is to determine a method for scaling the sequences $v_i(n)$ to ensure that no overflow occurs at any summation node, at the same time minimizing output roundoff noise variance. If we assume that there are k_j error sources input to the jth summation node (e.g., $k_1 = 3$, $k_2 = 4$ in the example of Fig. 5.19) and that each noise source is white, of uniform power spectrum density $Q^2/12$ where $Q = 2^{-b} =$ quantization step size, then the power spectrum of $e_j(n)$ is a white, uniform distribution of density $k_j(Q^2/12)$ by the assumption of uncorrelated sources. The output power density spectrum may then be obtained using linear system theory and is

$$N_y(e^{j\omega}) = \frac{Q^2}{12} \sum_j k_j |G_j(e^{j\omega})|^2 \tag{5.27}$$

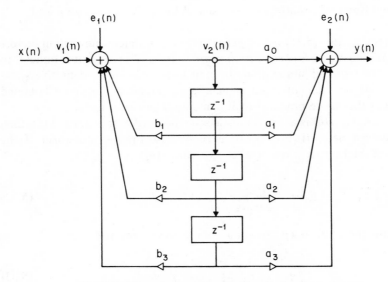

Fig. 5.19 Direct form realization of a third-order system.

If scaling is included in the filter design (and we use a prime to denote a scaled quantity), then Eq. (5.27) becomes

$$N_y(e^{j\omega}) = \frac{Q^2}{12} \sum_j k'_j |G'_j(e^{j\omega})|^2 \tag{5.28}$$

with $k'_j \geq k_j$ since scaling multipliers give additional error sources.

If we assume the input sequence $x(n)$ is bounded in magnitude by 1.0, it is easy to derive scale factors to guarantee $|v_i(n)| \leq 1$, $i = 1, 2, \ldots$. The sequence $v_i(n)$ satisfies the convolutional relation (assuming zero initial conditions and neglecting roundoff noise)

$$v_i(n) = \sum_{m=0}^{\infty} f_i(m)x(n - m) \tag{5.29}$$

Since $|x(n - m)| \leq 1$, this implies

$$|v_i(n)| \leq \sum_{m=0}^{\infty} |f_i(m)| \tag{5.30}$$

Thus to ensure $|v_i(n)| \leq 1$, it is sufficient that the scaled sequence $f'_1(m)$ must satisfy the relation

$$\sum_{m=0}^{\infty} |f'_i(m)| \leq 1 \tag{5.31}$$

It is straightforward to show the necessity of Eq. (5.31) to ensure $|v_i(n)| \leq 1$, for all n.

In practice, Eq. (5.31) is not generally used to determine scaling factors because it tends to be an *overly pessimistic bound* and because it is difficult to perform the required summation. Instead if one can make definitive statements about the class of input signals, e.g., bounded energy or bounded spectrum, other more practical scaling methods can be found.

If we assume $x(n)$ is a deterministic sequence with z transform $X(z)$, then we can determine $v_i(n)$ [Eq. (5.29)] as the inverse Fourier transform of the product of the Fourier transforms of $f_i(n)$ and $x(n)$; i.e.,

$$v_i(n) = \frac{1}{2\pi} \int_{-\pi}^{\pi} F_i(e^{j\omega})X(e^{j\omega})e^{j\omega n} \, d\omega \tag{5.32}$$

If we define the L_p norm ($p \geq 1$) of a Fourier transform $A(e^{j\omega})$ as

$$\|A\|_p = \left[\frac{1}{2\pi} \int_{-\pi}^{\pi} |A(e^{j\omega})|^p \, d\omega \right]^{1/p} \tag{5.33}$$

whenever the integral is finite, then the limit of Eq. (5.33) as $p \to \infty$ exists and is given by

$$\|A\|_\infty = \max_{-\pi \leq \omega \leq \pi} |A(e^{j\omega})| \qquad (5.34)$$

Thus the L_∞ norm of $A(e^{j\omega})$ is the peak value of $|A(e^{j\omega})|$ over all ω. Using L_p norms, it is now relatively easy to derive bounds on $|v_i(n)|$ from Eq. (5.32). For example, if $\|X\|_\infty \leq 1$, (i.e., the peak of the spectrum at the input is bounded), then [from Eq. (5.32)]

$$|v_i(n)| \leq \frac{1}{2\pi} \int_{-\pi}^{\pi} |F_i(e^{j\omega})| \, d\omega \qquad (5.35)$$

or, expressed in terms of norms,

$$|v_i(n)| \leq \|F_i\|_1 \|X\|_\infty \qquad (5.36)$$

Similarly, if $\|F\|_\infty$ is bounded, then

$$|v_i(n)| \leq \|F\|_\infty \|X\|_1 \qquad (5.37)$$

Using the Schwarz inequality on Eq. (5.32) gives

$$|v_i(n)|^2 \leq \left[\frac{1}{2\pi} \int_{-\pi}^{\pi} |F_i(e^{j\omega})|^2 \, d\omega\right]\left[\frac{1}{2\pi} \int_{-\pi}^{\pi} |X(e^{j\omega})|^2 \, d\omega\right] \qquad (5.38)$$

or

$$|v_i(n)| \leq \|F_i\|_2 \|X\|_2 \qquad (5.39)$$

In general it can be shown that

$$|v_i(n)| \leq \|F_i\|_p \cdot \|X\|_q \qquad (5.40)$$

with

$$\frac{1}{p} + \frac{1}{q} = 1 \qquad p, q \geq 1$$

If $F_i(\omega) = 1$, then $\|F_i\|_p = 1$, all $p \geq 1$, and Eq. (5.40) gives the special case

$$|v_i(n)| = |x(n)| \leq \|X\|_q = \|V_i\|_q \qquad q \geq 1 \qquad (5.41)$$

Therefore we can rewrite Eq. (5.40) in terms of spectral quantities as

$$\|V_i\|_1 \leq \|F_i\|_p \|X\|_q \qquad \frac{1}{p} + \frac{1}{q} = 1 \quad (5.42)$$

where $\|V_i\|_1$ is the mean absolute value of $V_i(e^{j\omega})$. Thus by Eq. (5.42) and (5.41) it is seen that the mean absolute value of $V_i(e^{j\omega})$ is bounded by $\|F_i\|_p \cdot \|X\|_q$ and this, in turn, provides a bound on $|v_i(n)|$.

Based on the arguments above, a sufficient condition on scaling can be given if one knows something about the L_q norm of the input. For example, assume $\|X\|_q \leq 1$ $(q \geq 1)$; then the L_p norm of the scaled $F_i(e^{j\omega})$ must satisfy the condition

$$\|F_i'\|_p \leq 1 \qquad p = \frac{q}{(q-1)} \tag{5.43}$$

The most important sets of (p, q) are $(1, \infty)$, $(2, 2)$ and $(\infty, 1)$. The case $p = 1$, $q = \infty$ corresponds to knowing the peak magnitude of the input spectrum and bounding the L_1 norm of $F_i(e^{j\omega})$. The case $p = 2$, $q = 2$ corresponds to placing an energy constraint on both input and transfer function $F_i(e^{j\omega})$. The case $p = \infty$, $q = 1$ corresponds to bounding the peak spectrum level of $F_i(e^{j\omega})$.

In the case of random input signals, Eq. (5.40) and (5.42) cannot be applied since Fourier transforms of random signals are not defined. Instead equivalent conditions can be obtained on the appropriate power spectrum densities and autocorrelation functions. Let $x(n)$ be a random signal with autocorrelation function $\varphi_x(n)$ and power density spectrum $\Phi_x(e^{j\omega})$, and let $v_i(n)$ be the resulting random signal at the ith branch node with auto-correlation $\varphi_{v_i}(n)$ and power density spectrum $\Phi_{v_i}(e^{j\omega})$. Then it can be shown that

$$\varphi_{v_i}(n) \leq \|F_i^2\|_p \|\Phi_x\|_q \tag{5.44}$$

or, equivalently,

$$\varphi_{v_i}(n) = \|F_i\|_{2p}^2 \|\Phi_x\|_q \tag{5.45}$$

with $1/p + 1/q = 1$ in both Eq. (5.44) and (5.45). Since $\sigma_{v_i}^2 = \varphi_{v_i}(0)$, Eq. (5.44) and (5.45) show the variance of $v_i(n)$ is similarly bounded. In the case $p = 1, q = \infty$, when the input power density spectrum is white (i.e., $\|\Phi_x\|_q = \sigma_x^2$), Eq. (5.45) gives

$$\sigma_{v_i}^2 \leq \sigma_x^2 \|F_i\|_2^2 \tag{5.46}$$

To ensure that $\sigma_{v_i}^2 \leq \sigma_x^2$, then $\|F_i'\|_2 \leq 1$; i.e., the energy in the scaled transfer function must be bounded by 1.0.

5.13 Dynamic Range Constraints in Direct Form Realizations

Figure 5.20 shows the direct form realization for an Nth-order filter with transfer function

$$H(z) = \frac{K' \sum_{i=0}^{N} a_i' z^{-i}}{1 + \sum_{i=1}^{N} b_i z^{-i}} = \frac{A(z)}{B(z)} \tag{5.47}$$

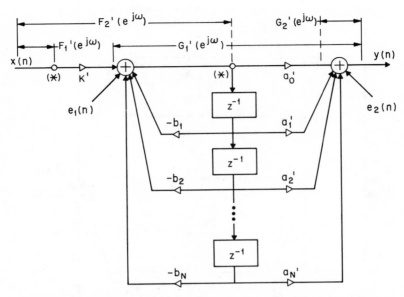

Fig. 5.20 Noise model for direct form realization of an Nth-order system. (After Jackson.)

The branch nodes are indicated by an asterisk in Fig. 5.20. The quantity K' is the only scaling multiplier (which is not identically 1.0) and is given by

$$K' = \frac{1}{\|1/B\|_p} \qquad (5.48)$$

[Eq. (5.48) is obtained from Eq. (5.43) using equality to give the greatest possible signal-to-noise ratio at the filter output.] Once a value for p is chosen (generally $p = 1, 2,$ or ∞), Eq. (5.48) gives K' and the filter specification (with respect to roundoff noise and dynamic range) is complete.

Given the value of K' of Eq. (5.48), the output noise power density spectrum is obtained as

$$N_y(e^{j\omega}) = \frac{Q^2}{12}(N+1)\left[1 + \left\|\frac{1}{B}\right\|_p^2 |H(e^{j\omega})|^2\right] \qquad (5.49)$$

The output noise variance is obtained from Eq. (5.49) as $\sigma_y^2 = \|N_y\|_1$.

5.14 Dynamic Range Considerations in Parallel Form Realizations

Figure 5.21 shows the parallel form realization for an Nth-order filter with transfer function

$$H(z) = \gamma_0 + \sum_{i=1}^{M} \rho_i \frac{(\gamma'_{1i}z^{-1} + \gamma'_{0i})}{(1 + \beta_{1i}z^{-1} + \beta_{2i}z^{-2})} = \gamma_0 + \sum_{i=1}^{M} \frac{\gamma_i(z)}{\beta_i(z)} \qquad (5.50)$$

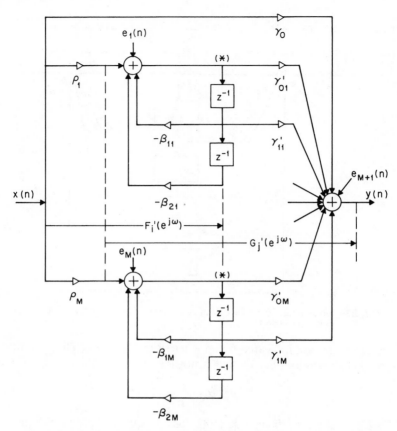

Fig. 5.21 Noise model for parallel realization of an Nth-order system. (After Jackson.)

The scaling multipliers (ρ_i) are of the form

$$\rho_i = \frac{1}{\|F_i\|_p} \tag{5.51}$$

where $F_i(e^{j\omega})$ is the transfer function to the ith branch node and is given by

$$F_i(e^{j\omega}) = \frac{1}{\beta_i(e^{j\omega})} \tag{5.52}$$

The spectrum of the roundoff noise at the output of the filter of Fig. 5.21 is

$$N_y(e^{j\omega}) = \frac{Q^2}{12}\left[(N+1) + \sum_{j=1}^{M} k_j' \left\|\frac{1}{\beta_j}\right\|_p^2 \left|\frac{\gamma_j(e^{j\omega})}{\beta_j(e^{j\omega})}\right|^2\right] \tag{5.53}$$

where k_j' is the total number of noise sources at the jth summation node (generally $k_j' = 3$).

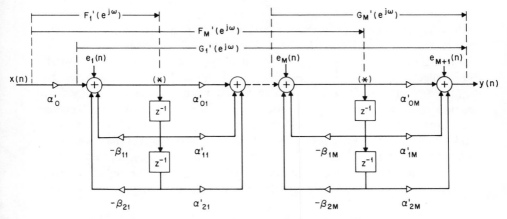

Fig. 5.22 Noise model for cascade realization of an Nth-order system. (After Jackson.)

5.15 Dynamic Range Considerations in Cascade Form Realizations

Figure 5.22 shows the cascade form realization of a digital filter with transfer function

$$H(z) = \alpha_0 \prod_{i=1}^{M} \frac{1 + \alpha_{1i}z^{-1} + \alpha_{2i}z^{-2}}{1 + \beta_{1i}z^{-1} + \beta_{2i}z^{-2}} = \alpha_0 \prod_{i=1}^{M} \frac{\alpha_i(z)}{\beta_i(z)} \qquad (5.54)$$

When scaling factors are included, Eq. (5.54) can be written as

$$H(z) = \alpha_0' \prod_{i=1}^{M} \frac{\alpha_i'(z)}{\beta_i(z)} = \alpha_0' \prod_{i=1}^{M} \frac{\alpha_{0i}' + \alpha_{1i}'z^{-1} + \alpha_{2i}'z^{-2}}{\beta_i(z)} \qquad (5.55)$$

The scaling multipliers are given by

$$\alpha_0' = \frac{1}{\|F_1\|_p}, \qquad \alpha_i'(z) = \frac{\|F_i\|_p}{\|F_{i+1}\|_p} \alpha_i(z) \qquad i = 1, 2, \ldots, M \qquad (5.56)$$

with $\|F_{M+1}\|_p = 1/\alpha_0$.

The resulting formula for the output noise variance (with the scaling above) is

$$\sigma_y^2 = \|N_y\|_1 = \frac{Q^2}{12}\left(k_{M+1}' + \alpha_0 \sum_{j=1}^{M} k_j' \left\| \frac{1}{\beta_j} \prod_{i=1}^{j-1} \frac{\alpha_i}{\beta_i} \right\|_p^2 \left\| \prod_{i=j}^{M} \frac{\alpha_i}{\beta_i} \right\|_2^2 \right) \qquad (5.57)$$

where k_j' is the total number of noise sources at the jth summation node (generally $k_j' = 5$).

5.16 Ordering and Pairing in Cascade Realizations

Equation (5.57) shows that the output noise variance is dependent on the sequential *ordering* of the M sections, as well as the exact way in which numerator and denominator sections are *paired* together. When one examines the expressions for the L_r norms ($r = 1, \infty$) of the output roundoff noise, it is seen that each term in the expressions corresponds to the noise contribution from one section of the filter. For each of these terms an expression involving the L_p norm as well as one involving the L_{r+1} norm is required. This reflects the fact that the sections preceding the given one determine the scaling for that section (and involve an L_p norm), while the succeeding sections filter the roundoff noise produced by that section (and involve the L_{r+1} norm). Therefore in the two most important cases (i.e., $p = 2$, $r = \infty$ and $r = 1$, $p = \infty$) the differences between orderings of the M sections is related to differences in the L_∞ and L_2 norms. Since the L_∞ norm is sensitive to the maxima of its argument, it is expected that a good ordering of sections is one which minimizes the peaked nature of the functions which are

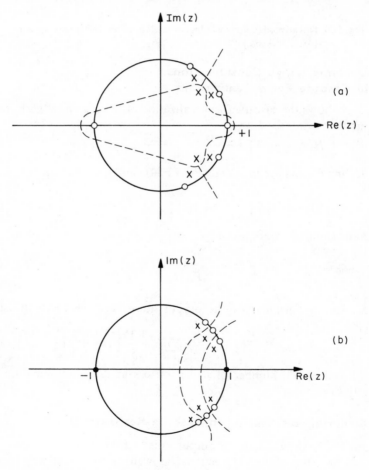

Fig. 5.23 Pole–zero pairing arrangements for two filters. (After Jackson.)

arguments of L_∞ norms. Therefore sections should be ordered from most peaked to least peaked or vice versa, depending on whether $r = \infty$ or $p = \infty$, respectively.

A reasonable measure of peakedness proposed by Jackson is the quantity

$$\rho_i = \frac{\|\alpha_i/\beta_i\|_\infty}{\|\alpha_i/\beta_i\|_2} \tag{5.58}$$

Thus if $p = 2$, $r = \infty$, sections should be ordered in terms of decreasing ρ_i; whereas if $r = 1$, $p = \infty$, sections should be ordered in terms of increasing ρ_i.

For the cases $p = 2$, $r = 1$ and $p = \infty$, $r = \infty$ there is just one norm involved in the output noise calculation. In these cases it can be argued that a good ordering should result from having ρ_i increase with increasing section number, although in these cases the effects of section order are not so significant as in the other cases.

The question of *pairing* of numerator and denominator sections still remains. Since numerator terms $\alpha_i(e^{j\omega})$ appear only in ratios in the expressions involving output roundoff noise, i.e., $\alpha_i(e^{j\omega})/\beta_i(e^{j\omega})$, it seems reasonable to minimize the L_∞ norm of the ratio; i.e., $\|\alpha_i/\beta_i\|_\infty$. In many cases this rule leads to a pairing of poles with zeros that are closest to them in the z plane. Figure 5.23(a) and (b) shows pole–zero pairing, as determined by the rules above, for a sixth-order bandpass filter, as well as a sixth-order band reject filter.

5.17 Summary of Roundoff Noise–Dynamic Range Interaction

The spectrum of the output roundoff noise from fixed point implementations has been shown to be of the form

$$N_y(e^{j\omega}) = \frac{Q^2}{12} \sum_j k'_j \, |G'_j(e^{j\omega})|^2 \tag{5.59}$$

where $G'_j(e^{j\omega})$ is the scaled transfer function from the jth summation node in the filter to the output and k'_j are the number of noise sources that input to the jth summation node.

If $F'_j(e^{j\omega})$ is the scaled transfer function to the ith branch node where dynamic range constraints must be met, then

$$\|F'_i\|_p \le 1 \qquad p \ge 1 \tag{5.60}$$

where $\|F'_i\|_p$ is the L_p norm of $F'_i(e^{j\omega})$, defined as

$$\|F'_i\|_p = \left[\frac{1}{2\pi} \int_{-\pi}^{\pi} |F'_i(e^{j\omega})|^p \, d\omega\right]^{1/p} \tag{5.61}$$

To preserve the largest signal-to-noise ratio, Eq. (5.60) is used with equality. This condition serves to define scale factors s_i as

$$F_i'(e^{j\omega}) = s_i F_i(e^{j\omega}) \tag{5.62}$$

or

$$s_i = \frac{1}{\|F_i\|_p} \tag{5.63}$$

Once the scaling factors s_i are known, the scaled transfer functions $G_j'(e^{j\omega})$ are readily determined and the output roundoff noise spectrum is given by Eq. (5.59). The output noise variance may be obtained as the L_1 norm of $N_y(e^{j\omega})$, i.e.,

$$\sigma_y^2 = \|N_y\|_1 \tag{5.64}$$

and the peak of the output power spectrum may be obtained as the L_∞ norm of $N_y(e^{j\omega})$, i.e.,

$$\max_{\omega} [N_y(e^{j\omega})] = \|N_y\|_\infty \tag{5.65}$$

Based on the above, formulas for output noise for direct, parallel, and cascade forms were given. Finally a discussion of section ordering and pole–zero pairing for the cascade form was given.

5.18 Additional Comments on Dynamic Range–Roundoff Noise Interactions

Several comments still are in order on roundoff noise–dynamic range interaction in recursive structures. The only type of quantization considered was rounding. If truncation is used instead of rounding, the main difference is that the equivalent noise sources at the summation nodes have a nonzero average value. Thus the output noise has a nonzero average value. Since the variances of the errors for truncation and rounding are identical, the output noise variance for truncation is the same as for rounding. The use of truncation has no effect (to first order) on the scaling parameters. Because of the nonzero average value of the output noise, rounding is generally preferred to truncation.

The third technique, that is, sign-magnitude truncation, gives an error source at each of the summation nodes that is correlated with the input signal (recall that the error is positive for positive signals and negative for negative signals). Therefore the assumptions leading to the output noise variance are not valid and the equations related to output noise variance are not valid in this case. Since this type of truncation is extremely difficult to analyze and because the error variance is greater than for rounding or straight truncation, this type of quantization is generally avoided in practice. Its use may be required in certain computer realizations, however, depending on the detailed instruction set of the machine.

Finally some comments should be made about what happens when the sum of the inputs to the adder in a digital filter exceeds the maximum signal level (i.e., 1.0) in spite of the scaling that has been performed. If 1's or 2's complement arithmetic is used, as generally is the case, the arithmetic has the desirable property that as long as the final sum of all the inputs to the adder is less than 1.0, partial sums can overflow and still cause no problems. Figure 5.24(a) illustrates this property graphically by presenting all the 2's-complement numbers in a circle [using $(b + 1)$ bits in the word]. Thus the biggest positive number $(1 - 2^{-b})$ is next to the biggest negative number (-1.0). Adding positive numbers leads to counterclockwise rotation around the circle, whereas negative numbers lead to clockwise rotation. Thus if the total sum is in the range $(-1, 1 - 2^{-b})$, it makes no difference if overflows occur on partial summations. Expressed in a slightly different way, the total

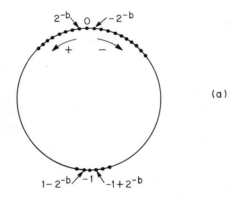

(a)

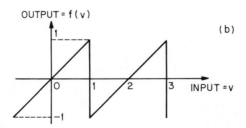

Fig. 5.24 Overflow characteristics of 2's-complement arithmetic. (After Ebert, Mazo, and Taylor.)

input–output characteristic of an adder is as shown in Fig. 5.24(b). Overflow occurs if the total input is out of the range $(-1, 1)$. Whenever overflow does occur, the resulting filter operates nonlinearly. In fact, Ebert, Mazo, and Taylor have shown that, as a result of overflow, self-sustained oscillations can occur within the filter. The question then occurs as to how to eliminate these highly undesirable oscillations. Ebert, et al. have shown that a simple but effective way of eliminating these oscillations is to replace the normal 2's-complement adder with a saturating adder. The total input–output characteristic of this adder is shown in Fig. 5.25. The use of this simple nonlinearity guarantees stable behavior when overflows occur. Of course, the filter is operating nonlinearly at these times. It can be argued heuristically, however, that saturation is perhaps the most reasonable thing to do in the case of overflow anyhow. As a result of these findings, many hardware digital filters use saturating adders to avoid the possibility of overflow oscillations.

5.19 Roundoff Noise in Nonrecursive Structures—Fixed Point Analysis

Because of their structural differences, it is worthwhile investigating separately quantization effects for the nonrecursive structures that are generally used to realize FIR filters, i.e., the direct and cascade forms. The techniques of the previous sections could be used by setting the denominator polynomial to 1. For the case of FIR filters, however, one is generally concerned with linear phase filters. The resulting constraints on the transfer function $H(z)$ [i.e., $H(z)$ is a mirror-image polynomial] change the realizations sufficiently so that it is not a trivial matter to apply directly the expressions developed for output noise in recursive structures to the nonrecursive structures. In Sec. 5.20 and 5.21 we consider the direct and cascade forms for linear phase FIR filters.

5.20 Roundoff Noise in Direct Form Nonrecursive Realizations

The transfer function for an FIR filter can be written in the form

$$H(z) = \sum_{n=0}^{N-1} h(n)z^{-n} \tag{5.66}$$

where $\{h(n)\}$ is the filter impulse response and is of duration N samples. (For convenience, N is assumed odd throughout this discussion.) The linear phase constraint implies

$$h(n) = h(N - 1 - n) \qquad 0 \le n \le N - 1 \tag{5.67}$$

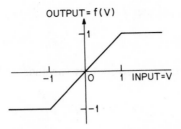

Fig. 5.25 Saturating adder input–output characteristic.

With this constraint, Eq. (5.66) can be written as

$$H(z) = \sum_{n=0}^{(N-3)/2} h(n)[z^{-n} + z^{-(N-1-n)}] + h\left(\frac{N-1}{2}\right)z^{-[(N-1)/2]} \quad (5.68)$$

Figure 5.26 shows the direct form realization of Eq. (5.68). It is seen from Eq. (5.68) and Fig. 5.26 that $[(N+1)/2]$ multipliers are required in this case rather than N multipliers in the standard direct form realization.

The statistics of the roundoff noise at the output depend on the location of points in the filter where rounding is performed. Generally two possibilities exist. First, if all multiplication products are represented exactly and rounding is performed only after they are summed, i.e., at the filter output, then only one noise source is present in the filter and it superimposes noise directly onto the output signal. In this case the output roundoff noise is uniformly distributed between $-Q/2$ and $Q/2$ with zero mean and variance of $Q^2/12$, where Q is the quantization step size.

In the second case, if all multiplication products are rounded before they are summed (in order to increase the speed at which the filter runs), then the output roundoff noise $e(n)$ is the sum of $[(N+1)/2]$ uncorrelated noise sources $e_i(n)$, $i = 0, 1, \ldots, (N-1)/2$, each uniformly distributed between

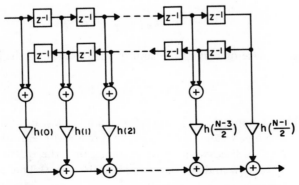

Fig. 5.26 Direct form realization of a linear phase FIR filter.

$-Q/2$ and $Q/2$, with zero mean and variance $Q^2/12$. Thus we can write

$$e(n) = \sum_{i=0}^{(N-1)/2} e_i(n) \tag{5.69}$$

The mean of $e(n)$ is clearly zero and its variance σ_0^2 is

$$\sigma_0^2 = \left(\frac{N+1}{2}\right)\frac{Q^2}{12} \tag{5.70}$$

The only other type of roundoff noise that is generally present in direct form realizations is A/D conversion noise. It is shown here that the output variance $\sigma_{A/D}^2$, due to A/D conversion noise, satisfies the relation

$$\sigma_{A/D}^2 \le \frac{Q^2}{12} \tag{5.71}$$

and hence is smaller than or equal to the variance of the roundoff noise produced by the multiplications within the filter. If we let $\varepsilon(n)$ denote the A/D noise sequence at the input to the filter [$\varepsilon(n)$ is uniform from $-Q/2$ to $Q/2$] and $\mathscr{E}(n)$ denote the output noise due to $\varepsilon(n)$, then $\mathscr{E}(n)$ satisfies the equation

$$\mathscr{E}(n) = \sum_{k=0}^{N-1} h(k)\varepsilon(n-k) \tag{5.72}$$

The mean of $\varepsilon(n)$ is clearly zero and its variance is

$$\sigma_{A/D}^2 = \frac{Q^2}{12}\sum_{n=0}^{N-1} h^2(n) \tag{5.73}$$

or, by Parseval's theorem,

$$\sigma_{A/D}^2 = \frac{Q^2}{12}\frac{1}{2\pi}\int_{-\pi}^{\pi} |H(e^{j\omega})|^2\, d\omega \tag{5.74}$$

Depending on how the scale factor is chosen for the filter (i.e., the L_p norm chosen), it can be shown that either

$$\sum_{n=0}^{N-1} h^2(n) \le 1 \tag{5.75a}$$

(as is valid in the case of scaling the sum of magnitudes of the impulse response) or

$$\frac{1}{2\pi}\int_{-\pi}^{\pi} |H(e^{j\omega})|^2\, d\omega \le 1 \tag{5.75b}$$

(as is valid for L_p scaling, $p \ge 2$). Therefore, in all cases, Eq. (5.71) is valid. Thus A/D noise can generally be neglected in direct form realizations of FIR filters. Note that the discussion above assumes that the quantization step size for A/D conversion and for rounding are the same. This need not be true in practice.

5.21 Roundoff Noise in Cascade Form Nonrecursive Realizations

For the cascade realization of linear phase FIR filters the problems of scaling and section ordering to maximize output signal-to-noise ratio are of prime importance. For cascade realizations, the filter's transfer function can be written in the form

$$H(z) = \prod_{i=1}^{N_s} (b_{0i} + b_{1i}z^{-1} + b_{2i}z^{-2}) = \prod_{i=1}^{N_s} H_i(z) \qquad (5.76)$$

where $N_s = (N - 1)/2$. For the case of linear phase filters, either the b_{ji} coefficients of Eq. (5.76) satisfy the relation

$$b_{0i} = b_{2i} \qquad (5.77a)$$

or some $j \neq i$ exists such that

$$\frac{b_{0i}}{b_{2j}} = \frac{b_{1i}}{b_{1j}} = \frac{b_{2i}}{b_{0j}} \qquad (5.77b)$$

Figure 5.27 shows a realization of the ith section of Eq. (5.76). From this figure and Eq. (5.77a), it is clear that there are either two or three multiplications per section. Therefore to model a section of the cascade, k_i noise sources are added to the output of the section (where $k_i = 2$ or 3) or, equivalently, one noise source of variance $k_i Q^2/12$ can instead be added. The value of k_i can be reduced to 1 by summing all products in each section before rounding but since this sacrifices speed of operation, it is generally not done.

If we define $G_i(z)$, the transfer function from the $(i + 1)$st section to the output, as

$$G_i(z) = \begin{cases} \prod_{j=i+1}^{N_s} H_j(z) & 0 < i \leq N_s - 1 \\ 1 & i = N_s \end{cases} \qquad (5.78)$$

and let $g_i(k)$ be the inverse z transform corresponding to Eq. (5.78), then roundoff noise in a cascade filter can be modeled as shown in Fig. 5.28(a)

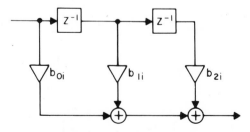

Fig. 5.27 Simple cascade section of an FIR linear phase filter.

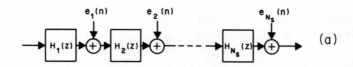

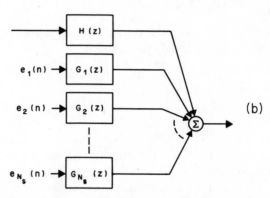

Fig. 5.28 Noise model of a cascade FIR filter.

and (b). The output noise variance is thus given by

$$\sigma^2 = \sum_{i=1}^{N_s} k_i \frac{Q^2}{12} \sum_k g_i^2(k) \tag{5.79}$$

Scaling multipliers must now be included between sections to prevent overflows, as in the case of recursive structures. If we define $H_i(z)$ of Eq. (5.76) as

$$H_i(z) = S_i \hat{H}_i(z) \tag{5.80}$$

where S_i is the scaling factor for ith section and $\hat{H}_i(z)$ is a normalized transfer function of the form

$$\hat{H}_i(z) = a_{0i} + a_{1i}z^{-1} + a_{2i}z^{-2} \tag{5.81}$$

with $a_{0i} \geq 0$ and

$$\sum_{j=0}^{2} |a_{ji}| = 1 \qquad i = 1, 2, \ldots, N_s \tag{5.82}$$

$$H(z) = \beta \prod_{i=1}^{N_s} \hat{H}_i(z) = \prod_{i=1}^{N_s} S_i \prod_{i=1}^{N_s} \hat{H}_i(z) \tag{5.83}$$

Given a value for β, $N_s - 1$ of the S_i's can be chosen arbitrarily.

If we make the following definitions

$$F_i(z) = \sum_{k=0}^{2i} f_i(k)z^{-k} = \prod_{j=1}^{i} H_j(z) \qquad 1 \leq i \leq N_s \qquad (5.84)$$

$$\hat{F}_i(z) = \sum_{k=0}^{2i} \hat{f}_i(k)z^{-k} = \prod_{j=1}^{i} \hat{H}_j(z) \qquad (5.85)$$

then if the input $x(n)$ is bounded in magnitude by 1.0, a necessary and sufficient condition on the scale factors to guarantee that the output of each section is bounded in magnitude by 1.0 is

$$\prod_{j=1}^{i} |S_j| \leq \left[\sum_{k=0}^{2i} |\hat{f}_i(k)| \right]^{-1} \qquad i = 1, 2, \ldots, N_s \qquad (5.86)$$

On the other hand if the input spectrum $X(e^{j\omega})$ satisfies the condition

$$\frac{1}{2\pi} \int_{-\pi}^{\pi} |X(e^{j\omega})| \, d\omega \leq 1 \qquad (5.87)$$

then a necessary and sufficient condition on the scale factors to guarantee that the output of each section is bounded in magnitude by 1.0 is

$$\prod_{j=1}^{i} |S_j| \leq \left[\max_{0 \leq \omega \leq 2\pi} |\hat{F}_i(e^{j\omega})| \right]^{-1} \qquad i = 1, 2, \ldots, N_s \qquad (5.88)$$

The rule of Eq. (5.86), satisfied with equality, is called *sum scaling;* whereas the rule of Eq. (5.88), satisfied with equality, is called *peak scaling.* As mentioned earlier, sum scaling has been found to be overly pessimistic in the case of recursive realizations of IIR filters. This is *not* the case for FIR filters because the conditions required on the input signal in order to attain equality on the scaling constraints are not overly restrictive as they apply only to a finite duration of signal. Thus, for example, the input need only have three consecutive samples of value 1.0 before the first section output attains magnitude of 1.0.

Thus, independent of the scaling rule chosen, all that remains to be done is to choose an ordering for the sections to minimize the resulting roundoff noise. As in the case of the recursive cascade structure, it is impossible to investigate all possible section orderings. It has been shown experimentally as well as heuristically, however, that a good ordering of sections is one for which the peak of the transfer function from the ith section to the output $(i = 2, 3, \ldots, N_s)$ is small and *does not vary significantly from section to section.* It has also been shown that the large majority of section orderings yield low output noise variances. Based on extensive study of practical design examples, a simple algorithm to order cascade sections is:

> Beginning with $i = N_s$, permanently build into position i the section that, together with the sections already built in, causes the smallest possible value for $\sum_k g_{i-1}^2(k)$ [recall $g_i(k)$ is the impulse response from the $(i + 1)$st section to the output].

The algorithm minimizes the output noise variances from individual sections rather than minimizing the sum of the output noise variances. However, because of the skewed distribution of output noise variances versus section orderings, *which is highly in favor of low noise variances*, the algorithm above has been found to yield section orderings very close to the optimum ordering in all cases tested.

In summary, the cascade realization of FIR filters involves choosing a scaling procedure (two of which were given) and ordering sections to minimize output noise variance using the proposed algorithm. When the ordering of sections has been done and scaling factors have been determined, it is then straightforward to determine output noise variance and hence the signal-to-noise ratio of the realization.

5.22 Roundoff Noise—Floating Point Realizations of Recursive Structures

In this section we discuss floating point realizations of recursive structures. As shown earlier, both additions and multiplications in floating point arithmetic lead to quantization noise that is proportional to the output of the operation. Following the development of Liu and Kaneko, the notation $fl[x]$ stands for the floating point number used to represent x. Thus addition and multiplication of two floating point numbers leads to the rules

$$fl[x + y] = (x + y)(1 + \varepsilon) \tag{5.89a}$$

$$fl[x \cdot y] = (x \cdot y)(1 + \delta) \tag{5.89b}$$

where ε and δ are uniform random variables distributed over the interval $-2^{-b} \leq \varepsilon$, $\delta \leq 2^{-b}$ [$(b + 1)$-bit mantissa] and ε and δ are independent of both x and y. The ideal (no quantization) second-order system

$$w(n) = b_0 x(n) - [a_1 w(n - 1) + a_2 w(n - 2)] \tag{5.90}$$

is actually computed as

$$y(n) = fl\{b_0 x(n) - [a_1 y(n - 1) + a_2 y(n - 2)]\} \tag{5.91}$$

where the products $a_1 y(n - 1)$, $a_2 y(n - 2)$, $b_0 x(n)$ are first computed and then the individual terms are summed to obtain $y(n)$. A flow diagram for these operations is given in Fig. 5.29. The individual roundoff error terms (e.g., $\delta_{n,0}$, $\varepsilon_{n,1}$) are assumed to be independent random variables. Thus Eq. (5.91)

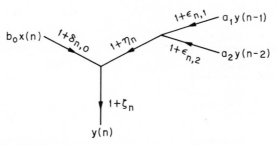

Fig. 5.29 Floating point quantization in a first-order system.

can be written in the form

$$y(n) = b_0 \theta_{n,0} x(n) - \sum_{k=1}^{2} a_k \Phi_{n,k} y(n - k) \qquad (5.92)$$

where

$$\theta_{n,0} = (1 + \delta_{n,0})(1 + \zeta_n)$$

$$\Phi_{n,1} = (1 + \varepsilon_{n,1})(1 + \eta_n)(1 + \zeta_n) \qquad (5.93)$$

$$\Phi_{n,2} = (1 + \varepsilon_{n,2})(1 + \eta_n)(1 + \zeta_n)$$

In order to determine the characteristics of the roundoff noise, the error signal $e(n) = y(n) - w(n)$ must be investigated.

Consider, for example, the direct form realization corresponding to the difference equation

$$y(n) = fl\left[\sum_{k=0}^{M} b_k x(n - k) - \sum_{k=1}^{N} a_k y(m - k)\right] \qquad (5.94)$$

The flow graph for this Lth-order filter is shown in Fig. 5.30. Equation (5.94) maybe written in the form

$$y(n) = \sum_{k=0}^{M} b_k \theta_{n,k} x(n - k) - \sum_{k=1}^{N} a_k \Phi_{n,k} y(n - k) \qquad (5.95)$$

where

$$\theta_{n,0} = (1 + \xi_n)(1 + \delta_{n,0}) \prod_{i=1}^{M} (1 + \zeta_{n,i})$$

$$\theta_{n,j} = (1 + \xi_n)(1 + \delta_{n,j}) \prod_{i=j}^{M} (1 + \zeta_{n,i}) \qquad j = 1, 2, \ldots, M$$

$$\Phi_{n,1} = (1 + \xi_n)(1 + \varepsilon_{n,1}) \prod_{i=2}^{N} (1 + \eta_{n,i}) \qquad (5.96)$$

$$\Phi_{n,j} = (1 + \xi_n)(1 + \varepsilon_{n,j} \prod_{i=j}^{N} (1) + \eta_{n,i}) \qquad j = 2, 3, \ldots, N$$

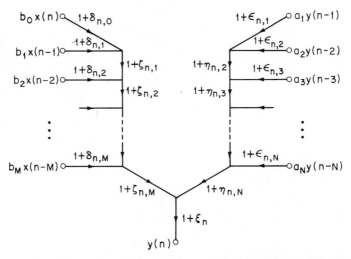

Fig. 5.30 Floating point quantization in an Nth-order system. (After Liu and Kaneko.)

Equation (5.95) cannot be used directly to solve for the statistics of the roundoff noise because it involves the time-varying random coefficients $\Phi_{n,k}$ and $\theta_{n,k}$. Instead a set of substitutions is made whereby averages of the random coefficients are used. The details of these substitutions are described in great detail by Liu and Kaneko. There is little insight gained by including all the details in obtaining the power spectral density of the output error sequence. Therefore the interested reader is referred to the paper by Liu and Kaneko who present results for the three most commonly used forms—i.e., the direct, parallel, and cascade forms in the case of floating point roundoff quantization.

5.23   Coefficient Quantization

Usually the coefficients of a digital filter are obtained by some theoretical design procedure that essentially assumes infinite precision representation of the filter coefficients. For practical realizations, the coefficients must be quantized to a fixed number of bits. As a consequence the frequency response of the actual filter which is realized deviates from that which would have been obtained with an infinite word length representation.

There are two general approaches to the analysis and synthesis of digital filters with finite precision coefficients. The first approach is to treat co-efficient quantization errors as intrinsically statistical quantites. In this case the effects of coefficient quantization can be represented by a stray transfer function in parallel with the corresponding ideal filter. Also by making certain assumptions about the coefficient errors, the expected mean square difference between the frequency responses of the actual and ideal filters can be readily evaluated.

The second approach to coefficient quantization is to study each individual filter separately. In this manner one can optimize the finite precision coefficients to minimize the maximum weighted difference between the ideal and actual frequency responses. Although one cannot make any general statements about coefficient quantization, this method has the benefit of yielding the best finite precision representation of the desired frequency response.

In the next sections we shall present examples of both these techniques and, whenever possible, illustrate their application via actual examples.

5.24   Coefficient Quantization in Recursive Structures

Kaiser was one of the first to investigate the effect of coefficient errors on filter performance. Studying direct form realizations, he derived an absolute accuracy bound for the filter coefficients that guaranteed the filter would remain stable. The error bound that was derived was overly pessimistic, however, because of the deterministic way in which Kaiser studied coefficient quantization. He was able to show convincingly, however, that for any reasonably complex filter with steep transitions between passband and stopband the use of the direct form should be avoided because of the *extreme* coefficient sensitivity problems.

Knowles and Olcayto formulated the coefficient quantization problem as a statistical one. The basis for their model is the assumption that the errors in the difference equation coefficients are uniformly distributed with zero mean. In Sec. 5.25 by way of example we consider the effect of coefficient quantization on the direct form of realization.

5.25   Coefficient Quantization in the Direct Form Realization

Let the actual (finite precision coefficients) digital filter being realized have as a transfer function

$$H(z) = \frac{A(z)}{B(z)} = \frac{\sum_{k=0}^{N} a_k z^{-k}}{1 + \sum_{k=1}^{N} b_k z^{-k}} \tag{5.97}$$

The filter coefficients may be written as

$$a_k = \bar{a}_k + \alpha_k$$
$$b_k = \bar{b}_k + \beta_k \tag{5.98}$$

where $\bar{a}_k$ and $\bar{b}_k$ are the infinite precision filter coefficients and the error quantities α_k and β_k are statistically independent, uniformly distributed random variables. If $x(n)$ denotes the actual filter input, $y'(n)$ denotes the actual filter output (in the absence of roundoff noise), and $y(n)$ denotes the ideal filter output with the same input sequence, then one can define an error quantity $e(n)$ as

$$e(n) = y'(n) - y(n) = \left[\sum_{k=0}^{N} a_k x(n - k) - \sum_{k=1}^{N} b_k y'(n - k) \right]$$
$$- \left[\sum_{k=0}^{N} \bar{a}_k x(n - k) - \sum_{k=1}^{N} \bar{b}_k y(n - k) \right] \quad (5.99)$$

$$e(n) = \sum_{k=0}^{N} \alpha_k x(n - k) - \sum_{k=1}^{N} \bar{b}_k e(n - k) - \sum_{k=1}^{N} \beta_k y(n - k) - \sum_{k=1}^{N} \beta_k e(n - k) \quad (5.100)$$

or, neglecting second-order quantities, one finds

$$e(n) = \sum_{k=0}^{N} \alpha_k x(n - k) - \sum_{k=1}^{N} \bar{b}_k e(n - k) - \sum_{k=1}^{N} \beta_k y(n - k) \quad (5.101)$$

Taking z transforms of Eq. (5.101) gives

$$0 = \alpha(z)X(z) - \beta(z)Y(z) - E(z)B_\infty(z) \quad (5.102)$$

where

$$\alpha(z) = \sum_{k=0}^{N} \alpha_k z^{-k} \quad (5.103a)$$

$$\beta(z) = \sum_{k=1}^{N} \beta_k z^{-k} \quad (5.103b)$$

$$A_\infty(z) = \sum_{k=0}^{N} \bar{a}_k z^{-k} \quad (5.103c)$$

$$B_\infty(z) = 1 + \sum_{k=1}^{N} \bar{b}_k z^{-k} \quad (5.103d)$$

$$E(z) = \sum_{k=0}^{\infty} e(k) z^{-k} \quad (5.103e)$$

Since

$$Y(z) = H_\infty(z)X(z) \quad (5.104)$$

Eq. (5.102) becomes

$$E(z) = \left[\frac{\alpha(z) - \beta(z)H_\infty(z)}{B_\infty(z)} \right] X(z) \quad (5.105)$$

One can solve for $Y'(z)$ from Eq. (5.99), (5.104), and (5.105) to give

$$Y'(z) = \left[H_\infty(z) + \frac{\alpha(z) - \beta(z)H_\infty(z)}{B_\infty(z)} \right] X(z) \quad (5.106)$$

Thus the actual filter may be represented as the *ideal filter in parallel with a stray transfer function*, as shown in Fig. 5.31.

One possible measure of the effects of coefficient quantization is the mean square error in the frequency response that may readily be determined from Eq. (5.106) as

$$\sigma^2 = E\left[\frac{1}{2\pi}\int_{-\pi}^{\pi}|H'(e^{j\omega}) - H_\infty(e^{j\omega})|^2\,d\omega\right] \qquad (5.107)$$

where $H'(e^{j\omega}) = Y'(e^{j\omega})/X(e^{j\omega})$ and E denotes expectation. Using the assumed statistical independence among α_k, β_k, and the input, Eq. (5.107) becomes

$$
\begin{aligned}
\sigma^2 = E\Bigg\{&\frac{1}{2\pi}\int_{-\pi}^{\pi}\left[\frac{\alpha(z) - \beta(z)H_\infty(z)}{B_\infty(z)}\right]\left[\frac{\alpha(z^{-1}) - \beta(z^{-1})H_\infty(z^{-1})}{B_\infty(z^{-1})}\right]\frac{dz}{z}\Bigg\} \\
= &\left(\sum_{k=0}^{N}\overline{\alpha_k^2}\right)\frac{1}{2\pi}\int_{-\pi}^{\pi}\frac{1}{B_\infty(z)B_\infty(z^{-1})}\frac{dz}{z} \\
&+ \left(\sum_{k=1}^{N}\overline{\beta_k^2}\right)\frac{1}{2\pi}\int_{-\pi}^{\pi}\frac{A_\infty(z^{-1})A_\infty(z)}{[B_\infty(z^{-1})]^2[B_\infty(z)]^2}\frac{dz}{z} \qquad (5.108)
\end{aligned}
$$

In the case of rounding, the quantities γ_k and β_k satisfy the relation

$$|\alpha_k| \le \frac{q}{2} \qquad (5.109a)$$

$$|\beta_k| \le \frac{q}{2} \qquad (5.109b)$$

$$\sum_{k=0}^{N}\overline{\alpha_k^2} = \mu\frac{q^2}{12} \qquad (5.109c)$$

$$\sum_{k=1}^{N}\overline{\beta_k^2} = \nu\frac{q^2}{12} \qquad (5.109d)$$

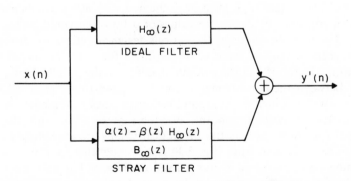

Fig. 5.31 Coefficient quantization model. (After Knowles and Olcayto.)

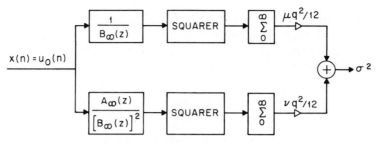

Fig. 5.32 Technique for measuring variance of error due to coefficient quantization. (After Knowles and Olcayto.)

where q is the size of the quantization step and μ and ν are the number of nonzero, nonunity coefficients in the numerator and denominator, respectively. Using Eq. (5.109c) and (5.109d), the quantity σ^2 of Eq. (5.108) can conveniently be measured by the system shown in Fig. 5.32 where the input is an impulse. Paraseval's relation has been used to convert the integrals in Eq. (5.108) to infinite summations. Of course a finite summation must be carried out in practice.

Similar formulas may be derived for coefficient roundoff noise variance for the parallel and cascade forms of realization. Rather than presenting these results we proceed to an experimental verification of the statistical model above.

5.26 Experimental Verification of Coefficient Quantization Model

The statistical model presented above was verified by Knowles and Olcayto for a twenty-second-order bandstop elliptic filter. The system of Fig. 5.33 was used to measure the value of σ^2 for various computer word lengths. The system $H_\infty(z)$ was implemented with 80-bit coefficients. Figure 5.34(a), (b), and (c) shows the values of σ both measured by the system of Fig. 5.33 and evaluated using the various design equations for direct, parallel, and cascade forms. In the case of the direct form, the predicted value of σ is within a factor of 2 of the computed value. In the parallel form, the agreement between theory and measurement is good down to word lengths of about 8 bits at which point second-order statistics of the coefficient quantization errors become significant. For the cascade form, only the measured values of σ are plotted as the theoretical values are exceedingly difficult to determine. It is seen from Fig. 5.34 that the parallel form has the smallest degradation in performance for a fixed number of bits and the direct form has, by far, the largest degradation, as expected.

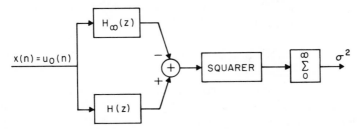

Fig. 5.33 Technique for measuring variance of the error due to coefficient quantization. (After Knowles and Olcayto.)

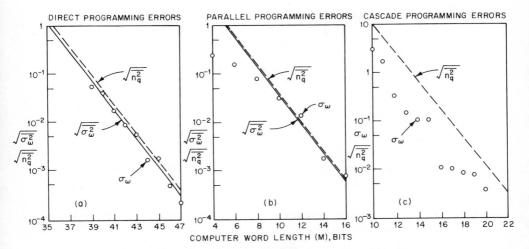

Fig. 5.34 Error variance for several recursive structures. (After Knowles and Olcayto.)

5.27   Optimization of Quantized Coefficients

The theory of coefficient quantization of the preceding sections presents a picture of coefficient quantization that is statistical in nature. Therefore, depending on the validity of the assumptions of the model, one can make probabilistic statements as to how closely the actual frequency response of the filter matches the ideal frequency response. Such statements are extremely useful for rules of thumb as they enable the designer to estimate the number of bits required to represent the filter coefficients, for a large class of filters, without specifically knowing the actual filter coefficients. In many cases, however, one wishes to optimize the finite precision filter coefficients to minimize some measure of the filter's performance.

Avenhaus and Schuessler have introduced as a measure of filter performance the index $\varepsilon(\omega)$, which is defined as

$$\varepsilon(\omega) = \frac{H_D(e^{j\omega}) - H(e^{j\omega})}{\delta(\omega)} \qquad (5.110)$$

where $H_D(e^{j\omega})$ is the desired frequency response of the filter, $H(e^{j\omega})$ is the actual filter frequency response (using finite precision coefficients), and $\delta(\omega)$ is a tolerance function. For example, in the case of a lowpass filter the quantities $H_D(e^{j\omega})$ and $\delta(\omega)$ may be defined as

$$H_D(e^{j\omega}) = \begin{cases} 1 & 0 \leq \omega \leq \omega_p \\ 0 & \omega_s \leq \omega \leq \pi \\ \Phi = \text{don't care} & \omega_p < \omega < \omega_s \end{cases}$$

$$\delta(\omega) = \begin{cases} \delta_1 & 0 \leq \omega \leq \omega_p \\ \delta_2 & \omega_s \leq \omega \leq \pi \\ \infty & \omega_p < \omega < \omega_s \end{cases}$$

Avenhaus and Schuessler proposed the measure max $\varepsilon(\omega)$ as a number that described the actual filter performance. If max $\varepsilon(\omega)$ was less than or equal to 1.0, the filter was acceptable as it met its tolerance bounds. If max $\varepsilon(\omega)$ was greater than 1.0, the filter was unacceptable and more bits had to be allotted to the filter coefficients.

Using an optimization procedure in a discrete parameter space to determine the finite precision coefficients that minimized the quantity max $\varepsilon(\omega)$, Avenhaus and Schuessler were able to make distinct improvements in filter performance over straight rounding of the filter coefficients, as done in the statistical model of Sec. 5.26. Figures 5.35 and 5.36 show the frequency responses and z-plane pole–zero positions of an eighth-order elliptic bandpass filter. For 36-bit coefficients, the figure of merit of the filter was max $\varepsilon_{36}(\omega) = 0.526054$. Rounding the filter coefficients to 8 bits ($Q = 2^{-6}$, the extra 2 bits were used for a sign bit and an integer bit since the filter coefficients can be as large as 2.0) gives a filter with max $\varepsilon(\omega) \approx 1.07$, an unacceptable filter. The minimum number of bits to which the 36-bit coefficients could be rounded with max $\varepsilon(\omega) \leq 1.0$ was 11 bits. Following optimization a set of 8-bit filter coefficients was found with max $\varepsilon(\omega) = 0.897$. Therefore the gain by optimization over straight rounding was 3 bits.

An interesting situation arises when the ideas above are tested on a variety of different filters. Since the quantity max $\varepsilon(\omega)$ must be less than 1.0 for the infinite precision filter (or else no finite precision case will be acceptable), one can always reduce the value of max $\varepsilon(\omega)$ by increasing the filter order, all other filter parameters remaining constant. Thus the question arises as to whether

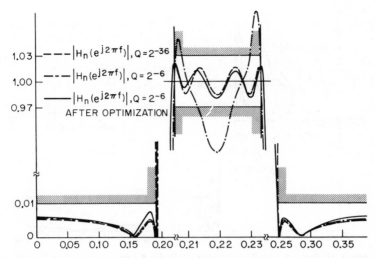

Fig. 5.35 Optimized filter coefficients and their effects on frequency response. (After Avenhaus and Schuessler.)

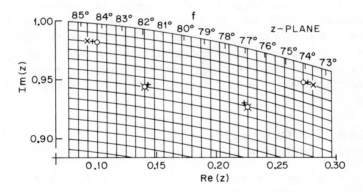

+   SOLUTION WITH k = 36
o   SOLUTION WITH k = 6 AFTER ROUNDING
x   SOLUTION WITH k = 6 AFTER OPTIMIZATION

Fig. 5.36 Pole–zero positions of optimized filters. (After Avenhaus and Schuessler.)

a higher-order filter can be quantized to a smaller number of bits than a lower-order filter because it initially begins with a smaller value of max $\varepsilon(\omega)$. Figure 5.37 illustrates this point for an elliptic lowpass filter. An eighth-order elliptic filter is the lowest-order filter that meets the given design specifications. After optimizing finite precision filter coefficients it is found that 15-bit coefficients are required to give max $\varepsilon(\omega) = 0.989$. A tenth-order elliptic filter also meets design specifications. For the tenth-order system the value of

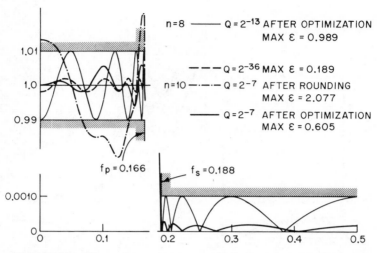

Fig. 5.37 The effects of quantization on the frequency response of one filter. (After Avenhaus and Schuessler.)

max $\varepsilon(\omega)$ for 36-bit coefficients is 0.189. Rounding the coefficients to 9 bits gives max $\varepsilon(\omega) = 2.077$; however, after optimization of the 9-bit coefficients the value of max $\varepsilon(\omega)$ is reduced to 0.605. The question then arises as to whether it is preferable to use four filter sections with 15-bit coefficients or five filter sections with 9-bit coefficients. The issues that come into play here are filter speed, roundoff noise considerations, cost of multiplier bits, etc.

5.28   Coefficient Quantization in the Implementation of a Pole-Pair

Since the second-degree section comprises the basic building block for higher-degree filters realized in either the cascade or parallel forms, it is worthwhile investigating its coefficient sensitivity in greater detail. For direct form implementation of a second-order section, the basic filter coefficients are $-r^2$ and $2r \cos \theta$. The poles are located at the points $z = re^{\pm j\theta}$. For a given quantization on the coefficients $-r^2$ and $2r \cos \theta$, the poles are constrained to an irregular grid of positions in the z plane defined by the intersection of concentric circles corresponding to quantization of r^2 and vertical lines corresponding to quantization of $r \cos \theta$. Figure 5.38 shows such a grid corresponding to six-bit quantization on both coefficients. The density of pole positions is highly nonuniform in the z plane. Although this need not be an undesirable feature, it does imply significantly more error in pole position due to finite precision coefficients for certain initial pole locations than for others.

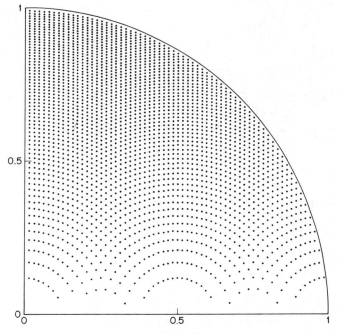

Fig. 5.38 Grid positions in the z plane for poles in a direct form second-order section.

An alternative method of realizing a second-order section is the coupled form proposed by Rader and Gold. The set of first-order difference equations corresponding to this case is given as

$$y_1(n) = r\,(\cos\,\theta)y_1(n-1) - r\,(\sin\,\theta)y_2(n-1) + x(n)$$

$$y_2(n) = r\,(\sin\,\theta)y_1(n-1) + r\,(\cos\,\theta)y_2(n-1)$$

$$y(n) = y_1(n)$$

Since the filter coefficients are $r\cos\theta$ and $r\sin\theta$, the poles lie on a rectangular grid in the z plane as shown in Fig. 5.39. As seen from Fig. 5.39 the coupled form realization of a second-order system has the property that the maximum error in pole position due to finite precision coefficients is the same for all initial pole regions in the z plane.

5.29   Coefficient Quantization in Nonrecursive Structures

In the case of nonrecursive structures for realizing linear phase FIR filters, somewhat stronger statements can be made as to the effects of coefficient quantization than in the case of the recursive structures of the previous section. For direct form realizations we shall derive statistical bounds on the

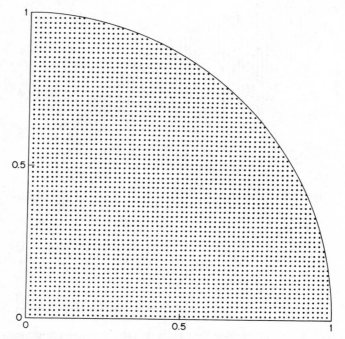

Fig. 5.39 Grid positions in the z plane for poles for coupled first-order sections.

error incurred in the frequency response due to coefficient rounding. For the cascade form, only certain observed properties of the sensitivities of this structure to rounded coefficients will be presented.

5.30 Coefficient Quantization in Direct Form Realizations of FIR Filters

The frequency response of a linear phase FIR filter can be written in the form

$$H(e^{j\omega}) = \left\{ \sum_{n=0}^{(N-3)/2} 2h(n) \cos \left[\left(\frac{N-1}{2} - n \right) \omega \right] + h \left(\frac{N-1}{2} \right) \right\} e^{-j\omega(N-1)/2}$$

(5.111)

where $\{h(n)\}$, $0 \leq n \leq N - 1$, is the impulse response of the filter and $h(n) = h(N - 1 - n)$. The factor $e^{-j\omega(N-1)/2}$ in Eq. (5.111) represents a pure delay of an integer number of samples (N odd) and is unaffected by quantization of the filter coefficients. Hence we can concentrate solely on the expression $\bar{H}(e^{j\omega}) = H(e^{j\omega})e^{j\omega(N-1)/2}$ with respect to coefficient quantization.

Let $\{h^*(n)\}$ be the sequence that results when $\{h(n)\}$ is rounded to a quantization step size Q. Then $h^*(n) = h(n) + e(n)$ and $h^*(n) = h^*(N - 1 - n)$ for $0 \leq n \leq (N - 1)/2$, where $e(n)$ is a random variable that is assumed to be uniformly distributed over the interval $[-Q/2, Q/2]$. Let $H^*(z)$ be the z transform of $\{h^*(n)\}$, and let $\bar{H}^*(e^{j\omega}) = H^*(e^{j\omega})e^{j\omega(N-1)/2}$. Finally, define an error function by

$$E_L(e^{j\omega}) = \bar{H}^*(e^{j\omega}) - \bar{H}(e^{j\omega}) \qquad (5.112)$$

or

$$E_L(e^{j\omega}) = \sum_{n=0}^{(N-3)/2} 2e(n) \cos\left[\left(\frac{N-1}{2} - n\right)\omega\right] + e\left(\frac{N-1}{2}\right) \qquad (5.113)$$

Equation (5.113) shows $E_L(e^{j\omega})$ to be the frequency response of a linear phase filter that has $\{e(n)\}$ as the first half of the impulse response and the second half is obtained as $e(N-1-n) = e(n)$, $0 \leq n \leq (N-1)/2$. As seen earlier the filter with quantized coefficients can be represented as a parallel connection of the infinite precision coefficient filter with a filter whose frequency response is $E_L(e^{j\omega})e^{-j\omega(N-1)/2}$.

A simple bound on $|E_L(e^{j\omega})|$ can be derived by observing that $|e(n)| \leq Q/2$. Thus from Eq. (5.113)

$$|E_L(e^{j\omega})| \leq \sum_{n=0}^{(N-3)/2} 2|e(n)| \left|\cos\left[\left(\frac{N-1}{2} - n\right)\omega\right]\right| + \left|e\left(\frac{N-1}{2}\right)\right|$$

$$\leq \frac{Q}{2}\left[1 + 2\sum_{n=1}^{(N-1)/2} |\cos(n\omega)|\right] \qquad (5.114)$$

or

$$|E_L(e^{j\omega})| \leq \frac{NQ}{2} \qquad (5.115)$$

The bound of Eq. (5.115) is overly pessimistic and therefore of little practical usefulness. If one assumes that errors due to the quantization of different coefficients are statistically independent, a more useful statistical bound than Eq. (5.115) can be derived. Even though, for any given filter, the quantization process is performed only once, after which the filter response is *exactly* determined, the aim of the statistical bound to be derived is to provide the filter designer a means of predicting how much accuracy is required for the coefficients to obtain a desired filter response, without a priori knowing the values of the filter coefficients.

From Eq. (5.113) an expression for $\overline{E_L(e^{j\omega})^2}$ can be written as

$$\overline{E_L^2(e^{j\omega})} = \sum_{n=0}^{(N-3)/2} \overline{4e(n)^2} \cos^2\left[\left(\frac{N-1}{2} - n\right)\omega\right] + \overline{e\left(\frac{N-1}{2}\right)^2} \qquad (5.116)$$

but since $\overline{e(n)^2} = Q^2/12$ for a uniform distribution, Eq. (5.116) can be written as

$$\overline{E_L^2(e^{j\omega})} = \frac{Q^2}{12}\left[1 + 4\sum_{n=1}^{(N-1)/2} \cos^2(\omega n)\right] \qquad (5.117)$$

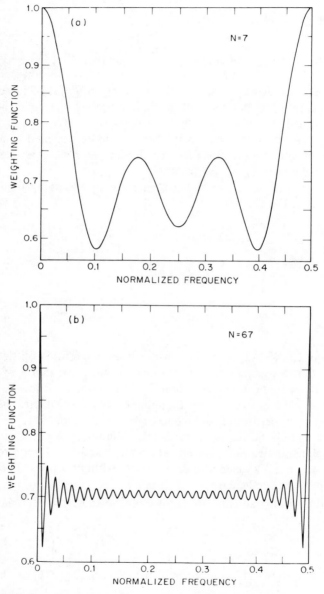

Fig. 5.40 Effects of quantized coefficients on the frequency response.

Defining the weighting function $W_N(\omega)$ as

$$W_N(\omega) = \left\{ \frac{1}{2N-1} \left[1 + 4 \sum_{n=1}^{(N-1)/2} \cos^2(\omega n) \right] \right\}^{1/2} \qquad (5.118)$$

the standard deviation of the error is given by

$$\sigma_{EL}(\omega) = [\overline{E_L^2(e^{j\omega})}]^{1/2} = \sqrt{\frac{2N-1}{3}} \frac{Q}{2} W_N(\omega) \qquad (5.119)$$

Clearly $0 < W_N(\omega) \leq 1$ and $W_N(0) = W_N(\pi) = 1$ for all N. Thus

$$\sigma_{EL}(\omega) \leq \frac{Q}{2} \sqrt{\frac{2N-1}{3}} \qquad (5.120)$$

It is easily shown that

$$\lim_{N \to \infty} W_N(\omega) = \frac{1}{\sqrt{2}} \qquad 0 < \omega < \pi \qquad (5.121)$$

Figure 5.40 shows plots of $W_N(2\pi f)$ versus f for $N = 7$ and $N = 67$. This figure shows how close $\sigma_{EL}(\omega)$ is to its bound in Eq. (5.120)—i.e., $\sigma_{EL}(\omega)$ is within a factor of 2 of its bound for all ω.

Since, for any ω, $E_L(e^{j\omega})$ is a sum of independent random variables whose probability density functions vanish outside some finite interval, $E_L(e^{j\omega})$ is essentially Gaussian for large N. Thus the mean and variance of $E_L(e^{j\omega})$ constitute an excellent statistical description of $E_L(e^{j\omega})$.

Chan and Rabiner have experimentally demonstrated the validity of the model of coefficient quantization above in the case of optimum lowpass filters. Based on this model and theoretical work of the relations between filter parameters in the infinite precision case, Chan and Rabiner were able to formulate a procedure for deciding on the number of bits required to realize a filter with given specifications on filter cutoff frequencies and passband and stopband ripples.

5.31 Coefficient Quantization in Cascade Realizations of FIR Filters

For cascade realizations of linear phase FIR filters, sections of the fourth order are generally required to preserve the linear phase characteristic with finite precision coefficients. Although zero location and frequency response sensitivities to coefficient changes can readily be obtained, no general statistical bounds, of the type given in Sec. 5.29 for the direct form, have been obtained to describe the behavior of this configuration. Experimental evidence shows that, in the case of band select filters, the cascade form is quite sensitive near $z = 1$ and quite insensitive near $z = 0$. Thus, in the case of lowpass filters, the cascade form is highly sensitive to inaccurate coefficients in the

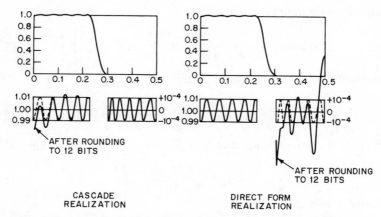

Fig. 5.41 Comparison of frequency responses of cascade and direct forms for coefficient quantization. (After Herrmann and Schuessler.)

passband, while the behavior in the stopband is much less sensitive. Figure 5.41 shows an example, due to Herrmann and Schuessler, of a linear phase FIR lowpass filter realized in the cascade form with coefficients rounded to 12 bits. The response with no rounding is also plotted. The response for the 12-bit case deviates from the ideal response significantly at the beginning of the passband, whereas no visible errors are seen anywhere in the stopband.

Since no general statistical bounds on the frequency response error are available for the cascade form, several coefficient optimization programs have been used to minimize the peak weighted frequency response error to aid in the use of this structure for specific cases.

5.32 Limit Cycle Oscillations

The discussion of roundoff noise in digital filters assumed that sample to sample changes in the input signal were large compared to the size of a quantization step. In this manner one could argue that the samples of the roundoff noise sequence were uncorrelated with each other, as well as with the input sequence. It is clear that there are many cases where the assumption above is no longer valid—e.g., constant (zero) input to a digital filter. Thus, for example, consider the difference equation

$$y(n) = 0.95y(n - 1) + x(n) \tag{5.122}$$

with input $x(n) = 0$ (i.e., the input is turned off) and initial condition $y(-1) = 13$. (The values of the variable y are expressed as integers times the quantization step Q. Thus the variable y can only take on integer values.) The table below shows a comparison between exact values of $y(n)$ as obtained from Eq. (5.122) using infinite precision arithmetic and rounded values of $y(n)$ as obtained from finite precision arithmetic. It is clear that even though the exact value of $y(n)$ decays exponentially to zero, the rounded value of $y(n)$ stays trapped at the value 10 and can never get any smaller. This is an example

n	$y(n)$—Exact	$y(n)$—Rounded
−1	13.0	13
0	12.35	12
1	11.7325	11
2	11.145875	10
3	10.58858125	10
4	10.05915219	10

of a limit cycle that occurs at the output of a recursive realization of a digital filter in response to zero input. The amplitude intervals in which such limit cycles are confined have been called deadbands by Blackman. In the example above, any value of $|y(-1)| \leq 10$ would give $y(n) = y(-1)$, $n \geq 0$, for $x(n) = 0$. Thus the deadband in the example above is the interval $[-10, 10]$.

Jackson has studied limit cycles in first- and second-order systems using an "effective value" model—i.e., realizing that limit cycles can only occur if the result of rounding effectively leads to poles on the unit circle. Thus in the case of the first-order difference equation

$$y(n) = x(n) - [\alpha y(n-1)]' \tag{5.123}$$

where $[\ ]'$ denotes the operation of rounding to the nearest integer with $x(n) = 0$, $n \geq 0$. The deadband in which limit cycles can exist is the region $[-k, k]$ with k the largest integer satisfying

$$k \leq \frac{0.5}{1 - |\alpha|} \tag{5.124}$$

As seen in the example above, if α is negative, the limit cycle is of constant magnitude and sign. If α is positive, the limit cycle is of constant magnitude and alternating sign. For all values of $y(n)$ in the deadband, the multiplier α has the effective value ± 1; i.e., $[\alpha y(n-1)]' = \pm y(n-1)$. Therefore the difference equation, Eq. (5.123), has an effective pole at $z = \pm 1$.

For the second-order system with difference equation

$$y(n) = x(n) - [\beta_1 y(n-1)]' - [\beta_2 y(n-2)]' \tag{5.125}$$

one can consider a deadband in which limit cycles can occur as the region $[-k, k]$ where k is the largest integer satisfying

$$k \leq \frac{0.5}{1 - \beta_2} \qquad (0 < \beta_2 < 1) \tag{5.126}$$

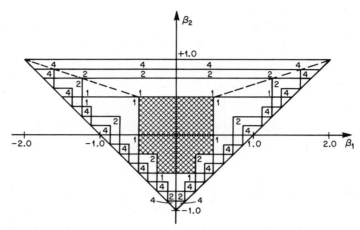

Fig. 5.42 Limit cycle amplitudes as a function of filter coefficients. (After Jackson.)

Equation (5.126) is analogous to Eq. (5.124) with α replaced by β_2 to ensure the poles of the filter are on the unit circle—i.e., the effective value of β_2 is 1.0. (Note $\beta_2 > 0$ for complex conjugate, stable poles.) The oscillation frequency of the limit cycle is controlled primarily by the value of β_1 but is also somewhat dependent on how the rounding affects the product $\beta_1 y(n-1)$ in Eq. (5.125).

Equation (5.126) shows that the minimum value of β_2 to achieve effective complex conjugate poles is $\beta_2 = 0.5$ for which $k \triangleq 1$. The next value for β_2 that has a larger value of k for its limit cycles is $\beta_2 = 0.75$ for which $k = 1$ or 2. For each value of β_2 there are only a finite number of ranges of β_1 for which distinct limit cycles occur. Figure 5.42 shows a diagram of the (β_1, β_2) plane for the second-order section of Eq. (5.125). The crosshatched area is the region where no limit cycles occur. The horizontal lines denote minimum values of β_2 for which distinct changes in the deadband interval occur. The numbers with each strip indicate the maximum magnitude of the limit cycle in that region of the (β_1, β_2) plane. The limit cycles indicated for $\beta_2 < 0.5$ are explained below.

The discussion of a second-order section above investigated limit cycles with effective complex conjugate poles. The other possibility in the second-order section is to have a real effective pole at $z = \pm 1$. The condition for a limit cycle of amplitude k to occur is

$$[\beta_2 k]' \pm [\beta_1 k]' = k \tag{5.127}$$

The regions of the (β_1, β_2) plane that satisfy Eq. (5.127) for various values of k are readily determined and are shown in Fig. 5.42.

There are two important reasons for studying limit cycles. In a communication environment the idle channel condition can lead to limit cycles that are extremely undesirable in that one would like to hear no signal over a line when no signal is put in. Thus when digital filters are used in the telephone plant, extensive care must be given to this problem. The second important reason for studying limit cycles is their potential application to the design of digital periodic waveform generators. By creating desirable limit cycles in a reliable manner, one can use the limit cycles as a source in digital signal processing.

Since Jackson's original treatment of limit cycles, there has been an extensive amount of study given to refining the bounds on the amplitudes and frequencies of the limit cycles. The details of such investigations are available elsewhere and will not be discussed here.

REFERENCES

General References

1. A. V. OPPENHEIM AND C. W. WEINSTEIN, "Effects of Finite Register Length in Digital Filters and the Fast Fourier Transform," *Proc. IEEE*, **60**, No. 8, 957–976, Aug., 1972.
2. B. GOLD AND C. M. RADER, *Digital Processing of Signals*, Chapter 4, McGraw-Hill Book Co., 1969.
3. B. LIU, "Effect of Finite Word Length on the Accuracy of Digital Filters—A Review," *IEEE Trans. Circuit Theory*, **CT-18**, 670–677, Nov., 1971.
4. W. R. BENNETT, "Spectra of Quantized Signals," *Bell Sys. Tech. J.*, **27**, 446–472, July, 1948.
5. C. M. RADER AND B. GOLD, "Effects of Parameter Quantization on the Poles of a Digital Filter," *Proc. IEEE*, **55**, No. 5, 688–689, May, 1967.

Roundoff Noise in Recursive Structures—Fixed Point Case

1. J. B. KNOWLES AND R. EDWARDS, "Effects of a Finite-Word-Length Computer in a Sampled-Data Feedback System," *Proc. Inst. Elec. Eng.*, **112**, 1197–1207, June, 1965.
2. B. GOLD AND C. M. RADER, "Effects of Quantization Noise in Digital Filters," *Proc. AFIPS 1966 Spring Joint Computer Conf.*, **28**, 213–219, 1966.
3. L. B. JACKSON, "On the Interaction of Roundoff Noise and Dynamic Range in Digital Filters," *Bell Sys. Tech. J.*, **49**, 159–184, Feb., 1970.
4. L. B. JACKSON, "Roundoff Noise Analysis for Fixed-Point Digital Filters Realized in Cascade or Parallel Form," *IEEE Trans. on Audio and Electroacoustics*, **AU-18**, 107–122, June, 1970.

Roundoff Noise in Nonrecursive Structures—Fixed Point Case

1. D. S. K. CHAN AND L. R. RABINER, "Theory of Roundoff Noise in Cascade Realizations of Finite Impulse Response Digital Filters," *Bell Sys. Tech. J.*, **52**, No. 3, 329–345, Mar., 1973.

2. D. S. K. CHAN AND L. R. RABINER, "An Algorithm for Minimizing Roundoff Noise in Cascade Realizations of Finite Impulse Response Digital Filters," *Bell Sys. Tech. J.*, **52**, No. 3, 347–385, Mar. 1973.

3. D. S. K. CHAN AND L. R. RABINER, "Analysis of Quantization Errors in the Direct Form for Finite Impulse Response Digital Filters," *IEEE Trans. on Audio and Electroacoustics*, **AU-21**, No. 4, 354–366, Aug., 1973.

Roundoff Noise in Recursive Structures—Floating Point Case

1. I. W. SANDBERG, "Floating-Point Roundoff Accumulation in Digital Filter Realization," *Bell Sys. Tech. J.*, **46**, 1775–1791, Oct., 1967.

2. T. KANEKO AND B. LIU, "Roundoff Error of Floating-Point Digital Filters," *Proc. 6th Annual Allerton Conf. on Circuit and System Theory*, 219–227, Oct., 1968.

3. C. WEINSTEIN AND A. V. OPPENHEIM, "A Comparison of Roundoff Noise in Floating Point and Fixed Point Digital Filter Realizations," *Proc. IEEE* (*Corresp.*), **57**, 1181–1183, June, 1969.

4. B. LIU AND T. KANEKO, "Error Analysis of Digital Filters With Floating-Point Arithmetic," *Proc. IEEE*, **57**, 1735–1747, Oct., 1969.

5. A. V. OPPENHEIM, "Realization of Digital Filters Using Block Floating-Point Arithmetic," *IEEE Trans. on Audio and Electroacoustics*, **AU-18**, 130–136, June, 1970.

Overflow Oscillations

1. P. M. EBERT, J. E. MAZO, AND M. G. TAYLOR, "Overflow Oscillations in Digital Filters," *Bell Sys. Tech. J.*, **48**, 3021–3030, Nov., 1968.

Coefficient Quantization—Recursive Structures

1. J. F. KAISER, "Some Practical Considerations in the Realization of Linear Digital Filters," *Proc. 3rd Annual Allerton Conf. on Circuit and System Theory*, 621–633, 1965.

2. C. M. RADER AND B. GOLD, "Effects of Parameter Quantization on the Poles of a Digital Filter," *Proc. IEEE* (Corresp.), **55**, 688–689, May, 1967.

3. J. B. KNOWLES AND E. M. OLCAYTO, "Coefficient Accuracy and Digital Filter Response," *IEEE Trans. Circuit Theory*, **15**, No. 1, 31–41, Mar., 1968.

4. E. AVENHAUS AND H. W. SCHUESSLER, "On the Approximation Problem in the Design of Digital Filters with Limited Wordlength," *Arch. Elek. Ubertragung*, **24**, 571–572, 1970.

Coefficient Quantization—Nonrecursive Structures

1. O. HERMANN AND H. W. SCHUESSLER, "On the Accuracy Problem in the Design of Nonrecursive Digital Filters," *Arch. Elek. Ubertragung*, **24**, 525–526, 1970.

2. D. S. K. CHAN AND L. R. RABINER, "Analysis of Quantization Errors in the Direct Form for Finite Impulse Response Digital Filters," *IEEE Trans. on Audio and Electroacoustics*, **AU-21**, No. 4, 354–366, Aug., 1973.

3. C. W. WEINSTEIN, "Quantization Effects in Frequency Sampling Filters," *NEREM Record*, 222, 1968.

Limit Cycles in Recursive Structures

1. R. B. BLACKMAN, *Linear Data-Smoothing and Prediction in Theory and Practice*, pp. 75–79, Addison-Wesley Pub. Co., Reading, Mass., 1965.
2. L. B. JACKSON, "An Analysis of Limit Cycles Due to Multiplication Rounding in Recursive Digital (Sub) Filters," *Proc. 7th Annual Allerton Conf. on Circuit and System Theory*, 69–78, 1969.
3. S. R. PARKER AND S. F. HESS, "Limit-Cycle Oscillations in Digital Filters," *IEEE Trans. Circuit Theory*, **CT-18**, 687–696, Nov., 1971.
4. I. W. SANDBERG, "A Theory Concerning Limit Cycles in Digital Filters," *Proc. 7th Allerton Conf. on Circuit and System Theory*, 63–67, 1969.
5. T. A. BRUBAKER AND J. N. GOWDY, "Limit Cycles in Digital Filters," *IEEE Trans. Automatic Control*, **17**, No. 5, 675–677, Oct., 1972.
6. I. W. SANDBERG AND J. F. KAISER, "A Bound on Limit Cycles in Fixed-Point Implementations of Digital Filters," *IEEE Trans. on Audio and Electroacoustics*, **AU-20**, No. 2, 110–112, June, 1972.

6

Spectrum Analysis and the Fast Fourier Transform

6.1 Introduction

There are many instances when signal processing involves the measurement of spectra. For example, in speech recognition problems, spectrum analysis is usually a preliminary to further acoustic processing. In speech bandwidth reduction systems, spectrum analysis is generally the basic measurement. In sonar systems, sophisticated spectrum analysis is required for the location of surface vessels and submarines. In radar systems, obtaining target velocity information generally implies the measurement of spectra.

It is important to realize that "spectrum analysis" encompasses a great variety of different measurements. A broad definition might be "a measurement that gives the exact or approximate value of the z transform of a discrete signal at selected values of z." Development of a comprehensive theory of spectrum analysis is made difficult by the fact that nearly all such measurements are taken over finite time intervals and the length of this interval is usually determined through intuition and experience. For example, the "spectrum" of a speech signal changes drastically as a function of time, more or less at a phonemic rate (about 10 times per second); despite this, the short-time speech spectrum is one of the most important speech measurements for many applications.

356

The set of algorithms known as the fast Fourier transform (FFT) consists of a variety of tricks for reducing the computation time required to compute a discrete Fourier transform (DFT). Since the DFT is the central computation in most spectrum analysis problems, it follows that the FFT implementation of the DFT, which, in some practical cases can improve performance by a factor of 100 or more over direct evaluation of the DFT, is crucially important and must be understood as part of any serious effort to utilize the digital signal processing techniques of spectral analysis. This chapter, therefore, begins with FFT theory including the well-known radix 2 decimation-in-time and decimation-in-frequency algorithms. It is then shown how the FFT can be presented as a single algorithm with many variations. The fact that a one-dimensional array of numbers can be represented as a two-dimensional array in more than one way accounts for much of the variability on FFT algorithms. This leads to a consideration of the mathematical manipulation that goes from one- to two-dimensional space as the central core of all FFT algorithms. From this unified viewpoint of the FFT, algorithmic variations can be derived in a relatively straightforward way. Many of these variations are treated in Chapter 10, on FFT hardware.

Given the FFT introduction, the stage is set to discuss the subject of the relationship between different measurement techniques. Included in this discussion are the questions of what portion of the z plane is of interest for various applications and how to treat these different cases, what the relationship is between spectrum analysis using the DFT as opposed to using a bank of digital filters, how to "improve" the spectrum analysis, and what the theoretical connections are between spectrum analysis and "chirp" filtering.

6.2   Introduction to Radix 2 FFT's

Although FFT algorithms are well-known and widely used, they are rather intricate and often difficult to grasp at first reading. There are several reasons for this difficulty. For one, the reader must learn many new terms; in addition, several different treatments of the basic results are present in the literature; also, there appear to be a great variety of different FFT algorithms. Finally, data shuffling in the FFT is intricate and usually best grasped using many examples.

The DFT of a finite duration sequence $\{x(n)\}, 0 \leq n \leq N - 1$, was defined in Chapter 2 as

$$X(k) = \sum_{n=0}^{N-1} x(n)e^{-j(2\pi/N)nk} \qquad k = 0, 1, \ldots, N - 1 \tag{6.1}$$

which may conveniently be written in the form

$$X(k) = \sum_{n=0}^{N-1} x(n)W^{nk} \tag{6.2}$$

where $W = e^{-j(2\pi/N)}$. It is easily seen that W^{nk} is periodic and of period N; i.e.,

$$W^{(n+mN)(k+lN)} = W^{nk} \qquad m, l = 0, \pm 1, \ldots \qquad (6.3)$$

We shall see that the periodicity of W^{nk} is one of the keys to the FFT. For emphasis the expression W_N is often used in place of W to explicitly give the period of W^{nk}.

Equation (6.1) shows that, in the case when $x(n)$ is a complex sequence, a complete direct evaluation of an N-point DFT requires $(N - 1)^2$ complex multiplications and $N(N - 1)$ complex additions. Thus for reasonably large values of N (on the order of 1000) direct evaluation of the DFT requires an inordinate amount of computation. The idea behind the FFT is to break the original N-point sequence into two shorter sequences, the DFT's of which can be combined to give the DFT of the original N-point sequence. Thus, for example, if N were even and the original N-point sequence were broken into two $(N/2)$-point sequences, it would require on the order of $[(N/2)^2 \cdot 2 = N^2/2]$ complex multiplications to evaluate the desired N-point DFT—a savings of a factor of 2 over direct evaluation. The factor $(N/2)^2$ represents the multiplications required for direct evaluation of an $(N/2)$-point DFT and the multiplier 2 represents the two DFT's that must be performed. The process above can be iterated so as to reduce the computation of an $(N/2)$-point DFT to two $(N/4)$-point DFT's (assuming $N/2$ is even) and thereby effect another savings of a factor of 2. The factor of 2 savings is only approximate because we have not accounted for a way of combining the smaller DFT's to give the desired N-point DFT.

We shall illustrate the process above for an N-point sequence $\{x(n)\}$ where N is assumed to be a power of 2. Define two $(N/2)$-point sequences $x_1(n)$ and $x_2(n)$ as the even and odd members of $x(n)$, respectively; i.e.,

$$x_1(n) = x(2n) \qquad n = 0, 1, \ldots, \frac{N}{2} - 1$$

$$x_2(n) = x(2n + 1) \qquad n = 0, 1, \ldots, \frac{N}{2} - 1 \qquad (6.4)$$

The N-point DFT of $\{x(n)\}$ can be written as

$$X(k) = \sum_{\substack{n=0 \\ n \text{ even}}}^{N-1} x(n)W_N^{nk} + \sum_{\substack{n=0 \\ n \text{ odd}}}^{N-1} x(n)W_N^{nk} \qquad (6.5)$$

$$= \sum_{n=0}^{N/2-1} x(2n)W_N^{2nk} + \sum_{n=0}^{N/2-1} x(2n + 1)W_N^{(2n+1)k} \qquad (6.6)$$

Recognizing that W_N^2 can be written as

$$W_N^2 = [e^{j(2\pi/N)}]^2 = e^{j[2\pi/(N/2)]} = W_{N/2} \qquad (6.7)$$

Eq. (6.6) can be put in the form

$$X(k) = \sum_{n=0}^{N/2-1} x_1(n)W_{N/2}^{nk} + W_N^k \sum_{n=0}^{N/2-1} x_2(n)W_{N/2}^{nk} \tag{6.8}$$

$$X(k) = X_1(k) + W_N^k X_2(k) \tag{6.9}$$

where $X_1(k)$ and $X_2(k)$ are seen to be the $(N/2)$-point DFT's of $x_1(n)$ and $x_2(n)$. Equation (6.9) shows that the N-point DFT $X(k)$ can be decomposed into two $(N/2)$-point DFT's that we combined according to the rule of Eq. (6.9). If the $(N/2)$-point DFT's were evaluated directly, it is readily seen that $(N^2/2 + N)$ complex multiplications are required to evaluate an N-point DFT. In the case of large N (i.e., $N^2/2 \gg N$) this represents a savings of 50% in computation time.

Since $X(k)$ is defined for $0 \leq k \leq N - 1$ and $X_1(k)$ and $X_2(k)$ are defined for $0 \leq k \leq N/2 - 1$, a rule must be given for how to interpret Eq. (6.9) for values of $k \geq N/2$. This rule is fairly straightforward and is given as[1]

$$X(k) = \begin{cases} X_1(k) + W_N^k X_2(k) & 0 \leq k \leq \dfrac{N}{2} - 1 \\[2mm] X_1\left(k - \dfrac{N}{2}\right) + W_N^k X_2\left(k - \dfrac{N}{2}\right) & \dfrac{N}{2} \leq k \leq N - 1 \end{cases} \tag{6.10}$$

Figure 6.1 illustrates, using a flow graph notation,[2] the processes involved in evaluating an eight-point DFT using two four-point transforms. The input sequence $x(n)$ is first shuffled into even and odd members to give $x_1(n)$ and $x_2(n)$, which are then transformed to give $X_1(k)$ and $X_2(k)$. The rule of Eq. (6.10) (see footnote 1, this page) is used to give $X(k)$.

The scheme above can be iterated to evaluate the $(N/2)$-point DFT's of Eq. (6.9) and (6.10). The sequences $x_1(n)$ and $x_2(n)$ are each divided into two sequences consisting of the even and odd members of the sequences. In a

---

[1] Equation (6.10) follows directly from the periodicity property of DFT. Notice, also, that $W_N^{k+N/2} = -W_N^k$ so that Eq. (6.10) can actually be rewritten as

$$X(k) = X_1(k) + W_N^k X_2(k) \qquad 0 \leq k \leq \frac{N}{2} - 1$$

$$= X_1\left(k - \frac{N}{2}\right) - W_N^{k-N/2} X_2\left(k - \frac{N}{2}\right) \qquad \frac{N}{2} \leq k \leq N - 1$$

[2] The right side of Fig. 6.1 defines the open circle ($\bigcirc$) as being an adder–subtracter; the sum always appears on top and the difference on the bottom. The arrow ($\rightarrow$) is defined as a multiplier with the value of the multiplier (a) being given above the arrow. In general, all variables are complex numbers. Notice that the open circle may also be interpreted as a two-point DFT. The nodes define the registers required to hold the inputs and outputs of the individual DFT's. This flow graph notation is consistent with the flow graph rules of linear system theory.

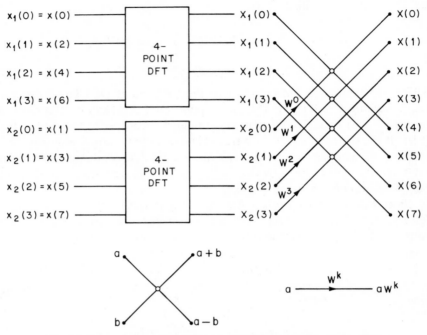

Fig. 6.1 Construction of an eight-point DFT from two four-point DFT's.

manner similar to the one described above, the $(N/2)$-point DFT's can be expressed as a combination of $(N/4)$-point DFT's; i.e., $X_1(k)$, $0 \le k \le N/2 - 1$, can be written as

$$X_1(k) = A(k) + W_{N/2}^k B(k) \tag{6.11}$$

or

$$X_1(k) = A(k) + W_N^{2k} B(k) \tag{6.12}$$

where $A(k)$ is the $(N/4)$-point DFT of the even members of $x_1(n)$ and $B(k)$ is the $(N/4)$-point DFT of the odd members of $x_1(n)$. Figure 6.2 shows the resulting flow graph when the four-point DFT's of Fig. 6.1 are evaluated as in Eq. (6.12).

The process above of reducing an L-point DFT (L is a power of 2) to $(L/2)$-point DFT's can be continued until you are left with two-point DFT's to evaluate. A two-point DFT, $F(k)$, $k = 0, 1$, may be evaluated (using no multiplications) as

$$F(0) = f(0) + f(1)W_8^0$$
$$F(1) = f(0) + f(1)W_8^4 \tag{6.13}$$

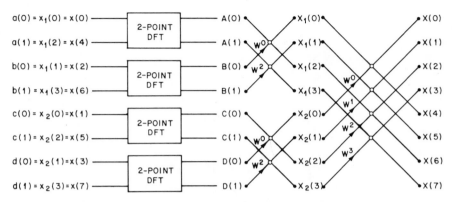

Fig. 6.2 Construction of an eight-point DFT from two four-point DFT's that are in turn constructed from four two-point DFT's.

where $f(n)$, $n = 0$, 1, is the two-point sequence being transformed. Since $W_8^0 = 1$ and $W_8^4 = -1$, no multiplications are required to evaluate Eq. (6.13). Thus the eight-point DFT of Fig. 6.1 and 6.2 finally reduces to the flow graph of Fig. 6.3.

6.3 Some Properties of Radix 2 Decimation-in-Time FFT

An analysis of Fig. 6.3 along with some general thoughts as to the mechanics of successively halving the size of the transforms shows that at each stage of the FFT (i.e., each halving) $(N/2)$ complex multiplications are required to combine the results of the previous stage. Since there are $(\log_2 N)$ stages, the number of complex multiplications required to evaluate an N-point DFT is approximately $N/2 \log_2 N$. The word *approximately* is used because it is seen that multiplications by factors such as W_N^0, $W_N^{N/2}$, $W_N^{N/4}$, and $W_N^{3N/4}$ are really just complex additions and subtractions. Thus, for example, in Fig. 6.3 it is seen that the first stage of the FFT consists solely of real additions and subtractions of complex quantities. Even the second stage consists entirely of complex additions and subtractions. In fact the flow chart of Fig. 6.3 indicates that there are only 2 nontrivial multiplications as opposed to the expected value of 12 ($4 \log_2 8$). For larger values of N, however, the actual number of nontrivial multiplications is well approximated by $N/2 \log_2 N$.

The algorithm described above has been called the decimation-in-time (DIT) algorithm since at each stage of the process the input sequence (i.e., time sequence) is divided into smaller sequences for processing—i.e., the input sequence is decimated at each stage. Another form of FFT algorithm, called the decimation-in-frequency (DIF) algorithm, also exists and will be

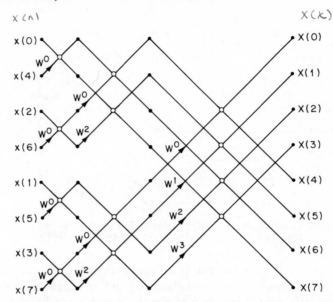

Fig. 6.3 Eight-point FFT obtained by successive splitting into two's.

described later in this section. First, however, it is both interesting and informative to discuss some of the general properties of methods for performing fast Fourier transforms.

The basic operation of the decimation-in-time algorithm is the so-called butterfly in which two inputs A and B are combined to give the two outputs X and Y via the operation

$$X = A + W_N^k B$$
$$Y = A - W_N^k B$$

(6.14)

Figure 6.4 shows a flow graph for the butterfly of Eq. (6.14). Careful examination of the flow graph of Fig. 6.3 shows that at each stage there are $(N/2)$ butterflies. For the case where W_N^k is a nontrivial multiplier, only one multiplication is required per butterfly since the quantity $B \cdot W_N^k$ can be computed and saved. The importance of the butterfly structure to the FFT flow graph is that only one additional memory location is required to transform an N-point sequence stored in memory. Thus one can compute

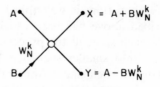

Fig. 6.4 FFT Butterfly.

intermediate stages of the FFT and store the results in the same locations in which the original data being transformed was stored. Therefore the array used to store the input sequence is also the same array into which the output sequence is stored. An algorithm that uses the same locations to store both the input and output sequences is called an in-place algorithm.

Figure 6.5 illustrates another way of viewing the radix 2 FFT. All the DFT's are two-point DFT's, requiring *no* multiplications. Combining two $(N/2)$-point DFT's to create an N-point DFT, however, requires on the order of $(N/2)$ multiplications. As exemplified in Fig. 6.3 an N-point FFT requires $(\log_2 N)$ stages and *all* multiplications are due to the combining operations. Since these combining multiplications are so important to the

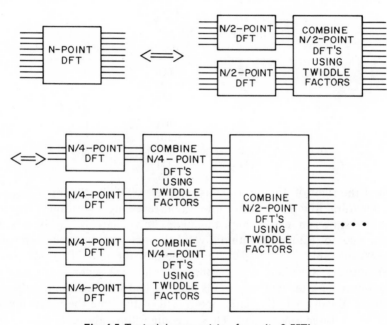

Fig. 6.5 Typical decomposition for radix 2 FFT's.

FFT and because they occur in different ways in the different FFT algorithms, they have been given a special name: *twiddle factors* or sometimes phase or rotation factors.

6.4   Data Shuffling and Bit Reversal

Another point worth noting about the decimation-in-time algorithm (as well as most other FFT algorithms) is that in order for the output sequence to be in natural order (i.e., $X(k)$, $k = 0, 1, \ldots, N - 1$), the input sequence had to be stored in a shuffled order. For the example in Fig. 6.3 the required order of the input sequence was $x(0)$, $x(4)$, $x(2)$, $x(6)$, $x(1)$, $x(5)$, $x(3)$, and finally $x(7)$. Although shuffled, the order of the input sequence can be determined in a relatively simple manner. It will be shown later that when N is a power of 2, the input sequence must be stored in bit-reversed order for the output to be computed in natural order. The definition of bit-reversed order is as follows: If one forms the L-bit binary representation of the natural order indices of the input sequence, where $N = 2^L$, and reverses the bits, the resulting number is the index of the input sequence that belongs in that position. Thus, in the case of $N = 8 = 2^3$, the natural-order indices are as shown at the left of Table 6.1, whereas the bit-reversed indices are as

Table 6.1

Index	Binary Representation	Bit-Reversed Binary	Bit-Reversed Index
0	000	000	0
1	001	100	4
2	010	010	2
3	011	110	6
4	100	001	1
5	101	101	5
6	110	011	3
7	111	111	7

shown at the right of Table 6.1. Thus in order to shuffle the input sequence from its natural to a bit-reversed order, a bit-reversing algorithm is required. Figure 6.6 shows an easily implementable flow chart for a bit-reversed counter due to Rader. Beginning with the first bit-reversed number X (e.g., 000 in Table 6.1) the algorithm generates the remaining bit-reversed indices in order. One-half the time the algorithm will generate the next bit-reversed number after two operations because half the time the most significant bit of X will be 1. Similarly one-fourth of the time the algorithm will terminate in three operations. Thus the algorithm above is exceedingly efficient.

It should also be clear from the discussion above that the input sequence can be shuffled to its bit-reversed order in place—i.e., pairs of numbers can be interchanged using only one temporary storage location. Thus for the case of Table 6.1, the shuffling is as graphically shown in Fig. 6.7.

A final point worth noting about the FFT is that the coefficients $\{W_N^k\}$, $k = 0, 1, \ldots, N - 1$, are required at various stages throughout the transform. There are several possible ways of obtaining these coefficients. The simplest way is to form a table of the coefficients that may then be referenced at various points during the computation. The only disadvantage of this method is that extra storage (on the order of N locations) is required to hold

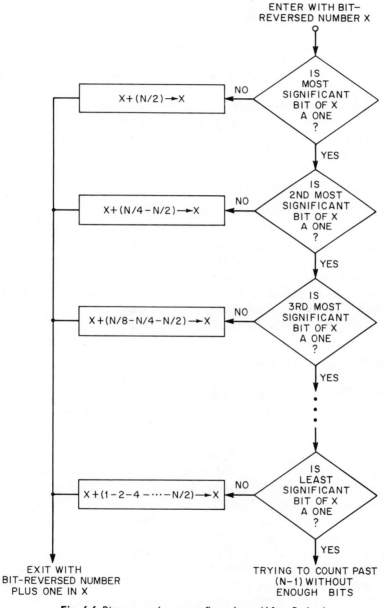

Fig. 6.6 Bit-reversed counter flow chart. (After Rader.)

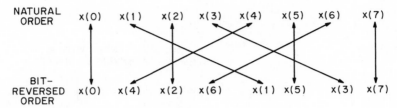

Fig. 6.7 In-place bit-reversed shuffling.

these coefficients. For large values of N, the extra storage may be prohibitive. A second method for obtaining the coefficients is to compute $W_N^k = \cos[(2\pi/N)k] - j \sin[(2\pi/N)k]$ directly, using standard sine and cosine routines, each time it is required. This method is a relatively slow one because sines and cosines generally require a fair amount of computation time. The third alternative is to use a simple recursion relation of the form

$$W_N^k = (W_N^{k-L}) \cdot W_N^L \qquad (6.15)$$

with an initial condition (e.g., $= W_N^0 = 1$) to start the algorithm. This recursion can be used because at each stage of the FFT the powers of W that are required are equally spaced. Thus, for the example of Fig. 6.3, the first stage requires W^0 and W^4; the second stage requires W^0, W^2, W^4, and W^6; and the third stage requires W^k, $k = 0, 1, \ldots, 7$. Thus the only quantities that need to be stored or computed are the multipliers W^4, W^2, and W in order to use Eq. (6.15) for each of the three stages.

6.5 FFT FORTRAN Program

In order to fix ideas about the FFT, it is worthwhile seeing a simple FORTRAN program that implements the decimation-in-time algorithm. One such program, due to Cooley, Lewis, and Welch, is given in Fig. 6.8. The input sequence is the complex array A of size up to 1024. The actual size of the FFT is $N = 2^M$, where M is read in via the calling statement. The coding from DO 7 I = 1, NM1 to statement 7 performs the in-place bit-reversing shuffling of the input data. The remainder of the coding performs the actual calculations for the FFT. Three nested DO loops are required for the entire algorithm. One loop keeps track of which of the M stages is being performed. Another loop keeps track of which butterfly within the stage is being calculated. The third loop performs the indexing on the powers of W as required by the different butterflies within a given stage. The algorithm of Fig. 6.8 uses the recursion relation of Eq. (6.15) to generate the required powers of W, using direct calculation of sines and cosines only for the evaluation of the increment in W. Cooley, Lewis, and Welch state that this procedure requires only about 15% more operations (complex additions or

```
      SUBROUTINE FFT(A,M,N)
      COMPLEX A(N),U,W,T
      N = 2**M
      NV2 = N/2
      NM1 = N - 1
      J = 1
      DO 7 I = 1,NM1
      IF(I .GE. J) GO TO 5
      T = A(J)
      A(J) = A(I)
      A(I) = T
    5 K = NV2
    6 IF(K .GE. J) GO TO 7
      J = J - K
      K = K/2
      GO TO 6
    7 J = J + K
      PI = 3.141592653589793
      DO 20 L = 1,M
      LE = 2**L
      LE1 = LE/2
      U = (1.0,0.)
      W = CMPLX(COS(PI/LE1),SIN(PI/LE1))
      DO 20 J = 1,LE1
      DO 10 I = J,N,LE
      IP = I + LE1
      T = A(IP) * U
      A(IP) = A(I) - T
   10 A(I) = A(I) + T
   20 U = U * W
      RETURN
      END
```

Fig. 6.8 FORTRAN code for a decimation-in-time, radix 2, in-place FFT. (After Cooley, Lewis, and Welch.)

multiplications) for $N = 1024$ than would be required for table storage of the entire W^k array. As Fig. 6.8 shows, it is indeed astonishing how remarkably simple the coding of an FFT algorithm can be.

Before proceeding to a second FFT algorithm, it is worthwhile re-emphasizing the gain in speed of an FFT algorithm (for N a power of 2) over direct evaluation of the DFT. For direct evaluation the number of operations (complex multiplications and additions) is on the order of N^2, while for the FFT it is on the order of $N \log_2 N$. Table 6.2 shows a comparison between

Table 6.2

N	N^2	$N \log_2 N$	$N^2/(N \log_2 N)$
2	4	2	2.0
4	16	8	2.0
8	64	24	2.7
16	256	64	4.0
32	1024	160	6.4
64	4096	384	10.7
128	16384	896	18.3
256	65536	1024	32.0
512	262144	4608	56.9
1024	1048576	10240	102.4
2048	4194304	22528	186.2

these numbers for various values of N from 2 to 2048. The gain in speed for $N \geq 1024$ is approximately two orders of magnitude. This gain in speed makes signal processing problems involving DFT's practical that were entirely impractical prior to the FFT.

6.6 Decimation-in-Frequency Algorithm

Another popular form of the FFT algorithm, for N a power of 2, is the so-called decimation-in-frequency algorithm (DIF). For this version of the FFT, the input sequence $\{x(n)\}$ is partitioned into two sequences each of length $(N/2)$ samples in the following manner. The first sequence $\{x_1(n)\}$ consists of the first $(N/2)$ points of $\{x(n)\}$, whereas the second sequence $\{x_2(n)\}$ consists of the last $(N/2)$ points of $\{x(n)\}$. Thus

$$x_1(n) = x(n) \qquad n = 0, 1, \ldots, \frac{N}{2} - 1$$

$$x_2(n) = x\left(n + \frac{N}{2}\right) \qquad n = 0, 1, \ldots, \frac{N}{2} - 1$$

(6.16)

The N-point DFT of $x(n)$ can now be written in the form

$$X(k) = \sum_{n=0}^{N/2-1} x(n)W_N^{nk} + \sum_{n=N/2}^{N-1} x(n)W_N^{nk} \tag{6.17}$$

$$= \sum_{n=0}^{N/2-1} x_1(n)W_N^{nk} + \sum_{n=0}^{N/2-1} x_2(n)W_N^{(n+N/2)k} \tag{6.18}$$

$$X(k) = \sum_{n=0}^{N/2-1} [x_1(n) + e^{-j\pi k}x_2(n)]W_N^{nk} \tag{6.19}$$

where we have used the fact that $W^{Nk/2} = e^{-j\pi k}$. If we now consider the even and odd samples of the DFT separately, we have the following relations:

$$X(2k) = \sum_{n=0}^{N/2-1} [x_1(n) + x_2(n)](W_N^2)^{nk} \tag{6.20}$$

$$= \sum_{n=0}^{N/2-1} [x_1(n) + x_2(n)]W_{N/2}^{nk} \tag{6.21}$$

$$X(2k + 1) = \sum_{n=0}^{N/2-1} [x_1(n) - x_2(n)]W_N^{n(2k+1)} \tag{6.22}$$

$$= \sum_{n=0}^{N/2-1} \{[x_1(n) - x_2(n)]W_N^n\}W_{N/2}^{nk} \tag{6.23}$$

Equations (6.21) and (6.23) show that the even- and odd-valued samples of the DFT can be obtained from the $(N/2)$-point DFT's of the sequences

$f(n)$ and $g(n)$, respectively, where

$$f(n) = x_1(n) + x_2(n) \qquad n = 0, 1, \ldots, \frac{N}{2} - 1$$

$$(6.24)$$

$$g(n) = [x_1(n) - x_2(n)]W_N^n \qquad n = 0, 1, \ldots, \frac{N}{2} - 1$$

Thus we have again reduced the problem of obtaining an N-point DFT to one of obtaining two $(N/2)$-point DFT's. Figure 6.9 illustrates the procedure for $N = 8$.

The procedure can now be iterated to express each $(N/2)$-point DFT as a combination of two $(N/4)$-point DFT's. Figures 6.10 and 6.11 show the results of expressing the four-point DFT's of Fig. 6.9 as two-point DFT's and finally evaluating the two-point DFT's directly.

Comparing Fig. 6.3 with Fig. 6.11, we note two apparent differences in the computational algorithms. First, for decimation-in-time (DIT), the input is bit-reversed while the output is in natural order; whereas the reverse is true for the DIF of Fig. 6.11. Both DIF and DIT can go from normal to shuffled data and vice versa, however, so this is not really a difference. Second, the DIF butterfly (Fig. 6.12) is slightly different from the DIT butterfly shown in Fig. 6.4, the difference being that the complex multiplication takes place *after* the add–subtract operation in DIF.

The similarities between the DIT and DIF algorithms cannot be overlooked. Both algorithms require on the order of $(N \log_2 N)$ operations to compute the DFT. Both algorithms can be done in place and both need to perform bit reversal at some place during the computation. Section 6.8

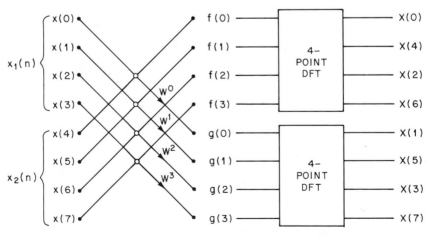

Fig. 6.9 Reduction of an eight-point DFT to two four-point DFT's by decimation-in-frequency.

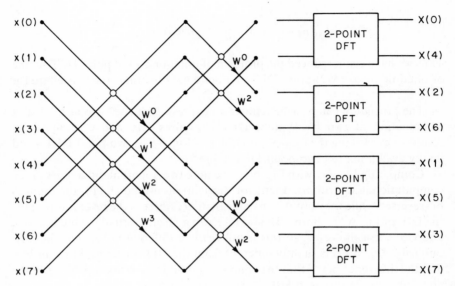

Fig. 6.10 Further reduction of Fig. 6.9.

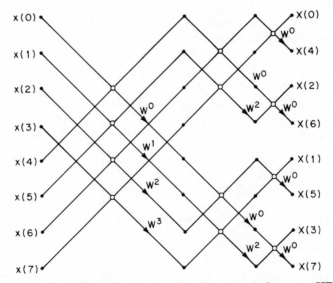

Fig. 6.11 Complete eight-point in-plane decimation-in-frequency FFT.

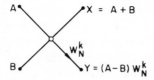

Fig. 6.12 Butterfly for decimation-in-frequency.

will show exactly why these similarities exist between seemingly disparate algorithms. A unified approach toward the FFT will be given for which the DIT and DIF algorithms will be special cases. This unified approach will also consider the case where N is a composite integer but not necessarily a power of 2.

6.7   Computing an Inverse DFT by Doing a Direct DFT

Before proceeding to the next section we shall show how an FFT algorithm can be used to compute an inverse DFT without *any* changes in the algorithm. The inverse DFT of an N-point sequence $\{X(k)\}$, $k = 0, 1,\ldots, N - 1$, is defined as

$$x(n) = \frac{1}{N}\sum_{k=0}^{N-1} X(k)W^{-nk} \tag{6.25}$$

If we take the complex conjugate of Eq. (6.25) and multiply by N, we find

$$Nx^*(n) = \sum_{k=0}^{N-1} X^*(k)W^{nk} \tag{6.26}$$

The right-hand side of Eq. (6.26) is recognized to be the DFT of the sequence $\{X^*(k)\}$ and may be computed using any FFT algorithm such as the ones described above. The desired output sequence $\{x(n)\}$ can then be obtained by complex conjugating the DFT of Eq. (6.26) and dividing by N to give

$$x(n) = \frac{1}{N}\left[\sum_{k=0}^{N-1} X^*(k)W^{nk}\right]^* \tag{6.27}$$

Thus a single FFT algorithm serves for evaluation of both the direct and inverse DFT's.

6.8   A Unified Approach to the FFT

There are a great variety of FFT algorithms. It turns out that they can all be derived from successive applications of a single operation, namely, that of representing a one-dimensional string of numbers as a two-dimensional array. In this section we present this unified approach. We begin with some terminology and discussion.

If we have an N-point sequence for which we desire to evaluate the N-point DFT, the integer N is either a prime or a composite number (until now we have assumed N was a highly composite number, i.e., a power of 2.) If N is prime, it cannot be expressed as the product of smaller integers. In this case the one-dimensional signal $x(0)$, $x(1), \ldots, x(N-1)$, *cannot* be expressed as a rectangular two-dimensional array so that no FFT algorithm exists for that array. Most practical problems are sufficiently malleable so that no difficulties occur when the sequence is artificially lengthened by augmentation with signal samples of zero value; the resulting spectrum will be an interpolated version of the nonaugmented spectrum.

As an example, choose $N = 60$. We can express 60 as a product of smaller numbers in a variety of ways: $60 = 3 \times 4 \times 5 = 4 \times 3 \times 5 = 5 \times 4 \times 3 = 12 \times 5 = 2 \times 2 \times 5 \times 3$, etc. Depending on the order and degree of decomposition, a different FFT variation may be obtained. The term *radix* is commonly used to describe the specific decomposition. The term *mixed radix* means that the factors of N are not all identical. For $N = 60$, all forms of the algorithm are mixed radices. If N can be expressed as a product of the same integer r, then the algorithm is called a radix r algorithm. For example, if $N = 64 = 2 \times 2 \times 2 \times 2 \times 2 \times 2$, we have the radix 2 algorithms as discussed in the preceding sections; but if N is expressed as $64 = 8 \times 8$, a radix 8 algorithm results.

With respect to the decomposition of a composite number, it is most important to note that successive decomposition of *one* number into *two* products can result in any and all possible decompositions. Thus, if $N = 60$, we can express 60 as 12×5 and then express 12 as 4×3. Alternately, we could have $60 = 3 \times 20$ and $20 = 5 \times 4$ and $4 = 2 \times 2$. Thus, careful attention to the properties of *one* into *two* decomposition should give us the necessary insight into the general case. Let us again take $N = 60$ as an example. One possible decomposition follows:

$$
\begin{array}{cccccccccccc}
0 & 1 & 2 & 3 & 4 & 5 & 6 & 7 & 8 & 9 & 10 & 11 \\
12 & 13 & 14 & 15 & 16 & 17 & 18 & 19 & 20 & 21 & 22 & 23 \\
24 & 25 & 26 & 27 & 28 & 29 & 30 & 31 & 32 & 33 & 34 & 35 \\
36 & 37 & 38 & 39 & 40 & 41 & 42 & 43 & 44 & 45 & 46 & 47 \\
48 & 49 & 50 & 51 & 52 & 53 & 54 & 55 & 56 & 57 & 58 & 59
\end{array}
$$

Here we have expressed the signal indices as a (5×12) matrix. Now since, the columns have 5 samples (a prime number), they can be decomposed no further; however, each 12-sample row can be expressed as a (3×4) matrix. For example, the first row is

$$
\begin{array}{cccc}
0 & 1 & 2 & 3 \\
4 & 5 & 6 & 7 \\
8 & 9 & 10 & 11
\end{array}
$$

and the other rows can be similarly expressed. Thus, the fundamental result we are after should tell us how manipulations on a two-dimensional array can lead to the DFT of the one-dimensional array from which it was derived.

To derive the central result, we wish to represent the entries in the matrix as a function of a row index and a column index. This can easily be done by replacing each entry by a pair of entries, for example, as follows:

0, 0	0 ,1	0, 2	0 ,3	0, 4	0, 5	0, 6	0, 7	0 ,8	0, 9	0, 10	0, 11
1, 0	1, 1	1, 2	1, 3	1, 4	1, 5	1, 6	1, 7	1, 8	1, 9	1, 10	1, 11
2, 0	2, 1	2, 2	2, 3	2, 4	2, 5	2, 6	2, 7	2, 8	2, 9	2, 10	2, 11
3, 0	3, 1	3, 2	3, 3	3, 4	3, 5	3, 6	3, 7	3, 8	3, 9	3, 10	3, 11
4, 0	4, 1	4, 2	4, 3	4, 4	4, 5	4, 6	4, 7	4, 8	4, 9	4, 10	4, 11

Now, letting the column index be m ($m = 0, 1, \ldots, 11$) and the row index be l ($l = 0, 1, \ldots, 4$) and if n is the original index ($n = 0, 1, \ldots, 59$), then we see that

$$n = Ml + m \qquad (6.28)$$

where, in this example, M ($= 12$) is the number of columns and L ($= 5$) is the number of rows.

Assuming we can indeed perform the DFT on our two-dimensional doubly indexed array, the result would also be a two-dimensional doubly indexed array. With m and l as the signal variables, let r and s be the transformed column and row variables, which when *recomposed* yield a single variable

$$k = Lr + s \qquad (6.29)$$

We are now in a position to express the DFT samples $X(k) = X(s, r)$ as the transform of $x(n) = x(l, m)$ by simply substituting Eq. (6.28) and (6.29) into the definition of the DFT [Eq. (6.2)] giving

$$X(k) = X(s, r) = \sum_{m=0}^{M-1} \sum_{l=0}^{L-1} x(l, m) W^{(Ml+m)(Lr+s)} \qquad (6.30)$$

Expanding $W^{(Ml+m)(Lr+s)}$, observing that $W^{MLlr} = W^{Nlr} = 1$, and properly associating indices with summation terms, we can rearrange Eq. (6.30) as

$$X(s, r) = \sum_{m=0}^{M-1} W^{Lmr} W^{ms} \underbrace{\sum_{l=0}^{L-1} x(l, m) W^{Msl}}_{q(s, m)} \qquad (6.31)$$

Equation (6.31), if correctly interpreted, is really all we need to establish the validity of our procedure for evaluating one-dimensional transforms via two-dimensional representations. Notice first that the L-fold interval sum is really the DFT of the mth column of the array having the kernel W^M. Thus, the first step in our computation procedure would be

1.   Compute the L-point DFT of each column.

The result is a function of s and m, with m going from 0 to $M - 1$. Call the result $q(s, m)$. Now Eq. (6.31) can be written

$$X(s, r) = \sum_{m=0}^{M-1} W^{Lmr} W^{ms} q(s, m) \tag{6.32}$$

From Eq. (6.32), step 2 in the procedure is deduced to be

2.   Obtain a new array $h(s, m)$ by multiplying every $q(s, m)$ by its *twiddle* factor W^{ms}.

Equation (6.32) now reduces to

$$X(s, r) = \sum_{m=0}^{M-1} h(s, m) W^{Lmr} \tag{6.33}$$

Equation (6.33) is recognized to be the M-point DFT of each row, with the row index x. Thus, the final step in the procedure is

3.   Compute the DFT of each row of the $h(s, m)$ matrix, with W^L as kernel.

The procedure above will be recognized to be *similar* to the computation of two-dimensional DFT's. In that case, one computes row DFT's followed by column DFT's to find the answer; step 2 is missing. The separability of the higher-dimensional kernel causes higher-dimensional DFT's to require less computation than one-dimensional DFT's with the same number of points.

But now the central result demonstrates that the two computations can be made equivalent by introducing the twiddles in step 2; and this step involves at most N additional multiplications for each decomposition of a one-into-two-dimensional array. More details on computation time will be discussed later in this section.

It should be noted that if the summations in Eq. (6.30) were performed in reverse order, i.e.,

$$X(s, r) = \sum_{l=0}^{L-1} W^{Mls} \underbrace{\overbrace{\sum_{m=0}^{M-1} x(l, m) W^{ms}}^{\text{step 1}} W^{Lrm}}_{\substack{\text{step 2}}}$$

step 3

then the computation procedure would become

1.   Twiddle the signal $x(l, m)$ by the factors W^{ms}.
2.   Compute the M-point DFT of each row.
3.   Compute the L-point DFT of each column.

Although the total computation is identical to the decomposition given previously, the detailed steps occur in a different order—i.e., the twiddles precede the row DFT's in the second procedure, whereas the twiddles follow the column DFT's in the first procedure. Such differences are reminiscent of and, in fact, related to the differences between the DIT and DIF algorithms for radix 2 FFT's as discussed previously in this chapter.

By observation of Eq. (6.30) we notice another very important effect of our procedure: m and r are column indices, while l and s are row indices. As m is incremented by 1, the signal index ($Ml + m$) is also incremented by 1. Incrementing the transformed column index r by 1, however, increments the argument of $x(s, r)$ by L. This implies that the row and column indices undergo permutation as a result of the transform. This point is so important that we shall illustrate the result for obtaining a 15-point DFT via a (3×5) decomposition. The original signal matrix can be written as

$$
\begin{array}{ccc}
x(0) = x(0, 0) & x(1) = x(0, 1) & x(2) = x(0, 2) \\
x(3) = x(0, 3) & x(4) = x(0, 4) & x(5) = x(1, 0) \\
x(6) = x(1, 1) & x(7) = x(1, 2) & x(8) = x(1, 3) \\
x(9) = x(1, 4) & x(10) = x(2, 0) & x(11) = x(2, 1) \\
x(12) = x(2, 2) & x(13) = x(2, 3) & x(14) = x(2, 4)
\end{array}
$$

whereas the resultant spectral matrix becomes

Table 6.3

$X(0) = X(0, 0)$	$X(3) = X(0, 1)$	$X(6) = X(0, 2)$	$X(9) = X(0, 3)$	$X(12) = X(0, 4)$
$X(1) = X(1, 0)$	$X(4) = X(1, 1)$	$X(7) = X(1, 2)$	$X(10) = X(1, 3)$	$X(13) = X(1, 4)$
$X(2) = X(2, 0)$	$X(5) = X(2, 1)$	$X(8) = X(2, 2)$	$X(11) = X(2, 3)$	$X(14) = X(2, 4)$

If we take a number like $N = 30 = 5 \times 6 = 5 \times 2 \times 3$, we see that the central theorem can be applied twice. We might start out with a (5×6) decomposition and then do the six-point DFT's by a (2×3) decomposition. It is instructive to proceed in this case by example; the reader can then try other N's. We begin by enumerating the 30 indices in the original signal matrix.

x Matrix—$x(l, m)$

$0 = (0, 0)$	$1 = (0, 1)$	$2 = (0, 2)$	$3 = (0, 3)$	$4 = (0, 4)$	$5 = (0, 5)$
$6 = (1, 0)$	$7 = (1, 1)$	$8 = (1, 2)$	$9 = (1, 3)$	$10 = (1, 4)$	$11 = (1, 5)$
$12 = (2, 0)$	$13 = (2, 1)$	$14 = (2, 2)$	$15 = (2, 3)$	$16 = (2, 4)$	$17 = (2, 5)$
$18 = (3, 0)$	$19 = (3, 1)$	$20 = (3, 2)$	$21 = (3, 3)$	$22 = (3, 4)$	$23 = (3, 5)$
$24 = (4, 0)$	$25 = (4, 1)$	$26 = (4, 2)$	$27 = (4, 3)$	$28 = (4, 4)$	$29 = (4, 5)$

We now perform our six five-point column DFT's and then twiddle the resulting matrix. The $h(s, m)$ matrix results:

h Matrix—$h(s, m)$

$$\begin{array}{cccccc}
h(0, 0) & h(0, 1) & h(0, 2) & h(0, 3) & h(0, 4) & h(0, 5) \\
h(1, 0) & h(1, 1) & h(1, 2) & h(1, 3) & h(1, 4) & h(1, 5) \\
h(2, 0) & h(2, 1) & h(2, 2) & h(2, 3) & h(2, 4) & h(2, 5) \\
h(3, 0) & h(3, 1) & h(3, 2) & h(3, 3) & h(3, 4) & h(3, 5) \\
h(4, 0) & h(4, 1) & h(4, 2) & h(4, 3) & h(4, 4) & h(4, 5)
\end{array}$$

Rather than performing the last step of the decomposition (i.e., the five six-point row transforms) directly, each six-point row is decomposed into a (2×3) matrix. Thus, for example, the first two rows of the h matrix become

Row 1

$$\begin{array}{ccc}
h(0, 0) & h(0, 1) & h(0, 2) \\
h(0, 3) & h(0, 4) & h(0, 5)
\end{array}$$

Row 2

$$\begin{array}{ccc}
h(1, 0) & h(1, 1) & h(1, 2) \\
h(1, 3) & h(1, 4) & h(1, 5)
\end{array}$$

Thus the required six-point DFT's can be obtained by first transforming the two-point columns, twiddling the results, and then transforming the three-point rows. When this has been completed for all five of the (2×3) arrays, we are done.

Let us review all operations required for this decomposition of an ($N = 30$)-point DFT into a $(5 \times 2 \times 3)$ result.

1. 6 five-point transforms.
2. 30 twiddles.
3. 15 two-point transforms.
4. 30 twiddles.
5. 10 three-point transforms.

From this example we can deduce how things go for $N = PLM$.

1. LM P-point transforms.
2. LMP twiddles.
3. MP L-point transforms.
4. LMP twiddles.
5. LP M-point transforms.

To obtain an estimate of the computational savings obtained through the use of the FFT, let us consider a few specific cases. First, take $N = LM$, where both L and M are primes larger than 2. By observation of Eq. (6.2), we can estimate that an M-fold DFT takes M^2 "operations," where an operation is a complex multiplication and addition. Applying this fact to the central result yields the number of operations for an $(N = LM)$ decomposition as

$$C_2 = ML^2 + LM^2 + LM = LM(L + M + 1) = N(M + L + 1)$$
$$(6.34)$$

The LM term comes from the twiddle function sandwiched between the row and column transforms. (Strictly speaking, the twiddle is only a complex multiplication and somewhat *less* than a full operation.) We quickly observe that if the N-point DFT were carried out by the slow method exclusively, it would take N^2 computations. Thus, Eq. (6.34) illuminates the savings intrinsic to the decomposition procedure. For example, if $N = 55$, $L = 11$, $M = 5$, the ratio of C_2 to the "slow" DFT time is $17:55$. It is also clear that more benefit is gained from larger values of N; for example, $N = 15$, $L = 5$, $M = 3$ yields a $9:15$ ratio.

Now let's assume that N can be expressed as the product of three prime integers, $N = PML$. By letting C_3 be the number of operations with the procedure enumerated above, we easily find

$$C_3 = P^2LM + PML^2 + PLM^2 + 2PLM = N(P + L + M + 2)$$
$$(6.35)$$

For example, if $P = 3$, $L = 5$, $M = 7$, the ratio of C_3 to slow DFT time is $17:105$ so we are getting close to an order of magnitude savings. To induce the general result, let's change our notation and let $N = N_1N_2N_3, \ldots, N_J$; then

$$C_J = N\left(\sum_{i=1}^{J} N_i + J - 1\right) \qquad (6.36)$$

All the results above apply fairly accurately to the case where each N_i is a prime (and not equal to 2) because in these cases it is quite fair to associate N_i^2 operations with an N_i-point transform. When N_i is not a prime or when $N_i = 2$, however, we have to be very careful. For example, for $N_i = 2$, as shown earlier, there are *no* multiplications associated with the two-point DFT. This is also true for $N_i = 4$; for $N_i = 8$ there are many fewer than 64 multiplications needed. Thus, a decomposition into numbers like 2, 4, and 8 will make the formulas above quite wrong. As an example, take N to be a power of 2, say $N = 2^J$. Then the term in Eq. (6.36), which refers to the two-point DFT, consists of a complex addition and a complex subtraction; whereas the $(J - 1)$ term refers to the twiddles. Equation (6.36) then becomes like an "apples and oranges" equation since it is not simple to compare the complexities of multiplications versus additions.

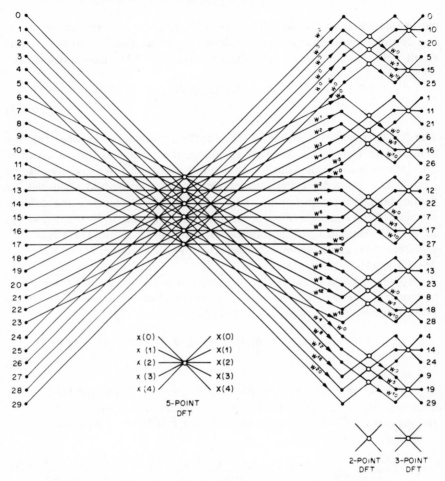

Fig. 6.13 Flow chart of a 30-point FFT obtained by successive decimation from one to two dimensions.

In order to derive a flow chart mechanism (similar to Figs. 6.3 and 6.11) for the more general decomposition, we need to extend the notation that we introduced for Figs. 6.3 and 6.11. This extension is most easily explained by reference to Fig. 6.13. Here we have decomposed a 30-point array into a five-row six-column by two-dimensional array, as shown below (with the numbers representing the indexes of the input sequence).

$$
\begin{array}{cccccc}
0 & 1 & 2 & 3 & 4 & 5 \\
6 & 7 & 8 & 9 & 10 & 11 \\
12 & 13 & 14 & 15 & 16 & 17 \\
18 & 19 & 20 & 21 & 22 & 23 \\
24 & 25 & 26 & 27 & 28 & 29
\end{array}
$$

The first FFT stage performs six five-point DFT's. Thus, an open circle in Fig. 6.13 represents a complete DFT, of length determined by the number of lines going into (and out of) the open circle. The nodes in these flow-graphs are the registers that hold the inputs and outputs of the DFT's. Following these DFT's, all outputs are twiddled, as shown by the 30 arrows with W's. The next step of the procedure is to do the row transforms. Since each row contains six elements, each row can be further decimated into a two-row by three-column matrix. In the example of Fig. 6.13, each six-point row is transformed by performing (1) three two-point DFT's, (2) twiddling, and (3) two three-point DFT's.

Another notational point: strictly speaking, every line coming out of an open circle should have an arrow with a W. If the twiddle factor happens to be $W^0 (= 1)$, it can be left out. In all these flow graphs (including Fig. 6.3 and 6.11), the authors have used "artistic license" in determining when to explicitly insert W^0 and when to omit W^0. To derive the twiddles in the second stage of Fig. 6.13, we recall that the W for that stage is really W_{30}^5. Thus, the twiddle matrix for each decimated row of the original five-row by six-column matrix is

$$W^0 \quad W^0 \quad W^0$$

$$W^0 \quad W^5 \quad W^{10}$$

Finally, we have to explain how the data shuffling in Fig. 6.13 comes about. If the 30 points had been decimated into a (5×6) matrix (and nothing more), we would have come out with a simple row–column permutation between the input data and the transformed data. Each row has been further deci-mated, however, so that there is an additional data shuffling within each row, leading to the result shown in Fig. 6.13.

6.9 Radix 2 Algorithms

At present, fixed radix FFT algorithms are by far more widely used than mixed radix algorithms. These cases (fixed radices) are simplest to analyze, to program, or to implement with hardware. In concept, there is no difference between fixed and mixed radices; given that $N = r^m$, where m is an integer, we proceed by first decomposing N into $r \times N/r$ and then further decomposing N/r into $r \times N/r^2$, etc. For example, if $N = 32$ and $r = 2$, we can proceed as follows.

1. Let samples 0 through 15 be the first row and samples 16 through 31 be the second row of the (2×16) decomposition. Begin by doing 16 column (two-point) DFT's, as in Fig. 6.14. Multiply the result by a 2-row by 16-column twiddle matrix.

2. Let each transformed row be arranged as a 2-row by 8-column matrix. Do the required column DFT's for both (2×8) matrices and twiddle

them with a (2×8) twiddle matrix based on W^2; the result is the second stage of Fig. 6.14.

3. Proceeding in the same manner, each eight samples of a row becomes a (2×4) matrix and each four samples becomes a (2×2) matrix, which is as far as we need to go in this case.

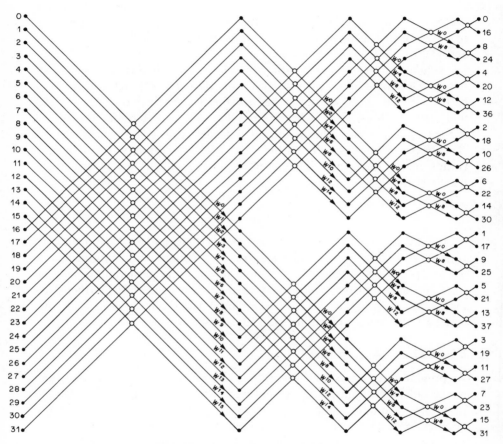

Fig. 6.14 32-Point, radix 2, in-place FFT.

As an exercise, the reader should try doing a 32-point DIT algorithm, using two-dimensional notions (or in any other way).

This completes the FFT discussion of this chapter. The main intent here is the development of algorithms and their theoretical meanings on the subject of spectrum analysis. Since it was felt that the use of the DFT in the remainder of the chapter would not be sufficiently motivated without a reasonable understanding of the FFT, we delved somewhat deeply into it but we have by no means exhausted that subject. We shall continue the development of FFT ideas in Chapter 10, which deals with the overall subject of FFT implementation.

6.10   Spectrum Analysis at a Single Point in the z Plane

In order to make measurements of the spectrum of a signal we must first inquire about what kind of signal we are working with and what we want to find out about the signal. Generally one can express the problem of spectral analysis as one of evaluating the z transform of a modified version of the signal over a region of the z plane. For example, Fig. 6.15 shows six possible regions

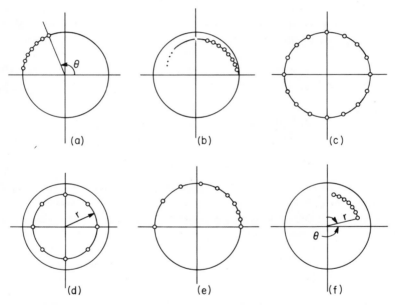

Fig. 6.15 Several possible regions in the z plane for spectral measurements.

of interest in the z plane for evaluating the spectrum of a signal. In Fig. 6.15(a), (c), and (e), the region of interest is a set of points on the unit circle in the z plane. In the remaining cases the measurements are made at a set of points off the unit circle. It is important to remember that there is no theoretical barrier to measuring the spectrum of a signal at any point z_1 in the z plane; the major practical considerations in such measurements are computation time and the effects of finite register length. With these points in mind, the (generalized) spectrum of a signal $x(n)$ can be defined as

$$S_n(z_1) = x(n) + x(n-1)z_1^{-1}$$
$$+ x(n-2)z_1^{-2} + \ldots + x(n-N+1)z_1^{-(N-1)} \quad (6.37)$$

or

$$S_n(z_1) = \sum_{m=n-N+1}^{n} x(m)z_1^{-(n-m)} \quad (6.38)$$

where N is the number of samples over which the spectrum is evaluated.

For many applications, i.e., when the signal spectrum is time-varying, we want to measure $S_n(z_1)$ for successive values of n; i.e., we want $S_0(z_1)$, $S_1(z_1)$, $S_2(z_1)$, etc. This type of measurement is called a *sliding* or *running* spectral measurement; what we do is move the spectral window (of duration N samples) one sample ahead and then repeat the measurement. An examination of Eq. (6.37) and (6.38) shows that a sliding spectral measurement at a single value of z, i.e., $z = z_1$, is equivalent to an FIR filtering operation. The equivalent impulse response of the filter is seen from Eq. (6.38) to be of the form

$$h(n) = z_1^{-n} \quad 0 \le n \le N-1 \quad (6.39)$$

Figure 6.16 shows a direct convolution realization of the spectral measurement of Eq. (5.37).

By considering two successive spectral measurements $S_{n-1}(z_1)$ and $S_n(z_1)$, a recursion relation can be obtained of the form

$$S_n(z_1) = z_1^{-1}S_{n-1}(z_1) + x(n) - z_1^{-N}x(n-N) \quad (6.40)$$

Equation (6.40) can be implemented as shown in Fig. 6.17. It should be noted that in both Figs. 6.16 and 6.17 the z's represent delay elements, whereas the z_1's are coefficient multipliers (in general, complex). Also the signals in the

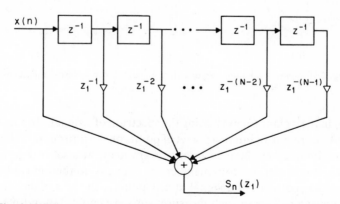

Fig. 6.16 FIR implementation of sliding spectrum analysis for a single point z_1.

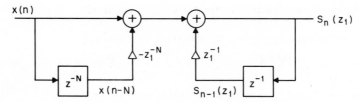

Fig. 6.17 Recursive implementation of sliding spectrum.

filters can be complex. Figure 6.17 shows that only two complex multiplications per sample are required to make a sliding measure of the signal spectrum at the point $z = z_1$.

6.11 Spectrum Analysis Using the FFT

So far we have discussed finding the spectrum at a single point in the z plane. To find the spectra at many points, say z_1, z_2, z_3, etc., we have to duplicate the computations above for each new value of z. In most practical cases we are interested in the behavior of the spectrum at some specialized but rather large assortment of points and our primary interest is in making these measurements with the fewest computations. Thus although the techniques discussed in Sec. 6.10 illustrate several fundamental facts about spectrum analysis, they are not generally used in practical spectrum analysis problems.

In a large percentage of applications, the problem of measuring the spectrum corresponds to finding the z transform of a finite record of the sequence at a large number of points equally spaced around the unit circle. These measurements correspond to measuring the DFT of the finite sequence and are generally most efficiently implemented by an FFT, as discussed earlier in this chapter. There are applications where it is desired to obtain a spectral measurement by evaluating the z transform of the sequence at equally spaced points *inside* the unit circle. For example, Fig. 6.15(d) shows a case where the point are uniformly spaced on a circle of radius r. It turns out that this computation can also be implemented as a DFT. The required measurement is

$$S[re^{j(2\pi/N)k}] = \sum_{n=0}^{N-1} \underbrace{[x(n)r^{-n}]}_{\hat{x}(n)} e^{-j(2\pi/N)nk} \qquad k = 0, 1, \ldots, N-1 \quad (6.41)$$

Equation (6.41) is the DFT of the sequence

$$\hat{x}(n) = x(n)r^{-n} \qquad (6.42)$$

Thus for this case the spectrum analysis is implemented by premultiplying the signal by r^{-n} and then performing an FFT.

6.12   Some Considerations in Spectrum Analysis

Two important considerations in spectrum analysis are

1. The number of frequencies at which a measurement of the spectrum is desired.
2. The "resolving" power of a spectral measurement.

Both these considerations can be discussed fruitfully using the equivalence between a spectral measurement and the output of a filter, as noted previously.

To illustrate some of the factors that enter into a determination of spectrum analysis parameters, consider the example of Fig. 6.15(c). For this case we want to compute the signal spectrum at 16 uniformly spaced points around the unit circle. Assume that N, the number of signal samples entering into the measurement, is chosen to be 16. The desired spectrum analysis can be obtained in two equivalent ways. Either a 16-point FFT or a 16-channel filter bank can be used to obtain the desired result. The impulse response of the kth filter, corresponding to evaluating the spectrum at the point $z_k = e^{j(2\pi k/N)}$, is of the form

$$h(n) = e^{-j(2\pi k n/N)} \qquad 0 \le n \le N - 1 \tag{6.43}$$

The z transform of the kth filter is therefore

$$H(z) = \sum_{n=0}^{N-1} e^{-j(2\pi k n/N)} z^{-n} = \frac{1 - z^{-N}}{1 - z^{-1} e^{-j(2\pi k/N)}} \tag{6.44}$$

Evaluating Eq. (6.44) on the unit circle gives

$$H(e^{j\omega}) = e^{-j\omega[(N-1)/2]} e^{j(\pi k/N)} \frac{\sin (N\omega/2)}{\sin (\omega/2 + \pi k/N)} \tag{6.45}$$

$$= e^{-j\omega[(N-1)/2]} e^{j(\pi k/N)} f_N(\omega, k) \tag{6.46}$$

Figure 6.18(a) shows a sketch of $f_N(\omega, k)$ for the example of Fig. 6.15(c), i.e., for $N = 16$ points uniformly spaced around the unit circle. Fig. 6.18(a) shows the filters with even values of k and the second line shows the odd values of k. Only the main lobes of these filters are shown (except for filter number 8) to avoid confusion. Examination of the resulting filter frequency responses shows that a sliding FFT corresponds to a fairly crude filter bank with rather high side lobes and appreciable overlap between adjacent filters.

Now assume that the number of data samples L is larger than the desired number of spectral points N. For example, assume there are $(L = 32)$ signal samples and $(N = 16)$ spectral points are desired. The easiest way to visualize the resulting frequency response is to first imagine that a 32-point sliding FFT was actually taken, but then we simply threw away alternate filters to arrive at the result in Fig. 6.18(b).

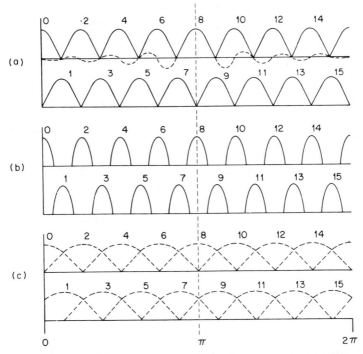

Fig. 6.18 Equivalent filter banks for sliding FFT's where the number of samples is equal to, greater than, and less than the number of desired frequencies.

Now let us assume that we still desire ($N = 16$) frequency points but have ($L = 8$) signal samples. This measurement is accomplished by augmenting the signal samples with ($N - L$) zero-valued samples to make the number of spectral samples and data samples equal and then doing an L-point FFT on these samples. Figure 6.18(c) shows the resulting filter frequency responses for the ($L = 8$, $N = 16$) example. The number of channels of the equivalent filter bank remains the same but each filter is wider. Thus the major change in the frequency response has been a loss in frequency resolution.

Implementation of the case where $N = 2L$ is almost immediately derivable from inspection of the FFT flow graph (e.g., Fig. 6.14). Notice that the even-numbered outputs appear at the top part of the diagram. This means that an FFT algorithm designed to compute only even-numbered outputs would require that the complete L points be at least partially processed to obtain the upper half samples of the processed first-stage output. As observed from Fig. 6.14, only that half is needed to obtain all even-numbered results. In fact, we can extend this result by noting, for example, that the upper set of eight outputs comprise every fourth output and can be obtained by doing half of the first stage, then half of the second stage, and processing these samples with an eight-point FFT. To generalize, let L, the number of time samples, obey the equation

$$L = NM \qquad (6.47)$$

where N is the desired number of frequency samples and M is an integer greater than 1. Then the transform we wish to compute is

$$X(k) = \sum_{n=0}^{N-1} \left[\sum_{m=0}^{M-1} x(n + mN) \right] e^{-j(2\pi/N)nk} \qquad k = 0, 1, \ldots, N-1 \quad (6.48)$$

Physically Eq. (6.48) can be described as presumming the input sequence at equidistant points N samples apart and then taking the DFT of the resulting sequence. This procedure makes sense if one recalls that sampling in the frequency domain (which is what we are doing) corresponds to aliasing or folding in the time domain [as in Eq. (6.48)].

6.13 Relation of "Hopping" FFT's to Filter Banks

Thus far we have shown that a sliding FFT is exactly equivalent to a specific filter bank; the question naturally arises as to which implementation is more computationally efficient. Without becoming involved in details we can see that a sliding FFT requires a complete FFT calculation for each signal sample thereby requiring $(N/2) \log_2 N$ butterfly computations per sample. For the filter bank implementation, which consists of a comb filter followed by a bank of digital resonators, as shown in Fig. 6.19, the number of complex multiplications is N. Thus for most cases of interest the filter bank implementation seems more efficient than the sliding FFT since the latter cannot take advantage of the iterative properties of the DFT. However, it will be shown below that the FFT computation can be made more efficient than an equivalent filter bank under the right conditions.

A standard method that allows one to use FFT's to compute running spectra approximately is to "hop" the FFT rather than slide it. Some examples of hopping are shown in Fig. 6.20. This figure shows three possible ways of hopping the FFT computation—i.e., in jumps of the size of the FFT, half the size of the FFT, or one-fourth the size of the FFT.

We can see, of course, that the efficiency of the FFT implementation rises proportionally to the "hop" size, but the question is, what happens to the resultant spectral measurement? The first thing to notice is that the hopped measurement is simply a *sampling* of the sliding measurement. Therefore, an equivalent result could have been obtained by sampling the filter bank outputs. The effects of such sampling can be treated by standard "folding" or

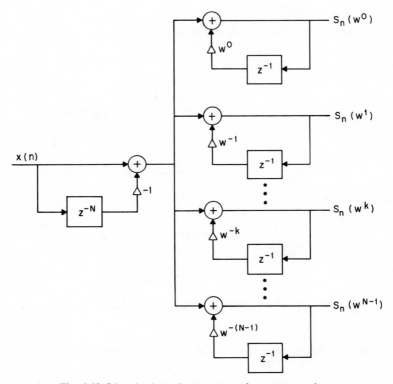

Fig. 6.19 Filter bank implementation of spectrum analyzer.

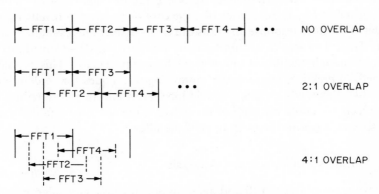

Fig. 6.20 Three examples of hopping FFT's.

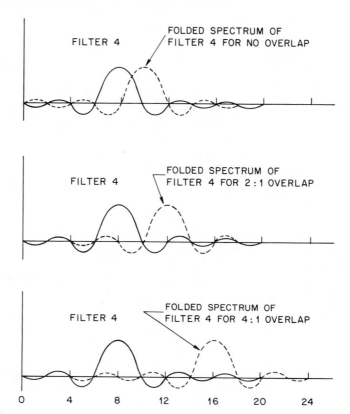

Fig. 6.21 Folding effects on spectral measurement of hopping FFT's or a sampled filter bank.

aliasing arguments. In Fig. 6.21 are shown the approximate effects of folding on the equivalent filter frequency response for the three cases of hopping FFT's given in Fig. 6.20. Notice that the case of no overlap results in severe folding of the filter response and thus can easily distort the resultant measurement. The 2:1 overlap case is substantially improved and the 4:1 overlap is even better, although significant amounts of fold-over remain. This problem is lessened, in practice, by using the windowing techniques (to be described in Sec. 6.14); this tends to diminish side lobe levels and therefore reduce fold-over effects. A crude measure of this effect is the size of the folded (dotted line) spectrum within the main lobe of the filter.

6.14 Windows in Spectrum Analysis

We have shown in Sec. 6.10 and again in Sec. 6.13 the equivalence between a spectrum measurement (defined in terms of the z transform) and the output of a filter (for a single point) or a bank of filters (for points spaced equally on the unit circle). The particular filter corresponding to our definition is a very special one and leads to a given spectral resolution. By varying the number of taps on the filter (equivalent to the number of terms in the z transform measurement) we can change the effective filter bandwidth, but thus far we have not described how the filter shape might be altered so as, for example, to reduce unwanted noise and signals outside the passband. This extension is accomplished by introducing a window $w(n)$ that multiplies each term in the selected portion of the waveform via the following formula:

$$\hat{S}_n(z_1) = x(n)w(0)$$
$$+ x(n-1)w(1)z_1^{-1} + \ldots + x(n-N+1)w(N-1)z_1^{-(N-1)} \quad (6.49)$$

or

$$\hat{S}_n(z_1) = \sum_{m=n-N+1}^{n} x(m)w(n-m)z_1^{-(n-m)} \quad (6.50)$$

Equation (6.49) is a windowed version of Eq. (6.37). The equivalent filter is obtained by replacing z_1^{-j} in Fig. 6.16 with $w(j)z_1^{-j}$. Since the choice of $w(n)$ is arbitrary, we now have a completely general filter equivalence to our modified spectral measurement [Eq. (6.49)].

The computational advantage of introducing the window in this way comes about most forcefully when doing a complete FFT spectrum analysis. If, as is true in many practical cases, all channels of the filter bank are identical except for a displacement in frequency, then multiplication of the signal section by the window (a total of N multiplications) "creates" the correct filter shapes. Processing this product via the FFT completes the spectrum analysis.

If two different resolution criteria are required in a spectrum analyzer, as in Fig. 6.22, it is probably simplest to repeat the FFT twice, using a different window prior to each FFT. In many cases, however, other tricks can be played. One method of particular interest is based on the equivalence between multiplication in the time domain and convolution in the frequency domain. Thus, for example, windowing can be accomplished by weighted adding of several adjacent (in frequency) values of $\hat{S}_n(z)$.

Mathematical formulation of windows has been treated at some length in Chapter 3. There it was also pointed out that other filter design techniques have been developed that are often more powerful than the window method. It is important to point out that the relative importance of the different design

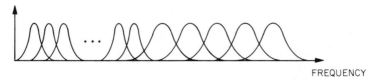

FREQUENCY

Fig. 6.22 Two resolution criteria in one spectrum analyzer.

approaches depends greatly on the implementation. For example, if we were to implement a spectrum analyzer by means of a bank of FIR filters, realized nonrecursively, then we are free to design each filter according to any criterion we desire with no consequent loss of computing efficiency. Thus, we might find it advantageous to obtain the filter coefficients by means of an equiripple design program. To take another example, if we were to use FIR filters realized recursively as in Fig. 6.19, a frequency sampling technique of taking a weighted sum of several adjacent filters to obtain a good composite filter could be the appropriate approach. Because of the many advantages of the hopping FFT implementation, however, this approach is of most interest and equivalent filter shapes are most efficiently obtained by windowing the data block prior to the FFT. We should also mention that the designer of an FFT spectrum analyzer can use weighted summing of FFT outputs to create a frequency domain equivalent of a window and if he wants to, he can use both time windows and frequency domain summing to obtain spectrum analyzers of varying resolution.

The FFT can actually be used to implement a filter bank approximately where each filter is specified to be a given IIR filter (such as Butterworth, Bessel, or elliptic) as shown in Sec. 5 of Chapter 12. The constraint on this design is that all the filters be equal bandwidth bandpass versions of a low-pass prototype IIR design. Then the notions of Sec. 6.12 combined with the window idea discussed in this section suffice. Let us say we wanted to measure the spectrum at N frequency points but were given $(L = MN)$ signal samples. If L were a large enough number so that the infinitely long IIR impulse response could be approximated by an L-point finite impulse response, then all one needs to do is to window the signal by this L-point approximation and then proceed to do a partial FFT as described in Sec. 6.12.

6.15 Measurement of the Spectrum over a Limited Angle in the z Plane by the FFT

Assume that we have L signal samples at our disposal and we wish to measure the z transform over a circular sector in the z plane. To make the example more specific, let $N = 64$ and let there be 16 points of interest as shown in Fig. 6.23. To begin with, for this example we can write

$$X(z_k) = \sum_{n=0}^{N-1} x(n)z_k^{-n} \qquad k = 0, \ldots, 15 \qquad (6.51)$$

where

$$z_k = re^{j[\theta+(2\pi k/N)]} \qquad (6.52)$$

so that

$$X(z_k) = \sum_{n=0}^{N-1} [x(n)r^{-n}e^{-jn\theta}]e^{-j(2\pi nk/N)} \qquad k = 0, 1, \ldots, 15 \qquad (6.53)$$

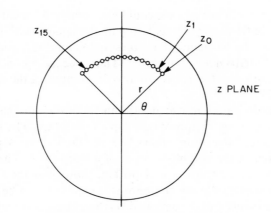

Fig. 6.23 Circular sector in the z plane.

Equation (6.53) says that premultiplication of $x(n)$ by $r^{-n}e^{-jn\theta}$ transforms the problem to one of finding the spectrum on an arc on the unit circle, as shown in Fig. 6.24.

In order for the spectrum to be efficiently evaluated by an FFT method, we would like to redistribute the spectral points to be uniformly spaced around the unit circle. This can be done by *sampling* the signal and combining spectra via the following procedure.

1. Perform a 16-point FFT on signal samples $0, 4, 8, \ldots, 60$ to find $X_0(k)$.
2. Perform a 16-point FFT on samples $1, 5, 9, \ldots, 61$ to find $X_1(k)$.
3. Repeat for $2, 6, 10, \ldots, 62$ and $3, 7, 11, \ldots, 63$ to find $X_2(k)$ and $X_3(k)$.
4. Add the transforms via the relation

$$X(k) = X_0(k) + W_N^k X_1(k) + W_N^{2k} X_2(k) + W_N^{3k} X_3(k) \qquad k = 0, 1, \ldots, 15$$

where $N = 64$ and $W_N = e^{-j(2\pi/N)}$.

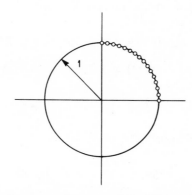

Fig. 6.24 Transformation of contour of Fig. 6.23 by premultiplication.

One point is worth noting. We could have done a 64-point transform and saved the first 16 frequency points to obtain the same result. Using the $(N/2)$ $\log_2 N$ formula, the 64-point FFT takes 192 butterflies, compared to 128 butterflies for four 16-point transforms. Thus, depending on the parameters of the system, we have at least two options for computing the spectrum on the arc of Fig. 6.23.

The next sections present two algorithms for evaluating the z transform of a finite sequence along an appropriate z-plane contour. The first algorithm, called the Bluestein algorithm, shows how the DFT of an N-point sequence can be measured as the weighted output of a chirp filter. The second algorithm, called the chirp z-transform algorithm, shows how the z-transform of a finite duration sequence along an appropriate contour can be expressed as a convolution and hence can be computed via the techniques of fast convolution.

6.16   Bluestein's Algorithm

The FFT algorithm significantly reduces computation time of an N-point DFT only if N is a highly composite number. Other efficient algorithms exist for evaluating the DFT of a sequence in a time approximately proportional to $N \log N$. The Bluestein algorithm is a digital filtering approach to measuring the DFT that is applicable to any value of N.

Consider a digital filter with impulse response $h(n)$ defined as

$$h(n) = \begin{cases} 0 & n < 0 \\ e^{j\pi n^2/N} & 0 \leq n \leq 2N - 1 \\ 0 & n \geq 2N \end{cases} \tag{6.54}$$

A filter with this impulse response is generally called a chirp filter because of its similarity to analog chirp filters. If the N-point input sequence $x(n)$ is the input to the filter [i.e., $x(n)$ is nonzero only in the interval $0 \leq n \leq N - 1$], then the filter output $y(n)$ in the interval $N \leq n \leq 2N - 1$ is

$$y(n) = \sum_{r=0}^{N-1} x(r)h(n - r) \qquad N \leq n \leq 2N - 1 \tag{6.55}$$

If we change indices in Eq. (6.55) by setting $k = n - N$, we obtain

$$y(k) = \sum_{r=0}^{N-1} x(r)h(k + N - r) \qquad 0 \leq k \leq N - 1$$

$$= \sum_{r=0}^{N-1} x(r)e^{j(\pi/N)(k+N-r)^2} \tag{6.56}$$

$$y(k) = e^{j\pi(k^2/N)}e^{j\pi N} \sum_{r=0}^{N-1} [\underbrace{x(r)e^{j\pi(r^2/N)}}_{w(r)}]e^{-j(2\pi/N)rk} \tag{6.57}$$

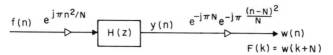

Fig. 6.25 An N-point FFT via linear filtering.

Equation (6.57) says that the output $y(k)$ is the weighted (by the factor $e^{j\pi k^2/N}e^{j\pi N}$) N-point DFT of the sequence $x(r)e^{j\pi r^2/N}$. Therefore if we desire the N-point DFT of the sequence $f(n)$, Eq. (6.57) shows that the sequence $x(n) = f(n)e^{-j\pi n^2/N}$ should be the input to the filter and the filter output $y(n)$ (for $N \leq n \leq 2N - 1$) weighted by the factor $e^{-j\pi(n-N)^2/N}e^{-j\pi N}$ is the desired N-point DFT. Figure 6.25 gives a summary of the operations required to evaluate an N-point DFT for the sequence $f(n)$. For certain values of N (e.g., N a perfect square) the digital filter may be realized by a number of computations proportional to $N^{1.5}$.

The major significance of the Bluestein algorithm is that it showed how the DFT of a sequence could be obtained by a linear filtering operation using a chirp filter.

6.17   The Chirp z Transform Algorithm

The chirp z transform (CZT) is an efficient algorithm for evaluating the z transform of a finite duration sequence along certain general contours in the z plane. Since the unit circle is one of the allowable contours, the CZT algorithm can be used to evaluate the DFT of the sequence efficiently, albeit not quite so efficiently as the FFT. The CZT eliminates many of the restrictions of the FFT, however, including the necessity that N, the number of samples in the sequence, be a composite number. The CZT algorithm can thus efficiently evaluate the DFT of a sequence even when N is a prime.

Let $\{x(n)\}$, $0 \leq n \leq N - 1$, be a given N-point sequence with z transform

$$X(z) = \sum_{n=0}^{N-1} x(n)z^{-n} \qquad (6.58)$$

The DFT of the sequence $X(k)$ is defined as (excusing the improper notation)

$$X(k) = X(z)\big|_{z=e^{j(2\pi/N)k}} \qquad k = 0, 1, \ldots, N - 1 \qquad (6.59)$$

Using the CZT, one can evaluate the z transform [Eq. (6.58)] along the more general contour

$$z_k = AW^{-k} \qquad k = 0, 1, \ldots, M - 1 \qquad (6.60)$$

where M is an arbitrary integer (not necessarily equal to N) and A and W are arbitrary complex numbers defined by

$$A = A_0 e^{j2\pi\theta_0} \qquad (6.61)$$

$$W = W_0 e^{j2\pi\varphi_0} \qquad (6.62)$$

Figure 6.26 shows a plot of the contour [Eq. (6.60)] in the z plane indicating the significance of A_0, W_0, θ_0, and φ_0.

As seen from Fig. 6.26 and Eqs. (6.60) to (6.62), if $A = 1$, $M = N$, and $W = e^{-j2\pi/N}$, then $z_k = e^{-j2\pi k/N}$ and the points along the contour are the points at which the N-point DFT of the sequence is evaluated.

The constant W_0 determines the rate at which the contour spirals in or out from a circle of radius A_0. If $W_0 > 1$, the contour spirals in, whereas if $W_0 < 1$, the contour spirals out. It is somewhat easier to interpret a spiraling contour in the z plane physically by examining the equivalent s-plane contours. If we make the substitution $z = e^{sT}$, then $z_0 = A = e^{s_0 T}$ or $s_0 = \ln A/T = \sigma_0 + j\omega_0 = \ln A_0/T + j2\pi\theta_0/T$. Further, since $z_k = e^{s_k T} = AW^{-k}$,

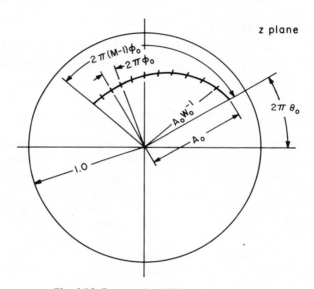

Fig. 6.26 Contour for CZT in the z plane.

we find $s_k = \ln (AW^{-k})/T = \ln A/T + \ln (W^{-k})/T = s_0 - k \ln (W)/T$. Figure 6.27 shows the equivalent s-plane contour to the contour of Fig. 6.26. Thus the spiraling contours in the z plane correspond to straight lines in the s plane. The rate of spiraling in the z plane determines the slope of the line in the s plane.

We now show how to evaluate the z transform of the sequence efficiently [Eq. (6.58)] along the contour of Eq. (6.60). If we let X_k be the z transform at $z = z_k$, i.e.,

$$X_k = \sum_{n=0}^{N-1} x(n)z_k^{-n} \qquad (6.63)$$

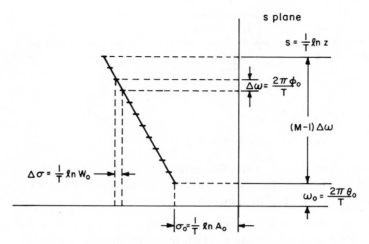

Fig. 6.27 Contour for CZT in the s plane.

then using Eq. (6.60) we find

$$X_k = \sum_{n=0}^{N-1} x(n) A^{-n} W^{nk} \qquad k = 0, 1, \ldots, M - 1 \qquad (6.64)$$

If we then make the substitution

$$nk = \frac{n^2 + k^2 - (k - n)^2}{2} \qquad (6.65)$$

into Eq. (6.64), we find

$$X_k = \sum_{n=0}^{N-1} x(n) A^{-n} W^{[n^2+k^2-(k-n)^2]/2}$$

$$= \sum_{n=0}^{N-1} [x(n) A^{-n} W^{n^2/2}] (W^{k^2/2}) W^{-(k-n)^2/2} \qquad (6.66)$$

$$X_k = W^{k^2/2} \sum_{n=0}^{N-1} y(n) v(k - n) \qquad (6.67)$$

where

$$y(n) = x(n) A^{-n} W^{n^2/2} \qquad (6.68)$$

$$v(n) = W^{n^2/2} \qquad (6.69)$$

Equation (6.67) shows that X_k may be evaluated as a weighted convolution of the sequences $y(n)$ and $v(n)$ and may thus be evaluated efficiently by high-speed convolution using the FFT. Figure 6.28 summarizes the operations of the CZT.

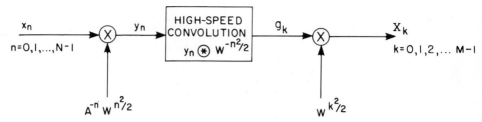

Fig. 6.28 Operations in the CZT.

Before proceeding to a detailed explanation of how the CZT may be carried out, it is worthwhile to summarize the advantages of the CZT algorithm over a standard FFT. These include:

1. N, the number of points in the input sequence, need not be the same as M, the number of points at which the z transform is evaluated.
2. Neither N nor M need be composite numbers—in fact, both can be primes.
3. The angular spacing of the z_k is arbitrary. Thus the resolution in frequency is arbitrary.
4. The contour need not be a circle in the z plane. As seen in Chapter 12, the use of a spiral contour offers some advantages in the analysis of speech.
5. The starting point in the z plane is arbitrary. This is useful for narrow-band analyses of data where the feature of high resolution (step 3) can be combined with arbitrary initial frequency.
6. If $A = 1$, $M = N$, $W = e^{-j(2\pi/N)}$, then the CZT can be used to evaluate the DFT, even when N is a prime.

In order to carry out the operations required to use the CZT algorithm, a series of steps must be performed. With the aid of some simple figures we summarize the steps required to program the CZT algorithm.

1. Choose L, the smallest integer greater than or equal to $(N + M - 1)$ and compatible with any available FFT algorithm. L denotes the size transforms that must be calculated in order to perform the required high-speed convolution of Fig. 6.28.
2. Form the L-point sequence $y(n)$ as

$$
y(n) = \begin{cases} A^{-n}W^{n^2/2}x(n) & n = 0, 1, \ldots, N - 1 \\ 0 & n = N, N + 1, \ldots, L - 1 \end{cases}
$$

Figures 6.29(a) and (b) show typical sequences $x(n)$ and $y(n)$.
3. Compute the L-point DFT of $y(n)$ using the available FFT routine; call this result Y_r. Figure 6.29(c) shows a typical sequence Y_r.

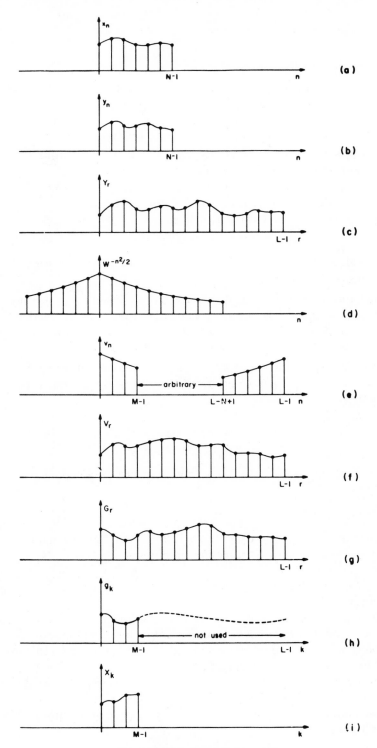

Fig. 6.29 Waveforms for the CZT.

4.   Define the L-point sequence $v(n)$ by

$$v(n) = \begin{cases} W^{-n^2/2} & 0 \le n \le M - 1 \\ W^{-(L-n)^2/2} & L - N + 1 \le n < L \\ \text{arbitrary} & \text{other } n, \text{ if any} \end{cases}$$

Figures 6.29(d) and (e) show the sequences $W^{-n^2/2}$ and $v(n)$.

5.   Compute the L-point DFT of $v(n)$; call it V_r [Fig. 6.29(f)].
6.   Multiply V_r by Y_r, point by point, to give $G_r = V_r Y_r$ [Fig. 6.29(g)].
7.   Compute the L-point IDFT of G_r; call this g_k [Fig. 6.29(h)].
8.   Multiply g_k by $W^{k^2/2}$ to give X_k defined as

$$X_k = g_k W^{k^2/2} \qquad k = 0, 1, \ldots, M - 1$$

The values of g_k for $k \ge M$ are not meaningful and are discarded. Figure 6.29(i) shows the sequence X_k.

It remains to be shown that the CZT is more efficient for evaluating the z transform of a sequence than a direct evaluation would be. To see that this is so, one merely has to count the number of complex multiplications in the CZT. An operations count for the CZT gives the following:

1.   To form $y(n)$ from $x(n)$ (step 2) requires N complex multiplications. The sequence $A^{-n}W^{n^2/2}$ can be prestored, or recursively generated, using the relation

$$A^{-(n+1)}W^{(n+1)^2/2} = (A^{-n}W^{n^2/2}) \cdot W^n W^{1/2}A^{-1} \qquad (6.70)$$

If we let

$$C_n = A^{-n}W^{n^2/2} \qquad (6.71)$$

$$D_n = W^n W^{1/2}A^{-1} \qquad (6.72)$$

then from Eq. (6.70) to (6.72) we find the recursion relations

$$D_{n+1} = D_n \cdot W \qquad (6.73)$$

$$C_{n+1} = C_n \cdot D_n \qquad (6.74)$$

with initial conditions $C_0 = 1$, $D_0 = W^{1/2}A^{-1}$. Thus an additional $2N$ complex multiplications are required if the sequence $A^{-n}W^{n^2/2}$ is generated recursively.

2.   The L-point FFT in step 3 requires on the order of $(L \log_2 L)$ complex multiplications.
3.   The sequence $v(n)$ is generally prestored, although it can be recursively generated (as it was above) requiring $2M$ complex multiplications.
4.   The L-point DFT to compute V_r (step 5) again requires on the order of $(L \log_2 L)$ complex multiplications (although if the CZT is performed repeatedly with the same values of M and N, this sequence need only be computed once and stored).
5.   The multiplication of the sequences V_r by Y_r (step 6) requires L complex multiplications.

6. The L-point FFT, in order to give g_k, again requires on the order of $(L \log_2 L)$ complex multiplications.
7. To form the M-point output, X_k requires M complex multiplications.

Thus from the above it is clearly seen that the dominant factors in evaluating the CZT are the three (or sometimes two) L-point FFT's that must be performed. Therefore the computation time for the CZT is again approximately proportional to $L \log_2 L$ (recall $L \geq N + M - 1$) where the constant of proportionality is on the order of two to three times larger than for the FFT. Thus the CZT is an efficient algorithm for evaluating the z transform of a sequence along certain general contours in the z plane. Applications of the CZT are discussed by Rabiner, Schafer, and Rader. (Recently the authors have been made aware that among other applications the CZT has been used in conjunction with a surface wave acoustic delay line to perform a DFT—private communication from Harper Whiteside.)

6.18 Power Spectrum for Noisy Signals

Spectral measurements on discrete random time functions are complicated by several factors. First, we are usually interested in the power spectrum (which has not yet been defined). Second, our measurement must be statistically useful in that whatever estimate we use must converge to some meaningful result as the measurement interval increases. A simple and useful way of defining *convergence* is by requiring that the variance of the estimate approaches zero as the measurement interval increases. In dealing with power spectra, it has long been known that the correlation function of the signal is an extremely useful tool. Thus, in this section, we briefly introduce the concepts of a correlation function and a power spectrum and formulate some of the relationships between them. Then we proceed to analyze the computational tradeoffs involved in two well-known methods of noise–power spectrum analysis.

Until this point we have used a rather general definition of *spectrum*, defining it as a z transform for any value or sets of values in the z plane. The following question arises: Can we define the power spectrum of a signal in the same general way? A reasonable approach appears to be as follows: If we think of a spectral density on some curve (not necessarily on the unit circle) in the z plane, then by integrating this density over the chosen curve we can obtain the total power in the signal. By invoking Parseval's theorem, we can convince ourselves that this approach works. Thus for a finite duration random sequence $x(n)$ of duration N samples, Parseval's theorem gives

$$\sum_{n=0}^{N-1} x^2(n) = \frac{1}{2\pi j} \oint X(z)X\left(\frac{1}{z}\right)z^{-1}\, dz \tag{6.75}$$

Equation (6.75) is a simplified version of the complex convolution theorem. The path of integration, when we have a finite duration sequence, can be any path enclosing the origin. Thus we can define a spectral density $S_N(z)$ as being proportional to $X(z)X(1/z)$ over any such path; i.e.,

$$S_N(z) \propto X(z)X(z^{-1}) \qquad (6.76)$$

Equation (6.76) implies that in order to measure the spectral density at the point $z = z_1$ in the z plane, the z transform of the signal must be computed at both $z = z_1$ and $z = 1/z_1$. As an example, if we wanted to measure spectral density on a circle of radius r, we would measure the z transforms of the sequence both on that circle and also on a circle of radius $1/r$. When $r = 1$, the two measurements become complex conjugates and we are, in essence, measuring $|X(e^{j\omega})|^2$ as the spectral density function.

To determine what the constant of proportionality should be, we resort to a statistical argument. Let us write

$$X(z)X\left(\frac{1}{z}\right) = \sum_{n=0}^{N-1} \sum_{m=0}^{N-1} x(n)x(m)z^{-(n-m)} \qquad (6.77)$$

Now, the mean value of this measurement is

$$E\left[X(z)X\left(\frac{1}{z}\right)\right] = \sum_{n=0}^{N-1} \sum_{m=0}^{N-1} E[x(n)x(m)]z^{-(n-m)} \qquad (6.78)$$

Let us choose the simple example of "white" noise, where the mean value of the product $x(n)x(m)$ is zero unless $n = m$, in which case we assume the mean value is K. This immediately tells us that

$$E\left[X(z)X\left(\frac{1}{z}\right)\right] = KN \qquad (6.79)$$

Since it is desirable that the mean value converge to a constant as N grows, it behooves us to define a spectral density as

$$S_N(z) = \frac{1}{N} X_N(z)X_N\left(\frac{1}{z}\right) = \frac{1}{N} \sum_{n=0}^{N-1} \sum_{m=0}^{N-1} x(n)x(m)z^{-(m-n)} \qquad (6.80)$$

As an example for $N = 5$, Eq. (6.80) gives

$$\begin{aligned}
S_5(z) = \tfrac{1}{5}[&x(0)x(0)z^0 + x(0)x(1)z^{-1} + x(0)x(2)z^{-2} + x(0)x(3)z^{-3} \\
&+ x(0)x(4)z^{-4} + x(1)x(0)z^1 + x(1)x(1)z^0 + x(1)x(2)z^{-1} \\
&+ x(1)x(3)z^{-2} + x(1)x(4)z^{-3} + x(2)x(0)z^2 + x(2)x(1)z^1 \\
&+ x(2)x(2)z^0 + x(2)x(3)z^{-1} + x(2)x(4)z^{-2} + x(3)x(0)z^3 \\
&+ x(3)x(1)z^2 + x(3)x(2)z^1 + x(3)x(3)z^0 + x(3)x(4)z^{-1} \\
&+ x(4)x(0)z^4 + x(4)x(1)z^3 + x(4)x(2)z^2 + x(4)x(3)z^1 \\
&+ x(4)x(4)z^0]
\end{aligned} \qquad (6.81)$$

Combining terms with the same power of z, we obtain

$$S_5(z) = \tfrac{1}{5}\{z^0[x(0)x(0) + x(1)x(1) + x(2)x(2) + x(3)x(3) + x(4)x(4)]$$
$$+ (z^1 + z^{-1})[x(0)x(1) + x(1)x(2) + x(2)x(3) + x(3)x(4)]$$
$$+ (z^2 + z^{-2})[x(0)x(2) + x(1)x(3) + x(2)x(4)]$$
$$+ (z^3 + z^{-3})[x(0)x(3) + x(1)x(4)] + (z^4 + z^{-4})[x(0)x(4)]\} \qquad (6.82)$$

From Eqs. (6.81) and (6.82) it is easy to induce the result for general N as

$$S_N(z) = \sum_{m=0}^{N-1} R(m)(z^{-m} + z^m) - R(0) = \sum_{m=-(N-1)}^{N-1} R(m)z^{-m} \qquad (6.83)$$

where

$$R(m) = \frac{1}{N} \sum_{n=0}^{N-1-m} x(n)x(m + n) \qquad 0 \le m \le N - 1 \qquad (6.84)$$

and $R(-m) \triangleq R(m)$. Equation (6.84) defines the autocorrelation function for the finite duration sequence $x(n)$.

Equation (6.83) shows the power density spectrum to be the z transform of the autocorrelation function of $x(n)$. Although we have tacitly assumed that the random signal is of finite duration, in general $x(n)$ is of infinite duration. In such cases the basic measurement of the autocorrelation function [Eq. (6.84)] is defined as

$$R(m) = R_x(m) = \lim_{N \to \infty} \frac{1}{N} \sum_{n=0}^{N-1} x(n)x(n + m) \qquad -\infty < m < \infty \qquad (6.85)$$

For most cases of practical interest, the power spectral density is generally required along the unit circle, in which case Eq. (6.83) generalizes to

$$S(z)\big|_{z=e^{j2\pi f}} = S_x(f) = \sum_{m=-\infty}^{\infty} R_x(m)e^{-j2\pi mf} \qquad (6.86)$$

i.e., $S_x(f)$ is the Fourier transform of $R_x(m)$. Equivalently $R_x(m)$ may be obtained as the inverse Fourier transform of $S_x(f)$; i.e.,

$$R_x(m) = \frac{1}{2\pi} \int_{-\pi}^{\pi} S_x(f)e^{j2\pi fm} \, df \qquad (6.87)$$

For the case when we desire to measure the cross-power spectrum or the cross-correlation function of two random waveforms $x(n)$ and $y(n)$, Eqs. (6.86) and (6.87) generalize to

$$R_{xy}(m) = \lim_{N \to \infty} \frac{1}{N} \sum_{n=0}^{N-1} x(n)y(n + m) \qquad -\infty < m < \infty \qquad (6.88)$$

$$S_{xy}(f) = \sum_{m=-\infty}^{\infty} R_{xy}(m)e^{-j2\pi mf} \qquad (6.89)$$

and

$$R_{xy}(m) = \frac{1}{2\pi} \int_{-\pi}^{\pi} S_{xy}(f)e^{j2\pi fm} \, df \qquad (6.90)$$

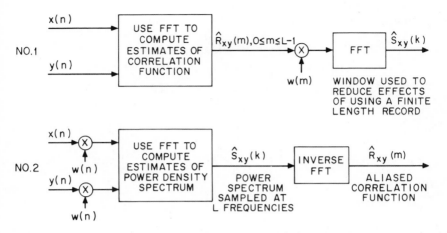

Fig. 6.30 Two methods for performing spectral measurements.

The cross-correlation measurement is generally of importance in cases when $y(n)$ is related to $x(n)$ by a filtering operation and a delay. For example, in a radar system, $y(n)$ may be a filtered, noisy, delayed version of $x(n)$. For such cases the peak of the cross-correlation function is a good indication of the delay of the input signal.

We have thus arrived at a definition of power spectral density, related it to the autocorrelation function, and shown that the mean value of these estimates converges, for large N, to some reasonable fixed number at least for the case of white noise input. As is well-known, however, the variance of $S_N(z)$ can only be made to converge to zero by performing some further averaging. In terms of FFT implementation, there appear to be two well-known methods of performing spectral noise measurements that satisfy the criterion of heading toward zero variance; one is based on correlation computation via the FFT and the other is a method of averaging successive directly measured power spectral densities. Figure 6.30 summarizes the basic ideas behind these two approaches.

For technique 1 the FFT is used directly to compute estimates of the cross-correlation[3] function $\hat{R}_{xy}(m)$ for L lags, where $2L$ is the size of the transform used. In order to estimate the power density spectrum $\hat{S}_{xy}(k)$ at a finite set of frequencies from $\hat{R}_{xy}(m)$, a smoothing window $w(m)$ must be used to reduce the undesirable effects of using a finite length record (i.e., L values of the correlation function) instead of the infinite correlation sequence. Using an L-point FFT, the power density spectrum may be estimated at L uniformly spaced frequencies around the unit circle.

---

[3] The descriptions above are simply modified to the case of measuring autocorrelations and power spectra by setting $y(n) = x(n)$.

Technique 2 uses the FFT to compute estimates of the power density spectrum directly rather than first estimating a correlation function as in technique 1. Thus each of the data sequences $x(n)$ and $y(n)$ must be smoothed by a window $w(n)$ prior to being transformed. The inverse FFT may then be used to estimate the correlation function from the power density spectrum. Since the power density spectrum is available only at a discrete set of L frequencies (rather than the infinite set of frequencies at which it must be theoretically determined), however, the use of the inverse FFT to produce the correlation function gives an aliased version of the desired result.

In the remainder of this section we discuss how techniques 1 and 2 may be carried out in practice. First, however, we show how the DFT can be used to perform the process of correlation. Consider the periodic sequences (of period L samples) $x_p(n)$ and $y_p(n)$ with L-point DFT's $X(k)$ and $Y(k)$ defined as

$$X(k) = \sum_{r=0}^{L-1} x_p(r)e^{-j(2\pi/L)rk} \tag{6.91}$$

$$Y(k) = \sum_{s=0}^{L-1} y_p(s)e^{-j(2\pi/L)sk} \tag{6.92}$$

We have already seen that multiplying $X(k)$ by $Y(k)$ corresponds to circularly convolving the periodic sequences $x_p(n)$ and $y_p(n)$; i.e.

$$X(k)Y(k) \leftrightarrow \sum_{m=0}^{L-1} x_p(m)y_p(n-m) \tag{6.93}$$

We now wish to show that circular correlation of $x_p(n)$ and $y_p(n)$ may be trivially obtained as

$$X(k)Y^*(k) \leftrightarrow \sum_{n=0}^{L-1} y_p(n)x_p(m+n) \tag{6.94}$$

To verify Eq. (6.94) we show that the inverse DFT of $X(k)Y^*(k)$ may be put in the form of the right side of Eq. (6.94). Using Eqs. (6.91) and (6.92), we find

$$\mathrm{DFT}^{-1}[X(k)Y^*(k)] = \frac{1}{L}\sum_{k=0}^{L-1} X(k)Y^*(k)e^{j(2\pi/L)mk}$$

$$= \frac{1}{L}\sum_{k=0}^{L-1}\left[\sum_{r=0}^{L-1} x_p(r)e^{-j(2\pi/L)rk}\right]$$

$$\times \left[\sum_{s=0}^{L-1} y_p(s)e^{j(2\pi/L)sk}\right]e^{j(2\pi/L)mk}$$

$$= \sum_{r=0}^{L-1}\sum_{s=0}^{L-1} x_p(r)y_p(s)\underbrace{\left[\frac{1}{L}\sum_{k=0}^{L-1} e^{j(2\pi/L)k(m-r+s)}\right]}_{\delta(m-r+s)}$$

$$= \sum_{s=0}^{L-1} y_p(s)x_p(m+s)$$

which is the desired result. [Note that we have assumed $x_p(n)$ and $y_p(n)$ are real in the derivation above.]

1. Technique 1—Correlation Method

Figure 6.31 shows one way of computing the first five points of an auto-correlation given an ($N = 20$)-point sequence. The sequence is shown on the top line of Fig. 6.31. The first (9×5) block of numbers corresponds to a circular correlation of two nine-point sequences, namely, $x_0 x_1 x_2 x_3 x_4 x_5 x_6 x_7 x_8$ with $x_0 x_1 x_2 x_3 x_4 0000$. The five useful results $R_0(0)$, $R_0(1)$, $R_0(2)$, $R_0(3)$, $R_0(4)$ are shown on the right. The second block begins with x_5 and the procedure is the same, resulting in $R_1(0)$ through $R_1(4)$. To compile the complete result for the first five correlations, we have to sum $R_i(m)$ over i, as follows.

$$R(m) = \sum_{i=0}^{3} R_i(m) \qquad m = 0, 1, 2, 3, 4 \qquad (6.95)$$

Our next step is to show how these computations can be made using FFT methods. For each of the four nine-point circular correlations we could have computed the result via the DFT by the following procedure.

1. Compute
$$U_0(k) = \text{DFT}\,\{x_0 x_1 x_2 x_3 x_4 x_5 x_6 x_7 x_8\}$$
   and
$$V_0(k) = \text{DFT}\,\{x_0 x_1 x_2 x_3 x_4 0000\}$$
$$U_1(k) = \text{DFT}\,\{x_5 x_6 x_7 x_8 x_9 x_{10} x_{11} x_{12} x_{13}\}$$
$$V_1(k) = \text{DFT}\,\{x_5 x_6 x_7 x_8 x_9 0000\}$$
$$U_2(k) = \text{DFT}\,\{x_{10} x_{11} x_{12} x_{13} x_{14} x_{15} x_{16} x_{17} x_{18}\}$$
$$V_2(k) = \text{DFT}\,\{x_{10} x_{11} x_{12} x_{13} x_{14} 0000\}$$
$$U_3(k) = \text{DFT}\,\{x_{15} x_{16} x_{17} x_{18} x_{19} \cdots\}$$
$$V_3(k) = \text{DFT}\,\{x_{15} x_{16} x_{17} x_{18} x_{19} 0000\}$$

2. Compute
$$X_i(k) = U_i(k) V_i^*(k) \qquad i = 0, 1, 2, 3$$

3. Compute
$$X(k) = \sum_{i=0}^{3} X_i(k) \qquad k = 0, 1, \ldots, 8$$

4. Compute
$$R(m) = IDFT\{X(k)\}$$

5. The first five points of $R(m)$ are the desired results and also satisfy the relation
$$R(m) = \sum_{i=0}^{3} R_i(m)$$

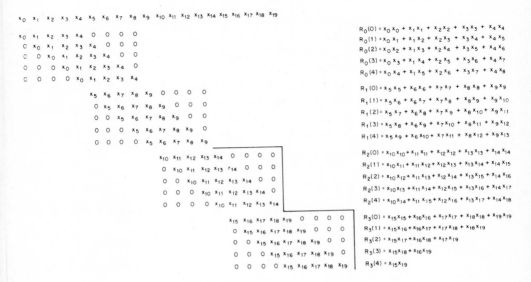

Fig. 6.31 Illustration of circular correlation.

The procedure just outlined is more efficient than first computing each partial correlation $R_i(m)$ by taking the inverse DFT of $X_i(k)$ and then directly performing the additions of the partial correlations. In the procedure outlined, two DFT's, $U_i(k)$ and $V_i(k)$, are performed for each (9×5) block in Fig. 6.31 and one "total" transform is derived by accumulating DFT results. The "total" correlation is computed from the "total" transform by means of a single DFT after accumulation. Thus, the total number of DFT's performed is $2K + 1$ where K is the number of "blocks" that have to be processed. If, however, accumulation of the final answer followed complete processing of each block as described above, we would have to do $3K$ DFT's.

From the procedure above we conclude that the size of the DFT's should be approximately twice the number of desired correlations and the number of DFT's is $2K + 1$ where $K = N/M$ (N is total number of samples and M is total number of desired correlations). At this point, a spectral estimate with any desired degree of frequency resolution can be obtained by augmenting the just computed $R(m)$ with zeros and performing a single DFT. Assuming that $R(m)$ is a good estimate with very low variance, then its transform should also be a good estimate.

A trick, due to Rader can be used for further improving the computational efficiency by noting that $U_i(k)$ in step 1 of our procedure can really be expressed as a function of $V_i(k)$ and $V_{i+1}(k)$.

The first step would be to change the sectioning of Fig. 6.31 slightly so that (1) a section contains an even number of samples, preferably a power of 2 and (2) the overlap in the measurement is exactly 2:1. This change is not particularly restrictive because most FFT algorithms require or desire such section lengths. The exact 2:1 overlap means that the transform size increases by 1 sample, certainly not a big computational loss. Figure 6.32 shows a 20-sample sequence with samples $x(0)$ through $x(19)$ divided into five sections. This figure also shows which sections are to be circularly autocorrelated. The equivalent FFT operations are shown on the right. The trick is to notice that any $U_i(k)$ can actually be written as a function of the $V_i(k)$'s as follows. If, for example,

$$V_3(k) = \text{DFT} \{x_{12}x_{13}x_{14}x_{15}0000\}$$

then

$$(-1)^k V_3(k) = \text{DFT} \{0000x_{12}x_{13}x_{14}x_{15}\}$$

the latter result following directly from the DFT property that rotation by $(N/2)$ samples is equivalent to multiplying the DFT by $(-1)^k$.

Therefore

$$U_2(k) = V_2(k) + (-1)^k V_3(k)$$

More generally,

$$\left.\begin{array}{l} U_i(k) = V_i(k) + (-1)^k V_{i+1}(k) \qquad i = 0, 1, 2, 3 \\[2mm] U_4(k) = V_4(k) \end{array}\right\} \tag{6.96}$$

with

Applying Eq. (6.93), we can now modify the procedure previously outlined.

1. Compute $V_0(k)$ through $V_4(k)$ where the $V_i(k)$ are the DFT's of the sequences shown in Fig. 6.32.
2. Compute

$$X_i(k) = V_i^*(k)[V_i(k) + (-1)^k V_{i+1}(k)]$$

with

$$X_4(k) = V_4(k)V_4^*(k)$$

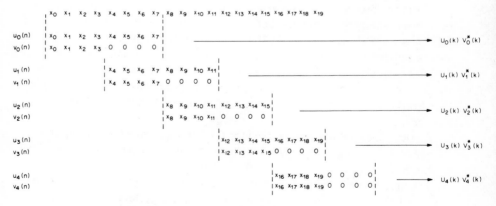

Fig. 6.32 Sectioning for autocorrelation estimate.

3. Compute

$$X(k) = \sum_{i=0}^{4} X_i(k) \qquad k = 0, 1, \ldots, 7$$

4. Compute

$$R(m) = \text{IDFT} \{X(k)\}$$

From this point, spectral estimates can be obtained by transforming $R(m)$. The important thing to notice is that the heart of the computation, namely, the number of DFT's, is now reduced to K where K is the number of sections.

In order to achieve convergence to zero variance, it is necessary that the argument m in $R(m)$ range over a small interval compared to the measurement interval. This implies that a spectral estimate is obtained by transforming a truncated version of $R(m)$ and, as we know, truncation in one domain can lead to excessive ripple in the transformed domain. Thus, a window is called for. Let us illustrate the above with an example: Assume that the total measurement interval contains 1024 samples and that we know that $R(m)$ will be small for $m > 16$ but we want spectral estimates at 64 frequency values. Therefore, the procedure would be as follows.

1. Section the 1024 samples into subsections of length 32.
2. Perform FFT's on these subsections with a 2:1 overlap; that is, do a new 32-point FFT every 16 samples.
3. Using the technique we have already described, obtain $R(m)$ for $m = 0$ through 16.
4. To obtain the spectral estimate, multiply $R(m)$ by an appropriate window $w(m)$. Augment with zeros from $m = 16$ through $m = 63$ and obtain the spectral estimate through a transform of the augmented $R(m)w(m)$.

For the general case in which we wish to compute L points of a correlation function, given N input points from which to make the measurement ($N \gg L$), technique 1 can be generalized in the following way. The function we wish to compute is

$$R_{xy}(m) = \sum_{n=0}^{N-1} x(n)y(n + m) \qquad m = 0, 1, \ldots, L - 1 \qquad (6.97)$$

A simple technique for computing Eq. (6.97), using $(P = 2L)$-point FFT's is shown in Fig. 6.33. The sequence $x(n)$ is partitioned into subsequences $x_i(n)$ by the rule

$$x_i(n) = x\left[n + \left(\frac{P}{2}\right)i\right] \qquad n = 0, 1, \ldots, P - 1$$

$$i = 0, 1, \ldots \qquad (6.98)$$

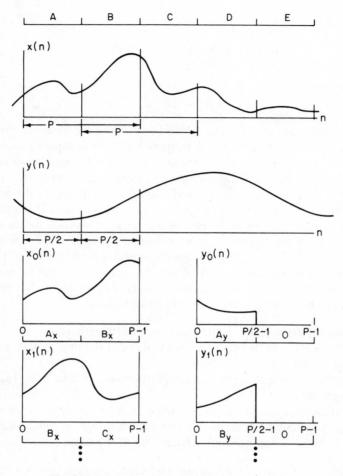

Fig. 6.33 Sectioning for computing cross correlations.

and $y(n)$ is partitioned into subsequences $y_i(n)$ by the rule

$$
y_i(n) = \begin{cases} y\left[n + \left(\dfrac{P}{2}\right)i\right] & n = 0, 1, \ldots, \dfrac{P}{2} - 1 \\[2mm] & i = 0, 1, \ldots \\[2mm] 0 & n = \dfrac{P}{2}, \ldots, P - 1 \end{cases} \tag{6.99}
$$

For each pair of P-point subsequences $x_i(n)$ and $y_i(n)$, the sequence $w_i(m)$ representing the periodic correlation of $x_i(n)$ and $y_i(n)$ is computed. In terms of the original sequences $x(n)$ and $y(n)$, $w_i(m)$ may be expressed as

$$w_i(m) = \sum_{n=0}^{P/2-1} x\left(n + \frac{iP}{2}\right) y\left(n + \frac{iP}{2} + m\right) \qquad m = 0, 1, \dots, \frac{P}{2} \quad (6.100)$$

[It should be noted that only $(P/2 + 1)$-points of the computed correlation function satisfy Eq. (6.100); the remaining $(P/2 - 1)$-points represent the undesired periodic correlation.] If each of the individual correction functions are added together, we find

$$z_i(m) = \sum_{j=0}^{i} w_j(m)$$

$$= \sum_{j=0}^{i} \sum_{n=0}^{P/2-1} x\left(n + \frac{jP}{2}\right) y\left(n + \frac{jP}{2} + m\right) \qquad (6.101)$$

$$= \sum_{n=0}^{(i+1)P/2-1} x(n)y(n + m) = \hat{R}_{xy}(m) \qquad (6.102)$$

which is identical to the desired result [Eq. (6.97)] for $N = (i + 1) \cdot P/2$. Thus, for any number of lags (L), correlation functions may be computed efficiently using the FFT.

As noted earlier, problems may occur in obtaining the power density spectrum from the correlation function by using a data window $w(m)$. Unless the transform of the window is entirely positive, there is a possibility that the computed power density spectrum may be negative—a highly undesirable result. This is due to the fact that the computed power density spectrum is the circular convolution of the window transform and the transform of the estimated correlation function and unless the transform of the window is positive for all frequencies, the possibility exists that due to statistical variation in estimating the correlation function, the resulting convolution may produce negative values for the power spectrum at some frequencies. There are windows whose transforms are entirely positive, e.g., the triangular window, and such windows should be used in cases where other windows lead to this undesirable result.

Another difficulty exists in estimating the cross-power density spectrum from a windowed cross-correlation using this method. As seen from Eq. (6.100), this technique measures the correlation function only for positive values of m. When using this technique in conjunction with estimating auto-correlation functions, the results for negative values of m are obtained trivially by symmetry. No such symmetry exists, however, for cross-correlation functions. Thus it is not clear how to window the cross-correlation function in order to obtain a good estimate of the cross-power density spectrum. Several possibilities exist and are illustrated in Fig. 6.34. The difficulty lies in not being able to easily attach a physical meaning to the value of the cross-correlation function at $m = 0$; e.g., the cross-correlation function is sensitive to delays between $x(n)$ and $y(n)$. Thus if $y(n) = x(n - N_0)$, then $\hat{R}_{xy}(m)$ would peak up at $m = N_0$. One would then wish to think of $m = N_0$ as some physical origin for $\hat{R}_{xy}(m)$. Figure 6.34 suggests three possible ways of windowing $\hat{R}_{xy}(m)$ by $w(m)$ prior to transforming to give a meaningful $\hat{S}_{xy}(k)$. In Fig. 6.34(a), it is assumed that a $(P/2)$-point FFT will be used to give $\hat{S}_{xy}(k)$ and the window (for this example, a triangular window) is centered at $m = P/4$. In Fig. 6.34(b), again a $(P/2)$-point FFT is used but here it is assumed that $m = 0$ corresponds to a meaningful physical time origin so the window is centered there. In Fig. 6.34(c), perhaps the most reasonable assumption is made. In this case both $\hat{R}_{xy}(m)$, $m = 0, 1, \ldots, P/2$, and $\hat{R}_{yx}(m) = \hat{R}_{xy}(-m)$, $m = 0, 1, \ldots, P/2$, are estimated. $\hat{R}_{yx}(m)$ is estimated by interchanging $x(n)$ and $y(n)$ and using the identical

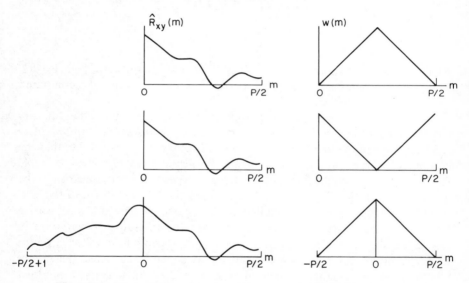

Fig. 6.34 Various windows for computing a cross spectrum from a cross correlation.

measuring technique as used for $\hat{R}_{xy}(m)$. In this case a P-point FFT is used and it is again assumed that $m = 0$ corresponds to the center of the window. Of course the computation time for this third possibility is twice that for either of the others but in many cases this extra computation is not unreasonable. Furthermore since a P-point FFT is used, this third possibility gives finer resolution in the estimate of the cross-power density spectrum.

Figures 6.35 to 6.37 illustrate the use of the correlation method for measuring the autocorrelation function and power density spectrum for colored noise—i.e., white noise that has been passed through an equiripple FIR lowpass filter with an impulse response duration of 39 samples. Thus, theoretically, the autocorrelation function of the lowpass filtered noise is

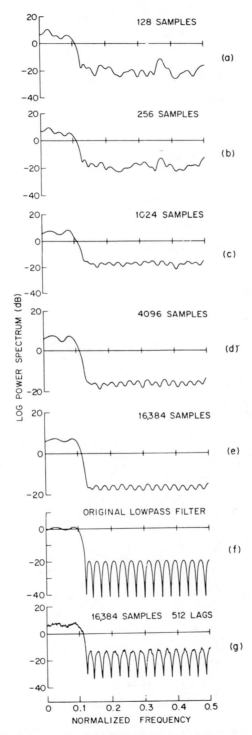

Fig 6.35 Power density spectra for lowpass filtered noise using power spectrum estimation technique #1.

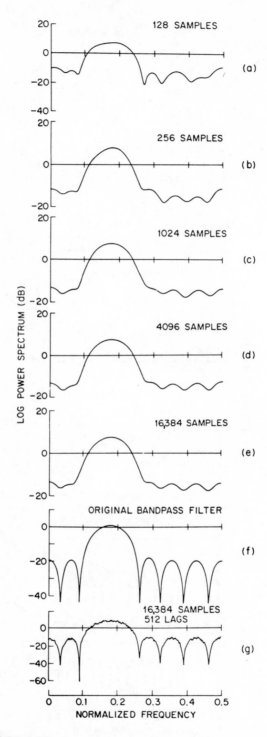

Fig 6.36 Power density spectra for bandpass filtered noise using power spectrum estimation technique #1.

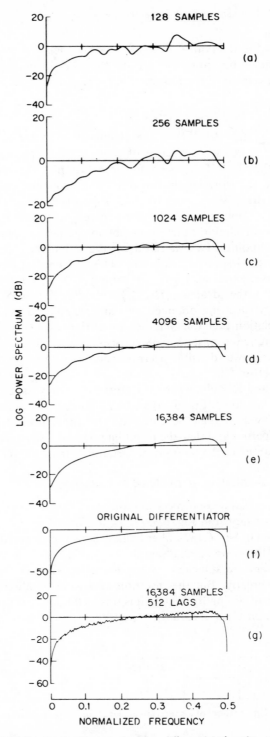

Fig. 6.37 Power density spectra for differentiated noise using power spectrum estimation technique #1.

nonzero only in the interval $-38 \leq m \leq 38$. In fact, the measured auto-correlation is small but nonzero outside this range. Similarly, the ideal power density spectrum of the lowpass filtered noise is

$$S_y(f) = S_x(f) \, |H(e^{j2\pi f})|^2$$

where $S_x(f) = \sigma_x^2$ and $H(e^{j2\pi f})$ is the frequency response of the lowpass filter. Figure 6.35 shows a sequence of plots of the measured log power spectrum as a function of N, the number of samples that enter into the measurement of the correlation function. The sequence of plots is for $N = 128$, 256, 1024, 4096, and 16,384, respectively, where the number of autocorrelation points is 59. In these cases, a Hamming window is used on the autocorrelation function to obtain the power spectrum density estimate. A 1024-point FFT was used on the augmented autocorrelation function to give good spectral resolution. This sequence of plots shows the rapidity of convergence of $\log S_y(f)$ to $\log|H(e^{j2\pi f})|$. For comparison, Fig. 6.35(f) shows the ideal $\log|H(e^{j2\pi f})|$. Finally, Fig. 6.35(g) shows the log power density spectrum obtained by transforming 512 measured points of the autocorrelation function weighted by a triangular window. Since the autocorrelation function past $m = 38$ consists of a random noise sequence, the measured log power density spectrum consists of a noise superimposed on the ideal $\log|H(e^{j2\pi f})|$.

Figures 6.36 and 6.37 show similar results for noise passed through a 15-point FIR bandpass filter (Fig. 6.36) and a 27-point FIR differentiator (Fig. 6.37). It is clearly seen that the correlation method has reasonably good convergence properties and that the resulting power density spectra are quite good approximations to the theoretical spectra for these simple cases.

2. Technique 2—Method of Modified Periodograms

Intuitively, we feel that a good spectral estimation can be obtained by means of a filter bank, as shown in Fig. 6.38.

Now, as shown in Sec. 6.13, a filter bank can be implemented by means of a sliding FFT. Also, if we insert a sampler at $y(n)$ in Fig. 6.38, then provided the bandpass filters are sufficiently narrow the statistics of the final result should not vary greatly. But the insertion of a sampler allows one to use hopping rather than sliding FFT's and it is this technique that has been called the *method of modified periodograms*.

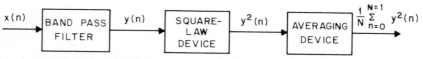

Fig. 6.38 One channel of a spectrum analyzer for noise measurements.

To illustrate how to apply the method we first consider the case where $y(n) = x(n)$, i.e., the method computes an estimate of the power density spectrum of the stationary process of which $x(n)$ is a typical sequence. Figure 6.39 shows the sequence $x(n)$ and its decomposition into subsequences $x_r(n)$ of length L samples. The subsequences $x_r(n)$ are spaced D samples apart where $D = L/3$ in Fig. 6.39. Thus $x_r(n)$ is related to $x(n)$ as

$$x_r(n) = x[n + (r - 1)D] \qquad r = 1, 2, \ldots, K \qquad (6.103)$$

where K is the number of subsequences that are used in the estimate of the power density spectrum. For each of the subsequences $x_r(n)$ the windowed FFT $X_r(k)$ is computed as

$$X_r(k) = \sum_{n=0}^{L-1} x_r(n)w(n)e^{-j(2\pi/L)nk} \qquad (6.104)$$

where $w(n)$ is an appropriate window function. The quantity $I_r(f_k)$, called a periodogram, is computed as

$$I_r(f_k) = \frac{1}{U} |X_r(k)|^2 \qquad (6.105)$$

where

$$f_k = \frac{k}{L} = \text{DFT frequency} \qquad (6.106)$$

and

$$U = \sum_{n=0}^{L-1} w^2(n) = \text{energy in window} \qquad (6.107)$$

The power density spectrum estimate $\hat{S}_x(f_k)$ is computed as

$$\hat{S}_x(f_k) = \frac{1}{K} \sum_{r=1}^{K} I_r(f_k) = \frac{1}{KU} \sum_{r=1}^{K} |X_r(k)|^2 \qquad (6.108)$$

Thus the power density spectrum estimate is a weighted sum of the periodograms of each of the individual subsequences. Welch has shown that the

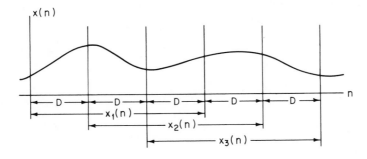

Fig. 6.39 Sectioning of the input sequence for computing correlations.

expectation of the estimate satisfies the equation

$$E[\hat{S}_x(f_k)] = \int_{-0.5}^{+0.5} H(f)S_x(f_k - f)\, df = H(f) * S_x(f) \tag{6.109}$$

where

$$H(f) = |W(e^{j2\pi f})|^2 \tag{6.110}$$

and

$$W(e^{j2\pi f}) = \sum_{n=0}^{L-1} w(n)e^{-j2\pi fn} \tag{6.111}$$

and $S_x(f)$ is the true power density spectrum of the process that generated $x(n)$. Equation (6.109) shows that the expected value of the estimate is the true power spectrum convolved with the magnitude-squared window transform. Welch has further shown that in the case when $x(n)$ is from a Gaussian random process and when $S_x(f)$ is reasonably flat over the frequency interval where the window transform is large, the variance of the estimate satisfies the equation

$$\text{Var}\,\{\hat{S}_x(f_n)\} = \frac{[S(f_n)]^2}{K}\left[1 + 2\sum_{j=1}^{K-1} \frac{K-j}{K}\,\rho(j)\right] \tag{6.112}$$

where

$$\rho(j) = \frac{\left[\sum_{k=0}^{L-1} w(k)w(k+jD)\right]^2}{\left[\sum_{k=0}^{L-1} w^2(k)\right]^2} \tag{6.113}$$

To see the applicability of Eq. (6.112), consider the case where $D = L$, in which case $\rho(j) = 0$ for $1 \leq j \leq K - 1$. In this case

$$\text{Var}^1\,[\hat{S}_x(f_n)] = \frac{[S(f_n)]^2}{K} \tag{6.114}$$

where $K = N/L$. Next, consider the case $D = L/2$, with $w(n)$ a triangular window as shown in Fig. 6.40. For this case $\rho(1) = \frac{1}{16}$ [from Eq. (6.113)] and $\rho(j) = 0, j \geq 2$. Thus

$$\text{Var}^2\,[\hat{S}_x(f_n)] = \frac{[S(f_n)]^2}{K'}\left(\frac{17}{16}\right) \tag{6.115}$$

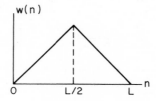

Fig. 6.40 Triangular window.

but $K' = N/(L/2) = 2K$. Hence Eq. (6.115) can be put in the form

$$\text{Var}^2\, \lfloor \hat{S}_x(f_n) \rfloor = \frac{[S(f_n)]^2}{K}\left(\frac{17}{32}\right) = \left(\frac{17}{32}\right)\text{Var}^1\,[\hat{S}_x(f_n)] \qquad (6.116)$$

and so the simple expedient of shifting subsequences half the window width rather than an entire window width (at the cost of double the computation) has reduced the variance of the estimate by almost a factor of 2.

For the case of cross-power spectra, both $x(n)$ and $y(n)$ are sectioned as in Fig. 6.39, giving subsequences $x_r(n)$ and $y_r(n)$. Each of these subsequences is windowed and transformed to give $X_r(k)$ and $Y_r(k)$. The periodogram is now defined as

$$I_r(f_k) = \frac{1}{U}\, X_r(k)Y_r^*(k) \qquad (6.117)$$

where U is as defined in Eq. (6.107). The cross-power spectrum estimate is obtained as

$$\hat{S}_{xy}(f_k) = \frac{1}{K}\sum_{r=1}^{K} I_r(f_k) = \frac{1}{KU}\sum_{r=1}^{K} X_r(k)Y_r^*(k) \qquad (6.118)$$

The expectation of the estimate may again be shown to be the convolution of the true cross-power density spectrum with the magnitude-squared transform of the window.

Figures 6.41 to 6.43 illustrate practical applications of the method of modified periodograms to estimation of power density spectra for the three-colored noise examples shown earlier in Figs. 6.35 to 6.37. For each of these cases the power density spectrum was estimated at 513 frequencies uniformly distributed from $f = 0$ to $f = 0.5$, i.e., an ($L = 1024$)-point FFT was used. A Hamming window was used as the data window on the input sequence. The

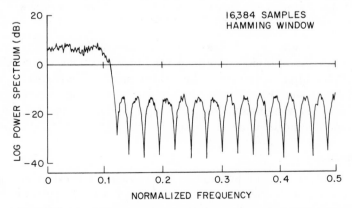

Fig. 6.41 Power density spectrum for lowpass filtered noise using power spectrum estimation technique #2.

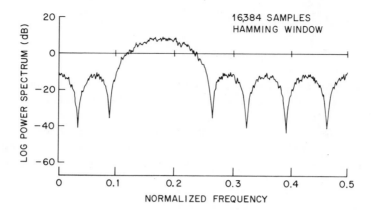

Fig. 6.42 Power density spectrum for bandpass filtered noise using power spectrum estimation technique #2.

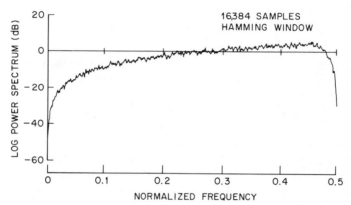

Fig. 6.43 Power density spectrum for differentiated noise using power spectrum estimation technique #2.

shift value used was $D = 512$ to minimize the variance of the estimate for a fixed number of input samples. A total of 32 sections was used—i.e., ($N = 16,384$) samples were used in the measurements. Figure 6.41 shows the measured log power spectrum for the lowpass filtered noise whereas Fig. 6.42 and 6.43 show measured log power spectra for the bandpass filtered and differentiated noise sequences, respectively.

3. Summary

In summary, two efficient procedures are available for estimating power spectra or correlation functions of random processes. Depending on the particular problem, either of these techniques may possess advantages over the other for that application. Both techniques have the desirable property that the mean of the estimate converges to a fixed value and the variance of the estimate tends to zero as the sample size tends to infinity.

6.19 Convolution and Correlation Using Number Theoretic Transforms[4]

The reader has now seen both theoretical and practical treatments of convolutions and correlations of data that are discrete in time. Quantized data, that is, data which are discrete in amplitude, have also been treated in Chapter 5, but there the point of view was to treat each sample as a sum of a continuous quantity and an error. The goal was to estimate the error, or to minimize it. In this section, we adopt a radically different point of view, namely that all of our data can only take on quantized (integer) values. We present an interesting method for computing lagged products (convolutions and correlations) of such quantities. Impressive computational savings can be demonstrated in a limited number of cases.

We have needed to come to grips with the issues of rounding and overflow only because computations without overflow or rounding often violate a fundamental requirement of any rigorous number system, namely that operations on numbers yield legitimate new numbers. Suppose we add together two 10-bit integers with a 10-bit word. The correct result of the addition could be expressed in an 11-bit word. With a 10-bit word, there are two possibilities. One possibility is that the most significant bit was a zero and the sum is indeed expressible in a 10-bit word. The other is that the most significant bit is a one, and the result cannot be stored in the register provided. The latter case is called an *overflow*. The mathematician would say that overflow indicates that the addition operation violates *closure* when only 10-bit integers are allowed in the set of possible numbers. Multiplication produces even greater problems of this sort.

Rounding and truncation were invented to restore closure. In effect the definition of addition and multiplication are modified to include a shifting and rounding operation. Now closure is satisfied, but *equality* under this modification can no longer be exact.

An entirely different approach would be to compute with modified definitions of addition and multiplication which preserve both closure and equality, at the expense of giving a result which we consider "wrong." If the *final* answer to a computational procedure is thus "wrong" we would regard the procedure as uninteresting. However, if the *final* answer to a procedure is guaranteed to be *correct*, it does not matter if the *intermediate* results are unfamiliar. An excellent example of this principle applies when a

[4] This section was written for this book by C. M. Rader, MIT Lincoln Laboratory, Lexington, Mass.

sum of signed integers is added together using a fixed point (one's or two's complement) representation. If the final sum is known a priori to be correctly representable in the wordlength being used, any overflows occurring along the way may be neglected. Since this principle is seen to be of use in one situation, we now shall examine its generalization in great detail. We consider the laws of addition and multiplication modulo M, an integer called the modulus.

The allowable numbers in this discussion are the integers $0, 1, 2, 3, 4, \ldots,$ $M - 1$. Two quantities, a and b, are called congruent modulo M if their difference is an exact multiple of M. Congruence is therefore not the same as equality unless one can show separately that the difference between a and b is also less than M. If that is the case, the only multiple of M less than M being zero, a and b would have to be equal. In the case of adding numbers using a k-bit word, an overflow created a situation in which the result differs from the "ordinary" result by 2^k. Therefore, using $M = 2^k$, such a result was congruent to the ordinary (many-bit) result. By induction, the final result must be congruent to the ordinary result in spite of possible multiple overflows. If the final ordinary result is known to be a representable k-bit number, then since the difference between any two k-bit numbers is less than 2^k, the ordinary result and the obtained result must be not only congruent but equal. We shall show how this principle can usefully apply with other moduli to situations more complicated than simple sums.

Let us review the laws relating to congruence.

1. Commutative law: $a + b \equiv b + a$; $ab \equiv ba$.
2. Associative law: $(a + b) + c \equiv a + (b + c)$; $(ab)c \equiv a(bc)$.
3. Distributive law: $a(b + c) \equiv ab + ac$.
4. Identity: $a + 0 \equiv a$; $a(1) \equiv a$.
5. Inverse: For every a there is an element $(-a)$ such that $a + (-a) \equiv 0$.
6. Closure: $a + b$, ab are legal elements.
7. Analogy: If $a \equiv b$ and $c \equiv d$, $a + c \equiv b + d$ and $ac \equiv bd$
        If $a \equiv b$ and $b \equiv c$, $a \equiv c$
        If $a \equiv b$ then $b \equiv a$
        .
        .
        .

Two comments are in order. First, closure is retained in computation modulo M because whenever the result of an arithmetic operation is larger than can be represented in the computing wordlength, we are allowed to subtract off any multiple of the modulus without affecting the result since it remains congruent to the result of the original arithmetic operation. Second, although there is an inverse to addition, permitting subtraction, there is no guarantee that there is an inverse to multiplication, so that division is not assured. This is also true of the integers with ordinary arithmetic. However, with the case of arithmetic modulo M, there *are* some numbers which have multiplicative inverses. For example, if $M = 14$, the integer $a = 2$ has no multiplicative inverse, but $a = 3$ and $b = 5$ are inverses of one another since $3 \cdot 5 = 15$ which is congruent to 1 modulo 14.

Since in what follows we are going to be using both congruences and equalities, it behooves us to define a notation. We represent a congruence by $a \equiv$. An equality is represented by the usual $=$. When a quantity is to be replaced by another quantity congruent to itself and in the set $0, 1, 2, \ldots$, $M - 1$, we use the symbol $((\cdot))$. Thus $((x)) = x + rM$ where r is an integer and

$$0 \leq ((x)) < M$$

The operation $((\cdot))$ is called residue reduction. It is the step by which we satisfy closure without sacrificing congruence when we are computing quantities with a finite wordlength. One possible mechanization of residue reduction is to divide by the modulus and keep only the remainder. Thus, with $M = 14$, we find that $((36)) = 8$ by

$$\tfrac{36}{14} = 2 \text{ rem } 8$$

If it were necessary to divide for residue reduction the operation would be quite costly. Fortunately there are simpler techniques.

We are interested in convolutions and correlations. Since we know that the ability to compute convolutions carries with it the ability to compute correlations, we will concentrate on convolutions. Since multiplication and addition are defined as modulo M, it is possible to talk about a convolution of two sequences of modulo M. Suppose the sequence c_n is given in ordinary arithmetic by

$$c_n = a_n * b_n$$

Applying the operation $((\cdot))$ during the computation might give us a sequence d_n. Therefore $d_n \equiv c_n$. We will mainly be interested in cases for which we can obtain c_n by computing d_n. Suppose that the largest possible value of c_n is c_u and the smallest possible of c_n is c_v. Then if M is greater than $c_u - c_v$ there can only be one number congruent to d_n which is equal to c_n. In that case, when d_n is computed, c_n can be trivially obtained. Of course we can always upper bound a convolution and choose a modulus to exceed this upper bound.

What we are now going to show is that there is a complete analogy to the method of computing convolutions by fast transform methods when the arithmetic is modulo M, although this statement is only true for certain choices of modulus M. To demonstrate this, we need to study the properties of an exponential sequence modulo M.

Consider the sequence $((a^n))$. There is an equation

$$((a^n)) = ((a \cdot ((a^{n-1}))))$$ (6.119)

or equivalently

$$a^n \equiv a \cdot a^{n-1} \qquad \text{mod } M$$ (6.120)

and therefore we recognize that we have a finite state machine; a may take on only M states and therefore $((a^n))$ must repeat periodically after at most $M - 1$ steps ($a = 0$ leads to its own, much shorter periodic sequence). This periodicity can be absolute, or can be with a preamble. We give two examples:

1. $M = 8, a = 2$
    $a^0 \equiv 1$, $a^1 \equiv 2$, $a^2 \equiv 4$, $a^3 \equiv 0$, $a^n \equiv 0$ for $n > 2$; periodic with a preamble.
2. $M = 9, a = 2$
    $a^0 \equiv 1$, $a^1 \equiv 2$, $a^2 \equiv 4$, $a^3 \equiv 8$, $a^4 \equiv 7$, $a^5 \equiv 5$, $a^6 \equiv 1$, $a^n \equiv a^{n-6}$; periodic.

The second situation is by far the more interesting. By the laws of exponentials we see that a^p and a^{6-p} have a product congruent to 1, and therefore they are multiplicative inverses of one another, in spite of the fact that there is no general guarantee in modulo arithmetic of a multiplicative inverse. In ordinary integer arithmetic, this cannot happen and therefore there is no analogy to the DFT using only ordinary integer arithmetic. In example 2, above, we have at least one example of a modulus and a sequence of powers exhibiting a periodic property analogous to the property $W^N = 1$ of the complex number field. Rather than search at random for cases of this type, we take note of a theorem:

> THEOREM   A quantity, p, has a multiplicative inverse, q, modulo M if p and M are mutually prime.
>
> The proof is intriguing. Since there are M possible values of q, we imagine trying them all, making a list

$$p \cdot 0 \equiv ?$$

$$p \cdot 1 \equiv ?$$

$$p \cdot 2 \equiv ?$$

$$\vdots$$

$$p \cdot (M - 1) \equiv ?$$

There are two possibilities. Either one of these M equations has a 1 in place of the question mark (which means that that equation was constructed using the correct quantity q), or none of the equations has a 1 in place of the question mark.

In the latter case, since there are M equations and only $M - 1$ values taken on (at most), there must be a repetition. Let the repetition be for $q = a$ and $q = b$. We assume $a > b$,

$$p \cdot a \equiv p \cdot b \qquad \text{modulo } M$$

$$p(a - b) \equiv 0 \qquad \text{modulo } M$$

Therefore M divides $p(a - b)$. Now $a - b$ is less than M so it cannot supply all the factors of M. By hypothesis, p had no factors in common with M. Therefore the latter possibility has led to a contradiction and we see that pq can be congruent to 1 mod M only if p and M have no factors in common. At the same time, we have seen that if p has no factors in common with M there can be no repetition in the list of equations and therefore one of the question marks must have the value 1. We have now demonstrated both sides of the theorem, the necessary and the sufficient condition.

It is a simple step from the above theorem to the conclusion that any number x, mutually prime to M, generates an exponential sequence $((x^n))$ with a period less than or equal to $M - 1$, in which each member has an inverse modulo M. More important, we can relate the period of the periodic sequence $((x^n))$ to the modulus M! Certainly this sequence can only generate members mutually prime to M (since x is itself mutually prime to M, which we have seen is a condition for periodicity in the first place). The number of integers less than M and mutually prime to M is called the *totient* function (a function of M), written $\phi(M)$. Clearly this is the maximum feasible period of the sequence $((x^n))$. The actual period of the sequence, we shall call N. A mathematician would say that x is of order N modulo M. It is an interesting result, due to Euler, that

$$((x^{\phi(M)})) = 1 \tag{6.121}$$

We prove this by a technique similar to the first proof, by listing (in our imagination) *all* the integers r_i less than M and mutually prime to M, and multiplying each by x modulo M. Since all the products must be different from one another and mutually prime to M, they constitute a permutation of the original set r_i. Now we multiply together all the members of each set.

$$\prod_{i=1}^{\phi(M)} ((xr_i)) \equiv \left(\prod_{i=1}^{\phi(M)} r_i \right) \quad \text{modulo } M \tag{6.122}$$

The result follows by cancelling πr_i leaving $x^{\phi(M)}$ congruent to 1. Therefore, the period, N, of the sequence $((x^n))$ must be a divisor of $\phi(M)$, which is a very powerful result indeed.

In many cases, the modulus is chosen to be a prime P. By definition, $\phi(P) = P - 1$, and any integer less than P is mutually prime to P so that

$$x^{P-1} \equiv 1 \text{ modulo } P \tag{6.123}$$

This result goes by the name Fermat's Theorem. No study of number theory goes very far without encountering some trace of the work of Fermat; we shall encounter his name again very shortly.

We seem to have deviated somewhat from our purpose by studying the properties of the sequence $((x^n))$, but we shall now return to a recognizable purpose. We are interested in demonstrating an analogy to FFT convolution modulo M. For this we need an analogy to the FFT, and thus an analogy to the DFT. We now define the Number Theoretic Transform of a sequence of N integers x_n with modulus M as another sequence of N integers X_k given by

$$X_k = \left(\left(\sum_{n=0}^{N-1} x_n \alpha^{nk}\right)\right) \tag{6.124}$$

where α is an integer mutually prime to M and with order N. This is like a DFT except that the role played in the DFT by $W = \exp(j2\pi/N)$ is played by an integer, α, whose Nth power is congruent to 1, and the computation is done modulo M. It will be recalled that in a derivation of the FFT algorithm, one makes use of the fact that $W^N = 1$. Therefore the Number Theoretic Transform can be computed by a similar algorithm, with multiplication by powers of W in the FFT replaced by multiplication by powers of α. As in the DFT, N must be a composite integer for the standard FFT algorithms to exist. However, unlike the DFT, there is no known use for the Number Theoretic Transform *by itself*. We shall use it only as a step in the computation of correlations and convolutions. This involves the multiplication of corresponding points of the Number Theoretic Transforms of two sequences, say x_n and h_n. The product sequence is *inverse transformed* to give the circular convolution of the original sequences. We therefore need to introduce the Inverse Number Theoretic Transform. By analogy to the DFT, we expect it to have a form

$$x_m = \left(\left(N^{-1} \sum_{k=0}^{N-1} X_k \alpha^{-mk}\right)\right) \tag{6.125}$$

The negative exponents of α in Eq. (6.125) have meaning because we have shown that the members of a periodic exponential sequence have inverses. We have tacitly assumed two things about Eq. (6.125). One is that N^{-1}, the multiplicative inverse of N, exists. We know that this only requires that N and M have no common factors. The other assumption is that the expression is indeed an Inverse Number Theoretic Transform. To show this we will substitute Eq. (6.124) in Eq. (6.125), and one more condition on N, M and α will emerge. Thus:

$$x_m = \left(\left(N^{-1} \sum_{k=0}^{N-1} \sum_{n=0}^{N-1} x_n \alpha^{nk} \alpha^{-mk}\right)\right) \tag{6.126}$$

Interchanging summations we find

$$x_m = \left(\left(\sum_{n=0}^{N-1} x_n \sum_{k=0}^{N-1} N^{-1} \alpha^{(n-m)k}\right)\right) \tag{6.127}$$

We can show that the two expressions are the inverse of one another if we can show that

$$\left(\left(\sum_{k=0}^{N-1}\alpha^{ik}\right)\right) = N\delta(i) \quad \text{where} \quad \delta(i) = \begin{cases} 1 & \text{if } i \equiv 0 \text{ modulo } N \\ 0 & \text{otherwise} \end{cases} \tag{6.128}$$

For $i \equiv 0 \bmod N$, $\alpha^i \equiv 1 \bmod M$ and Eq. (6.128) follows easily. For $i \not\equiv 0 \bmod N$, we multiply by $1 - \alpha^i$ giving

$$\left(\left((1 - \alpha^i)\sum_{k=0}^{N-1}\alpha^{ik}\right)\right) = ?$$

Adjacent terms in the sum cancel one another, leaving the first and the last, $((1 - \alpha^{Nk}))$, which also cancel one another. Therefore

$$\left(\left((1 - \alpha^i)\sum_{k=0}^{N-1}\alpha^{ik}\right)\right) = 0 \tag{6.129}$$

From this we can easily deduce the result desired, except for the inconvenient necessity to divide both sides of Eq. (6.129) by $1 - \alpha^i$. Division is not a meaningful operation. We are able to cancel $1 - \alpha^i$ *only if it has a multiplicative inverse*, in which case the full result holds. This gives our last constraint, that $1 - \alpha^i$ must have no factors in common with M for $i = 1, 2, \ldots, N - 1$.

We summarize what we have learned as follows:

1. The Number Theoretic Transform of a sequence of N points is defined by

$$X_k = \left(\left(\sum_{n=0}^{N-1}x_n\alpha^{nk}\right)\right)$$

where the modulus, M, and the sequence length, N, have no common factors and where N is a divisor of $\phi(M)$. α is chosen to be mutually prime to M and to have order N.

2. The inverse Number Theoretic Transform of a sequence of N points is defined as

$$x_m = \left(\left(N^{-1}\sum_{k=0}^{N-1}X_k\alpha^{-mk}\right)\right)$$

where $(1 - \alpha^k)$ is mutually prime to M for all k except values of k congruent to 0 modulo N.

It is possible to state all the conditions more compactly, as follows (without proof). One can find an invertible transform with α of order N if and only if N is a divisor of $P - 1$ for every prime factor, P of M.

It is straightforward to prove the convolution theorem for Number Theoretic Transforms. The mathematics is analogous to that which proves the result for the DFT, and it is left as an exercise to the reader. We shall now take it for granted that the product of Number Theoretic Transforms can be inverted to give a result congruent to the circular convolution of the original sequences. If N is composite, an FFT algorithm exists for the computation of the transforms and of the inverse, so that one may compute something congruent to a circular convolution of integers in very little time.

At this point, it seems worth trying a practical example. So as to make the mathematics clear, rather than to demonstrate savings, we shall choose a situation of extreme simplicity. Let us find the 4-point circular convolution of the sequence 1, 2, 0, 0 with itself. We choose the modulus $M = 17$ and $\alpha = 4$. N, of course, is 4, and we see that α is of order N modulo M. N^{-1} is 13. We list all the powers of α, for convenience: $\alpha^0 = 1$, $\alpha^1 = 4$, $\alpha^2 = 16$, $\alpha^3 = 13$.

Forward Transform, by FFT:

Original sequence	1	2	0	0
Bit reversed	1	0	2	0
After first stage				
(adds and subtracts)	1	1	2	2
After second stage	3	9	16	10

This is the transform of the original sequence. We square it mod 17.

transform of convolution 9 13 1 15

For the inverse, we multiply by 13, time reverse[5] and transform as before:

Transform times 13	15	16	13	8
Time reversed	15	8	13	16
Bit reversed	15	13	8	16
After first stage	11	2	7	9
After second stage	1	4	4	0

This is the answer to the circular convolution of the sequence with itself. Since the maximum value of the true answer is 4, the congruence modulo 17 is an equality. Note that during the computation, all results were expressed modulo 17 using the residue reduction whenever any computation gave a result outside the range 0 to 16.

Let us now assess the practical implications of all this. Unlike the FFT approach, convolutions obtained by use of the number theoretic transform are exact. Since the numbers modulo M form a finite set, there is no approximation at any time during the computation. Indeed if a mistake is made in even a single low order bit, the result would be completely useless, so it is fortunate that the number system does not require either rounding or truncation, no matter how involved the computation becomes. In applications, one is interested in an equality rather than a congruence, however, so

---

[5] As with the DFT, the same program can compute a forward or an inverse transform.

the terms of the convolution must be bounded by some other consideration in order that the congruence be replaced by an equality. For example, if two sequences of 64 points each are represented by 10-bit numbers, the result could require a 26-bit register for its exact representation. Such a computation using the number theoretic transform requires a modulus M greater than 2^{26}. Several ways have been devised for using smaller moduli in combination, but this trades one complication for another. Another point is that it is necessary to perform frequent residue reductions to avoid the increasing number of bits required to represent intermediate results. In fact, essentially every operation needs to be followed by a residue reduction. We have seen how this can be done by dividing the result by the modulus and retaining only the remainder. Division, however, is a fairly costly operation and we tend to avoid it, especially as a component of addition. Fortunately, there is no need to divide in practice, since we need only the remainder. Methods for extracting only the remainder depend in detail on the specific M for which the remainder is sought and the easiest and most important moduli are those of the form $M = 2^g \pm 1$.[6]

Arithmetic with a modulus $2^g - 1$ is well known. Each quantity can be represented with g-bits. Results of arithmetic operations (sums and products) which require more than g-bits can be residue reduced by noting that 2^g is congruent to 1. Thus the carry bit of an addition, with weight 2^g, is added into the least significant bit position. This is called *end-around carry*. A common name for the system of arithmetic is *one's complement*. In multiplication, more than one bit, in excess of the original g-bits, results. Just as the first extra bit has weight congruent to 1, the next has weight congruent to 2, the next bit has weight congruent to 4, etc. Therefore, the extra bits of a product, past the g least significant bits, form M a binary number which can be added to the binary number formed of the g least significant bits. For example, with $g = 8$, $M = 255$, the product $ab = 258 = 100000010$ is fractured up into two words: 00000001 and 00000010. Added together they give 00000011 = 3.

To date, an exploration of various choices of M of the form $2^g - 1$ has revealed some interesting choices, but none so interesting as have been found with moduli of the form $M = 2^g + 1$.

Arithmetic with a modulus $2^g + 1$ is less well known. To represent all the possible quantities, $g + 1$ bits are required. The representation is somewhat inefficient, since the most significant bit is a zero for all but one of the possible quantities, namely the quantity 2^g. Accepting this inefficiency, we note that there is a similarity between this and the former case. The bit whose weight is 2^g represents a quantity congruent to -1 modulo M. Therefore, instead of an end-around carry in addition, we need an end-around borrow. This is

---

[6] $M = 2^g$ has no interesting invertible transforms since N must equal 1.

different, but does not require more hardware than the end-around carry. In multiplication, a product usually has more than $g + 1$ bits, and if the product is fractured into the g least significant bits, and the other bits, the word formed of the most significant bits may be subtracted from the other. For example, with $g = 8$, $M = 257$, the product $ab = 258 = 100000010$ is fractured as before into 00000001 and 00000010, or 1 and 2; the difference is $2 - 1 = 1$. Thus residue reduction modulo $M = 2^g + 1$ is almost as easy as residue reduction modulo $2^g - 1$. Some caution is necessary because of the one extra quantity 2^g represented by a $g + 1$ bit word. The operations above will cause no trouble if two precautions are taken. The first precaution is to look out for the special case of a result congruent to -1. If (again for $M = 257$) we obtain a result 100000000, the fractured upper half word is $+1$ and the fracture lower half word is 0. End-around borrow for this case gives -1. This is the correct result, but recall that it is not representable with the 8-bit word. The second precaution to take is that for multiplication, the numerical value of the upper half word can exceed the numerical value of the lower half word so that after subtracting the upper half word from the lower half word, a negative quantity results. One must therefore add M to the negative quantity in order to obtain an answer in the range 0 to $M - 1$. After this addition, the former precaution, of course, applies.

Parenthetically, we mention the rule for changing the sign of a number, which is to complement the g-bits, and add $+2$ (again using the end-around borrow).

Admittedly, this arithmetic is somewhat more complicated than the systems of addition and multiplication we are used to; on the other hand, the need to divide in order to implement residue reduction has clearly been replaced by a much simpler operation. Therefore, we search among the possible moduli $M = 2^g + 1$ for a value of g with interesting transform possibilities, keeping in mind that there are constraining relations connecting M, N, and α. The simplest application of these constraints is when M is a prime. We ask, therefore, the question: for what values of g is $M = 2^g + 1$ a prime? We find that the answer, although not complete, is nevertheless surprisingly simple. Suppose g is odd, say $2r + 1$. Then $M = 2 \cdot 4^r + 1$, and M mod 3 is seen to be zero. This means that 3 divides M, and M is not prime.[7] Searching among even numbers g, we may as well consider searching over $M = 4^h + 1$. Now if h is odd, 5 will divide M because 4 is congruent to -1 modulo 5. Therefore M is not a prime.[8] Ranging over the cases with h even only, we might as well consider $M = 16^i + 1$. Here if i is odd, M has a factor 17. By iteration, we see that the only values of M which can be prime are those for which g is a power of two. Note that we have not proven that M is a

---

[7] Unless $M = 3$.
[8] Unless $M = 5$.

prime for g a power of two, but only that M is composite if g has any odd factor. The first few members of the remaining eligible numbers are $M = 3$, $M = 5$, $M = 17$, $M = 257$, $M = 65537$. All of these are primes. Fermat conjectured, but was unable to prove, that all numbers of the form $2^{2^m} + 1$ were prime. In fact, $2^{32} + 1$ is composite, having a factor 641. No other cases of primes of this type are known. Whether prime or not, integers of the form $2^{2^m} + 1$ are called *Fermat numbers*. It is important for our purposes to note a result of another famous number theorist, Lucas, who showed that any prime factor p of a Fermat number must be of the form $p = 2^{m+2}k + 1$. This tells us that if the modulus is a Fermat number there should exist Number Theoretic Transforms for any N which is a power of two less than or equal to 2^{m+2} for composite Fermat numbers, and for any N which is a power of 2, less than M, for prime Fermat numbers. These lengths of transform support FFT-like algorithms of the simplest type, i.e. radix-two algorithms.

The really exciting thing about the transforms based on M, a Fermat number, emerges when we investigate the choices of α. Since 2^g is congruent to -1 mod M, we see that 2^{2g} is congruent to 1. Very simple considerations show that no smaller positive power than $2g$ can be the order of 2 modulo M. In fact, for $M = 2^{2^m} + 1$, with $\alpha = 2$ and $N = 2^{m+1}$, all the conditions we have derived are satisfied. The transform for this extraordinary case is:

$$A_k = \left(\left(\sum_{n=0}^{N-1} a_n 2^{nk}\right)\right); \qquad N = 2g = 2^{m+1}$$

and the inverse is

$$a_n = \left(\left(2^t \sum_{k=0}^{N-1} A_k 2^{-nk}\right)\right)$$

where

$$t = 2^{m+1} - m - 1$$

EXAMPLE. $m = 3$, $2^m = 8$, $M = 257$, $N = 16$, $t = 12$. We call the special case of a transform modulo a Fermat Number, a Fermat Number Transform. Notice that the only multiplications called for in either the forward transform or the inverse transform are multiplications by powers of two. This remains true in the FFT implementation of the Fermat Number Transform. Furthermore, multiplication by a power of two, simple enough in the two's complement number system we are most used to, is reasonably simple in the Fermat Number system. One shifts the number to the left by a number of places equal to the power of two, and subtracts the upper half word from the lower half word. Since the upper word, after shifting, is one of the form 0000xxxx (illustrated for a shift of 4 in an 8-bit word) and the lower half is of the form yyyy0000, the subtraction is particularly simple, involving less logic, even, than a full adder. The actual operations needed are of the form:

$$y\,y\,y\,y \qquad \bar{x}\,\bar{x}\,\bar{x}\,\bar{x}$$
$$\underline{\qquad -1 \qquad\qquad +1 \qquad}$$

where the $+1$ is injected in the least significant bit of the shifted word (which is complemented), and the -1 is injected in the position where the least significant bit (1sb) of the original word ended up. Note that a carry from the $+1$ cancels the -1, while a borrow from the -1 adds another 1 into the 1sb.

At the time of this writing, the full logic design of a Fermat Number Transform arithmetic unit has not been completed, although Fermat Number Transforms have been programmed on IBM 360 computers and shown to be able to compute convolutions about three times as fast as the FFT implementation of the same convolution. The reasons for the speedup are:

1. The Fermat Number Transform requires no multiplications and, therefore, the implementation of a convolution requires only N multiplications for an N point convolution. The number of additions and subtractions (together) for a convolution is $2N \log_2 N$ and there are $N \log_2 N$ required "multiplications" by powers of two.
2. Only *real* operations are required. This buys about a two to one saving over the FFT requirement.
3. The Fermat Number Transform is able to compute an exact convolution, thus allowing a program to avoid the need for either floating point arithmetic or overflow checks or other precautions.

However, we should also consider the disadvantages associated with the technique. One is forced by the constraints on the modulus, the transform length, and the choice of α, into a rather narrow range of applications. The fact that the answer is exact forces us to use a large modulus to contain it, and this means that a long word is necessary throughout. In hardware applications, the storage cost may well exceed the multiplying cost, so that the elimination of nearly all the multiplications may be outweighed by the increased memory cost. An exact answer is almost never a requirement in a digital signal processing application.

Another factor operating to push up the word-length is that for Fermat Numbers and for $\alpha = 2$, the length of the circular convolution is only 2^{m+1}. This means that the wordlength must be about as long as the convolution length. In convolutions of interest, this may mean wordlengths of hundreds of bits, which is obviously beyond the range of current interest. We shall shortly see how to reduce this constraint somewhat.

Still other considerations emerge in practical problems. Signal processing to be done on a digital computer does not require the elimination of multiplications if the computer has a built-in hardware fast multiplication mechanism, while the Fermat Number operations, although simple, require several program steps apiece. Furthermore, the wordlength of a computer may greatly affect the ease with which a given residue reduction may be programmed. The reader may wish to consider for himself how he would program the residue reduction for a modulus $F = 2^{32} + 1$ on a computer with a 12-bit word, a 16-bit word, an 18-bit word, a 24-bit word, and a 32-bit word, assuming an analogous instruction set for each machine. Finally, although rotation of the bits in a word may seem to be a simpler operation than multiplication, it is not an *utterly* trivial operation in terms of time, especially on a very simple computer, for which shifts must frequently be performed one place at a time.

The constraint on a convolution length can be weakened in two ways. First, we note that the value of N can be doubled by using a value of α of order 2^{m+2} modulo M. Agarwal and Burrus have shown that the value of α given by Eq. (6.130) has order 2^{m+2} modulo $M = 2^{2^m} + 1$.

$$\alpha = 2^c - 2^d \quad \text{where } c = 3 \cdot 2^{m-2} \quad \text{and } d = 2^{m-2} \tag{6.130}$$

The reader may see that α^2 is equal to 2 modulo M, and therefore that even powers of α are powers of two, as before. The odd powers of α are easily seen to be different from 1, so that the order of α is twice the order of 2 modulo M. Multiplication by the powers of α given by Eq. (6.130) is slightly more complicated than was the case for powers of two, since here we require the tandem combination of several operations of the former type. This is still simpler than a multiplication, but the happiest circumstance is that such complexity is only· encountered in the first or last (not both) of the $\log_2 N$ stages of the fast transform implementation of a Fermat Number transform, since any odd powers of α are only encountered there.

A far more dramatic weakening on the constraint on transform length may be had, at somewhat greater cost, by the observation that any one-dimensional convolution may be computed using the algorithms for convolution of two-dimensional functions. This seemingly abstract statement is really only analogous to the observation that an ordinary convolution may be computed by the techniques for circular convolution; one arranges the data to be convolved into two dimensional arrays, using zeroes where necessary, and repetitions of the data where necessary, so that the result of the two dimensional convolution is the same numerical result as would have been obtained with a one-dimensional convolution. The reader should convince himself that the following one dimensional circular convolution:

$$\{a,b,c,d,e,f,g,h,i,j,k,l,m,n,p,q\} * \{A,B,C,D,E,F,G,H,I,J,K,L,M,N,P,Q\}$$

is obtained by convolving the following arrays:

$$\begin{pmatrix} m & n & p & q & a & b & c & d \\ a & b & c & d & e & f & g & h \\ e & f & g & h & i & j & k & l \\ i & j & k & l & m & n & p & q \end{pmatrix} * \begin{pmatrix} A & B & C & D & 0 & 0 & 0 & 0 \\ E & F & G & H & 0 & 0 & 0 & 0 \\ I & J & K & L & 0 & 0 & 0 & 0 \\ M & N & P & Q & 0 & 0 & 0 & 0 \end{pmatrix}$$

circularly in two dimensions. Note that half of the terms are correct answers and half of the terms are of no value. The use of a two dimensional convolution to compute a one-dimensional convolution means that one can make use of a two-dimensional Fermat Number Transform. When reflected back into the one dimensional situation, this means that the modulus which was limited to handling N point sequences is now able to handle sequences of $\frac{1}{2}N^2$ samples, unless, of course, the limitations of accuracy have intervened.

With an acquaintance with the advantages of Fermat Number Transforms and also with the disadvantages, it is possible to speculate on just what problems are likely to benefit from the new approach. We look for problems which have the following characteristics: 1) Fairly short sequences (about fifty lagged products), 2) Considerable accuracy needed, and 3) Multiplication very much more costly than addition.

Two situations come to mind. The first is the estimation of spectra of (many simultaneous) wideband signals. The theory of spectral estimation tells us that we should compute a correlation function only for a number of lags which is a fraction of the number of points available as data. Section 6.18 shows us an algorithm for computing an autocorrelation function of this sort by summing in the frequency domain. We can do the same sort of summing in the Fermat Number domain. Applying this sort of algorithm with 32-bit words, one could compute the 65 values of an autocorrelation function with $N = 128$, for as many points of data as we could handle without fear of ambiguity due to the modulus. For 10-bit signed words, this is over 2000 data. Apart from the Fermat Number additions and such which are required, this implementation would require only about one multiplication per input point, doing the work of 65 multiplications by the conventional method. Of course, the result is exact, which may be more of an advantage in this problem than in most. (It should be noted that the number theoretic techniques described in this section are equally well applied to signed as well as unsigned numbers.)

The second situation which comes to mind is two-dimensional finite impulse response filtering. Here we may consider an arbitrary $L \times L$ impulse response to be applied to a large picture. If L is in the range of 5–20 points, the FFT is not particularly attractive for convolution, although more so for two-dimensional convolution than for one-dimensional convolution.

The Fermat Number Transform is quite effective, however. For impulse responses in this size range, the number of multiplications is reduced by about two orders of magnitude over the direct method, in exchange for a number of additions which is not too different (usually less) than for the direct method. Thus we can expect that the Fermat Number Transform will play a part soon in the computations involved in the linear filtering of pictures.

Analogies to the Fast Fourier Transform in finite mathematical structures is an interesting field whose implications undoubtedly go beyond those of computing convolutions. We have seen that there are new computational considerations relating to the intimate connection between the wordlength, the modulus, the choice of α, and the tolerable ambiguity, if any. These considerations have greatly limited the use of a new set of techniques which seem initially to be exceedingly rewarding. Further research may uncover other mathematical structures with many of the advantages and none or few of the disadvantages of the structures we have been considering. It is perhaps more important that we have learned something about the method of high speed convolution which was unsuspected until we examined its analogy in another number system.

REFERENCES

FFT Algorithms

1. J. W. COOLEY AND J. W. TUKEY, "An Algorithm for the Machine Computation of Complex Fourier Series," *Math. Comp.*, **19**, 297–301, Apr., 1965.
2. G. D. BERGLAND, "A Guided Tour of the Fast Fourier Transform," *IEEE Spectrum*, **6**, No. 7, 41–52, 1969.
3. W. T. COCHRAN, J. W. COOLEY, D. L. FAVIN, H. D. HELMS, R. A. KAENEL, W. W. LANG, G. C. MALING, D. E. NELSON, C. M. RADER, AND P. D. WELCH, "What is the Fast Fourier Transform," *IEEE Trans. on Audio and Electroacoustics*, **15**, No. 2, 45–55, June, 1967.
4. J. W. COOLEY, P. LEWIS, AND P. D. WELCH, "The Finite Fourier Transform," *IEEE Trans. on Audio and Electroacoustics*, **17**, No. 2, 77–86, 1969.
5. J. W. COOLEY, P. LEWIS, AND P. D. WELCH, "Historical Notes on the Fast Fourier Transform," *IEEE Trans. on Audio and Electroacoustics*, **15**, No. 2, 76–79, June, 1967.
6. R. C. SINGLETON, "A Method for Computing the Fast Fourier Transform with Auxiliary Memory and Limited High Speed Storage," *IEEE Trans. on Audio and Electroacoustics*, **15**, No. 2, 91–98, June, 1967.
7. J. W. COOLEY, P. LEWIS, AND P. D. WELCH, "The Fast Fourier Transform Algorithm: Programming Considerations in the Calculation of Sine, Cosine, and Laplace Transforms," *J. Sound Vib.*, **12**, No. 3, 315–337, 1970.
8. R. C. SINGLETON, "An Algorithm for Computing the Mixed Radix Fast Fourier Transform," *IEEE Trans. on Audio and Electroacoustics*, **17**, No. 2, 93–103, June, 1969.

9. M. C. PEASE, "An Adaptation of the Fast Fourier Transform for Parallel Processing," *J. Assn. Comp. Mach.*, **15**, No. 2, 252–264, Apr., 1968.
10. C. M. RADER, "Discrete Fourier Transforms When the Number of Data Samples is Prime," *Proc. IEEE*, **56**, No. 6, 1107–1108, June, 1968.
11. J. W. COOLEY, P. LEWIS, AND P. D. WELCH, "The Fast Fourier Transform and Its Applications," *IEEE Trans. Education*, **12**, 27–34, Mar., 1969.
12. B. GOLD AND C. M. RADER, *Digital Processing of Signals*, McGraw-Hill Book Co., New York, 1969.
13. L. I. BLUESTEIN, "A Linear Filtering Approach to the Computation of Discrete Fourier Transform," *IEEE Trans. on Audio and Electroacoustics*, **18**, No. 4, 451–456, 1970.
14. L. R. RABINER, R. W. SCHAFER, AND C. M. RADER, "The Chirp z-Transform Algorithm and Its Application," *Bell Sys. Tech. J.*, **48**, No. 5, 1249–1292, May–June, 1969.

Statistical Spectrum Analysis Techniques

1. G. M. JENKINS AND D. G. WATTS, *Spectral Analysis and Its Applications*, Holden-Day, Inc., Pub., San Francisco, Calif., 1968.
2. P. D. WELCH, "The Use of the FFT for Estimation of Power Spectra: A Method Based on Averaging Over Short, Modified Periodograms," *IEEE Trans. on Audio and Electroacoustics*, **15**, No. 2, 70–73, 1967.
3. J. W. COOLEY, P. LEWIS, AND P. D. WELCH, "The Use of the Fast Fourier Transform Algorithm for the Estimation of Spectra and Cross Spectra," *Proc. of the 1969 Polytechnic Institute of Brooklyn Symposium on Computer Processing in Communications*, 5–20, 1969.
4. J. W. COOLEY, P. LEWIS, AND P. D. WELCH, "The Application of the Fast Fourier Transform Algorithm to the Estimation of Spectra and Cross-Spectra," *J. Sound Vib.*, **12**, 339–352, 1970.
5. C. M. RADER, "An Improved Algorithm for High Speed Autocorrelation with Applications to Spectral Estimation," *IEEE Trans. on Audio and Electroacoustics*, **18**, No. 4, 439–442, 1970.

Number Theoretic Transforms and Convolution

1. J. M. POLLARD, "The Fast Fourier Transform in a Finite Field," *Mathematics of Computation*, **25**, No. 114, 365–374, Apr, 1971.
2. C. M. RADER, "Discrete Convolution Via Mersenne Transforms," *IEEE Trans. on Computers*, **C-21**, No. 12, 1269–1273, Dec., 1972.
3. R. C. AGARWAL AND C. S. BURRUS, "Fast One-Dimensional Convolution by Multidimensional Techniques," *IEEE Trans. on Acoustics, Speech, and Signal Processing*, **ASSP-22**, No. 1, 1–10, Feb., 1974.
4. R. C. AGARWAL AND C. S. BURRUS, "Fast Convolution Using Fermat Number Transforms with Applications to Digital Filtering," *IEEE Trans. on Acoustics, Speech, and Signal Processing*, **ASSP-22**, No. 2, 87–97 Apr., 1974.

Appendix

Notational Conventions for FFT's

Several authors have presented material on the FFT using a variety of notational conventions. The Cooley–Tukey paper (Reference 6.1) used an algebraic notation and other authors have used matrix notation. Perhaps the most popular way of representing FFT's is via flow graphs (to our knowledge, this was first used by Rader and Stockham). In the present discussion we shall compare the notation we have used with this flow graph notation. It should be easy to absorb both notations and employ one or the other depending on the circumstances with little or no confusion.

In flow graph nomenclature, a summing point is represented by a node and a multiplication by an arrow; thus the radix 2 decimation-in-time butterfly can be shown as in Fig. A-6.1. Figure A-6.1, as drawn, is somewhat awkward when making a complete FFT flow graph and can be replaced by the flow graph of Fig. A-6.2. Figure A-6.2 is easier to use for drawing a complete FFT but suffers somewhat from the fact that its *appearance* suggests twice as many complex multiplications as are needed.

The notation used throughout this book makes use of a special symbol to represent an elementary DFT computation. Thus, for example, a DIT butterfly is shown in Fig. A-6.3, while a DIF butterfly is drawn in Fig. A-6.4. Figures A-6.5 and A-6.6 show a four-point DIT FFT done both ways. The reader can, of course, make his own judgment as to the relative aesthetic values of the two notations. We shall simply comment on the objective distinctions.

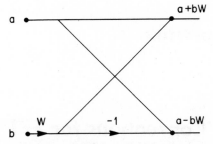

Fig. A-6.1 Radix 2 DIT butterfly using old notation.

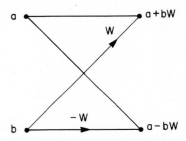

Fig. A-6.2 Improved notation for radix 2 DIT butterfly.

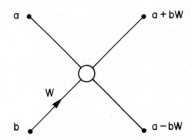

Fig. A-6.3 Radix 2 DIT butterfly using new notation.

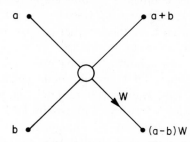

Fig. A-6.4 Radix 2 DIF butterfly using new notation.

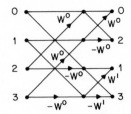

Fig. A-6.5 Four-point DIT FFT using old notation.

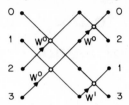

Fig. A-6.6 Four-point DIT FFT using new notation.

1. Notation 1 (as in Fig. A-6.5) is traditional flow graph notation, well-known to many, and of fairly widespread use.
2. Notation 1 can also be used to describe digital filters.
3. There are some general and powerful theorems about flow graphs that are applicable to FFT's.
4. Notation 2 has half as many lines for radix 2 (and $1/r$th as many lines for radix r).
5. Notation 2 can be used to draw a mixed radix FFT easily.
6. Notation 2 has the same number of multiplier symbols as there are multipliers for radix 2 FFT's.

7

An Introduction to the Theory of Two-Dimensional Signal Processing

7.1 Introduction

There are many signals that are inherently two-dimensional for which two-dimensional signal processing techniques are required. Included in this group of signals are photographic data, such as weather photos, air reconnaissance photos, and medical X rays, seismic records, gravity and magnetic data, and electron micrographs that are used to elucidate the spatial structure of the complex biological molecules. Although two-dimensional signals may be processed by one-dimensional systems, it is generally preferable to consider using two-dimensional systems. Many of the basic ideas of one-dimensional signal processing may be readily extended to the two-dimensional case. There are some very important concepts of one-dimensional systems, however, that are not directly extendable to two dimensions. It is the goal of this chapter to discuss the basic ideas and techniques of two-dimensional signal processing and to illustrate them in the context of two-dimensional filter design.

7.2 Two-Dimensional Signals

Let $x(n_1, n_2)$ be a two-dimensional sequence where n_1 and n_2 are integer variables. As in the one-dimensional case, the notation $x(n_1, n_2)$ is often shorthand for a sampled version of a continuous two-dimensional signal $x(s, t)$; i.e.,

$$x(n_1, n_2) = x(n_1 T_1, n_2 T_2) = x(s, t)\big|_{s=n_1 T_1, t=n_2 T_2} \qquad (7.1)$$

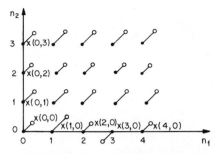

Fig. 7.1 Representation of the two-dimensional signal $x(n_1, n_2)$.

It is quite difficult to depict a two-dimensional sequence graphically so throughout this chapter we shall resort to perspective plots whenever appropriate. Thus, for example, the sequence $x(n_1, n_2)$, $0 \leq n_1, n_2 \leq \infty$, may be represented as shown in Fig. 7.1, where the value of $x(n_1, n_2)$ is given by the height of the sample at grid coordinates (n_1, n_2).

Some useful two-dimensional sequences are given in Figs. 7.2 to 7.4 and defined below. These include

1. Digital Impulse or Unit Sample

$$u_0(n_1, n_2) = \begin{cases} 1 & n_1 = n_2 = 0 \\ 0 & \text{elsewhere} \end{cases} \tag{7.2}$$

2. Digital Step

$$u_{-1}(n_1, n_2) = \begin{cases} 1 & n_1, n_2 \geq 0 \\ 0 & \text{otherwise} \end{cases} \tag{7.3}$$

3. Exponential

$$x(n_1, n_2) = \begin{cases} a_1^{n_1} a_2^{n_2} & n_1, n_2 \geq 0 \\ 0 & \text{otherwise} \end{cases} \tag{7.4}$$

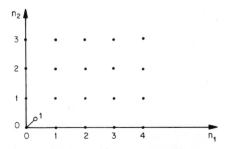

Fig. 7.2 A two-dimensional digital impulse.

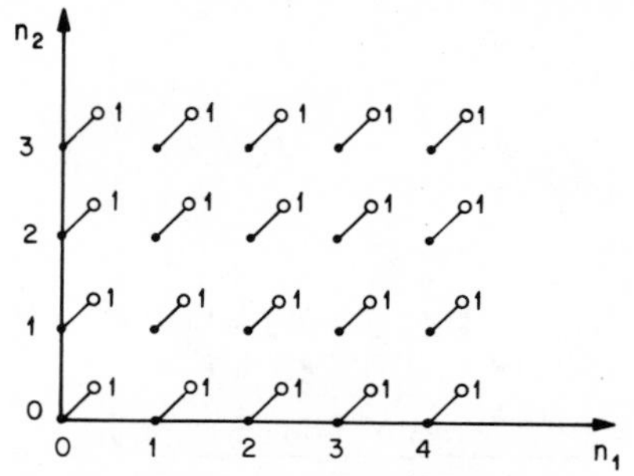

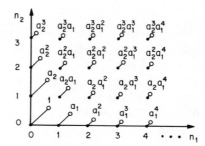

Fig. 7.3 A two-dimensional digital step.

Fig. 7.4 A two-dimensional exponential.

4. Sinusoid (complex)

$$x(n_1, n_2) = e^{j(\omega_1 n_1 + \omega_2 n_2)} \qquad -\infty \leq n_1, n_2 \leq \infty \tag{7.5}$$

As seen above, the two-dimensional step is related to the two-dimensional impulse by

$$u_{-1}(n_1, n_2) = \sum_{m_1=-\infty}^{n} \sum_{m_2=-\infty}^{n} u_0(m_1, m_2) \tag{7.6}$$

7.3 Two-Dimensional Systems

For linear time-invariant systems the basic convolution theorem is valid. Thus if $x(n_1, n_2)$ is input to an LTI system with impulse response $h(n_1, n_2)$, the output $y(n_1, n_2)$ may be determined as

$$y(n_1, n_2) = x(n_1, n_2) * h(n_1, n_2) \tag{7.7}$$

$$= \sum_{m_1=-\infty}^{\infty} \sum_{m_2=-\infty}^{\infty} h(m_1, m_2)x(n_1 - m_1, n_2 - m_2) \tag{7.8}$$

$$= \sum_{m_1=-\infty}^{\infty} \sum_{m_2=-\infty}^{\infty} x(m_1, m_2)h(n_1 - m_1, n_2 - m_2) \tag{7.9}$$

To illustrate the difficulty of applying Eq. (7.7) to (7.9) in all but the simplest of cases, consider the following example.

EXAMPLE I. Given an LTI system with impulse response

$$h(n_1, n_2) = a^{n_1 n_2} \qquad -\infty \leq n_1, n_2 \leq \infty$$

and input to this system of the form

$$x(n_1, n_2) = \begin{cases} 1 & 0 \leq n_1, n_2 \leq 2 \\ 0 & \text{otherwise} \end{cases}$$

determine $y(n_1, n_2)$, the output of the system.

Solution: Using Eq. (7.8) and (7.9), we can write

$$y(n_1, n_2) = \sum_{m_1=0}^{2} \sum_{m_2=0}^{2} a^{(n_1-m_1)(n_2-m_2)}$$

The difficulty in evaluating the right side of the equation above lies in the fact that $a^{m_1 m_2}$ cannot be factored into a product of the form $\alpha(m_1)\beta(m_2)$. Thus the two-dimensional sum must be carried out directly, term by term, to give $y(n_1, n_2)$. To complete this example we carry out the summation giving

$$\begin{aligned}
y(n_1, n_2) = {} & a^{n_1 n_2} + a^{n_1(n_2-1)} + a^{n_1(n_2-2)} \\
& + a^{(n_1-1)n_2} + a^{(n_1-1)(n_2-1)} + a^{(n_1-1)(n_2-2)} \\
& + a^{(n_1-2)n_2} + a^{(n_1-2)(n_2-1)} + a^{(n_1-2)(n_2-2)}
\end{aligned}$$

The difficulty of working with even simple examples is an omen of some of the inherent problems in working with two-dimensional systems.

7.4 Causality, Separability, Stability

A two-dimensional filter is said to be *realizable* or *causal*, if its impulse response satisfies the property

$$h(n_1, n_2) = 0 \qquad n_1 < 0 \quad \text{or} \quad n_2 < 0 \tag{7.10}$$

The filter is said to be *separable* if its impulse response can be factored into a product of one-dimensional responses; i.e.,

$$h(n_1, n_2) = g(n_1)f(n_2) \tag{7.11}$$

If Eq. (7.11) is not satisfied, the filter is said to be *nonseparable*. The advantage of separable filters is that the two-dimensional convolution [Eq. (7.8)] may be carried out as a sequence of one-dimensional convolutions. This is readily seen by rewriting Eq. (7.8) as

$$y(n_1, n_2) = \sum_{m_1=-\infty}^{\infty} \sum_{m_2=-\infty}^{\infty} g(m_1)f(m_2)x(n_1 - m_1, n_2 - m_2) \tag{7.12}$$

$$= \sum_{m_1=-\infty}^{\infty} g(m_1)\left[\sum_{m_2=-\infty}^{\infty} f(m_2)x(n_1 - m_1, n_2 - m_2) \right] \tag{7.13}$$

$$= \sum_{m_1=-\infty}^{\infty} g(m_1)a(n_1 - m_1, n_2) \tag{7.14}$$

where $a(n_1 - m_1, n_2)$ is a sequence of one-dimensional convolutions [obtained by evaluating the terms inside the bracket of Eq. (7.13) for each fixed value of m_1]. Equation (7.14) shows that $y(n_1, n_2)$ may be obtained by a second sequence of one-dimensional convolutions.

If both the input sequence $x(n_1, n_2)$ and the filter impulse response $h(n_1, n_2)$ are separable, then it is readily seen that the output sequence $y(n_1, n_2)$ is separable. In this case we obtain the result [using Eqs. (7.8) and (7.11)]

$$y(n_1, n_2) = \left[\sum_{m_1=-\infty}^{\infty} g(m_1)b(n_1 - m_1) \right] \left[\sum_{m_2=-\infty}^{\infty} f(m_2)c(n_2 - m_2) \right] \quad (7.15)$$

$$= \alpha(n_1)\beta(n_2) \quad (7.16)$$

where

$$x(n_1, n_2) = b(n_1)c(n_2) \quad (7.17)$$

A two-dimensional filter is said to be *stable* (in the sense that a bounded input produces a bounded output) if and only if its impulse response satisfies the constraint

$$\sum_{n_1=-\infty}^{\infty} \sum_{n_2=-\infty}^{\infty} |h(n_1, n_2)| < \infty \quad (7.18)$$

As in the one-dimensional case, Eq. (7.18) can be shown to be a necessary and sufficient condition for stability.

One problem with Eq. (7.18) is that it can be quite difficult to evaluate for arbitrary $h(n_1, n_2)$. Thus alternate measures of stability are desirable and necessary. We shall discuss such measures later in this chapter.

7.5 Two-Dimensional Difference Equations

As in the one-dimensional case, linear time-invariant two-dimensional filters can often be described by a constant coefficient, linear difference equation relating the output of the filter to its input. The most general form for such a difference equation, for a realizable filter, is

$$\sum_{m_1=0}^{M_1} \sum_{m_2=0}^{M_2} \alpha_{m_1,m_2} y(n_1 - m_1, n_2 - m_2) = \sum_{l_1=0}^{L_1} \sum_{l_2=0}^{L_2} \beta_{l_1,l_2} x(n_1 - l_1, n_2 - l_2) \quad (7.19)$$

where $\{\alpha_{m_1,m_2}\}$ and $\{\beta_{l_1,l_2}\}$ are the set of constant coefficients that characterize the particular filter. A set of initial conditions must also be given for the complete specification of the filter. Equation (7.19) can be written in the form of a recursion relation by solving for $y(n_1, n_2)$, the current output sample, in terms of past output samples as well as present and past input samples. Such a recursion (assuming $\alpha_{0,0} = 1$) is of the form

$$y(n_1, n_2) = \sum_{l_1=0}^{L_1} \sum_{l_2=0}^{L_2} \beta_{l_1,l_2} x(n_1 - l_1, n_2 - l_2)$$

$$- \sum_{m_1=0}^{M_1} \sum_{m_2=0}^{M_2} \alpha_{m_1,m_2} y(n_1 - m_1, n_2 - m_2) \qquad (7.20)$$
$$(m_1, m_2 \neq 0 \text{ simultaneously})$$

As an example, consider a filter described by the difference equation

$$y(n_1, n_2) = x(n_1, n_2) + \alpha_{1,0} y(n_1 - 1, n_2) + \alpha_{1,1} y(n_1 - 1, n_2 - 1) \qquad (7.21)$$

with initial conditions
$$y(n_1, n_2) = 0 \qquad \begin{cases} n_1 < 0 \\ n_2 < 0 \end{cases} \qquad (7.22)$$

If the input to this filter is a digital impulse, the output can be shown to be of the form

$$y(n_1, n_2) = \begin{cases} \binom{n_1}{n_2} (\alpha_{1,1})^{n_2} (\alpha_{1,0})^{n_1 - n_2} & \begin{array}{l} 0 \leq n_2 \leq n \\ 0 < n_1 < \infty \end{array} \\ 0 & \text{otherwise} \end{cases} \qquad (7.23)$$

where $\binom{n_1}{n_2}$ is the number of combinations of n_1 things taken n_2 at a time.

Equation (7.23) can readily be verified by manually computing $y(n_1, n_2)$ for a few values of (n_1, n_2) using Eqs. (7.21) and (7.22). With this procedure we find

$$y(0, 0) = x(0, 0) = 1$$
$$y(1, 0) = \alpha_{1,0} y(0, 0) = \alpha_{1,0}$$
$$y(2, 0) = \alpha_{1,0} y(1, 0) = (\alpha_{1,0})^2$$
$$y(0, 1) = 0$$
$$y(1, 1) = \alpha_{1,1} y(0, 0) = \alpha_{1,1}$$
$$y(0, 2) = 0$$
$$y(1, 2) = 0$$
$$y(2, 1) = \alpha_{1,0} y(1, 1) + \alpha_{1,1} y(1, 0) = 2 \cdot \alpha_{1,0} \cdot \alpha_{1,1}$$
$$y(3, 1) = \alpha_{1,0} y(2, 1) + \alpha_{1,1} y(2, 0) = 3(\alpha_{1,0})^2 \alpha_{1,1}$$

The techniques for solving difference equations in two dimensions are similar to those discussed in the one-dimensional case and will not be repeated here.

7.6 Frequency Domain Techniques

Two-dimensional sinusoids play an extremely important part in the analysis of linear, time-invariant two-dimensional systems because, as in the one-dimensional case, they are the eigenfunctions of such systems. Thus if the input to a two-dimensional system is the sinusoidal signal

$$x(n_1, n_2) = e^{j\omega_1 n_1} e^{j\omega_2 n_2} \qquad \begin{matrix} -\infty \le n_1 \le \infty \\ -\infty \le n_2 \le \infty \end{matrix} \qquad (7.24)$$

using the convolution theorem, the output of the system is

$$y(n_1, n_2) = \sum_{m_1=-\infty}^{\infty} \sum_{m_2=-\infty}^{\infty} h(m_1, m_2) e^{j\omega_1(n_1-m_1)} e^{j\omega_2(n_2-m_2)} \qquad (7.25)$$

$$= e^{j\omega_1 n_1} e^{j\omega_2 n_2} \sum_{m_1=-\infty}^{\infty} \sum_{m_2=-\infty}^{\infty} h(m_1, m_2) e^{-j\omega_1 m_1} e^{-j\omega_2 m_2} \qquad (7.26)$$

$$= x(n_1, n_2) H(e^{j\omega_1}, e^{j\omega_2}) \qquad (7.27)$$

where $H(e^{j\omega_1}, e^{j\omega_2})$ is the frequency response of the system and may be written as a two-dimensional Fourier series

$$H(e^{j\omega_1}, e^{j\omega_2}) = \sum_{m_1=-\infty}^{\infty} \sum_{m_2=-\infty}^{\infty} h(m_1, m_2) e^{-j\omega_1 m_1} e^{-j\omega_2 m_2} \qquad (7.28)$$

Since Eq. (7.28) represents a Fourier series, the filter coefficients $h(n_1, n_2)$ can be recovered using the well-known formula

$$h(n_1, n_2) = \frac{1}{(4\pi^2)} \int_{-\pi}^{\pi} \int_{-\pi}^{\pi} H(e^{j\omega_1}, e^{j\omega_2}) e^{j\omega_1 n_1} e^{j\omega_2 n_2} \, d\omega_1 \, d\omega_2 \qquad (7.29)$$

The frequency response ($He^{j\omega_1}, e^{j\omega_2}$) has several interesting properties worth noting. It is easily seen that $H(e^{j\omega_1}, e^{j\omega_2})$ is doubly periodic in frequency —i.e.,

$$H(e^{j\omega_1}, e^{j\omega_2}) = H[e^{j(\omega_1+l\cdot2\pi)}, e^{j(\omega_2+m\cdot2\pi)}] \qquad -\infty \le l, m \le \infty \quad (7.30)$$

If we restrict $h(n_1, n_2)$ to take on real values, then the frequency response satisfies the constraint

$$H(e^{j\omega_1}, e^{j\omega_2}) = H^*(e^{-j\omega_1}, e^{-j\omega_2}) \qquad (7.31)$$

Thus knowledge of the behavior of $H(^{j\omega_1}, e^{j\omega_2})$ in the region $0 \le \omega_1 \le \pi$, $0 \le \omega_2 \le \pi$, i.e., the first quadrant, implies knowledge of its behavior in the third quadrant and vice versa.

We now illustrate the application of Eq. (7.29) for two important examples of frequency response.

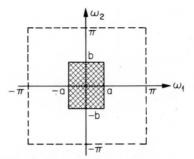

Fig. 7.5 Ideal rectangular lowpass filter frequency response specifications.

EXAMPLE 2. Find the Fourier series coefficients, $h(n_1, n_2)$ of the filter whose frequency response is

$$H(e^{j\omega_1}, e^{j\omega_2}) = \begin{cases} 1 & \begin{cases} -a \le \omega_1 \le a \\ -b \le \omega_2 \le b \end{cases} \\ 0 & \text{elsewhere} \end{cases}$$

i.e., $H(e^{j\omega_1}, e^{j\omega_2}) = 1$ in the crosshatched rectangle shown in Fig. 7.5.

Solution: From Eq. (7.29) we find

$$h(n_1, n_2) = \frac{1}{4\pi^2} \int_{-b}^{b} \int_{-a}^{a} e^{j\omega_1 n_1} e^{j\omega_2 n_2} \, d\omega_1 \, d\omega_2$$

$$= \left[\frac{\sin(an_1)}{\pi n_1} \right] \left[\frac{\sin(bn_2)}{\pi n_2} \right]$$

This example shows that if the frequency response can be decomposed into a product of a term in ω_1 and a term in ω_2, then the impulse response $h(n_1, n_2)$ can similarly be decomposed into a product of a term in n_1 and a term in n_2.

EXAMPLE 3. Find the Fourier series coefficients of the filter whose frequency response is

$$H(e^{j\omega_1}, e^{j\omega_2}) = \begin{cases} 1 & \omega_1^2 + \omega_2^2 \le R^2 \\ 0 & \text{otherwise} \end{cases}$$

i.e., $H(e^{j\omega_1}, e^{j\omega_2}) = 1$ in the crosshatched circle shown in Fig. 7.6.

Solution: It is easy to see that $H(e^{j\omega_1}, e^{j\omega_2})$ possesses circular symmetry—i.e., $H(e^{j\omega_1}, e^{j\omega_2}) = f(\sqrt{\omega_1^2 + \omega_2^2})$. Therefore it can be shown that the Fourier series coefficients also possess circular symmetry; i.e., $h(n_1, n_2) = g(\sqrt{n_1^2 + n_2^2})$

Thus an easy way to find $h(n_1, n_2)$ is to find $h(n_1, 0)$ and substitute $\sqrt{n_1^2 + n_2^2}$ for n_1. We obtain $h(n_1, 0)$ as

$$h(n_1, 0) = \frac{1}{(4\pi^2)} \int_{-R}^{R} e^{j\omega_1 n_1} \, d\omega_1 \int_{-\sqrt{R^2-\omega_1^2}}^{\sqrt{R^2-\omega_1^2}} d\omega_2$$

$$= \frac{1}{(4\pi^2)} \int_{-R}^{R} e^{j\omega_1 n_1} 2\sqrt{R^2 - \omega_1^2} \, d\omega_1$$

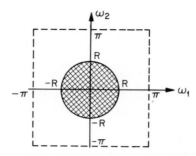

Fig. 7.6 Ideal circularly symmetric lowpass filter frequency response specifications.

If we use the substitution

$$\omega_1 = R \sin \varphi$$

$$d\omega_1 = R \cos \varphi \, d\varphi$$

then

$$h(n_1, 0) = \frac{1}{(4\pi^2)} \int_{-\pi/2}^{\pi/2} 2R^2 (\cos^2 \varphi) e^{jRn_1 \sin\varphi} \, d\varphi$$

$$= \frac{RJ_1(Rn_1)}{2\pi n_1}$$

where $J_1(x)$ is the first-order Bessel function. Therefore we find

$$h(n_1, n_2) = \frac{RJ_1(R\sqrt{n_1^2 + n_2^2})}{\sqrt{n_1^2 + n_2^2}}$$

7.7 Two-Dimensional z Transforms

A useful representation of a sequence $x(n_1, n_2)$ is its two-dimensional z transform, defined as

$$X(z_1, z_2) = \sum_{n_1=-\infty}^{\infty} \sum_{n_2=-\infty}^{\infty} x(n_1, n_2) z_1^{-n_1} z_2^{-n_2} \qquad (7.32)$$

where Eq. (7.32) is defined over some region of convergence—i.e., some range of values of z_1 and z_2. It is often exceedingly difficult to determine the convergence region of Eq. (7.32)—i.e., to find the locations of all the singularities of $X(z_1, z_2)$. We shall come back to this problem in our discussion of design of two-dimensional filters.

The inverse z transform is defined as

$$x(n_1, n_2) = \frac{1}{(2\pi j)^2} \int_{C_1} \int_{C_2} X(z_1, z_2) z_1^{n_1-1} z_2^{n_2-1} \, dz_1 \, dz_2 \qquad (7.33)$$

where C_1 and C_2 are suitable closed contours in the z_1 and z_2 planes.

To illustrate the application of Eq. (7.32) for determining a two-dimensional z transform, consider the following sequence:

$$x(n_1, n_2) = \begin{cases} K^{n_1}u_0(n_1 - n_2) & n_1, n_2 > 0 \\ 0 & \text{otherwise} \end{cases}$$

The z transform of this sequence is

$$X(z_1, z_2) = \sum_{n_1=-\infty}^{\infty} \sum_{n_2=-\infty}^{\infty} K^{n_1}u_0(n_1 - n_2)z_1^{-n_1}z_2^{-n_2}$$

$$= \sum_{n_1=0}^{\infty} K^{n_1}(z_1z_2)^{-n_1}$$

$$= \frac{1}{1 - Kz_1^{-1}z_2^{-1}}$$

which converges if

$$|Kz_1^{-1}z_2^{-1}| < 1$$

7.8   Finite Sequences

There is one class of sequences for which the z transform has the property that it converges everywhere in the z_1 and z_2 planes except possibly at $z_1 = 0$, $z_2 = 0$ or $z_1 = \infty$, $z_2 = \infty$. This class of sequences is the class of finite duration sequences that are bounded. Convergence of the z transform is guaranteed because Eq. (7.32) has finite limits on all the summations and since $x(n_1, n_2)$ is bounded, so is $x(n_1, n_2)z_1^{-n_1}z_2^{-n_2}$ bounded for finite values of z_1 and z_2. Since the z transform converges everywhere in the z_1 and z_2 planes for finite sequences, this guarantees that filters designed from finite sequences will be stable.

7.9   Convolution Property of z Transforms

It is easy to show that if a sequence $y(n_1, n_2)$ is the two-dimensional convolution of an input $x(n_1, n_2)$ and a filter impulse response $h(n_1, n_2)$, its z transform $Y(z_1, z_2)$ is the product of the z transforms of the input $X(z_1, z_2)$ and the filter $H(z_1, z_2)$. Thus we have the important relation

$$x(n_1, n_2) \leftrightarrow X(z_1, z_2) \tag{7.34}$$

$$h(n_1, n_2) \leftrightarrow H(z_1, z_2) \tag{7.35}$$

$$y(n_1, n_2) = x(n_1, n_2) * h(n_1, n_2) \tag{7.36}$$

$$Y(z_1, z_2) = X(z_1, z_2) \cdot H(z_1, z_2) \tag{7.37}$$

The derivation of Eq. (7.37) is left as an exercise for the reader. Thus using Eq. (7.37) the response $y(n_1, n_2)$ can be found as the inverse transform of the product of the transforms of $x(n_1, n_2)$ and $h(n_1, n_2)$. A simple example showing the use of this technique is given below.

EXAMPLE 4.   The input to a two-dimensional linear time-invariant filter is the sequence

$$x(n_1, n_2) = L^{n_1} u_0(n_1 - n_2) \qquad 0 \leq n_1, n_2 \leq \infty$$

The impulse response of the filter is of the form

$$h(n_1, n_2) = K^{n_1} u_0(n_1 - n_2) \qquad 0 \leq n_1, n_2 < \infty$$

Find the output $y(n_1, n_2)$ of this filter.

Solution: First we find the transforms of $x(n_1, n_2)$ and $h(n_1, n_2)$.

$$X(z_1, z_2) = \sum_{n_1=0}^{\infty} L^{n_1} z_1^{-n_1} z_2^{-n_1}$$

$$= \frac{1}{(1 - L z_2^{-1} z_2^{-1})} \qquad |z_1 z_2| > L$$

$$H(z_1, z_2) = \sum_{n_1=0}^{\infty} K^{n_1} z_1^{-n_1} z_2^{-n_2}$$

$$= \frac{1}{1 - K z_1^{-1} z_2^{-1}} \qquad |z_1 z_2| > K$$

Using Eq. (7.37) we find

$$Y(z_1, z_2) = \frac{1}{(1 - L z_1^{-1} z_2^{-1})(1 - K z_1^{-1} z_2^{-1})}$$

$$= \frac{L/(L - K)}{1 - L z_1^{-1} z_2^{-1}} + \frac{K/(K - L)}{1 - K z_1^{-1} z_2^{-1}}$$

The inverse transform can be solved by inspection to give

$$y(n_1, n_2) = \left(\frac{L}{L - K}\right) L^{n_1} u_0(n_1 - n_2) + \left(\frac{K}{K - L}\right) K^{n_1} u_0(n_1 - n_2)$$

7.10   Two-Dimensional DFT

We end this discussion on two-dimensional signals and systems with a brief mention of the two-dimensional DFT and how it may be evaluated efficiently using a one-dimensional FFT program.

A periodic signal in two dimensions is defined by the relations

$$x_p(n_1, n_2) = x_p(n_1 + m_1 N_1, n_2 + m_2 N_2) \tag{7.38}$$

where N_1 is the period along the first dimension, N_2 is the period along the second dimension, and m_1 and m_2 are any integers. The subscript p in Eq. (7.38) indicates the signal is periodic. Just as in one dimension, two-dimensional periodic signals can be exactly represented as a linear combination of a finite set of exponentials whose periods are subperiods of N_1 and N_2; i.e.,

$$x_p(n_1, n_2) = \frac{1}{N_1 N_2} \sum_{k_1=0}^{N_1-1} \sum_{k_2=0}^{N_2-1} X_p(k_1, k_2) e^{j(2\pi/N_1)n_1 k_1} e^{j(2\pi/N_2)n_2 k_2} \qquad (7.39)$$

where the periodic Fourier coefficients $X_p(k_1, k_2)$ represent the amplitude of $x_p(n_1, n_2)$ at the two-dimensional frequency $\omega_1 = (2\pi/N_1)k_1$, $\omega_2 = (2\pi/N_2)k_2$. $X_p(k_1, k_2)$ may be readily determined by evaluating the two-dimensional z transform over a period of $x_p(n_1, n_2)$ at the frequencies ω_1, ω_2 as above, giving

$$X_p(k_1, k_2) = X(z_1, z_2)\big|_{z_1 = e^{j(2\pi/N_1)k_1}, z_2 = e^{j(2\pi/N_2)k_2}} \qquad (7.40)$$

$$= \sum_{n_1=0}^{N_1-1} \sum_{n_2=0}^{N_2-1} x(n_1, n_2) e^{-j(2\pi/N_1)n_1 k_1} e^{-j(2\pi/N_2)n_2 k_2} \qquad (7.41)$$

Equation (7.39) is referred to as the two-dimensional inverse discrete Fourier transform, and Eq. (7.41) is the two-dimensional discrete Fourier transform.

Many properties of Eqs. (7.39) and (7.41) may be derived. Two important properties, however, concern the evaluation of Eqs. (7.39) and (7.41) as a series of one-dimensional DFT's and the extension of a DFT representation to finite duration signals. The first property may be seen by rewriting Eq. (7.41) as

$$X_p(k_1, k_2) = \sum_{n_1=0}^{N_1-1} e^{-j(2\pi/N_1)n_1 k_1} \left[\sum_{n_2=0}^{N_2-1} x_p(n_1, n_2) e^{-j(2\pi/N_2)n_2 k_2} \right] \qquad (7.42)$$

The bracketed term is a series of N_1 one-dimensional DFT's obtained by varying n_1 from 0 to $N_1 - 1$. If we call the result of each DFT $g_p(n_1, k_2)$, then Eq. (7.41) becomes

$$X_p(k_1, k_2) = \sum_{n_1=0}^{N_1-1} e^{-j(2\pi/N_1)n_1 k_1} g_p(n_1, k_2) \qquad (7.43)$$

which is again a series of N_2 one-dimensional DFT's as k_2 is varied from 0 to $N_2 - 1$. The extension of the argument above to the evaluation of the inverse DFT is straightforward.

The property that a DFT representation provides an exact representation for a finite duration sequence $x(n_1, n_2)$ may be readily seen by examining Eqs. (7.39) and (7.41) for the case when $x_p(n_1, n_2)$ is 0 outside the range $0 \leq n_1 \leq N_1 - 1$, $0 \leq n_2 \leq N_2 - 1$. We call this finite sequence $x(n_1, n_2)$. Although $x(n_1, n_2)$ is no longer periodic, its values within a period are identical to those of its periodic counterpart and thus the DFT, IDFT representation provides an exact and convenient means for obtaining Fourier transform coefficients.

The primary difference between $x(n_1, n_2)$ and the periodic sequence $x_p(n_1, n_2)$ is that the Fourier transform of $x(n_1, n_2)$ is a continuous function of ω_1 and ω_2, whereas the Fourier transform of $x_p(n_1, n_2)$ is strictly a line spectrum at the appropriate frequencies.

The applications of the DFT and IDFT for filtering have been discussed in the one-dimensional case and they apply equally well in two dimensions. Fast convolution using the FFT is probably the most important means for realizing two-dimensional filters.

7.11 Filter Design Considerations

A two-dimensional digital filter is characterized by the two-dimensional z transform

$$H(z_1, z_2) = \frac{\sum\limits_{i=0}^{p} \sum\limits_{j=0}^{q} a_{ij} z_1^{-i} z_2^{-j}}{\sum\limits_{i=0}^{p} \sum\limits_{j=0}^{q} b_{ij} z_1^{-i} z_2^{-j}} \tag{7.44}$$

where the set of a_{ij} and b_{ij} are constants and $b_{00} = 1$. The difference equation for this filter is of the form

$$y(n_1, n_2) = \sum_{i=0}^{p} \sum_{j=0}^{q} a_{ij} x(n_1 - i, n_2 - j)$$

$$- \sum_{\substack{i=0 \\ 0=i \neq j=0}}^{p} \sum_{j=0}^{q} b_{ij} y(n_1 - i, n_2 - j) \qquad \begin{array}{l} 0 \leq n_1 \leq \infty \\ 0 \leq n_2 \leq \infty \end{array} \tag{7.45}$$

where we have assumed the filter recursion is in the $(+n_1, +n_2)$ direction, i.e., the filter is causal. Equation (7.45) shows that a set of initial conditions is needed to solve for the output for $n_1, n_2 \geq 0$. The set of initial conditions that is required is shown as the shaded area in Fig. 7.7.

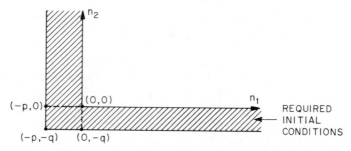

Fig. 7.7 Set of initial conditions required to solve a two-dimensional difference equation.

7.12 IIR Filters

To design infinite impulse response filters, we must be able to choose the filter coefficients a_{ij}, b_{ij} to approximate some desired response and then we must be able to ensure that the resulting filter is stable. Both these aspects of the design problem have met with considerable difficulty to the extent that no simple design technique is readily available. It is worth discussing the sources of difficulty before proceeding to the design of two-dimensional FIR filters. As far as the filter problem is concerned, an important property of one-dimensional z transforms is the ability to factor higher-order polynomials into a product of lower-order polynomials. Thus the polynomial

$$P(z) = z^4 + \alpha_1 z^3 + \alpha_2 z^2 + \alpha_3 z + \alpha_4 \qquad (7.46)$$

can be factored into

$$P(z) = (z - \beta_1)(z - \beta_2)(z - \beta_3)(z - \beta_4) \qquad (7.47)$$

The importance of this factorization cannot be minimized. It enables the designer to determine the stability of a high-order filter from simple criteria on the lower-order polynomials. Furthermore it provides a convenient means for realizing the filter as a cascade of lower-order filters, thereby greatly reducing certain coefficient sensitivity problems inherent in realizing high-order polynomials.

In two dimensions this factorization property is *not* preserved. Thus the general two-dimensional polynomial $P(z_1, z_2)$ has no simple lower-order factors. The main implication of this inability to perform factorization is that it is almost impossible to test the stability of an IIR filter, except for some simple cases. Furthermore it implies that an unstable filter cannot be stabilized by cascading an appropriate all-pass filter with the unstable filter. The final implication is that two-dimensional IIR filters cannot be realized using a cascade configuration of lower-order filters. Thus even if one could design a stable high-order IIR filter, it would be quite difficult to realize it in an efficient way.

In the following sections the approaches to determining the stability of an IIR filter by direct evaluation techniques will be discussed.

7.13 Stability Considerations

Several techniques have been proposed for determining stability of IIR filters directly from their two-dimensional z transforms. Although these techniques work in theory, they are often difficult and time-consuming to apply in practice. The following theorem due to Shanks relates to methods for determining the stability of IIR filters.

STABILITY THEOREM 1: A causal IIR filter with z transform $H(z_1, z_2) = A(z_1, z_2)/B(z_1, z_2)$ is stable *if and only if* there are no values of z_1 and z_2 such that $B(z_1, z_2) = 0$ and $|z_1| \geq 1$ and $|z_2| \geq 1$.

PROOF. Let

$$H(z_1, z_2) = \frac{A(z_1, z_2)}{B(z_1, z_2)} = \sum_{n_1=0}^{\infty} \sum_{n_2=0}^{\infty} h(n_1, n_2) z_1^{-n_1} z_2^{-n_2} \tag{7.48}$$

For a filter with impulse response $h(n_1, n_2)$ to be stable, the necessary and sufficient condition is

$$\sum_{n_1=0}^{\infty} \sum_{n_2=0}^{\infty} |h(n_1, n_2)| < \infty \tag{7.49}$$

Thus we want to show that Eq. (7.49) holds if and only if $H(z_1, z_2)$ is analytic in the region

$$D = \{(z_1, z_2); |z_1| \geq 1 \cap |z_2| \geq 1\} \tag{7.50}$$

The *if* part can be proved by noting that if $H(z_1, z_2)$ is analytic in D, we can find $\varepsilon > 0$ such that $H(z_1, z_2)$ is analytic in

$$D_1 = \{(z_1, z_2); |z_1| > 1 - \varepsilon \cap |z_2| > 1 - \varepsilon\} \tag{7.51}$$

which implies that

$$\sum_{n_1} \sum_{n_2} h(n_1, n_2) z_1^{-n_1} z_2^{-n_2}$$

be absolutely convergent in D_1 and therefore

$$\sum_{n_1} \sum_{n_2} |h(n_1, n_2)| < \infty$$

The *only if* part is shown by noting that if

$$\sum_{n_1} \sum_{n_2} |h(n_1, n_2)| < \infty$$

then by the M test (of series convergence)

$$\sum_{n_1} \sum_{n_2} h(n_1, n_2) z_1^{-n_1} z_2^{-n_2}$$

is absolutely convergent in D, which implies $H(z_1, z_2)$ is analytic in D.

The difficulty with this stability theorem lies in trying to apply it to a practical situation. To check stability the unit disk in the z_1 plane ($d_1 \equiv z_1$, $|z_1| \leq 1$) must be mapped into the z_2 plane by solving the *implicit* equation $B(z_1, z_2) = 0$. The filter is stable if and only if the image of d_1 in the z_2 plane does not overlap the unit disk in the z_2 plane.

This stability criterion requires that for a large set of values of $z_1 = \hat{z}_1$ ($0 \leq |\hat{z}_1| \leq 1$) we must solve the equation $B(\hat{z}_1, z_2) = 0$ for z_2. This is difficult to do in practice for all but the simplest of examples. To illustrate

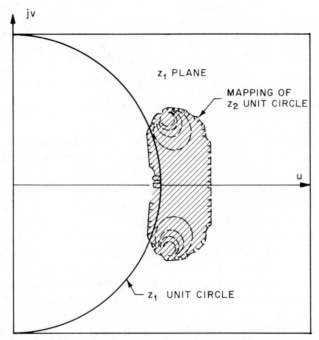

Fig. 7.8 Contour map of the absolute value of the z_2 roots as a function of the complex value of z_1 (After Shanks).

the use of this procedure, consider as an example the filter with z transform

$$H(z_1, z_2) = \frac{1}{B(z_1, z_2)}$$

$$B(z_1, z_2) = 1 - 0.8z_1^{-1} - 1.635z_2^{-1} + 1.272z_1^{-1}z_2^{-1} + 0.8z_2^{-2} - 0.64z_1^{-1}z_2^{-2}$$

Letting $z_1 = \hat{z}_1$ gives

$$B(\hat{z}_1, z_2) = (1 - 0.8\hat{z}_1^{-1}) + (-1.635 + 1.272\hat{z}_1^{-1})z_2^{-1} + (0.8 - 0.64\hat{z}_1^{-1})\hat{z}_2^{-2}$$

which is quadratic in z_2 (with complex coefficients) and hence can readily be solved to determine stability using the theorem above. Figure 7.8 shows a contour map given by Shanks of the absolute value of the z_2 roots as a function of the complex value of $z_1 = u + jv$. All contours for $|z_2| < 1$ have been suppressed. Therefore the shaded area is the mapping of the exterior of the z_2 unit circle onto the z_1 plane for $B(z_1, z_2) = 0$. Since the shaded area intersects the z_1 unit circle, the filter is unstable.

A simplified version of the stability criterion above (due to Huang) can be used to reduce computation greatly. Stated as a theorem it says:

STABILITY THEOREM 2:   A causal filter with z transform $H(z_1, z_2) = A(z_1, z_2)/B(z_1, z_2)$, where A and B are polynomials, is stable if and only if

1. The map of $\partial d_1 = (z_1: |z_1| = 1)$ in the z_2 plane according to $B(z_1, z_2) = 0$ lies inside $d_2 = (z_2: |z_2| \geq 1)$.
2. No point in $d_1 = (z_1: |z_1| \geq 1)$ maps into the point $z_2 = 0$ by the relation $B(z_1, z_2) = 0$.

The proof of this theorem is omitted here. The implications, however, are clear. The theorem says we need only consider the mapping $B(\hat{z}_1, z_2) = 0$ for $|\hat{z}_1| = 1$ and also solve $B(z_1, 0) = 0$ to see if there is any root with magnitude greater than 1.

Further refinements on methods of evaluating whether the stability criteria above are met have been given but we shall not discuss them here. Since in theory, if not in practice, stability of a two-dimensional filter can be ascertained, the more important questions are how do we determine filter coefficients (of a stable filter) to meet desired frequency response specifications, and how do we stabilize an unstable filter without affecting its magnitude response. Answers to these questions are, for the most part, not available.

A partial solution has been proposed by Shanks for the second question—that is, stabilizing an unstable filter. Let $H(z_1, z_2) = 1/B(z_1, z_2)$ be an unstable filter. To find a stable approximation to $H(z)$, first determine $C(z_1, z_2)$, a minimum mean square inverse of $B(z_1, z_2)$ defined by

$$C(z_1, z_2) = \sum_{n_1=0}^{a} \sum_{n_2=0}^{b} c(n_1, n_2) z_1^{-n_1} z_2^{-n_2} \tag{7.52}$$

where a and b are positive integers. Let

$$B(z_1, z_2) \cdot C(z_1, z_2) = 1 + d_{10} z_1^{-1} + d_{01} z_2^{-1} + d_{11} z_1^{-1} z_2^{-1} + \ldots \tag{7.53}$$

Then among all polynomials of order (a, b), C is the minimum mean square inverse of B if

$$\sum_{i=0}^{a} \sum_{j=0}^{b} (d_{ij})^2 \tag{7.54}$$

is minimized. Next determine $\hat{B}(z_1, z_2)$, the minimum mean square inverse of $C(z_1, z_2)$. It has been conjectured by Shanks that

$$\hat{H}(z_1, z_2) = \frac{1}{\hat{B}(z_1, z_2)} \tag{7.55}$$

is stable and that the magnitude of the frequency response of $\hat{H}$ is approximately equal to that of H. Figure 7.9 shows an example due to Shanks of the response of both an unstable filter and its "stabilized" double least

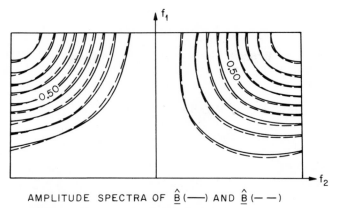

AMPLITUDE SPECTRA OF $\hat{\underline{B}}$ (——) AND $\underline{\hat{B}}$ (— —)

Fig. 7.9 Frequency responses of an unstable filter and its stabilized double least squares inverse (After Shanks).

squares inverse. In this figure are shown contour maps of the magnitude spectra of B and $\hat{B}$ that are very close to being identical.

7.14 FIR Filters

For two-dimensional FIR filters, problems of stability do not exist because the impulse response sequence is bounded and exists only for a finite time. Thus the stability condition

$$\sum_{n_1=0}^{N_1-1} \sum_{n_2=0}^{N_2-1} |h(n_1, n_2)| < \infty \tag{7.56}$$

is guaranteed for all finite N_1, N_2. (The impulse response is nonzero over the interval $0 \le n_1 \le N_1 - 1, 0 \le n_2 \le N_2 - 1$). The implications of Eq. (7.56) are that many of the one-dimensional filter design techniques are often directly extendable to two or more dimensions by appropriate modifications to the design procedure. In the remainder of this chapter we shall discuss, and illustrate with examples, the window method, frequency sampling techniques, and optimal filter design.

We shall assume throughout that the types of filters we are interested in are those whose frequency response approximates a circularly symmetric function; i.e.,

$$H(e^{j2\pi f_1}, e^{j2\pi f_2}) \approx \hat{H}(e^{j2\pi \sqrt{f_1^2+f_2^2}}) \tag{7.57}$$

There are many physical situations where filters of the type above are desired, such as in picture processing where there is no preferred spatial frequency axis. It is easily shown that, using a rectangular sampling grid, the requirement of circular symmetry cannot be met exactly. This is due to the rectangular periodicity of $H(e^{j2\pi f_1}, e^{j2\pi f_2})$; i.e.,

$$H(e^{j2\pi(f_1+m_1)}, e^{j2\pi(f_2+m_2)}) = H(e^{j2\pi f_1}, e^{j2\pi f_2}) \tag{7.58}$$

(m_1, m_2 are any integers) which is in conflict with Eq. (7.57). Very good approximations to Eq. (7.57) can be attained, however, in practice.

7.15   Two-Dimensional Windows

The technique of windowing is directly extendable to two dimensions. The desired frequency response $D(e^{j2\pi f_1}, e^{j2\pi f_2})$ (generally a circular symmetric function to within some limits) is expanded in a two-dimensional Fourier series to give the unwindowed impulse response $h(n_1, n_2)$ defined by

$$D(e^{j2\pi f_1}, e^{j2\pi f_2}) = \sum_{n_1=-\infty}^{\infty} \sum_{n_2=-\infty}^{\infty} h(n_1, n_2)e^{-j2\pi n_1 f_1}e^{-j2\pi n_2 f_2} \tag{7.59}$$

where

$$h(n_1, n_2) = \int_{-0.5}^{0.5} \int_{-0.5}^{0.5} D(e^{j2\pi f_1}, e^{j2\pi f_2})e^{j2\pi n_1 f_1}e^{j2\pi n_2 f_2} \, df_1 \, df_2 \tag{7.60}$$

To make the summation limits of Eq. (7.59) finite, and to improve the convergence of the truncated series at points of discontinuity of $D(e^{j2\pi f_1}, e^{j2\pi f_2})$, the Fourier series coefficients are multiplied by a finite series of window coefficients $w(n_1, n_2)$. Thus the final approximation to $D(e^{j2\pi f_1}, e^{j2\pi f_2})$ is obtained as

$$\hat{D}(e^{j2\pi f_1}, e^{j2\pi f_2}) = \sum_{n_1=-[(N_1-1)/2]}^{[(N_1-1)/2]} \sum_{n_2=-[(N_2-1)/2]}^{[(N_2-1)/2]} \hat{h}(n_1, n_2)e^{-j2\pi n_1 f_1}e^{-j2\pi n_2 f_2} \tag{7.61}$$

where

$$\hat{h}(n_1, n_2) = h(n_1, n_2)w(n_1, n_2) \tag{7.62}$$

It is easy to show that $\hat{D}(e^{j2\pi f_1}, e^{j2\pi f_2})$ is related to $D(e^{j2\pi f_1}, e^{j2\pi f_2})$ by the formula

$$\hat{D}(e^{j2\pi f_1}, e^{j2\pi f_2}) = D(e^{j2\pi f_1}, e^{j2\pi f_2}) * W(e^{j2\pi f_1}, e^{j2\pi f_2}) \tag{7.63}$$

where $*$ denotes circular convolution and

$$W(e^{j2\pi f_1}, e^{j2\pi f_2}) = \sum_{n_1=-[(N_1-1)/2]}^{[(N_1-1)/2]} \sum_{n_2=-[(N_2-1)/2]}^{[(N_2-1)/2]} w(n_1, n_2)e^{-j2\pi n_1 f_1}e^{-j2\pi n_2 f_2} \tag{7.64}$$

i.e., $W(e^{j2\pi f_1}, e^{j2\pi f_2})$ is the Fourier transform of $w(n_1, n_2)$. Equation (7.63) shows the filter frequency response is the desired response smeared by the response of the window.

The task for windowing is to choose an appropriate window function $w(n_1, n_2)$. The desired properties of the transform of the window are

1. $W(e^{j2\pi f_1}, e^{j\pi 2 f_2})$ should approximate a circularly symmetric function.
2. The volume under the main lobe of $W(e^{j2\pi f_1}, e^{j2\pi f_2})$ should be large.
3. The volume under the side lobes should be small.

It has been shown by Huang that good two-dimensional windows satisfying the properties above can be obtained from good one-dimensional windows via the relation

$$w(n_1, n_2) = \hat{w}(\sqrt{n_1^2 + n_2^2}) \tag{7.65}$$

where $\hat{w}$ is an appropriate one-dimensional (continuous) window sampled at the appropriate values. Thus, using Eq. (7.65), one can design approximations to circularly symmetric rectangular, Hamming, and Kaiser windows, for example.

7.16 Example of Window Design of a Lowpass Filter

This section will present a simple example of the design of a circular symmetric lowpass filter using both a rectangular window and a Kaiser window with $\beta = 5$. (The parameter β was defined in Chapter 3 on page 94).

Consider the frequency response of a circularly symmetric lowpass filter defined by

$$D(e^{j2\pi f_1}, e^{j2\pi f_2}) = \begin{cases} 1.0 & \sqrt{f_1^2 + f_2^2} \le 0.4 \\ 0.0 & \text{otherwise} \end{cases} \tag{7.66}$$

Direct application of Eq. (7.60) yields for $h(n_1, n_2)$

$$h(n_1, n_2) = \frac{0.8\pi J_1(0.8\pi\sqrt{n_1^2 + n_2^2})}{\sqrt{n_1^2 + n_2^2}} \quad \begin{matrix} -\infty \le n_1 \le \infty \\ -\infty \le n_2 \le \infty \end{matrix} \tag{7.67}$$

as shown in Example 3 on p. 445. [$J_1(x)$ is the first-order Bessel function.] A rectangular window is used to truncate the Fourier series. The rectangular window is a sequence defined by

$$w(n_1, n_2) = \begin{cases} 1 & \sqrt{n_1^2 + n_2^2} \le 14 \\ 0 & \text{otherwise} \end{cases} \tag{7.68}$$

A perspective plot of this window is given in Fig. 7.10(a). A perspective plot of $W(e^{j2\pi f_1}, e^{j2\pi f_2})$, the transform of this window, is given in Fig. 7.10(b). The large amount of overshoot in this window is apparent. Figure 7.11 shows the product $h(n_1, n_2) \cdot w(n_1, n_2)$. This sequence is the filter impulse response. The two perspective plots of Fig. 7.12(a) and (b) show the frequency responses of the resulting filter on linear and log scales. The large amount of overshoot and ripple in the region of the discontinuity is immediately apparent.

If a better window is used, the overshoot and ripple in the filter response can be significantly reduced. The next series of figures shows the results

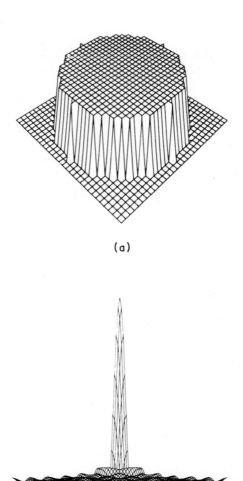

(a)

(b)

Fig. 7.10 Time and frequency responses of a two-dimensional rectangular window.

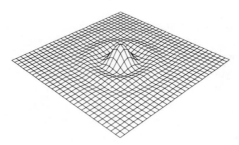

Fig. 7.11 The product of a rectangular window and an ideal lowpass filter response.

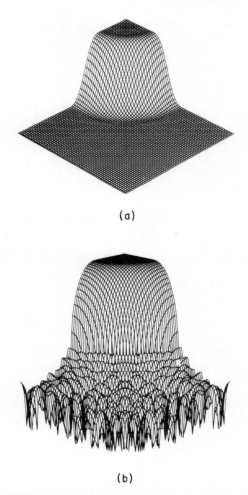

(a)

(b)

Fig. 7.14 Linear and log magnitude frequency responses of Kaiser window lowpass filter.

in filter response is dramatic over Fig. 7.12. Finally Fig. 7.15 shows a contour plot of the filter response obtained using the Kaiser window in the region $0 \leq f_1 \leq 0.5,\ 0 \leq f_2 \leq 0.5$. This figure shows the remarkable degree to which circular symmetry has been approximated with this window.

7.17 Frequency Sampling Filters

The basis for design of two-dimensional frequency sampling filters is the DFT, IDFT relations discussed earlier. Consider a filter with finite impulse

Fig. 7.15 Contour plot of Kaiser window lowpass filter.

$h(n_1, n_2)$ defined from $0 \le n_1 \le N_1 - 1$, $0 \le n_2 \le N_2 - 1$ with z transform

$$H(z_1, z_2) = \sum_{n_1=0}^{N_1-1} \sum_{n_2=0}^{N_2-1} h(n_1, n_2) z_1^{-n_1} z_2^{-n_2} \qquad (7.70)$$

The DFT relations for the filter may be obtained by evaluating Eq. (7.70) at the discrete set of values

$$\omega_{k_1} = k_1 \left(\frac{2\pi}{N_1} \right) \qquad k_1 = 0, 1, \ldots, N_1 - 1 \qquad (7.71)$$

$$\omega_{k_2} = k_2 \left(\frac{2\pi}{N_2} \right) \qquad k_2 = 0, 1, \ldots, N_2 - 1 \qquad (7.72)$$

giving

$$H(k_1, k_2) = \sum_{n_1=0}^{N_1-1} \sum_{n_2=0}^{N_2-1} h(n_1, n_2) \exp \left[-j2\pi \left(\frac{k_1 n_1}{N_1} + \frac{k_2 n_2}{N_2} \right) \right] \qquad (7.73)$$

The inverse DFT is readily obtained as

$$h(n_1, n_2) = \frac{1}{N_1 N_2} \sum_{k_1=0}^{N_1-1} \sum_{k_2=0}^{N_2-1} H(k_1, k_2) \exp \left[j2\pi \left(\frac{k_1 n_1}{N_1} + \frac{k_2 n_2}{N_2} \right) \right] \qquad (7.74)$$

A frequency interpolation formula from the DFT coefficients may be derived by inserting Eq. (7.74) into Eq. (7.70) giving

$$H(e^{j\omega_1}, e^{j\omega_2}) = \sum_{n_1=0}^{N_1-1} \sum_{n_2=0}^{N_2-1} \left\{ \frac{1}{N_1 N_2} \sum_{k_1=0}^{N_1-1} \sum_{k_2=0}^{N_2-1} H(k_1, k_2) \exp \left[j2\pi \left(\frac{k_1 n_1}{N_1} + \frac{k_2 n_2}{N_2} \right) \right] \right\}$$

$$\times \exp \left[-j(\omega_1 n_1 + \omega_2 n_2) \right] \qquad (7.75)$$

Interchanging orders of summation and summing over the n_1 and n_2 indices gives

$$H(e^{j\omega_1}, e^{j\omega_2}) = \sum_{k_1=0}^{N_1-1} \sum_{k_2=0}^{N_2-1} H(k_1, k_2) A(k_1, k_2, \omega_1, \omega_2) \tag{7.76}$$

where

$$A(k_1, k_2, \omega_1, \omega_2) = \frac{1}{N_1 N_2} \left(\frac{1 - \exp(-jN_1\omega_1)}{1 - \exp\{j[2\pi(k_1/N_1) - \omega_1]\}} \right)$$

$$\times \left(\frac{1 - \exp(-jN_2\omega_2)}{1 - \exp\{j[2\pi(k_2/N_2) - \omega_2]\}} \right) \tag{7.77}$$

Equation (7.76) serves as the basis for designing frequency sampling two-dimensional filters. As seen from Eq. (7.76), the continuous frequency response of the filter is a linear combination of shifted interpolating functions $[A(k_1, k_2, \omega_1, \omega_2)]$ weighted by the DFT coefficients $H(k_1, k_2)$. The DFT coefficients are called frequency samples as they exactly specify the value of the frequency response at uniformly spaced frequencies. For designing frequency sampling filters the majority of the frequency samples are given specific values depending on the frequency response being approximated. The remaining unspecified frequency samples are left as free variables to be optimized according to some minimization criterion.

For the specific case of linear phase frequency sampling filters, Eqs. (7.76) and (7.77) can be modified through use of the symmetry relations on the DFT coefficients, which are of the form

$$H(k_1, k_2) = |H(k_1, k_2)| e^{j\theta(k_1, k_2)} \tag{7.78}$$

where

$$|H(k_1, k_2)| = |H(k_1, N_2 - k_2)| = |H(N_1 - k_1, k_2)| \tag{7.79}$$

$$\theta(k_1, k_2) = \theta(k_1) + \theta(k_2) \tag{7.80}$$

$$\theta(k_1) = \begin{cases} -\dfrac{2\pi}{N_1} k_1 \left(\dfrac{N_1 - 1}{2} \right) & k_1 = 0, 1, \ldots, NU \\[3mm] \dfrac{2\pi}{N_1} (N_1 - k_1) \left(\dfrac{N_1 - 1}{2} \right) & k_1 = NU + 1, \ldots, N_1 - 1 \end{cases} \tag{7.81}$$

$$\theta(k_2) = \begin{cases} -\dfrac{2\pi}{N_2} k_2 \left(\dfrac{N_2 - 1}{2} \right) & k_2 = 0, 1, \ldots, NV \\[3mm] \dfrac{2\pi}{N_2} (N_2 - k_2) \left(\dfrac{N_2 - 1}{2} \right) & k_2 = NV + 1, \ldots, N_2 - 1 \end{cases} \tag{7.82}$$

and

$$NU = \begin{cases} \dfrac{N_1}{2} & N_1 \text{ even} \\[2ex] \dfrac{N_1 - 1}{2} & N_1 \text{ odd} \end{cases} \tag{7.83}$$

$$NV = \begin{cases} \dfrac{N_2}{2} & N_2 \text{ even} \\[2ex] \dfrac{N_2 - 1}{2} & N_2 \text{ odd} \end{cases} \tag{7.84}$$

Whenever N_1 or N_2 is even, the additional constraints

$$\theta\left(k_1, \frac{N_2}{2}\right) = \theta\left(\frac{N_1}{2}, k_2\right) = 0 \tag{7.85}$$

$$H\left(k_1, \frac{N_2}{2}\right) = H\left(\frac{N_1}{2}, k_2\right) = 0 \tag{7.86}$$

must be maintained.

Applying the constraints above into Eqs. (7.76) and (7.77) yields the following equations (after considerable arithmetic manipulations):

$$
\begin{aligned}
H(e^{j\omega_1}, e^{j\omega_2}) = {} & \exp\left\{-j\left[\omega_1\left(\frac{N_1 - 1}{2}\right) + \omega_2\left(\frac{N_2 - 1}{2}\right)\right]\right\} \cdot \frac{1}{N_1 N_2} \\[1ex]
& \times \Bigg[H(0, 0)\alpha(\omega_1, N_1)\alpha(\omega_2, N_2) \\[1ex]
& + \sum_{k_1=1}^{NU} |H(k_1, 0)|\, \alpha(\omega_2, N_2)\beta(\omega_1, k_1, N_1) \\[1ex]
& + \sum_{k_2=1}^{NV} |H(0, k_2)|\, \alpha(\omega_1, N_1)\beta(\omega_2, k_2, N_2) \\[1ex]
& + \sum_{k_1=1}^{NU}\sum_{k_2=1}^{NV} |H(k_1, k_2)|\, \beta(\omega_1, k_1, N_1)\beta(\omega_2, k_2, N_2) \Bigg]
\end{aligned} \tag{7.87}
$$

where

$$\alpha(\omega, N) = \frac{\sin(\omega N/2)}{\sin(\omega/2)} \tag{7.88}$$

$$\beta(\omega, k, N) = \frac{\sin\left[(\omega/2 - \pi k/N)N\right]}{\sin(\omega/2 - \pi k/N)} + \frac{\sin\left[(\omega/2 + \pi k/N)N\right]}{\sin(\omega/2 + \pi k/N)} \tag{7.89}$$

$$
NU = \begin{cases} \dfrac{N_1}{2} - 1 & N_1 \text{ even} \\[2ex] \dfrac{N_1 - 1}{2} & N_1 \text{ odd} \end{cases} \tag{7.90}
$$

$$
NV = \begin{cases} \dfrac{N_2}{2} - 1 & N_2 \text{ even} \\[2ex] \dfrac{N_2 - 1}{2} & N_2 \text{ odd} \end{cases} \tag{7.91}
$$

Equation (7.87), although cumbersome in appearance, is seen to consist basically of a sum of simple interpolating functions that are translated in frequency depending on k_1 and k_2. As such, Eq. (7.87) (without the leading linear phase terms) is the basic design equation for frequency sampling filters.

Figure 7.16 shows the desired set of frequency bands in the (ω_1, ω_2) plane for a circularly symmetric lowpass filter. The passband is the region where

$$
\rho(\omega_1, \omega_2) = (\omega_1^2 + \omega_2^2)^{1/2} \le R_1 \tag{7.92}
$$

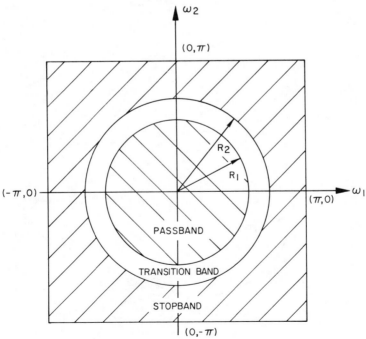

Fig. 7.16 Frequency bands for circularly symmetric lowpass filter.

The transition band is the region where

$$R_1 < \rho(\omega_1, \omega_2) < R_2 \tag{7.93}$$

and the stopband is the region where

$$\rho(\omega_1, \omega_2) \geq R_2 \tag{7.94}$$

With respect to this figure the DFT points, or frequency samples, are located on an $(N_1 \times N_2)$ grid of points in the (ω_1, ω_2) plane as shown in Fig. 7.17 for $N_1 = 9$, $N_2 = 9$. To design a frequency sampling approximation to the ideal circularly symmetric response

$$\hat{H}(e^{j\omega_1}, e^{j\omega_2}) = \begin{cases} 1 & \rho(\omega_1, \omega_2) \leq R_1 \\ 0 & \rho(\omega_1, \omega_2) \geq R_2 \end{cases} \tag{7.95}$$

the DFT values that occur inside the pa sband are given a value of 1.0; those that occur in the stopband are given a value 0.0; those that lie in the transition band are free variables whose value will be chosen by a minimization criterion.

To simplify both the equations and the computation involved in performing the desired minimizations, the following assumptions will be used. First we assume $N_1 = N_2 =$ odd integer. Next we assume

$$H(k_1, k_2) = H(k_2, k_1) \tag{7.96}$$

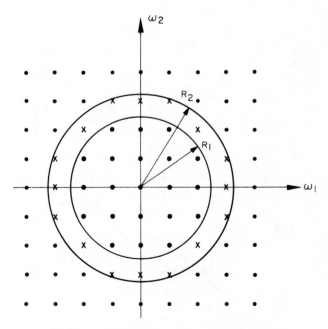

Fig. 7.17 Grid points for DFT of a (9 × 9) filter.

i.e., there is symmetry across the 45° diagonals in the (ω_1, ω_2) plane. This is a reasonable assumption for designing circularly symmetric filters, as in the ideal case of zero approximation error, Eq. (7.96) is valid. Finally we assume that the linear phase term in Eq. (7.87) can be disregarded in terms of the design procedure.

With these assumptions in mind, Eq. (7.87) can be simplified to the form

$$H(e^{j\omega_1}, e^{j\omega_2}) = H_F(\omega_1, \omega_2) + \sum_{m=1}^{M} T_m H_m(\omega_1, \omega_2) \qquad (7.97)$$

where $H_F(\omega_1, \omega_2)$ represents the contribution of the fixed DFT coefficients (the 1.0's), T_m represents the amplitude of the mth DFT coefficient in the transition band, $H_m(\omega_1, \omega_2)$ represents the interpolation function appropriate to the mth transition coefficient, and M is the total number of transition coefficients. The pertinent design equations can now be put in the form

$$1 - \alpha\delta \leq H(e^{j\omega_1}, e^{j\omega_2}) \leq 1 + \alpha\delta \qquad \text{for} \qquad \rho(\omega_1, \omega_2) \leq R_1 \quad (7.98)$$

$$-\delta \leq H(e^{j\omega_1}, e^{j\omega_2}) \leq \delta \qquad \text{for} \qquad \rho(\omega_1, \omega_2) \geq R_2 \quad (7.99)$$

where δ represents the peak approximation error in the stopband and $\alpha\delta$ (α any constant) represents the peak approximation error in the passband. The design goal is to choose the transition samples so as to minimize δ.

The problem above is readily seen to be a linear programming problem by evaluating Eqs. (7.98) and (7.99) at a dense set of points in the passband and stopband and by writing the equations explicitly as linear inequalities in the form

$$\left.\begin{aligned} \sum_{m=1}^{M} T_m H_m(\omega_1, \omega_2) - \alpha\delta \\ \leq 1 - H_F(\omega_1, \omega_2) \\ -\sum_{m=1}^{M} T_m H_m(\omega_1, \omega_2) - \alpha\delta \\ \leq -1 + H_F(\omega_1, \omega_2) \end{aligned}\right\} \quad \text{for all points with } \rho(\omega_1, \omega_2) \leq R_1 \quad (7.100)$$

$$\left.\begin{aligned} \sum_{m=1}^{M} T_m H_m(\omega_1, \omega_2) - \delta \\ \leq -H_F(\omega_1, \omega_2) \\ -\sum_{m=1}^{M} T_m H_m(\omega_1, \omega_2) - \delta \\ \leq H_F(\omega_1, \omega_1) \end{aligned}\right\} \quad \text{for all points with } \rho(\omega_1, \omega_2) \geq R_2 \quad (7.101)$$

Thus the minimization criterion is to chose $(T_1, T_2, \ldots, T_M)$ so as to minimize δ.

7.18   Two-Dimensional Frequency Sampling Lowpass Filters

To illustrate the application of the ideas of the preceding sections, the next set of figures shows the frequency responses of several lowpass filters designed using frequency sampling techniques. Figures 7.18 and 7.19 show linear and log magnitude responses of two lowpass filters with the parameters $N_1 = N_2 = N = 25$ and $R_1 = 5\pi/12.5$, $R_2 = 7\pi/12.5$ for Fig. 7.18; $R_1 = 3\pi/12.5$,

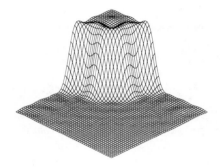

$R_1 = 5\,\pi\,/\,12.5$

$R_2 = 7\,\pi\,/\,12.5$

$N_1 = N_2 = 25$

AMPLITUDE   RESPONSE

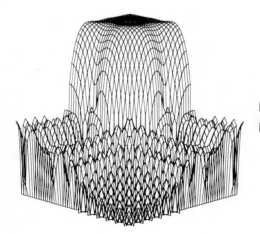

LOG   MAGNITUDE   RESPONSE

PEAK   ATTENUATION $= -41.8$ dB

Fig. 7.18 Linear and log magnitude responses of one frequency sampling lowpass filter.

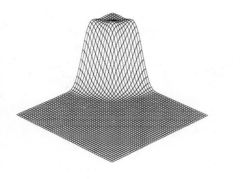

$R_1 = 3\pi / 12.5$
$R_2 = 6\pi / 12.5$
$N_1 = N_2 = 25$

AMPLITUDE RESPONSE

LOG MAGNITUDE RESPONSE
PEAK ATTENUATION = $-69.9\,$dB

Fig. 7.19 Linear and log magnitude responses of a second frequency sampling lowpass filter.

$R_2 = 6\pi/12.5$ for Fig. 7.19. For both these examples, $\alpha = 1000$, i.e., the in-band ripple was essentially left unconstrained. The minimum stopband attenuations were 41.8 dB for the filter of Fig. 7.18 and 69.9 dB for the filter of Fig. 7.19.

Figure 7.20 illustrates the effects of varying α (the ratio of peak in-band to peak out-of-band ripple) on the passband ripple and stopband attenuation. A surprising observation from this figure is that as α varies from 50 to 5, the peak passband ripple is reduced by a factor of 6; whereas the minimum stopband attenuation changes by only about 0.8 dB. On the basis of this result, it would seem reasonable to set α as close to the knee of the curve as possible.

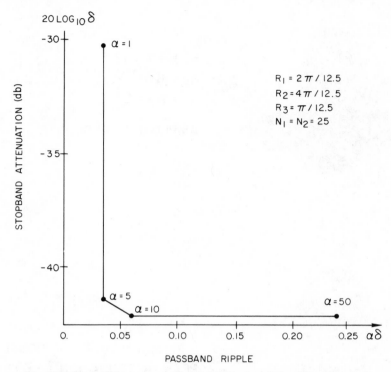

Fig. 7.20 Tradeoff between stopband attenuation and passband ripple for frequency sampling lowpass filters.

Finally, Figs. 7.21 and 7.22 illustrate the frequency response for the lowpass filter of Fig. 7.20 with $\alpha = 1$, as well as a contour plot for the passband and transition band of this filter. Figure 7.22 shows that the filter response (at least for $\alpha = 1$) bears a reasonable approximation to the desired circular symmetry.

7.19 Optimal (Minimax Error) Two-Dimensional Filter Design

As done in the case of one-dimensional filters, two-dimensional FIR filters can be designed that are optimal in the Chebyshev (minimax) sense. In such cases all the impulse response coefficients or, equivalently, all the DFT coefficients are treated as unknowns and are solved for by the optimization routine. Linear programming techniques may be used to solve for the resulting filter coefficients. Other one-dimensional algorithms for obtaining the Chebyshev solutions are not valid because the alternation theorem cannot be readily applied in two dimensions. (In Sec. 7.20 we discuss a procedure for

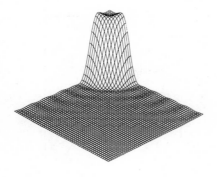

$R_1 = 2\pi / 12.5$

$R_2 = 4\pi / 12.5$

$R = \pi / 12.5$

$\alpha = 1$

$N_1 = N_2 = 25$

AMPLITUDE RESPONSE

LOG MAGNITUDE RESPONSE

IN-BAND RIPPLE = 0.03

PEAK ATTENUATION = – 30 dB

Fig. 7.21 Linear and log magnitude responses of a frequency sampling lowpass filter with a constrained passband.

transforming optimal one-dimensional filters to two-dimensional filters that are often optimal in two dimensions.)

Direct design of optimal two-dimensional filters is complicated by the number of variables and constraint equations that must be solved to obtain a reasonable solution. Thus a filter with an impulse response duration of (9 × 9) samples may involve thousands of constraint equations and up to 21 variables, thereby requiring on the order of 1 hour of computation time on a moderately high-speed computer. Thus the highest-order two-dimensional optimal filter that has been designed is on the order of a (9 × 9) filter. Figure 7.23 shows linear and log magnitude responses of one such optimal filter; Fig. 7.24 shows a contour plot of the magnitude response. As seen in Fig. 7.23, the peak magnitude error is equiripple—similar to the case of optimal one-dimensional filters. Figure 7.24 shows that in the passband and transition band the contours are approximately circles, as desired. In the

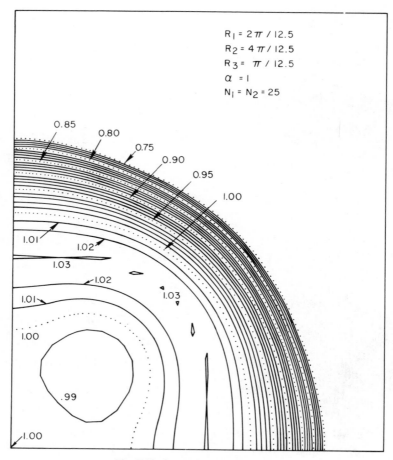

Fig. 7.22 Contour plot of Fig. 7.21.

stopband the contours are no longer circles, due to the nature of the rectangular sampling grid.

7.20   Frequency Transformations from One to Two Dimensions

As seen in the preceding sections, because of the nature of the design problem for two-dimensional filters, it is very difficult to design good approximations to most of the two-dimensional frequency responses of interest—especially so when one is interested in reasonably high-order filters. In this section we present a technique for mapping an optimal one-dimensional FIR filter into a two-dimensional FIR filter that, in many cases, preserves the optimality criterion.

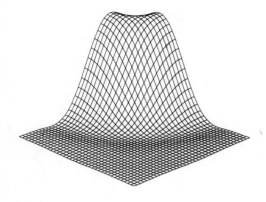

$R_1 = 1.5\,\pi/4.5$

$R_2 = 3\,\pi/4.5$

$N_1 = N_2 = 9$

AMPLITUDE RESPONSE

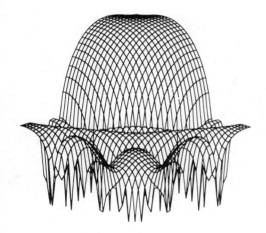

LOG MAGNITUDE RESPONSE
IN-BAND RIPPLE = 0.08
PEAK ATTENUATION = 32.5 dB

Fig. 7.23 Linear and log magnitude responses of a minimax (optimum) lowpass filter.

If we write the frequency response of a two-dimensional, linear phase, FIR filter in the form (assuming N_1 and N_2 are odd)

$$H(e^{j2\pi f_1}, e^{j2\pi f_2}) = e^{-j2\pi\{[(N_1-1)/2]f_1 + [(N_2-1)/2]f_2\}}$$

$$\times \sum_{n_1=0}^{(N_1-1)/2} \sum_{n_2=0}^{(N_2-1)/2} a(n_1, n_2) \cos(2\pi f_1 n_1) \cos(2\pi n_2 f_2) \quad (7.102)$$

then a real function $\hat{H}(e^{j2\pi_1}, e^{j2\pi f_2})$ can be defined as

$$\hat{H}(e^{j2\pi f_1}, e^{j2\pi f_2}) = H(e^{j2\pi f_1}, e^{j2\pi f_2})e^{j2\pi\{[(N_1-1)/2]f_1 + [(N_2-1)/2]f_2\}} \quad (7.103)$$

$$= \sum_{n_1=0}^{(N_1-1)/2} \sum_{n_2=0}^{(N_2-1)/2} a(n_1, n_2) \cos(2\pi f_1 n_1) \cos(2\pi f_2 n_2) \quad (7.104)$$

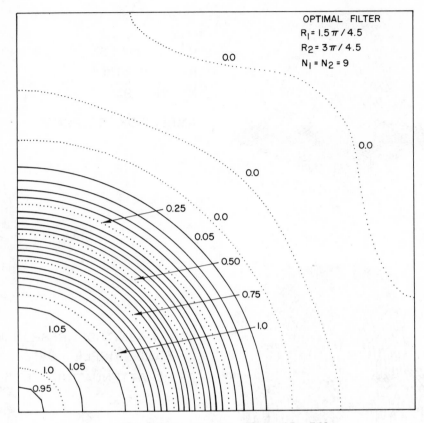

Fig. 7.24 Contour plot of filter of Fig. 7.23.

where $\{a(n_1, n_2\}$ is related to $\{h(n_1, n_2)\}$, the filter impulse response, by

$$a(0, 0) = h\left(\frac{N_1 - 1}{2}, \frac{N_2 - 1}{2}\right)$$

$$a(0, n_2) = 2h\left(\frac{N_1 - 1}{2}, \frac{N_2 - 1}{2} + n_2\right) \qquad 1 \leq n_2 \leq \frac{N_2 - 1}{2}$$

$$a(n_1, 0) = 2h\left(\frac{N_1 - 1}{2} + n_1, \frac{N_2 - 1}{2}\right) \qquad 1 \leq n_1 \leq \frac{N_1 - 1}{2}$$

$$a(n_1, n_2) = 4h\left(\frac{N_1 - 1}{2} + n_1, \frac{N_2 - 1}{2} + n_2\right) \qquad \begin{cases} 1 \leq n_1 \leq \dfrac{N_1 - 1}{2} \\[2mm] 1 \leq n_2 \leq \dfrac{N_2 - 1}{2} \end{cases}$$

In Chapter 3 it was shown that a linear phase FIR filter with an N-sample impulse response had the frequency response

$$G(e^{j2\pi f}) = e^{-j2\pi[(N-1)/2]f}\hat{G}(e^{j2\pi f}) \tag{7.105}$$

$$\hat{G}(e^{j2\pi f}) = \sum_{n=0}^{(N-1)/2} b(n)\cos(2\pi fn) \tag{7.106}$$

$$= \sum_{n=0}^{(N-1)/2} \hat{b}(n)[\cos(2\pi f)]^{n\cdot} \tag{7.107}$$

Equation (7.106) gives the frequency response of the filter as a trigonometric polynomial in f, whereas Eq. (7.107) expresses the frequency response as an ordinary polynomial in $\cos(2\pi f)$. By making the substitution (due to McClellan)

$$\cos(2\pi f) = A\cos(2\pi f_1) + B\cos(2\pi f_2) + C\cos(2\pi f_1)\cos(2\pi f_2) + D \tag{7.108}$$

Eq. (7.107) transforms to

$$\sum_{n=0}^{(N-1)/2} \hat{b}(n)\cos(2\pi f)^n \to \sum_{n_1=0}^{(N-1)/2}\sum_{n_2=0}^{(N-1)/2} \hat{a}(n_1, n_2)\cos(2\pi f_1)^{n_1}\cos(2\pi f_2)^{n_2} \tag{7.109}$$

which can be put in the form

$$\sum_{n_1=0}^{(N-1)/2}\sum_{n_2=0}^{(N-1)/2} a(n_1, n_2)\cos(2\pi f_1 n_1)\cos(2\pi f_2 n_2) \tag{7.110}$$

which is the desired form for a linear phase FIR filter as shown in Eq. (7.103). By letting A, B, C, D assume the values

$$A = B = C = -D = 0.5 \tag{7.111}$$

the mapping of Eq. (7.108) gives (solving for f_2 in terms of f and f_1)

$$f_2 = \frac{1}{2\pi}\cos^{-1}\left[\frac{\cos(2\pi f) + 0.5 - 0.5\cos(2\pi f_1)}{0.5 + 0.5\cos(2\pi f_1)}\right] \tag{7.112}$$

Figure 7.25 shows how the region $0 \leq f \leq 0.5$ maps to the (f_1, f_2) plane according to Eq. (7.112). For each value of f, there is a contour in the (f_1, f_2) plane along which the transformed two-dimensional frequency response is a constant equal to the value of the one-dimensional frequency response at frequency f. As f varies, a family of contours is generated (as shown in Fig. 7.25) that completely describes the transformed two-dimensional frequency response. Figure 7.25 shows that the contour lines distribute with approximately circular symmetry and hence are appropriate for the design of circularly symmetric filters.

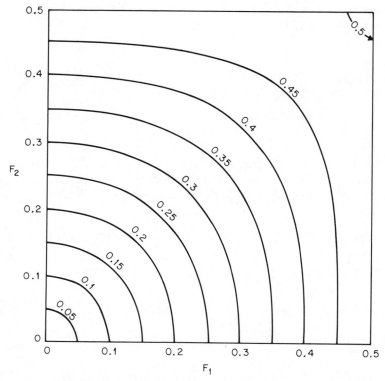

Fig. 7.25 The mapping in frequency from one to two dimensions (After McClellan).

To illustrate the method above, Figs. 7.26 to 7.28 show magnitude and contour responses of a low pass and a bandpass circularly symmetric filter. The magnitude response shown in Fig. 7.26 is for a circularly symmetric (9×9) lowpass filter with the passband and stopband cutoff frequencies of 0.1666 and 0.3333, respectively. Figure 7.27 shows a comparison of the cutoff frequency contours for the filter of Fig. 7.26 and a similar filter designed by linear programming techniques. As seen in this figure, the passband and stopband edges of the two filters are slightly different. In all other aspects (i.e., passband and stopband ripples), however, these filters are the same.

Figure 7.28 shows the magnitude response of a circularly symmetric (31×31) bandpass filter. One great advantage of this transformation method over standard two-dimensional designs is that filters with long impulse responses are readily obtained since the approximation problem need only be solved in one dimension. Furthermore it can be shown that, using the transformation of Eq. (7.108) with variables defined by Eq. (7.111), the

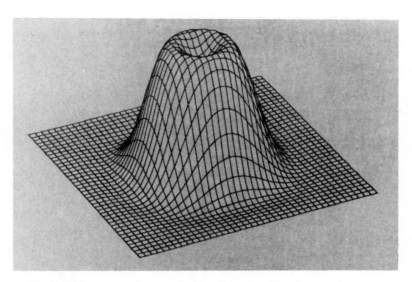

Fig. 7.26 Frequency response of a (9 × 9) lowpass filter (After McClellan).

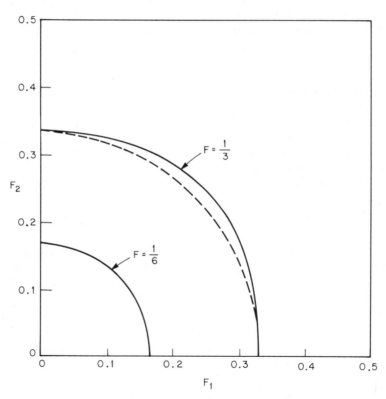

Fig. 7.27 Comparison of contours for two lowpass filters designed by independent methods (After McClellan).

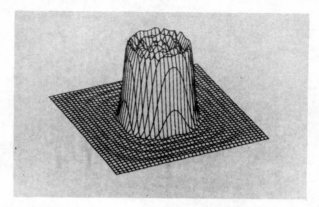

Fig. 7.28 Frequency response of a (31 × 31) bandpass filter (After McClellan).

resulting two-dimensional approximation is optimal in the Chebyshev sense. Thus the transformation technique appears to be the most useful design method for two-dimensional digital filters and can be applied in a variety of design problems.

7.21 Some Examples of Picture Processing

In this section we give two examples of the application of two-dimensional signal processing. The first example is primarily academic in that it shows the effect of lowpass and highpass filtering on a picture. The second example shows how nonlinear techniques can be used to enhance a picture.

Figures 7.29–7.31 show an original photograph, a lowpass filtered version of it, and a highpass filtered version of it. (These pictures were graciously supplied to the authors by Professor Leon Harmon and Alan Strelzoff of Case Western Reserve University.) The lowpass filter was a transformed two-dimensional frequency sampling filter (128 × 128 point impulse response) with a normalized cut off frequency of approximately 0.04 in both frequency dimensions. The highpass filter used was the complement of the lowpass filter. The pictures were made from an array of 384 × 384 points with 4 bits of intensity at each sample point. The lowpass filtered picture (Fig. 7.30) retains most of the picture information but lacks the sharp detail where the original picture changes intensity sharply. Conversely the highpass filtered picture (Fig. 7.31) preserves sharp contrasts, but loses the information within the constant intensity regions.

The next set of figures shows how nonlinear processing can enhance an image (these pictures were graciously supplied to the authors by Professor Thomas Stockham, University of Utah). Figure 7.32 shows an original picture

Fig. 7.29 An original photograph.

Fig. 7.30 A lowpass filtered version of Fig. 7.29.

Fig. 7.31 A highpass filtered version of Fig. 7.29.

Fig. 7.32 An original photograph.

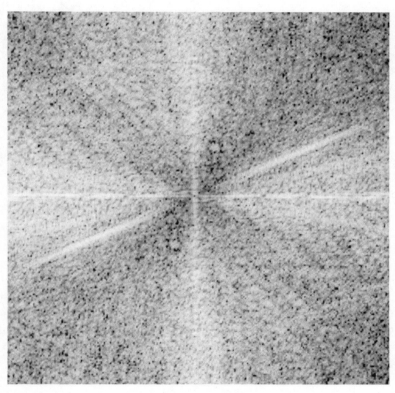

Fig. 7.33 The log spectrum of the picture in Fig. 7.32 after processing by a whitening filter.

Fig. 7.34 Nonlinearly processed and enhanced version of Fig. 7.32.

of a man inside a building. Figure 7.33 shows the spectral content of the log of the picture transform after it has been processed by a whitening filter. The high frequency components of the log images are clearly evident in this figure. Finally, Fig. 7.34 shows the result of processing the picture with a multiplicative system to reduce reflection and increase illumination, thereby greatly enhancing the picture. The multiplicative system used in the processing was a homomorphic system. We will discuss such systems in Chapter 12 in the context of nonlinear speech processing.

REFERENCES

1. S. TREITEL, J. L. SHANKS, AND C. W. FRASIER, "Some Aspects of Fan Filtering," *Geophysics*, **32**, No. 5, 789–800, Oct., 1967.
2. J. L. SHANKS, "Two Planar Digital Filtering Algorithms," *Proc. 5th Ann. Princeton Conf. on Information Sciences and Systems*, 48–53, Mar., 1971.
3. J. L. SHANKS, S. TREITEL, AND J. H. JUSTICE, "Stability and Synthesis of Two-Dimensional Recursive Filters," *IEEE Trans. on Audio and Electroacoustics*, **AU-20**, No. 2, 115–128, June, 1972.

4. T. S. HUANG, "Stability of Two-Dimensional Recursive Filters," *IEEE Trans. on Audio and Electroacoustics*, **AU-20,** No. 2, 158–163, June, 1972.
5. H. ANDREWS, A. TESCHER, AND R. KRUGER, "Image Processing by Digital Computer," *IEEE Spectrum*, **9,** No. 7, 20–32, July, 1972.
6. J. V. HU AND L. R. RABINER, "Design Techniques for Two-Dimensional Digital Filters," *IEEE Trans. on Audio and Electroacoustics*, **AU-20,** No. 4, 249–257, Oct., 1972.
7. T. S. HUANG, "Two-Dimensional Windows," *IEEE Trans. on Audio and Electroacoustics*, **AU-20,** No. 1, 88–89, Mar., 1972.
8. B. D. ANDERSON AND E. I. JURY, "Stability Test for Two-Dimensional Recursive Filters," *IEEE Trans. on Audio and Electroacoustics*, **AU-21,** No. 4, 366–372, Aug., 1973.
9. J. H. MCCLELLAN AND T. W. PARKS, "Equiripple Approximation of Fan Filters," *Geophysics*, **37,** No. 4, 573–583, Aug., 1972.
10. J. H. MCCLELLAN, "On the Design of One-Dimensional and Two-Dimensional FIR Digital Filters," Ph.D. Thesis, Rice Institute, Houston, Texas, Apr., 1973.
11. A. V. OPPENHEIM, R. W. SCHAFER, AND T. G. STOCKHAM, "Nonlinear Filtering of Multiplied and Convolved Signals," *Proc. IEEE*, Vol. 56, 1264–1291, Aug., 1968.

8

Introduction to Digital Hardware

8.1 Introduction

We begin by restating one of the basic aims of this book, which is to help the engineer to specify, design, and implement algorithms oriented toward digital signal processing tasks. By *implementation* we usually mean that the algorithm is either programmed on an existing general-purpose computer or is realized as a piece of digital hardware. Also included are situations that may call for the design of a *new* general-purpose computer plus the programs needed to implement the algorithms.

In principle, one might argue that writing a program and designing hardware are similar; at the same time, any practicing engineer realizes the many important differences between the two disciplines. Let us first discuss similarities. A general-purpose computer is, after all, a piece of hardware that some other engineer designed, while a special-purpose digital hardware device can be thought of as a computer with a wired program (this is precisely the control function in a piece of digital hardware). Thus, in either case, one is dealing with *computers* that realize the desired algorithms through *programs*. We believe that keeping this fundamental similarity in mind will help the reader comprehend the next few hardware-oriented chapters.

On the other hand, the design of a hardware device requires the learning of a whole new language, the gaining of many new insights, and the acquisition of a bag of tricks not needed by programmers. Unfortunately, there is no shortcut to understanding digital hardware; what we hope to accomplish in the next few chapters is to give the reader a "feel" for digital signal processing hardware issues. Our examples are necessarily drawn from today's component technology, but the reader should keep in mind that this technology has undergone very rapid changes and undoubtedly will continue to evolve but in a rather explosive way. We might note that Maxwell's equations remain pertinent for many years, whereas vacuum tube technology is no longer of prime interest. Analogously, multiplication algorithms interest us more than the chemistry of silicon devices.

8.2 Discussion of Design Procedures for Digital Signal Processing Hardware

The authors' experiences have been in research and development and not in manufacturing, so our remarks will no doubt be colored by this fact. From this point of view, four steps in the design procedure emerge:

1. Choice of a logic family or logic families.
2. Choice of a construction methodology.
3. Development of a hardware structure.
4. Simulation of all or part of the proposed system to help determine parameters.

Choice of a logic family is similar to the decision making needed to buy a new general-purpose computer. The answer is almost never clear-cut because the job can almost always be done in a variety of ways, each having their advocates. As an example, a digital frequency synthesizer built at the Lincoln Laboratory was done more or less as an exercise; the idea seemed novel, the required time and money seemed modest, and there was an engineer who was enthusiastic about the project. At the time, this engineer was not familiar with the fastest available logic so although it would have been more elegant to demonstrate a super-fast device, we settled for the well-known and moderately fast TTL (transistor–transistor logic) line.

As another example, it was decided that a large radar digital signal processor should be built with the fastest available components. The sponsor viewed this project as a step in the evolution of even higher performance systems and an essential step in this process was attaining mastery of the construction methodology of the high-speed circuits. Also, the project was relatively well-funded so that engineering time could be spared toward advancement of high-speed techniques.

In telephony, there is need for large numbers of filters to help route signals to their correct destination. In such a case where the long-range goal is large-scale production, the approach would be to evolve custom integrated circuit packages. Thus, we can imagine an iterative procedure of breadboard performance evaluation based on available components followed by attempts to integrate the breadboard.

A given logic family has distinct properties, such as energizing voltages, signal levels, impedances, and temperature sensitivities, which usually make them incompatible with other logic families. Often these problems can be solved through appropriate interfacing hardware. Sometimes it is well worth mixing logic families; for example, a given subsystem might have very moderate speed requirements, while the rest of the system requires high-speed logic. In such a case, one might consider how to design the subsystem in question with highly integrated, low power components. In our experience, engineers have tended to stay with a single logic line in planning signal processing systems.

Given a logic family with its speed capability and power requirements allows us next to get a handle on the construction methodology. This includes consideration of the card size and card construction, back plane wiring, ground and power distribution, availability of check points, etc. Figure 8.1 shows an example of construction using printed circuit techniques. This is a card with the conducting paths on both sides etched in. One difficulty with this technique is that of fixing mistakes; thus the layout for such a card should be done with great care. Once finished, however, the printed circuit is an economical method for production since duplicates can be easily photoetched.

Figure 8.2 illustrates another technique, called *wire-wrap*. This figure illustrates a complete construction of a small but nontrivial device using 2-nsec ECL (emitter-coupled logic). Shown in this figure are the two boards for mounting the integrated circuits. One of the boards is swiveled to show the wiring. Also visible are the power supply on the right, the cooling system, and part of the bundle of interconnection wires between boards. Figure 8.3 is a rather awesome blowup of part of the board wiring. It is obvious that to build this board required either a huge amount of manual work or an enormously complicated machine. In practice there is usually a mixture of manual and automatic procedures for building and testing a board like this. Despite its complexity, wire-wrapping is a more flexible and thus usually less time-consuming job than printed circuit techniques for one-of-a-kind digital devices. Further, mistakes in wire-wrapping are easily correctable.

In Figure 8.3, notice the large number of discrete components (resistors and capacitors); these are needed primarily to avoid noise and extra propagation delay. When the transistor circuits start switching at under 1 nsec,

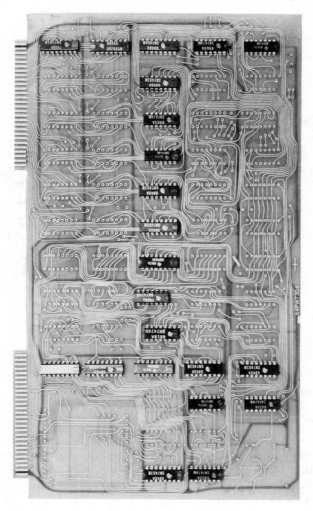

Fig. 8.1 Printed circuit card for a digital frequency synthesizer.

wire-wrap techniques are no longer manageable and printed circuits using transmission line techniques must be employed.

Working within a logic family and a specific (if tentative) construction framework makes it possible for the designer to begin evolving an appropriate structure for the hardware. For example, an airborne processor has construction and power constraints quite different from a ground-based processor. In competitive manufacturing, marketing considerations may cause engineering constraints (such as limited size, weight, and power). But even neglecting these "imposed" conditions, there is a natural desire on the part of the professional hardware designer to construct an "elegant"

Fig. 8.2 Construction using wire-wrap techniques.

system using the minimum number of components, to devise a mechanical structure within which check-out and alterations are simple, and (above all) to build a reliable piece of hardware. It is clear that the desire for excellence often runs head on into schedule and marketing constraints. We shall leave this point with the thought that digital hardware design, like most engineering, is based on the intuitive attitude of the designers to a large number of rather ill-defined tradeoffs. This seems to put engineering in the category of the "soft" sciences, such as psychology and social work. The difference, we think, is this: Once the engineer has dealt with the intuitive questions of choosing a logic family, a construction methodology, and a computer architecture as well as several items discussed below, the rest of the work is pretty clearly mapped out and the main requirements become dogged attention to masses of details, the careful planning of schedules, and the creation of important checkpoints along the road to completion of the project.

In many digital signal processing systems, considerations of register lengths, arithmetic number systems, and memory configurations will have important effects on the cost and performance of the system. Register lengths for such items as digital filters and FFT processors are best estimated through

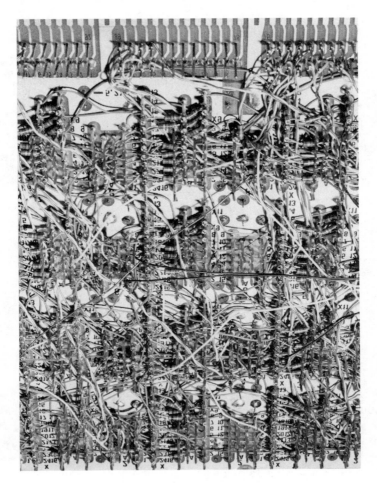

Fig. 8.3 Close-up of a wire-wrap board underside.

computer simulation of these devices. Both performance and complexity are correlated to the type of arithmetic system (such as fixed point or floating point number representation, overflow checking, and modules of the number system). A wide variety of system parameters can be adjusted through experience with a simulation program both before and, to a lesser extent, during the design and construction of the hardware.

In addition to simulation, it is also useful to design and build a few subsystems with a view toward estimating speeds, integrated circuit package count, cost, and time. At this point, specific components from manufacturers' handbooks are used and some logic designs are carried out using Boolean nomenclature. Many inexperienced hardware designers put too much

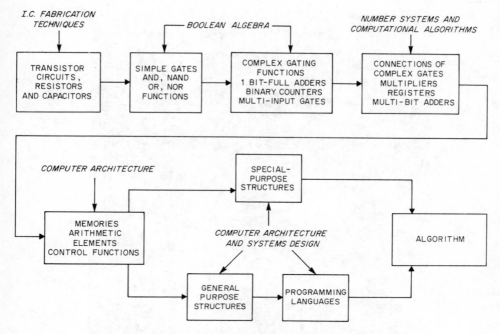

Fig. 8.4 Hierarchy of hardware leading to signal processing algorithms.

emphasis on the logic equations, just as novice programmers worry about the individual instructions of a program. In our experience these are relatively minor stages in the overall planning and execution of a complete digital signal processing design.

Figure 8.4 is a sketch of the hierarchy of circuits, subsystems, and systems that lead to an algorithm, along with a description of the disciplines needed for these steps. This sketch really applies to the development of any digital system so it is worth a short discussion of some of the major preoccupations associated with digital signal processing algorithms.

Computational speed is a very important factor in the design of digital signal processing algorithms. Increased speed may permit real-time operation, less parallelism (and thus less complexity) of the computer structure, savings of general-purpose computer facility time for production runs, and a greater degree of multiplexing. Because of the emphasis on speed, certain logic lines such as TTL (transistor–transistor logic) and ECL (emitter-coupled logic) will be of greater interest than RTL (resistor–transistor logic).

Multiplication and addition are the primary arithmetic operations of digital signal processing. In particular, the design of high-speed, compact, cheap multipliers is of direct and substantial benefit to many hardware or software implementations of signal processing algorithms.

8.3   Boolean Nomenclature; Examples of Simple Logic Nets

From a hardware point of view, the universal language of digital devices is binary arithmetic; the algebra of this arithmetic is Boolean algebra. Thus, the functional description of any algorithm can be expressed by a set of Boolean equations. A manufacturer's description of his logic family always includes the input–output relations expressed via a series of these equations. Not all manufacturers use precisely the same notation so we have arbitrarily chosen one that is popular.

All digital functions can be derived from the elementary AND, OR, and invert functions. Let x and y be two binary variables, and let u be a binary function of x and y. Then, if $u = 1$ *only* when both $x = 1$ and $y = 1$, u is defined as the AND of x and y and written as

$$u = xy \tag{8.1}$$

The complete functional relations among x, y, and u can also be expressed via a *truth table*, as follows:

x	y	u	
0	0	0	
0	1	0	(T.T.1)
1	0	0	
1	1	1	

The truth table is a tabulation that tells us what u is for every conceivable combination of x and y. Equations (8.1) and (T.T.1) are exact equivalents. Very often, an algebraic expression such as Eq. (8.1) can be derived from (T.T.1) and then reduced to simpler form using the rules of Boolean algebra.

If $u = 1$ when either $x = 1$ *or* $y = 1$, then u is called the OR of x and y, represented as

$$u = x + y \tag{8.2}$$

More complex functions can be formed. For example, let x_1, x_2, and x_3 be binary variables and define

$$u = x_1 x_2 + x_3 \tag{8.3}$$

Equation (8.3) states that if both $x_1 = 1$ *and* $x_2 = 1$ *or* if $x_3 = 1$, then $u = 1$. As an exercise, the reader should construct the truth tables corresponding to Eqs. (8.2) and (8.3).

By introducing complements, manipulation of Boolean equations can be made more flexible. Thus $\bar{x}$ represents the complement of x; if $x = 0$, $\bar{x} = 1$ and vice versa. The equations

$$u = \overline{xy}, \qquad u = \overline{x + y} \tag{8.4}$$

are called NAND (not AND) and NOR (not OR). The so-called exclusive OR (also called addition modulo 2) is defined as

$$u = x\bar{y} + \bar{x}y = x \oplus y \tag{8.5}$$

The five Boolean functions OR, NOR, AND, NAND, and exclusive OR form the nucleus of more complex logic. Figure 8.5 shows the convention we are using to represent the logic diagrams corresponding to either the Boolean equations or the truth tables. (The small circles in this and succeeding figures denote inversion.)

In Fig. 8.6 we see how AND and OR gates can be combined to produce the exclusive OR function. It is important to note that Fig. 8.6 requires *two* levels of gating; thus, we would expect the switching time of such a gate to be longer than the four elementary functions shown in Fig. 8.5. The switching time of a given Boolean function is a fundamental parameter of that logic. Delays of commercial integrated circuit packages are specified according to "typical" delay and maximum delay. These delays are also functions of temperature and fan-out (the number of gate packages driven by a circuit). Higher fan-out puts a heavier load on the circuit and increases propagation delay.

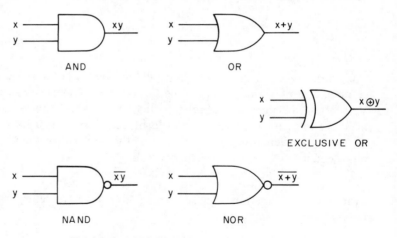

Fig. 8.5 Simple Boolean functions.

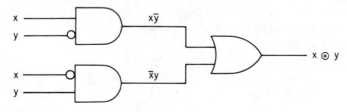

Fig. 8.6 Exclusive OR in terms of AND and OR.

Fan-out can be divided into ac fan-out and dc fan-out. The ac fan-out problems are caused by the fact that every logic circuit input has some intrinsic capacitance; thus if the output of a given gate is forced to drive many gate inputs, the capacitive loading will slow down the switching action. The dc fan-out, on the other hand, causes a change in the transfer characteristic of the transistors, reducing the voltage swing and thus aggravating the noise problem.

As an exercise, the reader should try the following:

1. Design an eight-bit parity tree by generating a logical *one* for even parity and a logical *zero* for odd parity.
2. Design an eight-bit data selecter; that is, select one of eight input lines depending on a 3-bit input code.
3. Determine the largest of two 3-bit numbers. Assume the number system deals with only positive numbers so that, for example, $010 = 2$ and $111 = 7$.

One of the most useful Boolean functions for digital signal processing is the one-bit adder, which accepts as inputs three bits (two sum bits and a carry input bit) and generates a sum bit and a carry output bit (and their complements, in a more comprehensive package), as shown in Fig. 8.7.

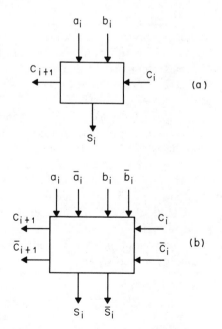

Fig. 8.7 One-bit adder inputs and outputs.

The input–output relations can be depicted by (T.T.2) shown below.

| | | | Arithmetic | | |
a	b	c_i	Sum	c_0	s
0	0	0	0	0	0
0	0	1	1	0	1
0	1	0	1	0	1
0	1	1	2	1	0
1	0	0	1	0	1
1	0	1	2	1	0
1	1	0	2	1	0
1	1	1	3	1	1

(T.T.2)

(T.T.2) completely defines the logic equations. By direct inspection of (T.T.2) the Boolean expression for c_0 and s as functions of a, b, and c_i can be written. Figure 8.8 shows a specific commercial package that is a particular implementation of the logic function. As exercises, the reader should try the following:

1. Find the truth table and logic function for a one-bit subtractor, using two's-complement notation.
2. A four-bit adder is depicted in Fig. 8.9. The problem here is to determine the approximate "settling" time of the adder; i.e., assuming that all inputs are simultaneously applied at zero time, how long before all sum and carry logic levels will be sufficiently stable to be usable by other logic components in the system. To do this problem we need to know all the gate propagation times for each input to every output. We choose the following numbers: $\tau(a_i \rightarrow s_i) = \tau(b_i \rightarrow s_i) = 10$ nsec, $\tau(c_i \rightarrow s_i) = 15$ nsec, $\tau(a_i \rightarrow c_{i+1}) = \tau(b_i \rightarrow c_{i+1}) = 20$ nsec, and $\tau(c_i \rightarrow c_{i+1}) = 5$ nsec. Compute the worst-case propagation time for these numbers.
3. A parallel adder, consisting of eight of the one-bit adder packages of the preceding problem, turns out to be too slow. See if you can work out the logic equations of several carry look-ahead schemes so that the carries propagate more rapidly. Try to come up with a good scheme that you are willing to defend. Assume basic gate propagation delays of 3 nsec in the look-ahead logic.
4. In Fig. 8.8, find the logic equations for $\bar{S}_i$ and $\bar{C}_{i+1}$. Also, prove that alternative expressions for S_i and C_i are

$$S_i = A_i \oplus B_i \oplus C_i$$

$$C_{i+1} = A_i B_i + (A_i \oplus B_i)C_i$$

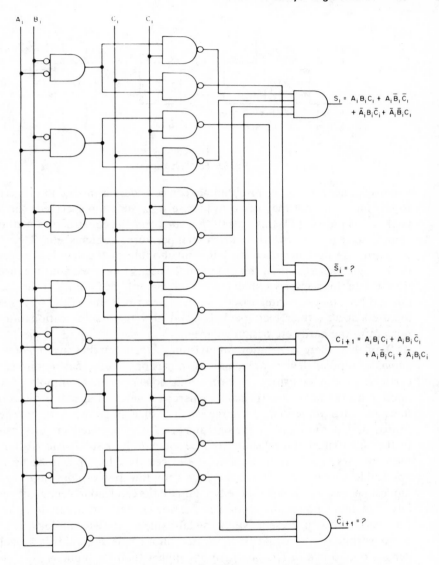

Fig. 8.8 Logic diagram of one-bit adder package (After Motorola Reference Handbook).

8.4 The Major Logic Families

We have already indicated that designing hardware has certain similarities to writing a computer program. In the latter case one tries to implement an algorithm based on the instruction set of a given general-purpose computer. Analogously, one usually builds hardware for an algorithm using the Boolean functions available from a given logic family. A key difference between the

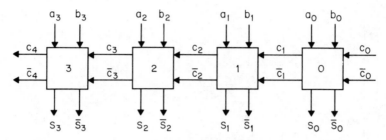

Fig. 8.9 Four-bit adder.

two disciplines is this: The programmer can consider his instruction repertoire to be "ideal" in that they perform precisely the function specified; the logic package will also fulfill this "ideal" *provided that* the construction method is sound, noise is kept low, there are no temperature problems, etc. Therefore, the hardware designer needs to know much more of the detailed properties of the logic packages than the programmer needs to know about his instruction set. In this section we shall discuss the major logic families of the present era and how they are categorized; we shall describe some of the ways manufacturers specify their logic packages and how these specifications are important to the hardware design engineer.

To begin, there is a major subdivision of logic families into *bipolar* and *unipolar*. Bipolar devices are standard transistor circuits that require charge carriers of both polarities (electrons and holes), whereas unipolar devices operate on the field-effect principle whereby a gate voltage controls current flow, in either direction, of charge carriers of a single polarity between a source and a drain. Both major categories of integrated circuits can be further subdivided and we shall describe some of these subdivisions after one more remark, namely, that unipolar devices thus far appear to be most applicable to high density packaging, with low power consumption and modest speeds, while most of the bipolar circuits can lead to very great speeds but with more power dissipation and a lower level of circuit integration. For example, with present MOS (metal-oxide-silicon) technology, a fairly complex digital filter could be fabricated on a single chip; this is a packing density one to two orders of magnitude higher than for bipolar.

Bipolar devices can be further categorized as being comprised of transistor circuits wherein the transistors are driven into that part of their transfer characteristic that causes current saturation (saturated logic) during the switching epoch or wherein the transistors always operate in the linear portion of their transfer characteristic (linear logic). We first describe two saturated logic families and then one linear logic family.

A four-input resistor–transistor logic (RTL) gate is shown in Fig. 8.10. If the four input levels x_1, x_2, x_3, and x_4 are low (0), none of the transistors conduct and no current flows through R_2 so that the output y is at voltage V.

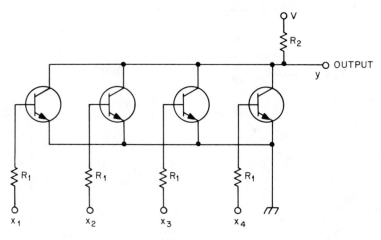

a) TRANSISTOR CIRCUIT OF 4-INPUT GATE

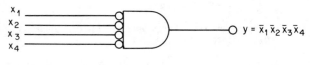

b) LOGIC DIAGRAM

Fig. 8.10 Transistor circuit and logic diagram or our-input AND gate (After Garrett).

If any of the inputs is raised to a sufficiently high level, current flows and the output voltage drops. Thus, Fig. 8.10 is a switching circuit with the Boolean equation

$$y = \bar{x}_1 \bar{x}_2 \bar{x}_3 \bar{x}_4 \tag{8.6}$$

or, in words, $y = 1$ when, and only when, all x_i's $= 0$.

In this example we have used the convention that a high voltage means logical 1 and a low voltage means logical 0. But what are the actual voltages, what are the required power supply values, what kind of transistor do we use, and how do we choose the resistor values? All this plus the ability of the fabrication process to fabricate sufficiently precise parameter values will determine speed, reliability, noise, etc. To obtain such information we must study the manufacturer's data sheets. For example, Fig. 8.11 shows the input voltage–output voltage transfer function of an RTL circuit. RTL transistors are designed with high amplification β so that a relatively small input level change will saturate the output voltage. Figure 8.11 explicitly shows why this transistor design causes switching behavior and also explains the name *saturated logic*. Notice also the specification "fan-out = 1" and then observe

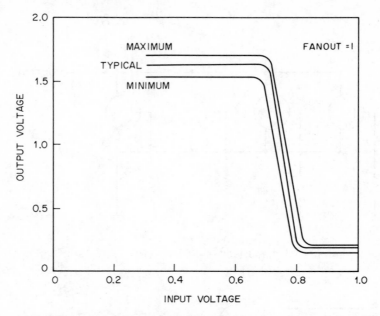

Fig. 8.11 Input voltage–output voltage transfer characteristic for an RTL circuit (After Garrett).

Fig. 8.12, which shows the transfer characteristic of the same circuit for "fan-out = 5"; these are dc fan-out specifications. Thus, a gate that is loaded down by driving five other gates has a somewhat poorer switching behavior than the more lightly loaded gate, the voltage swing being reduced by about 30%. Ultimately, too large a fan-out will deteriorate performance too much to be usable.

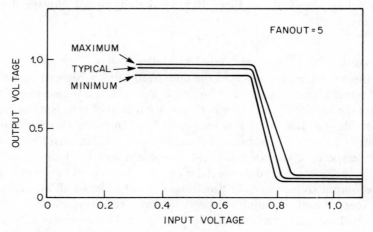

Fig. 8.12 Transfer characteristic for an RTL circuit with a fan-out of 5 (After Garrett).

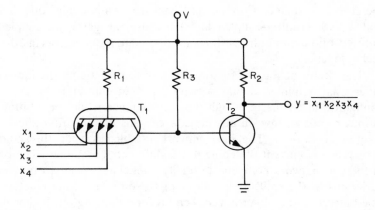

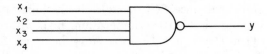

Fig. 8.13 Simplified TTL circuit and logic diagram (After Garrett).

Figure 8.13 indicates how a simplified TTL (transistor–transistor logic) circuit operates. If x_1, x_2, x_3, or x_4 is low (zero), current flows in R_1 and T_1 (transistor 1), thus turning off T_2 and causing the output to be high (1). Therefore, the logic equation is

$$\bar{y} = x_1 x_2 x_3 x_4 \quad \text{or} \quad y = \overline{x_1 x_2 x_3 x_4} \tag{8.7}$$

Notice that T_1 is a multiemitter transistor. Figure 8.13 is not an actual commercial circuit but illustrates the switching action and helps explain the nomenclature TTL. In TTL, the inputs are coupled into the circuit via transistors, as opposed to RTL where the coupling is via resistors. There is also DTL (diode–transistor logic) wherein coupling is via diodes. For reasons too numerous and specialized to discuss here, the TTL arrangement has led to the fastest and most versatile logic line of these three. Typical switching times for medium-speed TTL are 10 to 12 nsec. Higher-speed TTL [HTTL (high-speed TTL) switches at 6 nsec and STTL (Shottky TTL) switches at 4 nsec] are also commercially available.

TTL, DTL, and RTL circuits are all saturated logic families. Faster switching times are made possible by keeping some of the transistors in the linear region of their transfer characteristic. This avoids the capacitive effects due to charge depletion in the saturated region. This corresponds to class A operation; i.e., current is always flowing in the circuit and more power is therefore dissipated. Commercial packages embodying this feature are called ECL (emitter-coupled logic) because the gates are coupled by the emitters of the output transistors, as opposed to the collectors in other bipolar technologies.

The circuit of Fig. 8.14 is divided into four sections: the input circuit, a differential amplifier, a bias network, and an emitter follower. The input configuration results in a high input impedance while the emitter follower output results in low output impedance; this combination allows for substantial dc fan-out. The bias network establishes a dc threshold for the occurrence of current switching in the differential amplifier. The two diodes in the bias network compensate for dc level changes caused by temperature variations in the emitting junctions of the differential amplifier transistors. The differential amplifier is very helpful for balancing out higher harmonics of switching transients. It can be seen that ECL circuits have been designed with emphasis on good high-frequency performance.

We mentioned before that since ECL operates in the linear part of the transistor, power dissipation is high (by implication, high with respect to other logic such as TTL). That is not the whole story, however. Since ECL is linear, power dissipation is independent of switching speed. With TTL, however, current is drawn during switching times so that the faster the system speed, the more "on" time there is in a TTL circuit so that power rises with

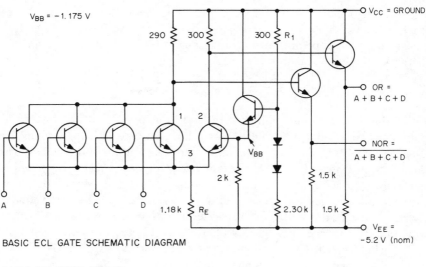

BASIC ECL GATE SCHEMATIC DIAGRAM

BASIC ECL GATE SYMBOL

Fig. 8.14 ECL circuit and logic diagram (After Garrett).

switching rate. This is clearly in evidence in the illustrative curve of Fig. 8.15, which shows gate power versus switching rate for TTL and ECL components.

Although currently TTL is more popular, the advantages of ECL should gradually emerge as engineers become used to handling fast, complex systems. We predict that ECL will make great inroads into the integrated circuit market and compete favorably with TTL in digital signal processing applications, where speed is often essential.

For any high-speed logic family, overall construction methodology assumes great importance. The additional problems that arise at high speeds include time delays through the interconnections, waveform distortion due to the transmission line behavior at these speeds of even short leads, greater cross talk between adjacent signal leads, and thermal problems caused by the high power dissipation. To solve these problems, the engineer must pay special attention to the clock and power distribution systems and thermal management (heat sinking and good mounting techniques to allow uniformly distributed airflow). Much of this can be solved through the design of multi-layer printed circuit boards that can lead to good ground planes and reduce interconnection problems, yielding a higher packing density and minimizing delays.

Up to now, we have discussed bipolar devices. The unipolar devices are also comprised of transistors but the physical principles of operation are different. Unipolar devices depend on the field effect wherein the charge flow between two semiconductors is controlled by varying the voltage on a metal "gate." This gate is isolated from the semiconductors by a layer of silicon

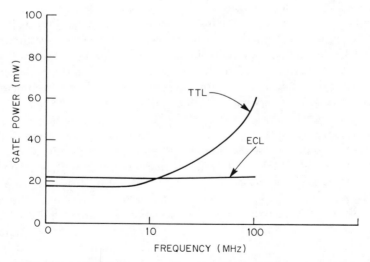

Fig. 8.15 Power versus speed for saturated logic (TTL) as compared to linear logic (ECL).

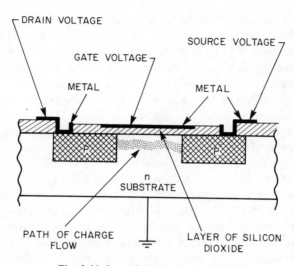

Fig. 8.16 Simplified sketch of an MOS chip.

dioxide, as seen in the sketch in Fig. 8.16. Given the metal gate, the oxide insulating layer, and the (silicon) semiconductors, these devices are called MOS (metal-oxide-silicon). Figure 8.16 shows one variation of such a device; we begin with an *n*-type silicon substrate and diffuse two *p* regions, which correspond to the *source* and the *drain*. On top of this an insulating layer of silicon oxide is applied over which a metal control element, the gate (usually made of aluminum), is laid. Finally metal contacts to the two *p* regions are made. A sufficiently large negative voltage applied to the gate will cause the *n*-type silicon to invert by attracting "holes" to its surface; eventually this effect, in conjunction with the two *p*-type regions, causes the combination to behave like a single *p*-type semiconductor, with consequent current flow. The threshold from no current to full current flow is quite sharp; hence the device is effectively a switch. At present there are a number of commercially available variations of this basic MOS technology but, in addition, a great deal of research is still going on and several of the newer technologies promise rather fantastic results. Thus, in our attempt to tabulate the various advantages and disadvantages of the different technologies, we shall restrict ourselves to the items that are commercially available. Given that, we can make a few comparative statements between unipolar (MOS) and bipolar technology. The chart shown in Fig. 8.17 shows 1973 technology. MOS has achieved a much higher degree of circuit integration, while bipolar circuits have achieved a much higher speed. MOS circuits are clearly more amenable to LSI; the reason lies in the intrinsic simplicity of the fabrication process and in the smaller size and lower dissipation of the MOS transistors, which in turn greatly simplifies the interconnection problem.

MOS   BIPOLAR TTL

	MOS	BIPOLAR TTL
MEMORY	1000 - 4000 BITS 200 - 400 nsec	64 - 1000 BITS 30 - 100 nsec
LOGIC	2000 GATES 1 μ sec	50 GATES 2 nsec

Fig. 8.17 Present-day capabilities of integrated circuits (After Raffel).

From Fig. 8.17, it seems clear that for high speed digital signal processing applications the speed-complexity tradeoff favors bipolar over MOS circuits. It is tempting, however, to realize that with present MOS technology a complete digital filter of reasonable complexity can be fabricated on a single chip! At the present time, the main application of MOS technology to digital signal processing appears to be in shift–register memories; with 4:1 multiplexing techniques, data rates of up to 20 MHz have been attained. As will be seen in Chapter 9, digital filter hardware configurations can make excellent use of shift–register memory.

Table 8.1

COMPARISON TABLE OF MAJOR LOGIC FAMILIES
(From Garrett)

	RTL	*DTL*	*TTL*	*ECL*	*P-MOS*	*CMOS*
Typical Output Impedance	kilohms	kilohms	10–17 ohms	6–15 ohms	kilohms	kilohms
Fan-out	5	8	10	10–25	20	50
Typical Power Dissipation per Gate	2.5–12 mW	8–12 mW	12–22 mW	40–55 mW	0.2–10 mW	1 mW (at 1 MHz)
Immunity to External Noise	fair	good	very good	good	fair	very good
Noise Generation	low-medium	medium	medium-high	low-medium	medium	low-medium
Propagation Delay per Gate (n sec)	12–25	30	6–12	1–4	300	70
Typical Clock Rate for Flip-Flop MHz	2.5–8	12–30	15–60	60–400	2	5
Versatility	good	fair	very good	good	low but growing	low but growing

At the bottom of p 503 we give a table (edited) adapted from Garrett's papers on the major logic families. Let us now summarize this section; first of all, the only components of present interest to digital signal processing are TTL, ECL, and MOS. Although CMOS (complementary MOS; last column in table) has not been discussed by us, the table indicates that it is a definite improvement over P-MOS (P-channel MOS). The most popular logic at present is TTL, which is highly versatile, quite fast, and well understood by hardware engineers. On the other hand, ECL is rapidly gaining acceptance and appears to be most suited for digital signal processing applications. Undoubtedly, all three logic families will find great use as this application area expands.

8.5 Commercial Logic Packages: Gates, Multiplexers, and Decoders; Flip-Flops; Arithmetic Units; Memories

When a manufacturer evolves a new logic family, his biggest problem is to figure out how to penetrate the market. Even when the new logic family has many advantages over existing families, such penetration is difficult and usually takes several years. For such reasons, a manufacturer may start out by introducing only a few functions to see if they "catch on" and then gradually expand his repertoire if the new devices appear potentially profitable. Once a family is established, it ought to be fairly self-sufficient, which means logic versatility. At the present time, in the bipolar world, this implies that a wide variety of gating functions and flip-flop types exist; that useful, highly

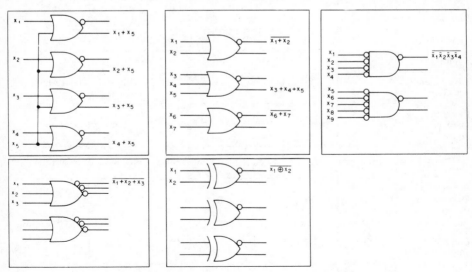

Fig. 8.18 Examples of multiple gating functions of commercial logic packages.

integrated arithmetic circuits exist, and that memory elements compatible with the logic be available. In addition, it is useful to have decoding and multiplexing packages for higher level control functions. In this section, we give examples of these functions. The reader may be surprised by the apparent arbitrariness of the functions chosen. One point worth remembering is the structure of a package; in most cases they are "16-pin DIP's" (dual-in-line packages) or "flat-packs," which also have a limited number of pins. The same logic family will often offer large packages of 24-pin DIP's when the level of circuit integration is too high for the simpler chip of the 16-pin DIP. Thus, (for the 16-pin DIP) if, say, there are 3 power leads, this leaves 13 signal leads. The reader can verify, from Fig. 8.18, that the number of signal inputs and outputs is always 12 or 13, for some typical logic chips.

An important result from Boolean algebra is that equivalent logic functions can be obtained using NOR or AND and also NAND and OR. Figure 8.19 shows an example of a three input NOR gate that can be replaced by a three-input AND gate provided that the inputs are complemented prior to AND gating. In one manufacturer's data sheets, shown in (T.T.3), a set of Boolean equations can be derived that should aid the reader in following the various nomenclatures.

		AND	NAND	OR	NOR				
u	x	y	uxy	$\overline{uxy}$	$u+x+y$	$\overline{u+x+y}$	$\bar{u}\bar{x}\bar{y}$	$\overline{\bar{u}\bar{x}\bar{y}}$	$\bar{u}+\bar{x}+\bar{y}$	$\overline{\bar{u}+\bar{x}+\bar{y}}$

u x y	uxy	$\overline{uxy}$	$u+x+y$	$\overline{u+x+y}$	$\bar{u}\bar{x}\bar{y}$	$\overline{\bar{u}\bar{x}\bar{y}}$	$\bar{u}+\bar{x}+\bar{y}$	$\overline{\bar{u}+\bar{x}+\bar{y}}$
0 0 0	0	1	0	1	1	0	1	0
0 0 1	0	1	1	0	0	1	1	0
0 1 0	0	1	1	0	0	1	1	0
0 1 1	0	1	1	0	0	1	1	0
1 0 0	0	1	1	0	0	1	1	0
1 0 1	0	1	1	0	0	1	1	0
1 1 0	0	1	1	0	0	1	1	0
1 1 1	1	0	1	0	0	1	0	1

(T.T.3)

From (T.T.3) we observe that

$$\left.\begin{array}{r} \bar{u}\bar{x}\bar{y} = \overline{u+x+y} \\ \overline{\bar{u}\bar{x}\bar{y}} = u+x+y \\ uxy = \overline{\bar{u}+\bar{x}+\bar{y}} \\ \overline{uxy} = \bar{u}+\bar{x}+\bar{y} \end{array}\right\} \qquad (8.8)$$

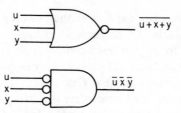

Fig. 8.19 A 3-input NOR gate and an equivalent 3-input AND gate.

Thus, an AND relation between variables can be written as the complement of an OR relation between the complements of the variables and vice versa.

The basic logic functions shown in Figs. 8.18 and 8.19 can be combined in many ways to help implement a given system. With relatively few different packages, any system can be designed but size reduction and overall system speed depend greatly on the versatility and driving capabilities of the available functions and also on the availability within single packages of specialized functions (such as addition). As an exercise the reader should try using the logic functions of Fig. 8.18 to design a logic net that detects the presence of all zeros in a 16-bit register. Assume the register information is available as 16 lines.

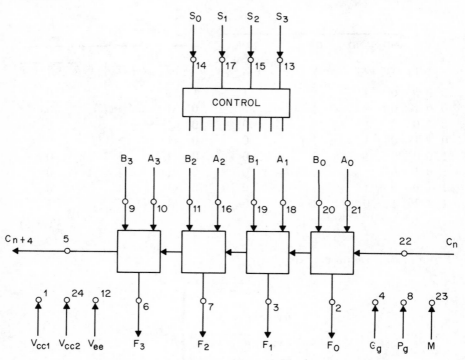

Fig. 8.20 Input–output connections of MC10181 four-bit adder.

As an example of an arithmetic package, we have extracted information from a data book published by Motorola on the MC10181, which is a high-speed four-bit arithmetic–logic unit (ALU). This package has 62 gates, which is somewhat higher than our estimates, but this package is a 24-pin DIP with considerably more package area than the 16-pin DIP's. The MC10181 takes a function F_0, F_1, F_2, F_3 of two four-bit inputs, A_0, A_1, A_2, A_3 and B_0, B_1, B_2, B_3. As shown in Fig. 8.20 there are three pin connections to power supplies in addition to the 12 just mentioned. Also, there are four control bits S_0, S_1, S_2, S_3 that determine the specific function; this means that this ALU can generate up to 16 different functions of the two four-bit inputs. The pins, labeled G_g, P_g, and C_{n+4}, are associated with the carry operation. For straightforward carry-ripple addition, C_n is the input carry to the first adder bit in the four-bit chain, while C_{n+4} is the output carry of the fourth adder bit. G_g and P_g are used in conjunction with other logic packages for fast carry look-aheads. M is another control bit that determines whether the package operates as an arithmetic or Boolean logic unit. Figure 8.21 is a table showing the different functions attainable with the MC10181. Since M is a control bit, there are really 32 control functions available; the second column shows the 16 logic functions, while the third column shows the arithmetic functions. We have used the $+$ notations to mean logical OR, while arithmetic operations are *plus* or *minus*. Thus we see

S3	S2	S1	SO	LOGIC FUNCTION	ARITHMETIC OPERATION
0	0	0	0	$F = \bar{A}$	$F = A$ MINUS 1
0	0	0	1	$F = \overline{A + B}$	$F = A$ PLUS $(A + \bar{B})$
0	0	1	0	$F = \bar{A} \cdot B$	$F = A$ PLUS $(A + B)$
0	0	1	1	$F = 0$	$F = 2A$
0	1	0	0	$F = \overline{A \cdot B}$	$F = (A \cdot B)$ MINUS 1
0	1	0	1	$F = \bar{B}$	$F = (A \cdot B)$ PLUS $(A + \bar{B})$
0	1	1	0	$F = A \oplus B$	$F = A$ PLUS B
0	1	1	1	$F = A \cdot \bar{B}$	$F = A$ PLUS $(A \cdot B)$
1	0	0	0	$F = \bar{A} + B$	$F = (A \cdot \bar{B})$ MINUS 1
1	0	0	1	$F = A \odot B$	$F = A$ MINUS B MINUS 1
1	0	1	0	$F = B$	$F = (A \cdot \bar{B})$ PLUS $(A + B)$
1	0	1	1	$F = A \cdot B$	$F = (A \cdot \bar{B})$ PLUS A
1	1	0	0	$F = 1$	$F = $ MINUS 1(2'S COMPLEMENT)
1	1	0	1	$F = A + \bar{B}$	$F = (A + \bar{B})$ PLUS O
1	1	1	0	$F = A + B$	$F = (A + B)$ PLUS O
1	1	1	1	$F = A$	$F = A$ PLUS O

Fig. 8.21 Table of control functions of the MC10181 (After Motorola Reference Handbook).

that the MC10181 can do many Boolean operations on a bit-by-bit basis, such as AND, OR, and exclusive OR, as well as 16 variations on straightforward addition. This means that a computer using MC10181's need not have separate hardware for, say, AND and arithmetic addition, but simply has to actuate the appropriate control bits on the MC10181's to obtain the desired function. This concept, of building much flexibility into a single package, appears to be very useful for the design of general-purpose computers. On the other hand, for specialized hardware, many of the available functions will not be needed so that the MC10181 could turn out to be somewhat wasteful.

In addition to the logic diagram (Fig. 8.22), the manufacturer supplies tables specifying dc test conditions and various propagation delays. The delay tables give the designer a feel for the overall package capabilities and help him estimate the speeds of systems, such as long adders or multipliers that use a collection of MC10181's.

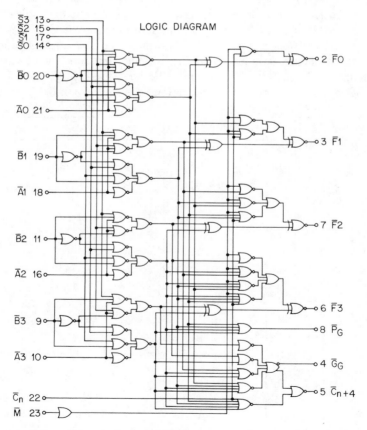

Fig. 8.22 Logic diagram of MC10181 (four-bit adder) (After Motorola Reference Handbook).

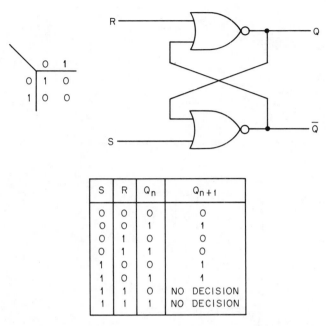

S	R	Q_n	Q_{n+1}
0	0	0	0
0	0	1	1
0	1	0	0
0	1	1	0
1	0	0	1
1	0	1	1
1	1	0	NO DECISION
1	1	1	NO DECISION

Fig. 8.23 Arrangement of two NOR gates to realize a flip-flop.

Let us move on to flip-flops. Before describing a commercial flip-flop, let us derive the concept of a flip-flop from Boolean logic. Figure 8.23 shows two NOR gates with their outputs connected back to an input of the opposite gate. The table on the left shows the NOR logic, while the table on the bottom shows the overall logic. If R (the reset input) goes to 1, the output Q goes to 0, independent of the original state. Similarly, if S goes to 1, Q goes to 1. If R and S remain at 0, the outputs will not change; thus the flip-flop is a one-bit memory element. Bringing R and S simultaneously to 1 causes both Q and $\bar{Q}$ of Fig. 8.23 to be 0; thus this combination is illegal as the basis for a flip-flop.

An important application of flip-flops is their use in binary counters. For such applications, the flip-flops should be able to *commute*, that is, keep complementing its state in response to the same clock input at successive intervals. The flip-flop of Fig. 8.24 has this property. The operation of the flip-flop can be understood by reference to Fig. 8.24 and the associated table. The two gates are three-input AND gates and are the same as described previously (i.e., Fig. 8.23). Remembering that $R = 1$ means *reset* to 0 and $S = 1$ means *set* to 1, we can see that, for example, setting $J = 1$ will set $S = 1$ via the AND gate if $Q = 1$ (the clock should be a short pulse which acts as the third AND input when it is up) with a consequent switching of Q to the 1 state. On the other hand, if Q is already 1, neither gate will be

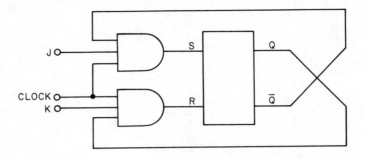

Q_n	J	K	R	S	Q_{n+1}
O	O	O	O	O	O
O	1	O	O	1	1
1	1	O	O	1	1
1	O	1	1	O	O
O	O	1	1	O	O
O	1	1	O	1	1 $\Big\}\bar{Q}_n$
1	1	1	1	O	O
1	O	O	O	O	1

Fig. 8.24 Operation of a J-K flip-flop.

turned on and Q will remain at 1. Thus J and K act like S and R of the previous example. Now, however, if both J and K are set to 1, the flip-flop will flip, as shown by the two lower rows of the table. Note that whereas R, S, J, and K are input levels, the clock can be a relatively short pulse. In this way, new data can be "clocked" into the flip-flop in a synchronous manner if desired.

Flip-flops may be configured in other ways, for example, as latches wherein the flip-flop is set to the value of a given information bit during the clock "on" time. Commercial packages may contain one, two, four, or even eight flip-flops. One interesting method of configuring a package containing a number of flip-flops is to permit parallel readin and serial readout or vice versa. As applications expand in a logic family, it becomes apparent that certain complex logic functions within a single package are called for. An example is shown in Fig. 8.25, the MC10164, an eight-channel data selector and multiplexer, whereby a three-bit code ABC determines which one of eight signals, $X0$ through $X7$, is routed to the output. An obvious use of such logic is in a busing scheme, whereby one or two of many registers may be connected to an adder or other arithmetic element. As an exercise the reader should design a 16:1 data selector using two MC10164's.

Another very useful module is shown in Fig. 8.26. This is the MC10162, which, as a function of a three-bit code ABC, causes one of the eight output

TRUTH TABLE

ENABLE	ADDRESS INPUTS C	B	A	Z
L	L	L	L	X0
L	L	L	H	X1
L	L	H	L	X2
L	L	H	H	X3
L	H	L	L	X4
L	H	L	H	X5
L	H	H	L	X6
L	H	H	H	X7
H	φ	φ	φ	L

φ = DON'T CARE

t_{pd} = 3.5 ns typ (DATA TO OUTPUT)

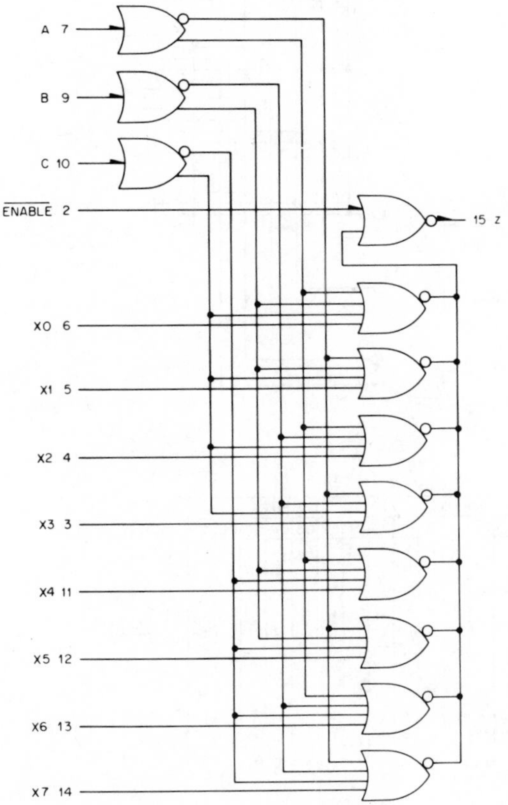

A 7
B 9
C 10
ENABLE 2
15 Z
X0 6
X1 5
X2 4
X3 3
X4 11
X5 12
X6 13
X7 14

Fig. 8.25 Eight-channel data selector (After Motorola Reference Handbook).

POSITIVE LOGIC

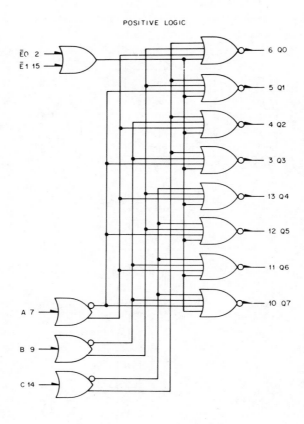

TRUTH TABLE

INPUTS					OUTPUTS							
$\overline{E0}$	$\overline{E1}$	C	B	A	Q0	Q1	Q2	Q3	Q4	Q5	Q6	Q7
L	L	L	L	L	H	L	L	L	L	L	L	L
L	L	L	L	H	L	H	L	L	L	L	L	L
L	L	L	H	L	L	L	H	L	L	L	L	L
L	L	L	H	H	L	L	L	H	L	L	L	L
L	L	H	L	L	L	L	L	L	H	L	L	L
L	L	H	L	H	L	L	L	L	L	H	L	L
L	L	H	H	L	L	L	L	L	L	L	H	L
L	L	H	H	H	L	L	L	L	L	L	L	H
H	φ	φ	φ	φ	L	L	L	L	L	L	L	L
φ	H	φ	φ	φ	L	L	L	L	L	L	L	L

φ DON'T CARE

$t_{pd} = 4.0$ ns typ

Fig. 8.26 Three-bit address decoder (After Motorola Reference Handbook).

512

lines to be zero while the other output lines stay high. This package is very useful for addressing a random access memory, as will be seen directly.

Memories are of great importance to all workers in digital hardware for the simple reason that a large part of the cost of a system so often resides in the memory. In our experience, digital signal processing hardware is no exception. At the present time, a profound transition is taking place in memory technology; namely, that integrated circuit (IC) memories are gradually replacing ferrite core memories as the major source of memories for new computers. Because of the great geometric regularity of an I.C. memory chip, a higher density can be attained for memories on a single chip as compared to other logic elements. At the present time, high-speed random access bipolar memories have as many as 256 bits per chip with early predictions of 1024 bits per chip, while lower-speed MOS shift–register memories will soon attain 4096 bits per chip.

Memories can also be categorized as *static* or *dynamic*. A static memory retains its information indefinitely as long as power is on, while a dynamic memory needs to be refreshed. Bipolar memories are static, while MOS memories can be static or dynamic. In addition, memories can be categorized by their *writability* (changeability): read-only memories (ROMS) can only be read and not written, programmable ROMS (PROMS) can be written once by the user; read-mostly memories can be written slowly compared to the read times and read–write memories can be both read and written, usually at about the same speeds. Memories can also be *random access* or *sequential;* in the first case, successive addresses can be chosen arbitrarily (so that random access memories should really be called arbitrary access memories), whereas sequential memories (shift registers) are functionally similar to discs, drums, or delay lines where the contents of geometrically adjacent bits appear at the output in sequence.

To see how an arbitrarily addressable memory chip operates, consider the 16-bit chip shown in Fig. 8.27. The bits (which are implemented as transistor flip-flops) are arranged as a (4 × 4) array. To access one of the 16 bits requires a 4-bit address. The address word is split into two subwords, one for the rows and one for the columns. Each 2-bit subword is decoded by logic similar to that shown in Fig. 8.26 and one of the four decoder lines is activated; the coincidence of the row and column activated lines determine the active bit. In addition to the four address bits, there is an input line to accept new data in the write mode and an output line to deliver the information in the chosen bit in the read mode, the mode being determined by the write enable control line. Finally, there is a chip select control when several chips are combined to create a memory with more registers. The chip in our example corresponds to a memory of word length 1 bit and 16 registers. To extend the word length to n bits requires n such chips, all controlled by the same address, Din, Dout, WE, and CS lines. To extend the number of

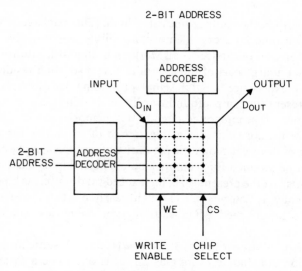

Fig. 8.27 Block diagram of a 16-bit memory chip.

registers, consider the example of Fig. 8.28, which shows a 64-register by 1-bit memory. The Din, Dout, and WE lines are all connected together. Addressing can now be split into three fields, going to the rows and columns and CS lines of each chip, as shown in Fig. 8.28. In this way a single bit from only one of the four chips is selected. By adding chips horizontally the word length can be increased 1 bit at a time; by adding bits vertically, for our example, the number of registers can be increased 16 at a time.

8.6 Multipliers

Because of their critical importance in digital signal processing, multipliers will be treated in some detail in this section. From a logic design viewpoint, we can divide multipliers into two major categories, clocked multipliers and array multipliers. In both types, the product is effectively obtained through successive additions, the difference being that sufficient parallelism exists in the array multiplier so that the final answer is obtained without registering intermediate results. The number system used also determines the multiplier's hardware configuration. We begin with a brief discussion of number systems as they apply to multipliers and then proceed to descriptions of the various clocked and array multiplier algorithms. The reader is referred to Chapter 5 for a more general number system discussion.

In Chapter 5 we discussed the sign-magnitude, 1's-complement, and 2's-complement number systems. In comparing the last two systems, it was argued that 2's complement was most suitable for high-speed performance so here we shall discuss only sign-magnitude and 2's-complement systems.

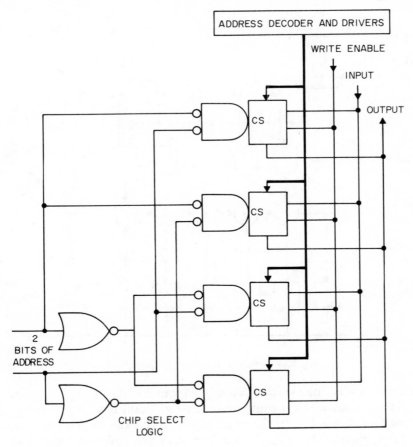

Fig. 8.28 A 64-register by 1-bit memory using 16-bit chips.

It is probably true that sign-magnitude multiplication is conceptually slightly simpler than 2's-complement multiplication; thus our early examples will be multiplication of positive numbers. Since *addition* of two sign-magnitude numbers is more awkward than 2's-complement addition, however, 2's-complement techniques warrant much attention.

A sketch of a simple (5 × 4) add–shift multiplier is shown in Fig. 8.29. This device successively accumulates the rows shown below if the corresponding bit of the y column is 1; if the bit is 0, a row of 0's is gated into the adders.

$$x_4 x_3 x_2 x_1 x_0 \quad y_0$$

$$x_4 x_3 x_2 x_1 x_0 \quad y_1$$

$$x_4 x_3 x_2 x_1 x_0 \quad y_2$$

$$x_4 x_3 x_2 x_1 x_0 \quad y_3$$

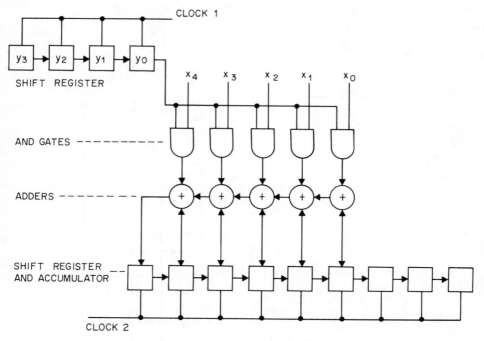

Fig. 8.29 Sketch of add–shift multiplier.

Each time a row is gated in, the accumulated sum is shifted one place to the right so that successive rows contribute more significant bits, as dictated by the algorithm. Clock 1 shifts the y bits into the AND gates so that the x bits or 0's are applied to the adders. After the adder levels have stabilized, clock 2 strobes the adder outputs into the accumulator flip-flops, which, after settling down, can then be shifted one place to the right. This multiplier, although simple, is quite slow. Since there is no look-ahead logic in the adders, each accumulation must allow for the carry bits to propagate through the complete adder. An approximate worst-case settling time for an n-bit adder is $(n-1)\tau_c + \tau_s + \tau_r$ where τ_c is the carry propagation time, τ_s is the sum propagation time for a one-bit adder, and τ_r is the settling time for the y-shift register. If n bits are multiplied by m bits, total multiplication time is about $m(n-1)\tau_c + m\tau_s + \tau_r$.

Figure 8.29 illustrates an important hardware fact about many add–shift multiplier configurations, namely, that the y bits are kept in a serial shift register. We shall see in Chapter 9 that the serialization of the data stream is very useful for digital filter structures and leads to memory cost savings. Given this structure, the question now arises as to how the add–shift multiply can be implemented when dealing with signed 2's-complement numbers instead of just the positive numbers of Fig. 8.29. This is a good place to review 2's-complement representation and multiplication. A useful arithmetic representation of a 2's-complement number is given by

$$X = -2^n x_n + \sum_{m=0}^{n-1} 2^m x_m = -2^n x_n + 2^{n-1} x_{n-1} + \ldots + 2^0 x_0 \qquad (8.9)$$

where x_m are the bits associated with the number X, the sum in Eq. (8.9) being arithmetic, not logical. x_n is the sign bit. As an exercise, the reader should use Eq. (8.9) to derive an algorithm for negating a positive 2's-complement number.

From Eq. (8.9) we can derive an arithmetic expression for the product XY of a pair of 2's-complement numbers where X is an $(n + 1)$-bit number and Y is an $(m + 1)$-bit number.

$$XY = 2^{n+m} x_n y_m - 2^n x_n \sum_{j=0}^{m-1} 2^j y_j - 2^m y_m \sum_{k=0}^{n-1} 2^k x_k + \sum_{j=0}^{m-1} \sum_{k=0}^{n-1} 2^j 2^k x_j x_k$$

$$(8.10)$$

The first and fourth expressions on the right of Eq. (8.10) are positive numbers, while the second and third expressions are negative. The fourth expression is the complete result when dealing with exclusively positive numbers and corresponds to the successive additions implemented in Fig. 8.29. Equation (8.10) tells us, however, that sums must be subtracted as well as added if either x_n or y_m is 1. A convenient logical extension of the add–shift multiplier to take account of the sign bit is known as *Booth's algorithm*. It can be explained by writing the number Y as follows:

$$Y = (y_{m-1} - y_m)2^m + \sum_{j=0}^{m-1} 2^j (y_{j-1} - y_j) \qquad y_{-1} = 0 \qquad (8.11)$$

Now, if the multiplier (Y) multiplies the multiplicand (X) on a bit-by-bit basis, it will operate on successive rows of x_n's according to these prescribed steps.

1. If $y_{j-1} = y_j$, the multiplicand is not accumulated but a row of 0's is injected.
2. If $y_{j-1} = 1$ and $y_j = 0$, the multiplicand is added to the accumulator.
3. If $y_{j-1} = 0$ and $y_j = 1$, the multiplicand is *subtracted* from the accumulator.

As an exercise, the reader should use Booth's algorithm to multiply X ($= -6$) by Y ($= -27$). Also, modify the logic of Fig. 8.29 to perform Booth's algorithm.

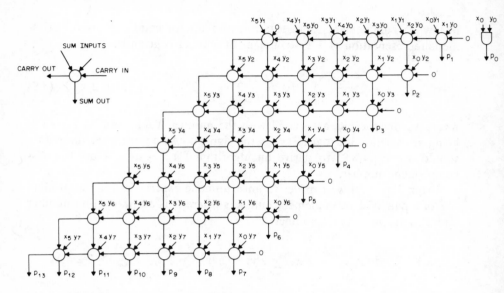

Fig. 8.30 Array multiplication of $(x_5\,x_4\,x_3\,x_2\,x_1\,x_0)$ by $(y_7\,y_6\,y_5\,y_4\,y_3\,y_2\,y_1\,y_0)$, both positive numbers.

Booth's algorithm can be speeded up by a factor of 2 by shifting two bits at a time but checking the logic of three successive bits to determine how to accumulate the multiplicand; we leave this as an exercise for interested readers.

The fastest multipliers consist of a two-dimensional array of one-bit adders and are referred to as array multipliers. Unlike the clocked multiplier of Fig. 8.29, the array multiplier is a complete memoryless logic net that only requires a prescribed settling time after the application of the input signals before the product can be used. There are many array configurations but they can be fairly well categorized according to the method of connecting the adders and according to the way negative numbers are handled. Figure 8.30 shows a not very efficient array for multiplying positive numbers. Each little circle is a one-bit adder. We can see that each row of adders is a full ripple–carry adder and each row of adders gives its partial sum to the row below. To estimate the settling time of such a multiplier, consider that the sum bit p_7 requires m sum delays and n carry delays but to reach p_{13} requires an additional n carry delays; thus the total settling time τ_t is given by

$$\tau_t = 2n\tau_c + m\tau_s \tag{8.12}$$

This is appreciably less than the successive addition scheme of Fig. 8.29. The total worst-case delay can be reduced, however, by propagating the carry bits diagonally, as shown in Fig. 8.31. It makes no difference into which row the carry from a previous column is entered. Thus, Fig. 8.31 yields the same final result as does Fig. 8.30 but it has the advantage that carries and

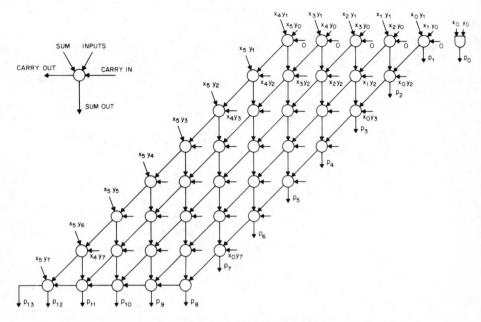

Fig. 8.31 Faster array structure for multiplication of Fig. 8.30.

sums generated from lower level adders propagate *simultaneously* toward the higher level adders. Such a scheme is called *carry-save* and is applicable for multipliers other than array multipliers. For example, the output p_8 is not available via column 8 until five sum times have elapsed *or* until eight carry times have elapsed, whichever takes longer. If these two times happen to be equal, then the sum time can be ignored. We must now include the total carry time in the final row of adders. Thus in Fig. 8.31 the worst case might be p_{13}, which would take $12\tau_c$, *provided* that $8\tau_c \leq 5\tau_o$. If this were not true then total time would be $5\tau_s + 5\tau_c$. In general,

$$\begin{aligned}\tau_t &= [(m-1) + (n-1)]\tau_c && \text{if } m\tau_c > (n-)1\tau_s \\ &= (n-1)(\tau_c + \tau_s) && \text{if } m\tau_c < (n-1)\tau_s\end{aligned} \tag{8.13}$$

If one were using an adder element with τ_s appreciably greater than τ_c, it becomes desirable to find a structure wherein special attention is paid to reduction of the number of τ_s times needed. The idea for one such structure can be noted via Fig. 8.32, wherein the array of adders is arranged as a "tree." Shown in Fig. 8.32 is an $n \times 16$-bit multiplier where each plus symbol represents n bits. In this arrangement, only four sum times are necessary. The same number of adders are required as before but more rows of adders are

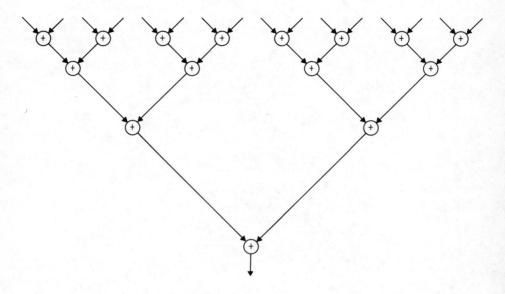

Fig. 8.32 Tree idea for array multiplier.

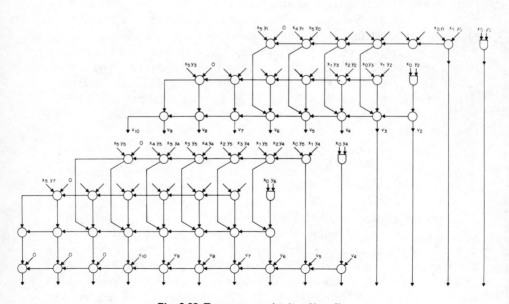

Fig. 8.33 Tree array multiplier (6 × 8).

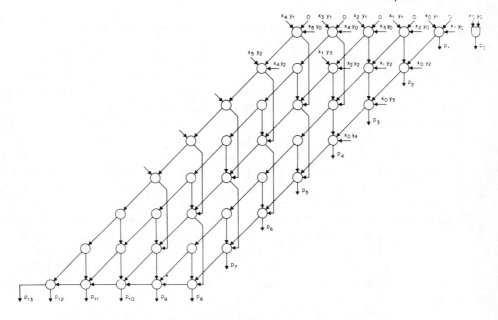

Fig. 8.34 Array multiplier with diagonal carry and tree arrangement for sums.

operating in parallel. In Fig. 8.33, a realization is given for our (6×8) multiplier in which the carries within each adder remain in the same row. Thus, the total add time for each adder row is really the carry-ripple time or $(n - 1)\tau_c$. Since there are four levels, each of time $(n - 1)\tau_c$, we obtain $4(n - 1)\tau_c$ as the total time. Thus, this scheme is not so efficient as that of Fig. 8.31, unless τ_s is a very large number.

The idea of the tree and the diagonal propagation of the carries can be combined to yield perhaps the fastest configuration. An example is shown in Fig. 8.34.

Since the array structure of Fig. 8.34 appears to be the fastest of the structures thus far discussed, one may question the need to describe the other structures. The answer sheds some light on an important problem in system design using I.C. components. Usually, a given component is designed and produced for a variety of applications. Thus, for example, the designer of an array multiplier must choose the most appropriate of the *available* components with which to build his multiplier. An interesting package that appeared commercially in 1971 was the MC10181 four-bit adder package with fast carry look-ahead logic. Such a component could be arranged along a diagonal, as in Fig. 8.31, or horizontally, as in Fig. 8.30 or 8.33. While Fig. 8.31 leads to a faster multiplier, it also requires extra packages for gating the x_j and y_k bits (formation of the partial products); these extras would not be required if horizontal carries are propagated since then one of the control bits can be used to gate all four inputs bits. The lesson to be learned is this; that the "best" method depends on the I.C. hardware being used and the speed–cost tradeoffs.

Signed 2's-complement multiplication can be explained by the basic representation of Eq. (8.9). The implications of this equation are as follows; each bit except the sign bit is interpreted as possessing the *arithmetic* values 0 or 1, while the sign bit is interpreted as 0 or -1. By performing the arithmetic addition shown in Eq. (8.9), the correct number is always generated from the representation if we use the interpretation above.

A one-bit adder element can be defined, as we have already done, in terms of logic equations or in terms of an arithmetic nomenclature. If each of the three inputs can be (arithmetic) 0 or 1, then the two-bit output can assume the values 0, 1, 2, or 3. We shall realize this function by associating the value 0 or 2 to the carry and 0 or 1 to the sum bit and defining the output in terms of these possibilities. (T.T.4) shows the truth table of an adder using this nomenclature, where A, B, and C are the inputs, D is the sum or low level output, and E is the carry. A little thought will show that there is no intrinsic need to express the two possible values of a given output as either 0 or 1; it could be $+$ or $-$, 0 or 2, 0 or 100, etc. By using (T.T.4) nomenclature, we obtain a natural connection between the binary outputs and the intuitively appealing arithmetic outputs.

A	B	C	D	E	*Arithmetic*	
0	0	0	0	0	0	A, B, C = inputs
0	0	1	1	0	1	D = sum (Boolean)
0	1	0	1	0	1	E = carry = arithmetic sum $-$
0	1	1	0	2	2	Boolean sum (T.T.4)
1	0	0	1	0	1	(I)
1	0	1	0	2	2	(IV) is sign reversal of (I)
1	1	0	0	2	2	
1	1	1	1	2	3	

This nomenclature may now be extended to describe logic functions that can perform arithmetic addition on negative as well as positive numbers. For example, consider a three-input circuit wherein one of the inputs takes as the value -1 or 0, while the other two are 0 or 1. Clearly, the four possible arithmetic sums are -1, 0, $+1$, 2; again, these sums are obtainable from a two-bit output circuit in which the low level bit assumes the values 0 or -1, while the high level bit is 0 or 2.

The truth table for this logic component is shown in (T.T.5) and again we see that the arithmetic sum, defined as the arithmetic sum of the two outputs, also corresponds to the arithmetic sum of the three inputs. Similarly, (T.T.6) and (T.T.7) are truth tables for two other cases.

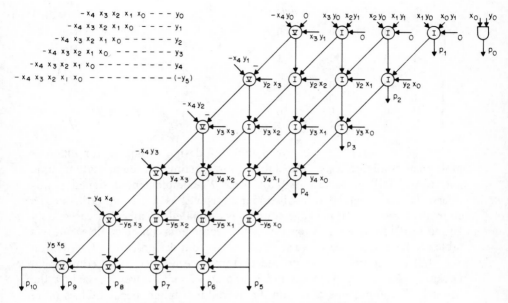

Fig. 8.35 A 2's-complement signed array multiplier (6 × 5).

Armed with this notation, it is now possible to express a 2's-complement signed array multiplier as successive additions, provided that logic packages capable of summing negative numbers are included. Figure 8.35 is an example of such a multiplier, where packages corresponding to an ordinary adder in (T.T.4) and packages corresponding to (T.T.6) and (T.T.7) are used.

A	B	C	D	E	Arithmetic		
0	0	0	0	0	0		
0	0	1	−1	2	1		
0	1	0	−1	2	1		
0	1	1	0	2	2		
−1	0	0	−1	0	−1	(II)	(T.T.5)
−1	0	1	0	0	0		
−1	1	0	0	0	0		
−1	1	1	−1	2	1		

A	B	C	D	E	Arithmetic		
0	0	0	0	0	0		
0	0	1	1	0	1		
0	−1	0	1	−2	−1		
0	−1	1	0	0	0		(T.T.6)
−1	0	0	1	−2	−1	(III)	
−1	0	1	0	0	0		
−1	−1	0	0	−2	−2		
−1	−1	1	1	−2	−1		

A	B	C	D	E	Arithmetic
0	0	0	0	0	0
0	0	−1	1	−2	−1
0	−1	0	1	−2	−1
0	−1	−1	0	−2	−2
1	0	0	1	0	1
1	0	−1	0	0	0
1	−1	0	0	0	0
1	−1	−1	1	−2	−1

$$(V) \qquad\qquad (T.T.7)$$

At the top of p.525 we give a table of design estimates for different multiplier configurations. The table shows estimated computation time and 16-pin DIP package count for three 2's-complemented signed multipliers. The commercial four-bit adder of row 2 is a 24-pin DIP, so the equivalent 16-pin DIP package count was estimated; in all other cases, only 16-pin DIP's were used. A few interesting comments can be made about this table. Referring to row 1, we see that the custom design (to be described in Sec. 8.8) has appreciably improved performance relative to all commercial packages. Referring to rows 2 and 3 (and also 6 and 7), we can see how the faster ECL components reflect on the overall multiplier speed; ECL is more than twice as fast. Referring to rows 1 and 5, we can see how the configuration influences both the speed and package count; somewhat surprisingly the array has the best tradeoff between speed and package count. One of the reasons for this is the extreme simplicity of the control functions in an array as compared to a clocked multiplier (rows 5, 6, 7). Also it is rather interesting that the two-bit (16-pin DIP) adder package (row 1) is more efficient in overall package count than the four-bit (24-pin DIP) adder package (rows 2 or 3). This says that a higher level of integration may be *harmful* in some configurations if the chip requires more real estate.

8.7 Dividers and Floating Point Hardware

Digital signal processing systems are necessarily comprised of the same elements as other computing systems, with major emphasis on the speed of these elements. Up to now we have focused on the adders and multipliers, which are undoubtedly the most important elements. In this section we shall describe design techniques for performing fast division and floating point addition. As digital filtering and FFT algorithms are applied to increasingly sophisticated problems, there is greater need for a good understanding of how to implement computation other than adders and multipliers. It turns out that the unclocked array concept can be put to good use for all the cases considered in this section; high speeds can be attained and the cost is not great. We anticipate that the continuing advance of component technology will make these statements even truer in the near future.

Table 8.2
MULTIPLIER COMPARISON TABLE

Configuration	Type of Basic Package	Component Technology	Computation Time (nsec)			Package Count		
			16×12	16×8	9×9	16×12	16×8	9×9
Fast trapezoid (Fig. 8.35)	Custom two-bit adder package	ECL	32	28	22	112	82	45
Tree	Commercial four-bit adder package	ECL	62	48	40	160	120	85
Tree	Commercial four-bit adder package	TTL	155	100	85	160	120	85
	(2×4)-bit (commercial) array package	TTL	261	195	135	80	58	30
Two bits at a time add–shift	Custom two-bit adder package	ECL	130	120	100	40	36	25
Two bits at a time add–shift	Commercial two-bit adder package	ECL	210	140	116	40	36	25
Two bits at a time add–shift	Commercial two-bit adder package	TTL	450	300	250	40	36	25

The authors thank P. Blankenship and A. Huntoon for permission to use this table, which is based on their design work.

I. Four-Quadrant Divider Array

Let us now consider a four-quadrant divider comprised of a combinational array of adder and subtracter logic elements. The network accepts a 24-bit word as a dividend (numerator), the word being interpreted as a signed 2's-complement number. The divisor (denominator) is a 12-bit 2's-complement word. The array produces a 12-bit quotient and a 12-bit remainder. The divisor and dividend may be considered to be integer, fractional, or fixed numbers; in our discussion we shall think of the numbers as fractional with the binary point directly to the right of the sign bit; the overflow logic is consistent with that interpretation.

The underlying operating principle of the array is that of nonrestoring binary division. The procedure is most easily undertood by considering the divisor and dividend to be positive fractional quantities wherein the divisor is larger than or equal to the dividend. The quotient will be a positive fraction in this instance. To obtain the first quotient data bit, a trial divisor equal to half the actual divisor is subtracted from the dividend, yielding a partial dividend. If the partial dividend is positive, then a 1 is entered as the quotient bit. If negative, however, the trial divisor did not "go into" the dividend and a 0 must be entered as the quotient bit. In normal (restoring) division it would be necessary at this point to add the trial divisor to the partial dividend. This would restore the original dividend before an attempt is made to subtract a new trial divisor. Nonrestoring division makes use of the fact that each successive trial divisor is half the preceding one. Thus the addition (or restoration) of a trial divisor followed by subtraction of one-half that very same trial divisor is nothing more than a net *addition* of half the trial divisor. Returning to the example, if the first data bit of the quotient is a 1, the previous trial divisor is halved and subtracted from the partial dividend to produce the next quotient bit. If the first quotient bit is a 0, the trial divisor is halved and added to the partial dividend to produce the next quotient bit. This procedure continues until all desired quotient data bits have been produced. The algorithm has the distinct advantage of being realizable as an unclocked array. There are no feedback loops; the process flows unconditionally from beginning to end without any "backup" steps.

Realizing this division procedure in practice, for the four-quadrant case (all sign combinations of divisor and dividend possible), requires some manipulation. The heart of the divider array consists of a series of adder/subtracter stages. Each of the stages will either add or subtract the appropriate divisor from the appropriate partial dividend depending on the *sign bit* of the partial dividend in question and the sign bit of the divisor proper. The array will accept any combination of dividend and divisor signs; however, the set of quotient data bits produced by the array must be corrected at the end for certain sign combinations.

The topmost portion of a diagram for the procedure (Fig. 8.36) depicts generation of the quotient data bits. The bottom section depicts the end correction. The rules for generating a quotient bit at any given stage of the array are

1. If the present partial dividend is positive, enter a 1 as the concomitant quotient bit. If not, enter a 0.
2. If the divisor *and* the present partial dividend have the same sign, *subtract* the next trial divisor.
3. If the divisor and the present partial dividend have differing signs, *add* the next trial divisor.

The conventions above imply two interesting facts: First, a positive dividend with either a positive or a negative divisor will yield the proper quotient *magnitude*. Second, a negative dividend with either a positive or a

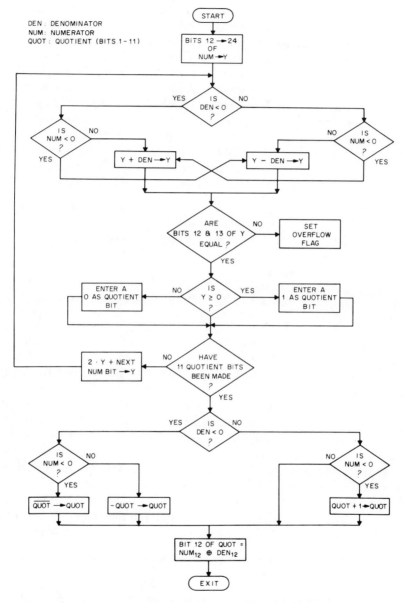

Fig. 8.36 Conceptual flow chart of the extraction of four bits of ratio via the nonrestoring algorithm (After Blankenship).

527

negative divisor will yield the *complement* of the magnitude of the quotient. A correction based on the actual signs of the operands must be applied to derive a correct 2's-complement quotient representation. For example, suppose the divisor is positive and the numerator is negative. The quotient bits generated will turn out to be a 1's-*complement* representation of the correct negative result. Thus the result must be incremented by 1 to yield the proper 2's-complement representation. As a further example, suppose both the divisor and dividend are negative. Clearly the quotient ought to be positive. The array produces, however, the *complement* of the correct results and a pure inversion of the quotient bits is necessary. All of these cases are dealt with via an extra adder/subtracter stage at the very bottom of the array that performs the actual correction. Only one case needs no correction: positive divisor and positive dividend. The correction rules are:

1. If both the divisor and dividend are positive, assign the quotient sign bit the value 0 and *do nothing* to the quotient data bits.
2. If both the divisor and dividend are negative, assign the quotient sign bit the value 0 and *complement* the quotient data bits.
3. If the divisor is positive and the dividend negative, assign the sign bit the value 1 and *increment* the quotient data bits *by 1*.
4. If the divisor is negative and the dividend is positive, assign the sign bit the value 1. *Complement* the quotient data bits and *increment* by 1.

Conceptually, all array additions and subtractions involving an N-bit signed divisor can be carried out on an $(N + 1)$-bit basis. The difference (or sum) between any given partial dividend and its trial divisor should also be representable in *no more* than N bits [really $(N - 1)$ bits plus sign]. Thus, for this case, bits 12 and 13 of any partial dividend, being *both* presumably sign bits, should always be in agreement. If not, an overflow indication is rendered. This indication implies, in terms of fractional operands, that the dividend was greater in magnitude than the divisor. The condition is illegal because the quotient would have to be greater than 1; hence it could not be represented as a signed fraction and the overflow flag is set.

The partial dividend that determined the last quotient bit (i.e., the result of the last add/subtract stage) is considered to be the remainder. It is a 12-bit entity (11 bits plus sign) and is related to the other operands by the equation

$$\text{Dividend} = (\text{quotient}) \times (\text{Divisor}) + \text{Remainder}$$

It can be used in conjunction with more dividend bits to derive an extended precision quotient. The actual formation of the extended dividend is involved but it can be done and the signed remainder is necessary.

Figure 8.37 shows an actual hardware realization of a divider that accepts a 4-bit divisor and an 8-bit dividend yielding a 4-bit quotient and a 4-bit remainder. The realization can be extended in a straightforward manner to

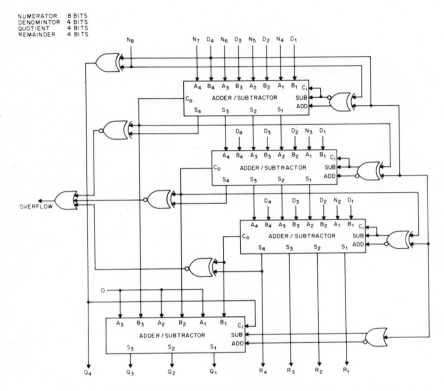

Fig. 8.37 A (4 × 8) nonrestoring divider array with overflow detection and end sign correction (After Blankenship).

the 24-bit/12-bit case. The adder/subtracters represented can be realized with the 2-nsec ECL 4-bit, ALU package (MC10181). The unit can be programmed either to add or to subtract in response to a control. The actual subtraction is accomplished by changing the sign of the B input and adding. This operation, in effect, requires that the B input be inverted and incremented by 1. The package does the complementation internally but the 1 must be supplied at the C_i (carry "in" 0) input wherever a subtraction is to occur.

It might be expected that since the divisor is a 4-bit entity, all add/subtracts ought to be done on a 5-bit basis as was inferred earlier. It can be shown via some manipulation, that the $(N + 1)$st-bit can be simply realized as nothing more than the carry out of the Nth bit with a slight change in rules. The simplified rules now can be stated succinctly:

1. *First stage*—If the sign bits of the divisor and dividend differ, then add. If not, subtract.

2. *All subsequent stages*—If the carry out (C_o) of the Nth bit of the previous adder/subtracter is a 1, enter 1 as the quotient digit. Also, if the carry out is different from the divisor sign bit, set the present stage to subtract. Add otherwise.
3. *Overflow*—If the carry out of the Nth bit for any given stage is the *same* as the Nth bit out of that stage, signal an overflow.
4. *End correction*—This can best be summarized in tabular form:

Sign of Divisor (S_D)	Sign of Dividend (S_N)	Quotient Correction	Sign of Quotient (S_Q)	Add	Subtract	Carry In (C_i)
+	+	(None)	+	√		
+	−	$Q + 1$	−	√		√
−	+	$\overline{Q} + 1$	−		√	√
−	−	$\overline{Q}$	+		√	

Logic equations are easily derived:

$$S_Q = C_i = S_N + S_D$$

$$\text{SUB} = S_D$$

$$\text{ADD} = \overline{S}_D$$

These are seen as the controls implemented in the figure for the end correction stage. Notice that the A input to this stage is necessarily a hard-wired zero.

The actual 24-bit/12-bit divider is realized using 12 12-bit stages. The first 11 derive the quotient bits, the last does the correction. Each stage requires three MC10181 packages with a look-ahead carry generator arranged to feed bit 9. Thus 36 MC10181 units are required. Each stage is capable of producing all necessary partial dividend bits in 13 nsec. Therefore the entire operation will require $12 \times 13 = 156$ nsec. The number of packages required to synthesize controls and overflow functions and to perform data distribution is incidental. Thus, the entire unit is smaller in terms of packages than a (12×12) multiplier using the same logic family.

2. Floating Point Hardware

In systems planning, dynamic range is very often an important subject. Floating point number systems greatly increase the available dynamic range. In intricate number crunching problems, such as inversion of large matrices and linear programming, the magnitudes of the numbers generated during the course of the computation may vary greatly; for problems like these, floating point arithmetic is normally used.

There are good reasons for resorting to floating point in the implementation of digital filters and FFT boxes. Recursive digital filters with poles close to the unit circle exhibit high gains at frequencies near the poles so that the output signal levels may fluctuate greatly for different input signals. An FFT stage can have a $2:1$ gain in signal level so that an N-point FFT may have a gain of $2^{\log_2 N} = N$.

Floating point does not appear to worsen roundoff problems in filtering and the FFT and often improves performance. The disadvantage of floating point is the greater expense of its implementation. Many general-purpose computers have no floating point hardware so that programming floating point means longer programs with longer execution times. In hardware, both floating multiplication and floating addition are more complex algorithms than their fixed point counterparts.

The rule for floating multiplication is to add exponents and do a fixed point multiplication. Since the decimal equivalent of the mantissa is always of magnitude greater than or equal to 0.5 and less than 1, the product of two mantissas is between 1 and 0.25. This means that the product will either be left justified[1] or within one bit of left justification. As an exercise, the reader should design the (four-bit) exponent logic for a floating multiplier.

The steps needed to perform floating additions are

1. Determine the larger of the two exponents.
2. Subtract the smaller from the larger exponent.
3. Right shift the mantissa corresponding to the smaller exponent by the difference obtained in step 2.
4. Perform the addition of the two mantissas.
5. Determine the number of leading 0's (or 1's) in the sum.
6. Left justify the sum, entering the number of leading 0's in the exponent.

As we shall see in a moment, the hardware design for the above is appreciably more complicated than fixed point addition. Furthermore, there is a fair amount of decision making and therefore sequencing of operations, which tends to slow down the overall algorithm. For this reason, despite the computational advantages of floating point, the choice of fixed or floating is often a nontrivial and possibly agonizing decision.

From steps 3 and 6 it is seen that a basic hardware component is a device to perform shifting by an arbitrary amount. An obvious way to do this is

---

[1] Left justification means that the mantissa bits are shifted as far to the left as they can be shifted without overflow. A synonym for left justification is *normalization*. If the mantissa is considered to be a signed decimal fraction, left justification moves the decimal point just to the right of the sign bit.

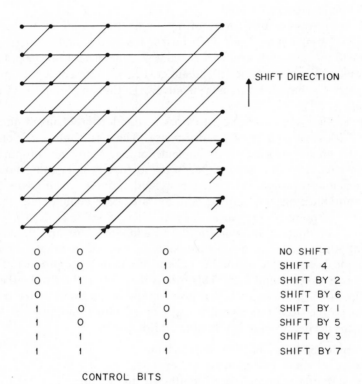

O	O	O	NO SHIFT
O	O	1	SHIFT 4
O	1	O	SHIFT BY 2
O	1	1	SHIFT BY 6
1	O	O	SHIFT BY 1
1	O	1	SHIFT BY 5
1	1	O	SHIFT BY 3
1	1	1	SHIFT BY 7

CONTROL BITS

Fig. 8.38 Shifting array for shifts up to eight.

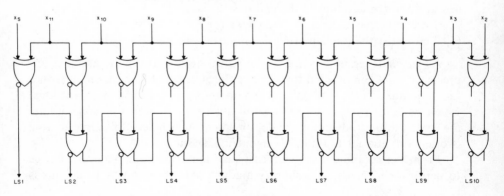

Fig. 8.39 Network to count the number of left shifts necessary to normalize (After Blankenship).

to shift the register one bit at a time until a control counter runs down; this is a cheap but slow method. Another way would be to wire in all possible shifts with a logic array; this is a fast but expensive method, probably more costly than array multiplication. A compromise method is shown in Fig. 8.38. Each dot represents a switch wherein one of the two output lines is selected to be active via a control bit. By connecting the lines as shown in Fig. 8.38, it can be seen that three levels of switching logic lead to all the possible eight shifts; more generally, $\log_2 N$ logic levels are needed for a general N-bit shift network.

From step 5 we see the need for a logic network that must count the number of leading 0's or 1's in a register. Figures 8.39 and 8.40 show networks which perform this function for the example of a 12-bit register. Figure 8.39 is a network that accepts 11 input bits and produces 10 output bits. The number of 1's in the output is determined by the number of left shifts needed to left justify (normalize) the input. Figure 8.40 is a network of full adders to sum the outputs of Fig. 8.39.

8.8 An Illustrative Example; Design of a Fast Array Multiplier

We have previously pointed out that the design of a complete digital system from commercial logic packages is usually inefficient, due to the fact that the logic packages were designed with multipurpose uses in mind. It is therefore instructive to consider a practical case where the basic circuit was designed specifically for the system requirements. This will also allow us to trace some of the hierarchy of circuits, logic, components, and systems referred to in Fig. 8.4, p. 490.

First we inquire as to the system considerations in building a fast multiplier. From the discussion in Sec. 8.6, it seems clear that the array is the fastest known method. Furthermore, the structure of Fig. 8.35 appears to be the fastest variation of the many possible array configurations. Notice, however, that three different logic elements are needed; this raises a problem that three separate circuits will have to be designed and three distinct logic packages appear to be necessary. The solution to this problem was obtained by designing the circuit so that very minor changes in interconnections of the same circuit led to all variations; furthermore, the chip was constructed with two metallization layers and only the top one needed to be altered to accommodate the three circuit types.

The basic logic package was a two-bit adder. From systems considerations, the most crucial design parameter is the two-bit carry propagation time. Thus, the circuit was designed to minimize the two-bit carry time.

Given a set of specifications on the package, let us list the steps leading to its eventual construction.

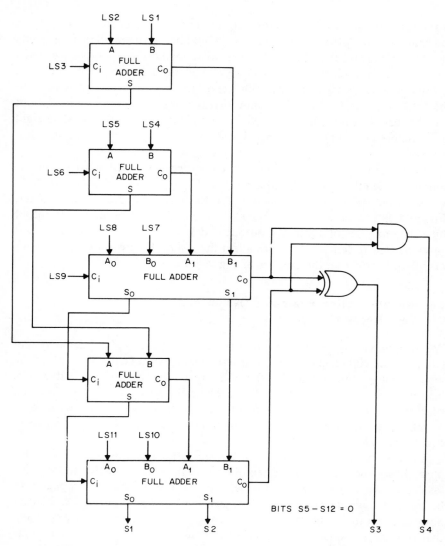

Fig. 8.40 Network to map left-shift count into four-bit binary number (After Blankenship).

1. Circuit design.
2. Design of masks.
3. Fabrication and testing of the package.

   A complete schematic of the two-bit adder is shown in Fig. 8.41. There is no need for the reader to understand the circuit details but, by observing the dotted lines and then referring to Fig. 8.42, we can see how the same circuit was altered to give the three versions (L101, L102, and L103) of the

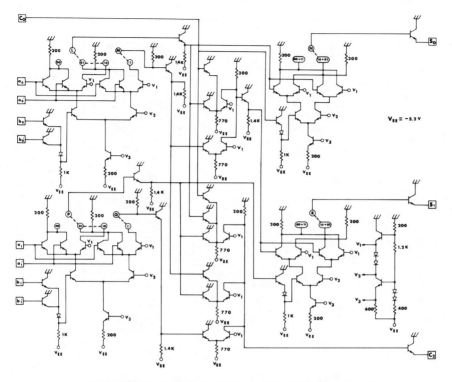

Fig. 8.41 Schematic of the L101 two-bit adder (After Pezaris).

two-bit adder packages required for Fig. 8.35. The important point, from a fabrications viewpoint, was that all variations could be realized by changing only the second-level metal interconnections.

The mask design for these packages was done manually because at the time the computer-aided mask-making facility was not sufficiently flexible. At this point it might be worth interjecting a few comments on mask making. Much of the fabrication depends on photographic likenesses of the actual chip layout and the computer can aid greatly in the transition from the circuit to the layout. In Fig. 8.43 an example is shown of a magnified chip corresponding to the ECL circuit below. The picture shown indicates regions on the chip corresponding to transistors and resistors. For example, in the lower left-hand corner are shown the three transistors and the common collector junction of Q_1, Q_2, and Q_3. The resistors correspond to the long paths terminated at each end by junctions.

Figure 8.44 illustrates the interactive computer program that allows the user to make a very precise layout. Photographs can be produced from this layout that then serve as masks for the various diffusion processes. Figure 8.45 shows the completed L101 chip. Finally, Fig. 8.46 shows a picture of a

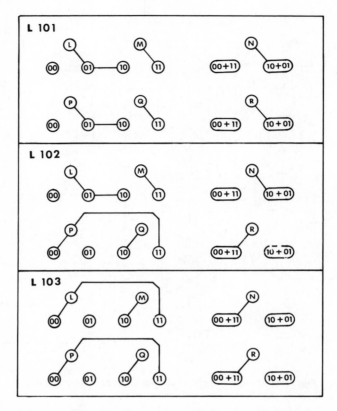

Fig. 8.42 Second-level metal interconnections that cause the basic two-bit adder chip to become an L101, L102, or L103 adder (After Pezaris).

completed (17×17) array multiplier using these logic packages on a conventional four-layer printed circuit card. Multiplication time is 40 nsec.

8.9 Summary

This chapter appears to depart from the previously presented material in this book. Up to now emphasis has been on fundamental theory and design techniques for digital filters and spectrum analyzers. Also, the previous chapters have been oriented toward fairly advanced methods, whereas this chapter is introductory. Probably engineers well versed in digital hardware will find the material too elementary and far from a complete introduction to digital hardware. Despite these shortcomings, we strongly felt the need for such a chapter as a link between the analysis–synthesis ideas of the previous chapters and the implementation-oriented later chapters. Our experience

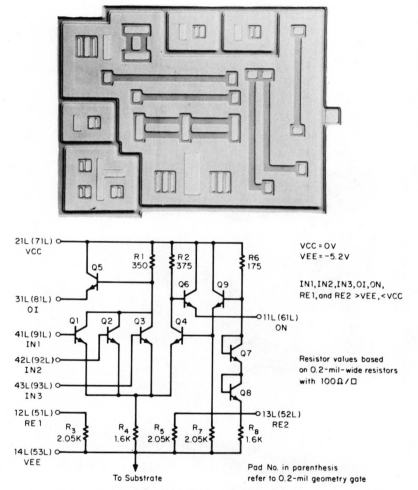

VCC = 0 V
VEE = −5.2 V

IN1, IN2, IN3, OI, ON,
RE1, and RE2 >VEE, <VCC

Resistor values based
on 0.2-mil-wide resistors
with 100 Ω/□

Pad No. in parenthesis
refer to 0.2-mil geometry gate

3-INPUT GATE SCHEMATIC UNDER NORMAL BIAS

Fig. 8.43 ECL circuit and chip (After Pezaris).

537

Fig. 8.44 Mask making using the *TX*-2 computer and an interactive tablet–scope drafting program.

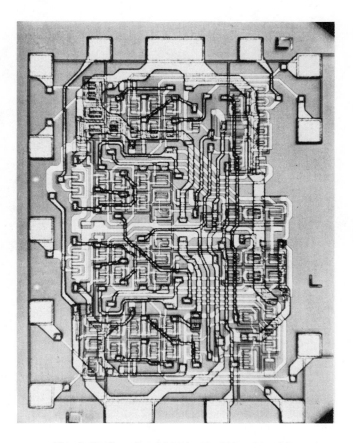

Fig. 8.45 Completed L101 chip (After Pezaris).

Fig. 8.46 The 40-n sec, (17 × 17) array multiplier (After Pezaris).

strongly indicates the need for sophistication in high-speed logic methods and we were unable to find any hardware reference work that accented these methods. This explains our preoccupation with ECL logic and array multipliers. Other methods of multiplication are more readily studied in a host of books.

REFERENCES

1. LANE S. GARRETT, "Integrated Circuit Digital Logic Families," Parts I, II, and III, *IEEE Spectrum*, Oct., Nov., and Dec., 1970.
2. EUGENE R. HNATEK, *A User's Handbook of Integrated Circuits*, John Wiley & Sons, Inc., New York, 1973.
3. *MECL Integrated Circuits Data Book*, Nov., 1972, published by Motorola, Inc.
4. *MECL System Design Handbook*, Oct., 1971, published by Motorola, Inc.
5. PEZARIS, S., "A 40 Nanosecond 17 × 17 Array Multiplier," *IEEE Trans. on Computers*, C-20, 442–447, No. 4, Apr., 1971.
6. P. BLANKENSHIP, B. GOLD, P. MCHUGH, AND C. WEINSTEIN, "Design Study of the Advanced Signal Processor," *Lincoln Lab. Technical Note*, Apr., 1972.

9

Special-Purpose Hardware for Digital Filtering and Signal Generation

9.1 Introduction

In preceding chapters we have been concerned primarily with software realizations of digital signal processing algorithms. Here and in Chapters 10 and 11 we shall be concerned with special-purpose digital hardware for the realization of these algorithms with the goal of speeding up their execution and making possible "real-time" simulations of several desired digital systems.

We have already discussed the basic building blocks of digital hardware including delay networks, adders, multipliers, general-purpose memories, and the various logic circuit types (and their resulting speed, cost, and power consumption tradeoffs) that are commercially available. In the realization of digital systems using special-purpose hardware, the concepts of parallelism, multiplexing, and pipelining are of great importance in achieving a maximum value of performance-to-cost ratio for the particular application being considered. In this chapter we discuss, in particular, various digital hardware realizations of digital filters including direct and cascade FIR realizations and direct, cascade, and parallel IIR realizations. We then shall discuss several practical examples of digital systems which incorporate digital filters in their realization and which have been built in hardware. Finally a discussion of a digital frequency synthesizer and a digital noise generator will be given.

9.2   Direct Form FIR Hardware

Let us consider realizations for the direct form FIR filter shown in Fig. 9.1. First, we shall define a basic control structure that is valid for a variety of realizations. From this basis we shall be able to derive the variations in terms of the parallelism used. This parallelism will encompass not only multi-arithmetic elements but also memory parallelisms.

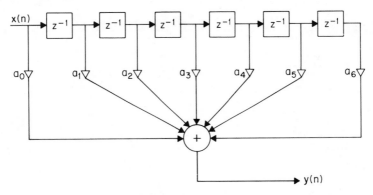

Fig. 9.1 Direct form FIR filter.

Figure 9.2 shows a simple structure for realizing the filter of Fig. 9.1 using a single computational element (consisting of a multiplier and an adder), a shift register to hold the filter states, and a ROM for the coefficients. The important element of this figure is the manner in which the shift register is controlled. By means of a multiplier and an accumulator, a single output sample can be computed by successive addition as the shift register circulates. During the first computation [of $a_6 x(n - 6)$], the new input $x(n)$ enters the shift register while $x(n - 6)$ is shifted off the end. Afterward, each iteration includes a circulation of one datum around the shift register, as indicated in the chart accompanying Fig. 9.2. When $y(n)$ is obtained, it is sent on while the accumulator is cleared; then the next major cycle begins.

In Fig. 9.2, we have as yet not specified in detail the format of the shift register memory. The first interesting point has to do with components technology. It has turned out that LSI techniques have proved quite successful with MOS shift register elements. At the current time, up to several thousand bits can be arranged serially on a single chip. This indicates that for many FIR filters a single package contains all the necessary state memory, provided that the data are arranged serially and, of course, fed serially into

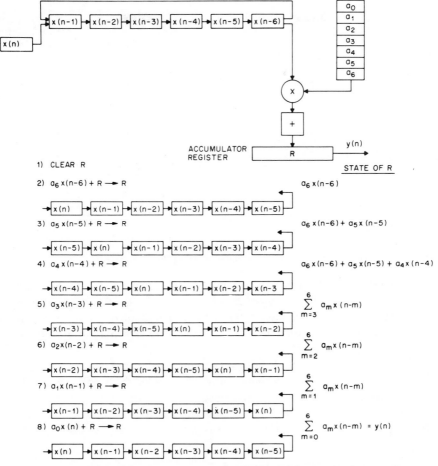

Fig. 9.2 Structure, program, and states for implementation of direct form FIR filter.

the computational element. Such potential economy in the memory elements tempts us to contrive structures for direct form FIR filters that use this component. The simplest such realization is shown in Fig. 9.3 wherein each product is performed by successive gating and adding from each bit of the state variable [$x(n-6)$, in this case]. All states of Fig. 9.3 are in serial form, although only $x(n-6)$ is explicitly shown that way.

In performing the multiplication (via the shifts and adds), we must recall that in dealing with signed numbers the precise algorithm varies depending on the sign bit. For example, if sign-magnitude convention is used, ordinary summing takes place and the sign of the product is determined by comparing the sign bit of the state with the sign bit of the coefficient. This is certainly an

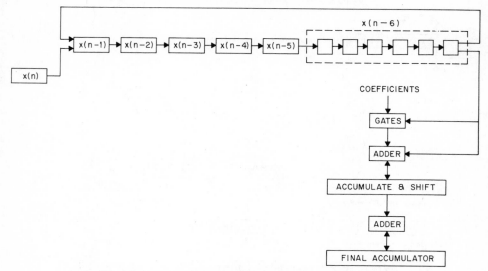

Fig. 9.3 Implementation of FIR filter using Booth algorithm.

efficient method of add–shift multiplication; however, the final accumulation of all the products is more complicated since addition of sign-magnitude numbers requires an adder–subtracter with proper control to subtract the smaller from the larger number. To avoid this, the sign magnitude can be changed to 2's complement.

Rather than sign magnitude, the states and coefficients can be stored as 2's-complement numbers. This means that control of the smaller adder in Fig. 9.3 is needed. A recommended procedure for implementing the algorithm of Fig. 9.3 is Booth's algorithm (as discussed in Chapter 8), which, by looking at two successive bits, controls whether to add or subtract.

9.3 Parallelism for Direct Form FIR

If the data word length is b bits and the filter requires N multiplications per iteration and if it is assumed that the shift register shift time per bit is equal to the add time, then the highest data rate R that can be processed by the realization discussed in Sec. 9.2 (e.g., Fig. 9.3) is $1/(Nb\tau)$ where τ is the shift time per bit. For example, if $\tau = 100$ nsec, $N = 16$, and $b = 16$, then

$$R = \frac{10^9}{256 \times 100} = 39,062.5 \text{ Hz}$$

For applications where R must be greater, we can devise various forms of parallelism for both memory and arithmetic. By using two or more shift register memories, speed can be proportionally increased; similarly, by connecting two or more computational elements, we can also gain speed.

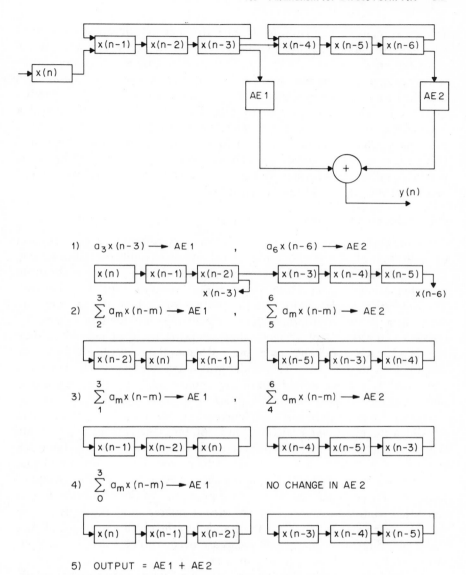

Fig. 9.4 Structure, program, and states for FIR implementation using two parallel AE's.

Let us look carefully at these variations to make sure we understand the extra control functions and arithmetic hardware needed.

Figure 9.4 shows the elements and the paths of a system having two arithmetic elements operating in parallel. Figure 9.4 also shows the program for using the structure in this figure and the accompanying shift register state after each multiply-accumulate operation. This parallel structure yields a

2:1 speed increase compared to the one arithmetic element (AE) system of Fig. 9.3. AE1 and AE2 are presumed to have the same hardware as the computational element of Fig. 9.3. This type of parallelism is straightforwardly extended to any number of AE's; however, it should be kept in mind that the high level of circuit integration is maintained only if each subsection of the shift register contains many bits. Thus, for example, if we had a twentieth-order filter with 16-bit states, using a single AE would mean we could pack 320 bits on a chip. Using two AE's for the same order filter would require a tapped shift register, which with presently available components would mean two chips, although there is no reason that a special tapped shift register of this sort could not be fabricated.

9.4 Cascade FIR Filter

The same basic structure (as in Fig. 9.4) can be extended to the cascade form of realization of FIR filters. Figure 9.5 shows an example of a three-section filter, each section being fourth order. Figure 9.6 shows how to construct this filter using a single AE and one serial shift register for state memory. The only trick needed is to save the outputs of each of the sections and enter them at the appropriate clock time in the shift register. Figure 9.7 traces a complete set of 12 clock times showing the contents of the state memory at each clock time and a single program that controls the hardware and implements the filter. The reader will notice that $x(n-4)$, $u(n-4)$, and $v(n-4)$ must be shifted out of the memory while the three new inputs must be shifted in as is shown in Fig. 9.7.

Figure 9.5 can also be easily implemented via parallel arithmetic hardware for which there are two straightforward options. One is to have a separate AE for each of the fourth-order sections, which leads to the three-AE system of Fig. 9.8. As seen by studying the accompanying program in Fig. 9.8, all three AE's proceed to compute successive partial sums of products until we have compiled the sum of four products. In the fifth cycle, $u(n)$ is computed from the first AE, the new $x(n)$ is entered, and the registers are shifted. Given $u(n)$ as input, it can be added to the result in the second AE, etc. Thus, three additional clock cycles are needed to complete the computation, with each AE working in sequence.

An interesting exercise for the reader is to create an appropriate cascade FIR structure using four AE's and using enough parallelism such that a complete section is completed in one cycle.

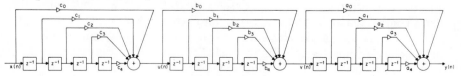

Fig. 9.5 Cascade FIR filter.

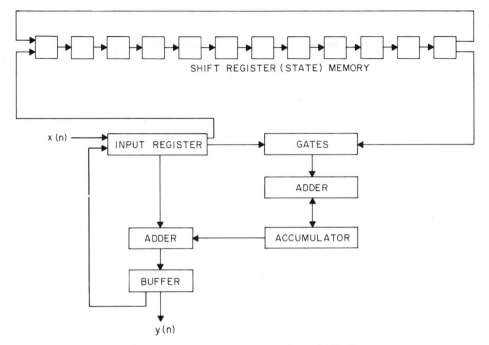

Fig. 9.6 Single AE implementation of cascade FIR filter.

	v(n-1)	v(n-2)	v(n-3)	v(n-4)	u(n-1)	u(n-2)	u(n-3)	u(n-4)	x(n-1)	x(n-2)	x(n-3)	x(n-4)
1)	——	v(n-1)	v(n-2)	v(n-3)	v(n-4)	u(n-1)	u(n-2)	u(n-3)	u(n-4)	x(n-1)	x(n-2)	x(n-3)
2)	x(n-3)	——	v(n-1)	v(n-2)	v(n-3)	v(n-4)	u(n-1)	u(n-2)	u(n-3)	u(n-4)	x(n-1)	x(n-2)
3)	x(n-2)	x(n-3)	——	v(n-1)	v(n-2)	v(n-3)	v(n-4)	u(n-1)	u(n-2)	u(n-3)	u(n-4)	x(n-1)
4)	x(n-1)	x(n-2)	x(n-3)	——	v(n-1)	v(n-2)	v(n-3)	v(n-4)	u(n-1)	u(n-2)	u(n-3)	u(n-4)
5)	x(n)	x(n-1)	x(n-2)	x(n-3)	——	v(n-1)	v(n-2)	v(n-3)	v(n-4)	u(n-1)	u(n-2)	u(n-3)
6)	u(n-3)	x(n)	x(n-1)	x(n-2)	x(n-3)	——	v(n-1)	v(n-2)	v(n-3)	v(n-4)	u(n-1)	u(n-2)
7)	u(n-2)	u(n-3)	x(n)	x(n-1)	x(n-2)	x(n-3)	——	v(n-1)	v(n-2)	v(n-3)	v(n-4)	u(n-1)
8)	u(n-1)	u(n-2)	u(n-3)	x(n)	x(n-1)	x(n-2)	x(n-3)	——	v(n-1)	v(n-2)	v(n-3)	v(n-4)
9)	u(n)	u(n-1)	u(n-2)	u(n-3)	x(n)	x(n-1)	x(n-2)	x(n-3)	——	v(n-1)	v(n-2)	v(n-3)
10)	v(n-3)	u(n)	u(n-1)	u(n-2)	u(n-3)	x(n)	x(n-1)	x(n-2)	x(n-3)	——	v(n-1)	v(n-2)
11)	v(n-2)	v(n-3)	u(n)	u(n-1)	u(n-2)	u(n-3)	x(n)	x(n-1)	x(n-2)	x(n-3)	——	v(n-1)
12)	v(n-1)	v(n-2)	v(n-3)	u(n)	u(n-1)	u(n-2)	u(n-3)	x(n)	x(n-1)	x(n-2)	x(n-3)	——
13)	v(n)	v(n-1)	v(n-2)	v(n-3)	u(n)	u(n-1)	u(n-2)	u(n-3)	x(n)	x(n-1)	x(n-2)	x(n-3)

(a)

Fig. 9.7 (a) Successive states of memory for cascade FIR filter.

547

1) $c_4 x(n-4)$ $\longrightarrow$ ACC.

2) $c_4 x(n-4) + c_3 x(n-3)$ $\longrightarrow$ ACC.

3) $\displaystyle\sum_{m=2}^{4} c_m x(n-m)$ $\longrightarrow$ ACC.

4) $\displaystyle\sum_{m=1}^{4} c_m x(n-m)$ $\longrightarrow$ ACC.

5) $u(n) = \displaystyle\sum_{m=0}^{4} c_m x(n-m)$: ENTER $x(n)$ INTO STATE MEMORY :

   ENTER $u(n)$ INTO INPUT REGISTER : $b_4 u(n-4)$ $\longrightarrow$ ACC.

6) $b_4 u(n-4) + b_3 u(n-3)$ $\longrightarrow$ ACC.

7) $\displaystyle\sum_{m=2}^{4} b_m u(n-m)$ $\longrightarrow$ ACC.

8) $\displaystyle\sum_{m=1}^{4} b_m u(n-m)$ $\longrightarrow$ ACC.

9) $v(n) = \displaystyle\sum_{m=0}^{4} b_m u(n-m)$: ENTER $u(n)$ INTO STATE MEMORY :

   ENTER $v(n)$ INTO INPUT REGISTER : $a_4 v(n-4)$ $\longrightarrow$ ACC.

10) $a_4 v(n-4) + a_3 v(n-3)$ $\longrightarrow$ ACC.

11) $\displaystyle\sum_{m=2}^{4} a_m v(n-m)$ $\longrightarrow$ ACC.

12) $\displaystyle\sum_{m=1}^{4} a_m v(n-m)$ $\longrightarrow$ ACC.

13) $y(n) = \displaystyle\sum_{m=0}^{4} a_m v(n-m)$: OUTPUT $y(n)$

   ENTER $v(n)$ INTO STATE MEMORY : ENTER NEW $x(n)$
   INTO INPUT REGISTER

(b)

Fig. 9.7 (b) Program for cascade FIR filter.

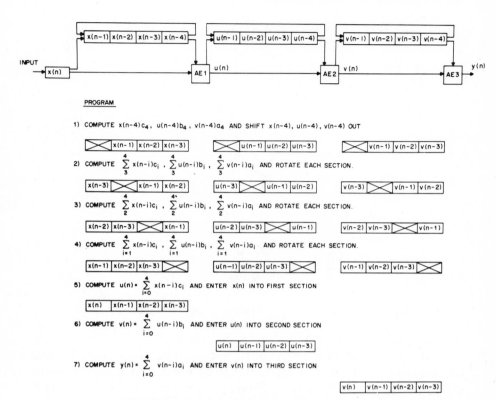

Fig. 9.8 Implementation of cascade FIR filter with three AE's.

9.5 Highly Parallel FIR Direct Form

As we have pointed out, the basic structure of a shift register can accommodate parallelism in both arithmetic computation and state memory flow. The natural question to ask is, how far can we go in parallelism before the system becomes too awkward? To investigate such questions further, let us look at a rather natural seeming parallelism, namely, the use of N arithmetic elements for an Nth-order system. Such a system is shown in Fig. 9.9, wherein each state variable has its own hardware multiplier. Note that in this case there is no need for shift register circulation and the shift register is used just like an analog tapped delay line. Note also that the advantages of present LSI memory elements disappear because of the need for taps. We remark that the shift registers, just as in the previous examples, can be run bit-wise in parallel. It can also be seen that there are N adders in addition to the N multipliers. Assuming that multiplication time is matched to the time needed to shift from one state register to the next, there seems to be an added time needed to accumulate the products. This extra time can be avoided by

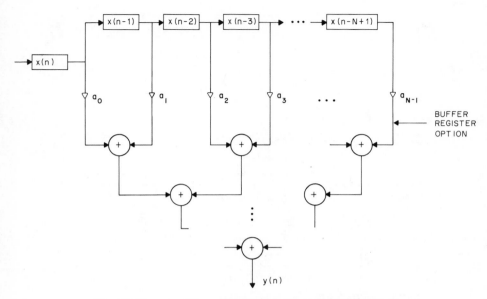

Fig. 9.9 The use of N parallel elements for an Nth-order FIR filter.

inserting buffers after the multipliers at the indicated place in Fig. 9.9. The insertion of such buffers is a form of pipelining. If we assume that the total accumulation time is exactly equal to the multiply time, then a single sample's delay is introduced in the output. Figure 9.9 is not the only way of using N arithmetic elements for realizing an Nth-order parallel FIR filter. Another configuration is shown in Fig. 9.10.

The equations of this network are as follows:

$$x_1(n) = a_0 x(n - 1) + a_1 x(n)$$

$$x_2(n) = x_1(n - 1) + a_2 x(n) = a_0 x(n - 2) + a_1 x(n - 1) + a_2 x(n)$$

$$x_3(n) = x_2(n - 1) + a_3 x(n) = \sum_{i=0}^{3} a_i x(n - 3 + i)$$

$$y(n) = x_3(n - 1) + a_4 x(n) = \sum_{i=0}^{4} a_i x(n - 4 + i)$$

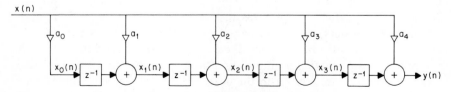

Fig. 9.10 Another realization for an FIR filter.

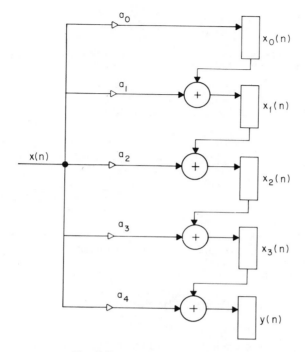

Fig. 9.11 An implementation of an FIR filter.

We see that our result is a convolution, just as before, although the indices appear scrambled. By setting $b_0 = a_3$, $b_1 = a_2$, $b_2 = a_1$, $b_3 = a_0$, we find the usual form. Figure 9.11 shows a simple implementation of the network of Fig. 9.10.

Figure 9.11 suggests one neat way of building the hardware, provided that multiplication is done by treating $x(n)$ as a serial bit stream, each bit successively multiplying the coefficients. For this scheme, the products can be made available serially and the adders can be serial adders. Thus, while register $x_i(n)$ is shifted into an adder, it is receiving the new $x_i(n)$ in a serial version from the product $a_i x(n)$ plus $x_{i-1}(n)$.

9.6   Direct Form IIR Filters

Up to now, the hardware configurations have all corresponded to FIR filters. Relatively minor changes in these basic structures will also accommodate IIR filters. As an example we shall do a hardware design of an all-pole filter using the direct form realization as shown in Fig. 9.12. In this case we assume that coefficients enter the filter as parallel words, while the state memory is serial, just as in the FIR case. The difference equation for the all-pole IIR

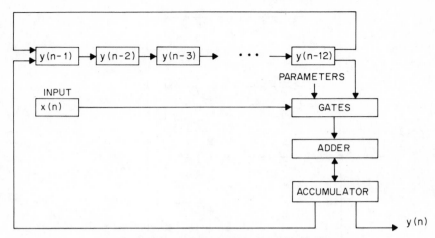

Fig. 9.12 Structure for direct form IIR filter.

filter of Fig. 9.12 is of the form

$$y(n) = \sum_{j=1}^{12} y(n-j)a_j + bx(n)$$

where the a_j's and b are all known.

The states $y(n-1)$ and the input $x(n)$ are presented as a serial bit stream to the gates. During the multiplication $y(n-12)a_{12}$, the state variable $y(n-12)$ is shifted off the end of the shift register. During the remaining 11 multiplications $y(n-11)a_{11}$ through $y(n-1)a_1$, the shift register circulates. Thus, at the instant before $x(n)$ is applied, the shift register state is as shown in Fig. 9.13.

Until $bx(n)$ is computed, $y(n)$ is not available for entry into the shift register. Therefore, no circulation takes place during that computation. The final cycle includes shifting the accumulated $y(n)$ into the shift register, entering a new $x(n)$ and then clearing the accumulator just prior to beginning the next cycle.

In typical applications, a reasonable sampling rate for the output $y(n)$ is 8 kHz, corresponding to a 125-μsec sampling interval. Assuming that each state variable is 16 bits long, then the total number of shift register 1-bit

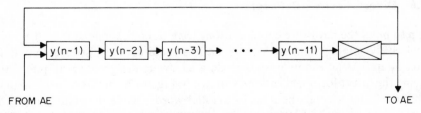

Fig. 9.13 State memory of IIR filter prior to application of input $x(n)$.

clocking cycles is $16 \times 14 = 224$, which means that clock pulses must appear at $\frac{125}{224}$ µsec ($= 0.55$ µsec). Thus, the shift register must circulate at about a 2-megabit rate and the adder must add at a 2-megabit rate. Both these numbers are well within moderate speed component technology and can lead to LSI fabrications of such devices.

9.7 Cascade Form IIR Filters

Figure 9.14 shows an IIR filter realized as a cascade of three second-order sections, each containing two poles and two zeros. The individual second-order sections are realized using the canonic direct form—i.e., sharing common delays. A simple two-AE realization of the cascade structure is

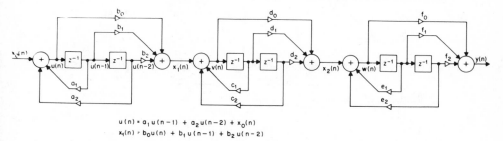

$$u(n) = a_1 u(n-1) + a_2 u(n-2) + x_0(n)$$
$$x_1(n) = b_0 u(n) + b_1 u(n-1) + b_2 u(n-2)$$

Fig. 9.14 IIR filter realized as a cascade of three second-order sections.

shown in Fig. 9.15, along with the necessary sequence of instructions (the control) for the specific filter of Fig. 9.14. This structure is similar to those discussed previously in that the state memory is stored in serial shift registers and the multiplier is an add–shift, one-state bit-at-a-time device. The two-AE system is convenient for computing partial sums for the feed-forward and the feed-backward portion of each second-order section in the cascade. Since the feed-backward portion of the calculation must be completed before $u(n)$, $v(n)$, and $w(n)$ can be used in the feed-forward calculation, two cycles of shifting state variables into the AE's are followed by two internal AE calculations—one to obtain the remaining input for the feed-forward path and the other to generate the input to the next section.

The program in Fig. 9.15 notes when shifts occur in the shift registers and when no shifts are to take place. The sixth-order filter is seen to require a total of seven shifts and six times when no shift is to take place. Scaling (i.e., simple shifting) of $x_0(n)$, $x_1(n)$, and $x_2(n)$ can also take place when they are added into AE1 in the calculation.

9.8 Multiplexing

Few practical systems are as simple as the ones we have discussed up to now. More common are configurations using filter banks with either IIR or FIR filters where each filter may itself be a cascade or parallel form. The greater

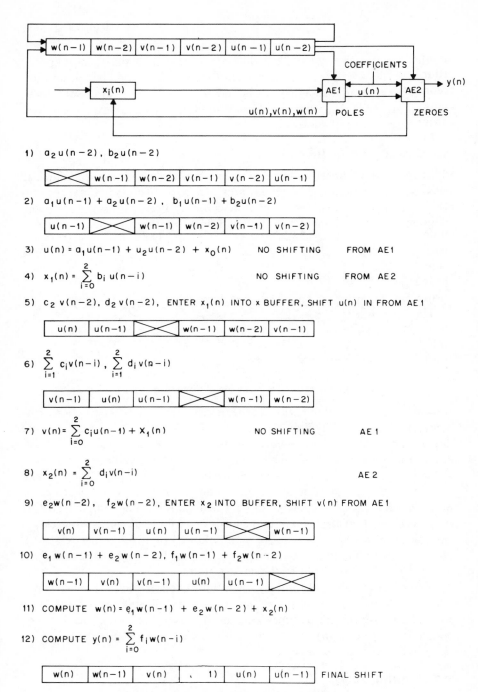

Fig. 9.15 Two AE realization of cascade IIR filter.

the number of filters to be multiplexed, the more beneficial a digital realization seems to be since the LSI developments in serial memories can easily accommodate the additional states. Caution is desirable in trying to assign too many diverse functions to a system using common memory and arithmetic hardware, however, because the control hardware will tend to blow up.

Figures 9.16 to 9.18 show an example of how multiplexing can be used to realize three separate FIR filters with a single computational element. In Fig. 9.16 it is seen that the three independent inputs are $x_1(n)$, $x_2(n)$, and $x_3(n)$ with outputs $y_1(n)$, $y_2(n)$, and $y_3(n)$, respectively. The appropriate filter coefficients are also given in this figure. Figure 9.17 shows a simple hardware configuration for realizing all three FIR filters in which the inputs are multiplexed and processed independently by the arithmetic element. A distributor is responsible for sending the correct output to the appropriate channel. Finally, Fig. 9.18 gives a program that describes the control for the hardware and gives the state of the circulating memory at various points during the computational cycle.

The example of Figs. 9.16 to 9.18 is not the only possible type of multiplexing for digital hardware. A second form of multiplexing is when a single input is filtered simultaneously by a number of different filters as would occur in a filter bank spectrum analyzer. The major consideration in multiplexing

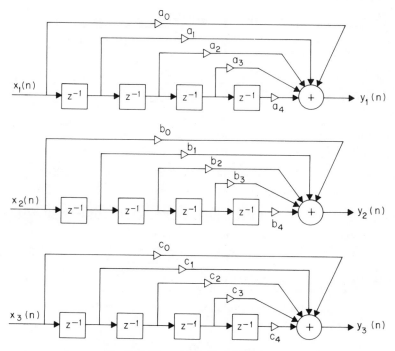

Fig. 9.16 Illustration of multiplexing.

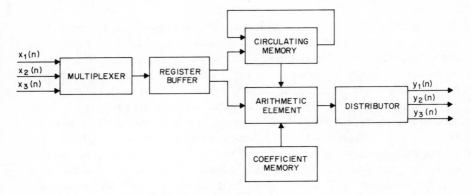

Fig. 9.17 Hardware configuration for multiplexed FIR filter.

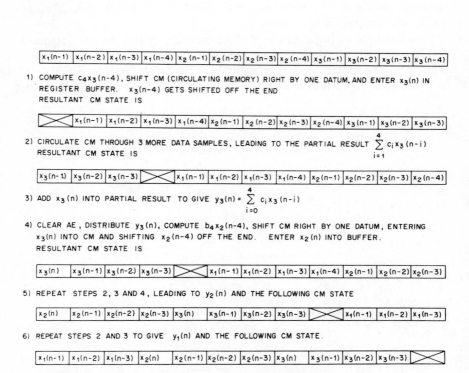

| $x_1(n-1)$ | $x_1(n-2)$ | $x_1(n-3)$ | $x_1(n-4)$ | $x_2(n-1)$ | $x_2(n-2)$ | $x_2(n-3)$ | $x_2(n-4)$ | $x_3(n-1)$ | $x_3(n-2)$ | $x_3(n-3)$ | $x_3(n-4)$ |

1) COMPUTE $c_4 x_3(n-4)$, SHIFT CM (CIRCULATING MEMORY) RIGHT BY ONE DATUM, AND ENTER $x_3(n)$ IN REGISTER BUFFER. $x_3(n-4)$ GETS SHIFTED OFF THE END
   RESULTANT CM STATE IS

| ✕ | $x_1(n-1)$ | $x_1(n-2)$ | $x_1(n-3)$ | $x_1(n-4)$ | $x_2(n-1)$ | $x_2(n-2)$ | $x_2(n-3)$ | $x_2(n-4)$ | $x_3(n-1)$ | $x_3(n-2)$ | $x_3(n-3)$ |

2) CIRCULATE CM THROUGH 3 MORE DATA SAMPLES, LEADING TO THE PARTIAL RESULT $\sum_{i=1}^{4} c_i x_3(n-i)$
   RESULTANT CM STATE IS

| $x_3(n-1)$ | $x_3(n-2)$ | $x_3(n-3)$ | ✕ | $x_1(n-1)$ | $x_1(n-2)$ | $x_1(n-3)$ | $x_1(n-4)$ | $x_2(n-1)$ | $x_2(n-2)$ | $x_2(n-3)$ | $x_2(n-4)$ |

3) ADD $x_3(n)$ INTO PARTIAL RESULT TO GIVE $y_3(n) = \sum_{i=0}^{4} c_i x_3(n-i)$

4) CLEAR AE, DISTRIBUTE $y_3(n)$, COMPUTE $b_4 x_2(n-4)$, SHIFT CM RIGHT BY ONE DATUM, ENTERING $x_3(n)$ INTO CM AND SHIFTING $x_2(n-4)$ OFF THE END. ENTER $x_2(n)$ INTO BUFFER.
   RESULTANT CM STATE IS

| $x_3(n)$ | $x_3(n-1)$ | $x_3(n-2)$ | $x_3(n-3)$ | ✕ | $x_1(n-1)$ | $x_1(n-2)$ | $x_1(n-3)$ | $x_1(n-4)$ | $x_2(n-1)$ | $x_2(n-2)$ | $x_2(n-3)$ |

5) REPEAT STEPS 2, 3 AND 4, LEADING TO $y_2(n)$ AND THE FOLLOWING CM STATE

| $x_2(n)$ | $x_2(n-1)$ | $x_2(n-2)$ | $x_2(n-3)$ | $x_3(n)$ | $x_3(n-1)$ | $x_3(n-2)$ | $x_3(n-3)$ | ✕ | $x_1(n-1)$ | $x_1(n-2)$ | $x_1(n-3)$ |

6) REPEAT STEPS 2 AND 3 TO GIVE $y_1(n)$ AND THE FOLLOWING CM STATE.

| $x_1(n-1)$ | $x_1(n-2)$ | $x_1(n-3)$ | $x_2(n)$ | $x_2(n-1)$ | $x_2(n-2)$ | $x_2(n-3)$ | $x_3(n)$ | $x_3(n-1)$ | $x_3(n-2)$ | $x_3(n-3)$ | ✕ |

7) FINAL STEP: SHIFT IN $x_1(n)$ AND SHIFT OUT RIGHT-MOST "GARBAGE" DATUM. BY CHANGING n TO n-1 WE ARE NOW BACK TO INITIAL STATE AND READY FOR THE NEXT ITERATION.

Fig. 9.18 Control program for multiplexed filter.

is the overall data rate required by the total number of multiplexed filters to be realized. As long as this data rate is within the speed limitations of the components, the overall system can generally be realized in a multiplexed format.

In the next sections we shall discuss several practical systems which have been implemented in digital hardware and which employ multiplexing to make the digital realization more attractive.

9.9 Digital Touch-Tone Receiver (TTR)

An excellent illustrative example of the application of digital hardware to practical systems is the all-digital touch-tone receiver that was designed and constructed at Bell Telephone Laboratories and described by Jackson, Kaiser, and McDonald. Figure 9.19 shows the multiplexed hardware used in the realization and Fig. 9.20 shows the digital TTR illustrating the multiplexed filters as well as the nonlinear processing elements. The basic arithmetic unit, which is capable of realizing a cascade IIR filter, has four multipliers and two three-input adders. A read-only coefficient

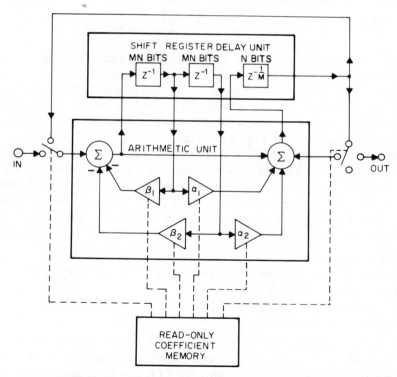

Fig. 9.19 Multiplexed hardware for digital touch-tone receiver (After Jackson, Kaiser, and McDonald).

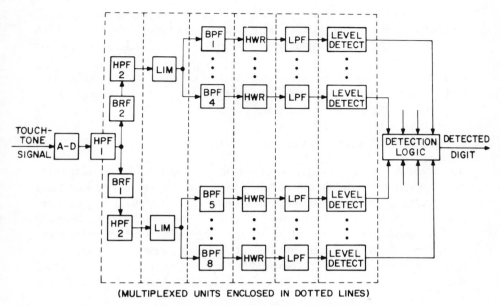

(MULTIPLEXED UNITS ENCLOSED IN DOTTED LINES)

Fig. 9.20 Digital touch-tone receiver, showing multiplexed filters and nonlinear units (After Jackson, Kaiser, and McDonald).

memory is used to store the filter coefficients and a shift register delay line is used to store the filter state variables. In the block diagram of Fig. 9.20, the highpass filters (HPF) are third order; the band-rejection filters (BRF) are each sixth order; the bandpass filters (BPF) are each second order; and the lowpass filters (LPF) are each first order. All these filters are realized using the multiplexed hardware of Fig. 9.19. The nonlinear processing required to implement the TTR includes limiters (LIM), half-wave rectifiers (HWR), and level detectors. All the nonlinear processing is also implemented digitally.

A multiplexing factor of 8:1 is used to realize all the units within dotted lines in Fig. 9.20 in single multiplexed units. Thus the linear filter operations require two multiplexed second-order sections and one multiplexed first-order section. In particular some parameters of the TTR are as follows:

Sampling rate: 10 kHz.
Input quantization (following A/D conversion): 7 bits per sample.
Data word length: 10 bits per sample.
Filter coefficients: 6 bits fractional part.
Multiplexing factor: 8:1.

Thus the overall data rate which is computed as sampling rate times data word length times multiplexing factor equals $800K$ bits per second. The major componentry required to realize the TTR was 40 serial adders and 400 bits of shift register storage.

9.10 Digital Time Division Multiplexing (TDM) to Frequency Division Multiplexing (FDM) Translator

An excellent practical example of the use of digital filter hardware in a large system is given by Freeny, et al. in their description of a TDM to FDM translator. In the telephone system, messages are often formated in either a TDM or an FDM format; thus there is need of a translator that converts one format to another. Figure 9.21 shows the desired translation for a series of 12 TDM speech inputs, each sampled at 8 kHz, for which we wish to assign a different portion of the spectrum to each channel for FDM transmission. In particular the band from 56 to 112 kHz is chosen for the 12 channels. Each of the 12 speech signals is stored in single sideband format in the FDM system. Only 4 kHz is alloted to each band, or a total of 48 kHz is required for all 12 channels. The additional 56 kHz of bandwidth (from 0 to 56 kHz) is alloted to cross talk between channels. Thus the TDM to FDM translation consists of a method of single sideband heterodyning to move the spectrum the appropriate amount.

To avoid cross talk (i.e., interference) between adjacent channels, each signal must be sharply filtered at 4 kHz. A special complication occurs when one contemplates building an all-digital version of this translator in that the spectrum of sampled speech, being periodic, is spread over all frequencies. Thus a simple heterodyning technique is inappropriate. Another complication results from the desire to implement the system at as low a rate as possible, thus avoiding the need for unduly fast computational elements. If not for these considerations, a simple-minded approach for implementing

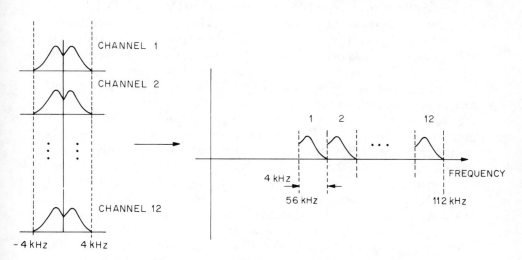

Fig. 9.21 Translation from 12 audio channels to a single FDM channel.

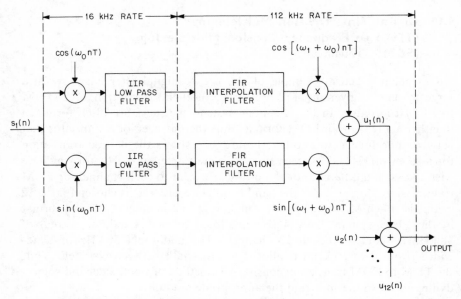

Fig. 9.22 Block diagram of TDM to FDM translator (After Freeny, Kieburtz, Mina and Tewksbury).

the translator would be to run each digitized speech waveform through an appropriate bandpass filter. This approach requires a sampling rate of over 100 kHz (112 kHz to be specific), however, and thus requires expensive hardware. To avoid these difficulties, Freeny, et al. devised a scheme whereby the majority of the computation takes place at a low sampling rate (16 kHz), the remainder of the computation requiring a 112-kHz rate.

Figure 9.22 shows a block diagram of the TDM to FDM translator implemented by Freeny, et al., while Fig. 9.23 shows spectral representations of the signals at various points in the system. The original signal spectrum [of $s_1(n)$] is shown in Fig. 9.23(a) and is seen to be periodic with a period of 8 kHz. The signal $s_1(n)$ is modulated by a 2-kHz ($\omega_0 = 4000\pi$) cosine and sine to give the signal spectra, shown in Fig. 9.23(b) (for the cosine) and Fig. 9.23(e) (for the sine). (Note that the signals being drawn may be complex and that we are only demonstrating the processing pictorially.) The frequency response of the 16-kHz sampling rate IIR lowpass filter is shown in Fig. 9.23(c). The lowpass filter has a 2-KHz passband. Figures 9.23(d) and (f) show the spectra at the outputs of the IIR lowpass filters. The sampling rate of the system is then increased to 112 kHz (seven times the previous rate of 16 kHz) by filling in six zero-valued samples between each 16-kHz sample; the augmented sequence is filtered by an FIR interpolating lowpass filter with the frequency response shown in Fig. 9.23(g). It should be noted that since six out of every seven input samples to the FIR filter are exactly zero, only

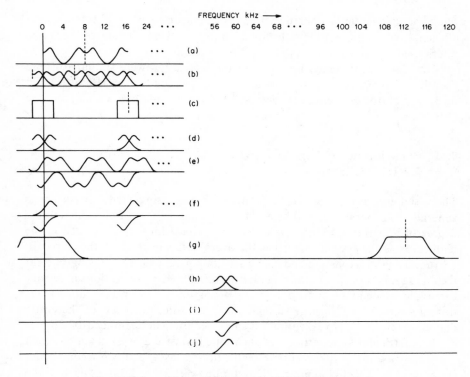

Fig. 9.23 Spectral transformations in TDM to FDM translator.

one-seventh of the number of multiplies per output sample is required to process the sequence. Thus for a 21-point impulse response filter ($N = 21$), as was actually used, only three multiplications per sample were required to realize the filter. The final modulation by the cosine and sine functions shifts the signal spectrum to the desired high-frequency band as shown in Fig. 9.23(h) and (i). Finally, the addition of the quadrature signals gives the desired single sideband spectrum as shown in Fig. 9.23(j). By adding all 12 such channels, the desired FDM signal is obtained.

It should be noted that this scheme depends on precision removal of the undesired aliased spectrum by signal subtraction. This kind of operation is quite appropriate for digital implementation and is seen to work quite well. One other tricky aspect of this system is the complementary filtering operations of the 16-kHz IIR lowpass filter and the 112-kHz FIR interpolating lowpass filter. Clearly all filtering could have been done at the high rate of 112 kHz but the extra computation would have made the entire approach impractical.

Some particular aspects of the digital filters are that the IIR lowpass filter used was a ninth-order design that was realized with 8-bit coefficients and a 22-bit word length for the filter variables. The FIR interpolating lowpass filter was a twentieth-order design that was realized with 10-bit filter co-efficients and 18-bit word lengths for the filter variables. Details of the actual filters used are given in the paper by Freeny, et al.

The actual arithmetic units used in the realization were similar to those used by Jackson, et al. in the digital TTR of Sec. 9.9 for the IIR filters and fairly straightforward circuitry was used for the modulators, adders, and the FIR filter.

9.11 Partitioning of Digital Filters for IC Realization

The ultimate criterion for realization of systems using digital hardware is generally an economic one. The cost of a digital system is often related to the degree of integration obtainable in the individual chips that are used in the realization of the system, as well as the overall bit rate at which the chips can be run. Heightley has studied one partitioning scheme that, using current technology, is capable of realizing all the digital filtering operations necessary for a digital channel bank consisting of 24 fourth-order IIR filters. Partitioning means dividing the storage and processing functions of the entire digital system for LSI realization on small economical beam lead IC chips.

The particular system that was partitioned consisted of 24 channels, each containing a fourth-order recursive realization (a cascade of two second-order systems) of an IIR digital filter. The sampling rate at the input of each of the 24 channels was 32 kHz even though the input speech was sampled at 8 kHz. The interpolation to 32 kHz was used so that an inexpensive binary rate multiplier–delta modulator could be used as a digital-to-analog converter. The word length of the digital filter variables was 16 bits and the filter coefficient word length was 12 bits. Thus a total of $(16 \times 24 \times 4 = 1536)$ bits of storage were required for the filter variables. Each multiplexed second-order section had four multipliers for a total of $(4 \times 12 = 48)$ bits of multi-plication per second-order section.

The overall bit rate at which the chips could run was 25 MHz. Heightley partitioned the system as follows:

Storage: 128 bits per chip	$\rightarrow$	12 chips
Multipliers: 4 bits per chip	$\rightarrow$	12 chips
Three input adders: 2 required	$\rightarrow$	2 chips
Overflow detectors: 4 required	$\rightarrow$	2 chips
Two's complementors: 4 required	$\rightarrow$	2 chips
Control logic: 1 required	$\rightarrow$	10 chips

This gives a total of 40 chips for the entire 24-channel bank using currently available LSI technology. Heightley's estimate was that by the late 1970's only about 12 to 14 chips would be required for this system and that they would run at twice the rate and require half the power of currently available chips.

9.12 Hardware Realization of a Digital Frequency Synthesizer

Synthesis of very accurately defined sinusoidal frequencies has received much attention and has resulted in a lively competition among various instrument manufacturers. At the present writing, all but one of these commercial devices uses analog techniques, whereby a crystal controlled source is fed into a system of mixers and filters to produce a large number of frequencies. In this section, we shall first describe the principles behind digital frequency synthesis and then describe one of the constructed pieces of hardware.

Digital sinusoids can be generated exclusively from a table look-up scheme, from a recursion, or from a combination of table look-up and a computation. Let us first consider the exclusive table look-up as illustrated in Fig. 9.24.

The register which addresses the sine table may have more bits than $\log_2 M$ (where M is the number of sine table entries) which are needed to address every register in the sine table. The reason for this is that the smallest increment determines the lowest possible generated frequency. For example, the length of the sine table M may be 1024, while the address register R may be 20 bits. Now, if the increment is 1, the address in R will dwell on the same sine table register for 1024 successive sampling times before moving to the next table value. For these numbers, a very inexact digital sinusoid is

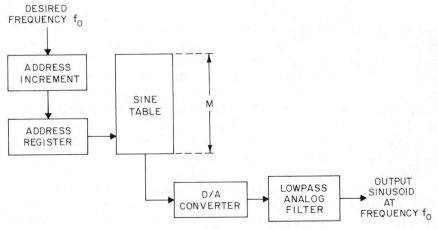

Fig. 9.24 A digital frequency synthesizer.

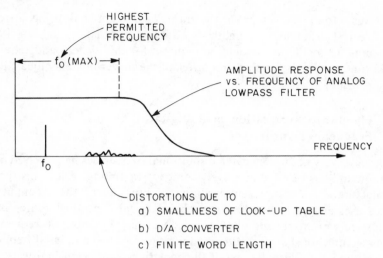

Fig. 9.25 Distortion in digital frequency synthesizer.

produced and this will cause distortion harmonics to appear in the spectrum that cannot readily be filtered out by the lowpass filter.

The situation described above is depicted in Fig. 9.25. In order to obtain as pure a sinusoid as possible, it is necessary that distortions of digital waveform generation be at frequencies above f_c, the cutoff of the lowpass analog filter.

Instead of a table look-up, a simple recursion can be used, as follows: Let $x(n)$ be the complex exponential $\exp[j(2\pi nk/NT)]$. Then, implementation of the formula

$$x(n) = \exp\left(j\,\frac{2\pi k}{NT}\right)x(n-1) \tag{9.1}$$

will generate the desired complex exponential whose real part is a cosinusoid and whose imaginary part is a sinusoid at frequency $f = k/(NT)$. In this manner, neglecting quantizing effects, a perfectly pure digital sine wave can be generated without a table look-up, as shown in Fig. 9.26. The system is set into motion by an applied unit impulse. Changing the frequency implies changing the value of k in the exponential multiplier, with the option of reapplying the impulse to reset the phase or using the last previous output as the new initial condition.

At the current time, a frequency synthesizer using this idea has not been built. Because of quantization noise, the fear exists that too much undesired noise may build up in the system. On the other hand, the theory of limit cycles indicates that a stable oscillation will always occur but whether this will be suitable for producing a very pure analog sinusoid is questionable. Another reason this device has not been built is that coefficient quantization yields unequally spaced frequencies.

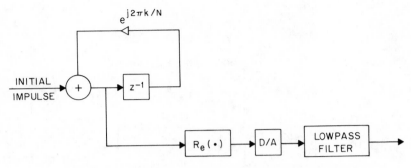

Fig. 9.26 Recursive technique for digital frequency synthesis.

One technique that has been implemented in digital hardware is a hybrid scheme, using both a table look-up plus a multiplier to produce the desired result. The design specifications were (1) 2^{15} frequencies, (2) 409.6-kHz bandwidth, (3) 12.5-Hz frequency spacing (or lowest frequency), and (4) 70-dB spectral purity (defined as ratio of power in the desired frequency to power in any other 100-Hz band). The algorithm was based on the simple trigonometric identities

$$\sin (x + y) = \sin x \cos y + \cos x \sin y$$
$$\cos (x + y) = \cos x \cos y - \sin x \sin y \qquad (9.2)$$

which corresponds to the complex exponential recursion described before. The difference is, however, that the x and the y are always generated from look-up tables so that no recursive quantization noise or limit cycle effect enters into the system. For the desired specifications we allow x to correspond to rough increments (of 2^8 possible frequencies) and y to the fine increments (2^{15} possible increments), the latter being treated, via Eq. (9.2), as a fine grain interpolation on the x increments. Using the additional information that, for any sine or cosine table, only one-quarter of the period be stored, we are reduced to two table look-up memories, each of length 64. The complete digital frequency synthesizer is shown in Fig. 9.27.

The rough increments are stored in 11 bits of a 16-bit by 64-word read-only memory. For the fine increments, the ($\cos y$) term can be approximated by unity to within 14 bits, while the ($\sin y$) term for such small values of y is well-represented by a 5-bit number. Thus, it turns out that two multiplications, each of size 5×8, must be performed as well as two 12-bit additions.

9.13 Techniques for Generating Pseudo-Random Numbers

Although a large number of techniques for generating a sequence of uniformly distributed pseudo-random numbers have been described in the literature,

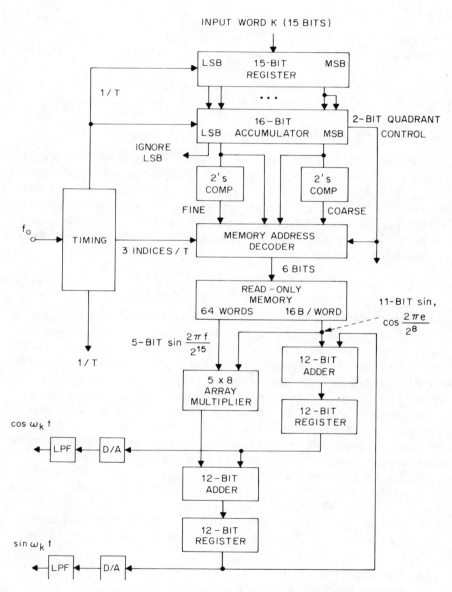

INPUT WORD K (15 BITS)

LSB   15-BIT   MSB
        REGISTER

1/T

· · ·

16-BIT
LSB   ACCUMULATOR   MSB

2-BIT QUADRANT
CONTROL

IGNORE
LSB

2's
COMP

2's
COMP

FINE

COARSE

f_0

TIMING

3 INDICES / T

MEMORY ADDRESS
DECODER

6 BITS

READ - ONLY
MEMORY
64 WORDS   16 B / WORD

11-BIT sin,
$\cos \dfrac{2\pi e}{2^8}$

1/T

5-BIT $\sin \dfrac{2\pi f}{2^{15}}$

5 x 8
ARRAY
MULTIPLIER

12 - BIT
ADDER

12 - BIT
REGISTER

$\cos \omega_k t$

LPF   D/A

12 - BIT
ADDER

12 - BIT
REGISTER

$\sin \omega_k t$

LPF   D/A

Fig. 9.27 Block diagram of a digital frequency synthesizer (After Tierney).

566

we shall briefly describe three of these techniques. One of the oldest, and most popular, techniques for generating pseudo-random numbers is the so-called congruential method whereby the random number $x(n)$ is generated from the preceding random number $x(n-1)$ by the rule

$$x(n) = [A \cdot x(n-1)] \text{ modulo } p \tag{9.3}$$

where p is a large prime and A is a suitably chosen constant. For appropriate values of A, this rule will generate integers in the range 1 to $p-1$ in seemingly random order before repeating. The advantages of the method are its simplicity and the small amount of storage required to carry out the computation. The disadvantages are that it is a slow method because one multiplication (and usually one division) must be performed each iteration and the method is extremely sensitive to values of A and p.

A second popular method for generating random sequences is shown in Fig. 9.28. For this method it is assumed that initially k random numbers in the range $(-1/2, 1/2)$ (as obtained from a table of random numbers) are stored in the k registers. The new random number $x(n)$ is generated by the rule

$$x(n) = [x(n-1) + x(n-k)] \text{ modulo } (\tfrac{1}{2}) \tag{9.4}$$

where modulo $\frac{1}{2}$ is interpeted as shown in the box of Fig. 9.28—i.e., if the result of the addition is greater than $\frac{1}{2}$, 1 is subtracted; whereas if the result of the addition is less than $-\frac{1}{2}$, 1 is added. In this manner $x(n)$ is always in the interval $(-\frac{1}{2}, \frac{1}{2})$. Figure 9.29 shows that if $x(n-1)$ and $x(n-k)$ are uniformly distributed over the interval $(-\frac{1}{2}, \frac{1}{2})$, as was assumed, then $x(n)$ will also be uniformly distributed over the same interval. It is seen from this figure that the triangular distribution that one obtains from summing two uniform distributions is converted back to a uniform distribution by taking

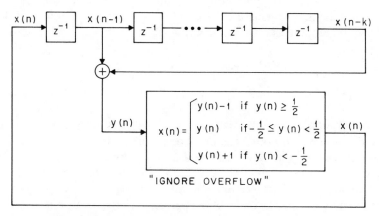

Fig. 9.28 Technique for generating uniform random numbers.

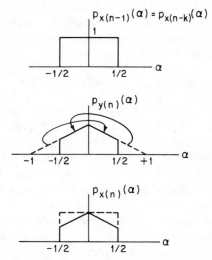

Fig. 9.29 Explanation of overflow wraparound for uniform noise generator.

the sum modulo $\frac{1}{2}$. Thus, when implemented using 2's-complement arithmetic, if the largest machine word is considered to be $\frac{1}{2}$, Eq. (9.4) can be implemented as a 2's-complement addition where overflows are disregarded. Thus one of the advantages of this method is that the only computation required is one addition per iteration. Additionally, k storage registers are required to save $x(n-1)$ to $x(n-k)$ where k is generally on the order of 50. Measurements of the statistics of the pseudo-random sequences produced by this technique show the distribution is approximately uniform and the power density spectrum is approximately white. Thus this generator is highly recommended for signal processing applications.

A third method for generating random numbers is illustrated in Fig. 9.30. The rule used to generate the current L-bit random number $x(n)$ from the previous outputs $x(n-1)$ and $x(n-2)$ is

$$x(n) = T_p[x(n-1) + x(n-2)] \tag{9.5}$$

where $T_p[\cdot]$ indicates a cyclic rotation of P places to the right and $+$ denotes exclusive OR.

In Fig. 9.30 the value of P is 1. This algorithm is given here because it is fast and efficient. The only computation required is an L-bit exclusive OR operation. The only storage required is L bits to store $x(n-1)$ and L bits to store $x(n-2)$. For certain values of L, the output of the generator has been shown to be approximately uniform and the power density spectrum to be approximately white. For $L = 19$ the period of the generator is 14, 942, 265, or approximately 1500 sec of output at a 10-kHz sampling frequency.

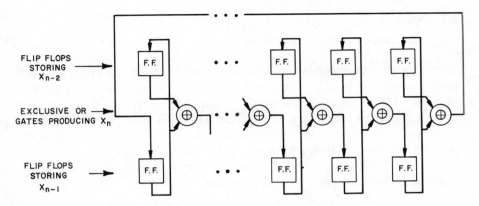

Fig. 9.30 Configuration for uniform pseudo-random number generator.

Because of its inherent simplicity, this generator has been built in digital hardware for $L = 19$. Figure 9.31 shows a histogram of approximately 50,000 samples of the output that was normalized to lie in the range $(-32,768, 32,767)$. The dashed line shows the theoretical frequency of occurrence for a uniform generator. (The cells on each end of the histogram were slightly smaller than the other cells. Thus they had proportionally fewer occurrences.) The approximation to a uniform distribution is reasonably good.

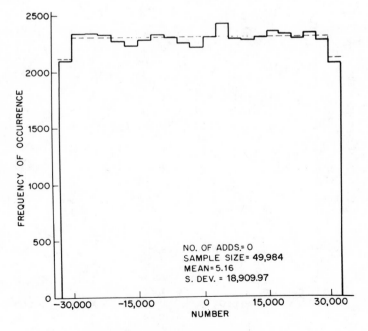

Fig. 9.31 Histogram of uniform generator.

9.14   Techniques for Generating Gaussian Random Numbers

The techniques described in Sec. 9.13 dealt exclusively with uniformly distributed numbers. It is often important to generate a sequence of pseudo-random numbers that are Gaussian distributed. Perhaps the simplest way to do this is to use the Central Limit Theorem, which states that, in the limit as N tends to infinity, the sum of N identically distributed, independent random variables tends to a Gaussian. Thus one can convert a sequence of independent, uniformly distributed numbers $\{x(n)\}$ to a sequence of Gaussianly distributed numbers $\{y(n)\}$ by the rule

$$y(n) = \frac{1}{N}\sum_{i=0}^{N-1} x(nN - i) \tag{9.6}$$

when N is sufficiently large. For all practical cases, a value of N on the order of 10 yields a reasonably good approximation to a Gaussian distribution. Figure 9.32 shows a measured distribution for the noise generator of Fig. 9.30 ($L = 19$) for $N = 12$. The agreement between the measured and the theoretical Gaussian distribution is reasonably good.

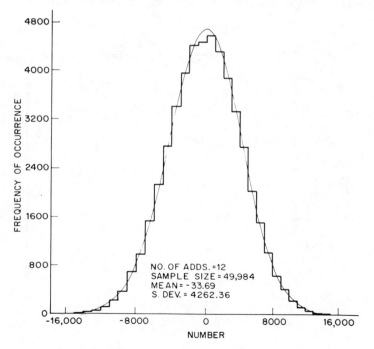

NO. OF ADDS. = 12
SAMPLE SIZE = 49,984
MEAN = -33.69
S. DEV. = 4262.36

Fig. 9.32 Histogram of Gaussian generator.

A variation on the technique above was devised by Rader where a sequence of L independent, uniform random variables was converted to a sequence of L uncorrelated, Gaussian random variables through use of a Hadamard matrix. Each of the L Gaussian variables was effectively the sum (or difference) of L uniform variables. Thus for values of L greater than about 16, the approximation to a Gaussian distribution was reasonably good. The method above is exceedingly efficient since L uniform variables were used to give L Gaussian variables rather than L/N as in Eq. 9.6. The outputs are not independent, however, although they can be made nearly so by slight modifications.

There is a straightforward theoretical way of converting a pair of uniformly distributed random variables to a pair of Gaussian random variables. If we let $\{x(n)\}$ be a sequence of uniform random variables, distributed between 0 and 1, and define $y(n)$ as

$$y(n) = \sqrt{2\sigma^2 \ln\left(\frac{1}{x(n)}\right)} \qquad (9.7)$$

then $y(n)$ is Rayleigh distributed; i.e.,

$$P_y(y_0) = \frac{y_0}{\sigma^2} \exp\left(\frac{-y_0^2}{2\sigma^2}\right) \qquad (9.8)$$

If we then form the two random variables $w(n)$ and $w(n+1)$ as

$$w(n) = y(n) \cos\left[2\pi x(n+1)\right] \qquad (9.9)$$

$$w(n+1) = y(n) \sin\left[2\pi x(n+1)\right] \qquad (9.10)$$

then $w(n)$ [and $w(n+1)$] are Gaussian distributed with 0 mean and variance σ^2. Furthermore, $w(n)$ and $w(n+1)$ are uncorrelated which, for Gaussian variables means they are independent. Although the technique above works quite well in practice, it is exceedingly time-consuming because of the need to compute logs, square roots, sines, and cosines and hence is often avoided when large numbers of Gaussian numbers need to be generated.

REFERENCES

1. L. B. Jackson, J. F. Kaiser, and H. S. McDonald, "An Approach to the Implementation of Digital Filters," *IEEE Trans. on Audio and Electroacoustics*, **16**, No. 3, 413–421, Sept., 1968.
2. J. D. Heightley, "Partitioning of Digital Filters for Integrated-Circuit Realization," *IEEE Trans. on Communication Tech.*, **COM-19**, 1059–1063, Dec., 1971.
3. S. L. Freeny, R. B. Kieburtz, K. V. Mina, and S. K. Tewksbury, "Design of Digital Filters for an All Digital Frequency Division Multiplex-Time Division Multiplex Translator," *IEEE Trans. Circuit Theory*, **CT-18**, 702–711, Nov., 1971.

4. S. L. FREENY, R. B. KIEBURTZ, K. V. MINA, AND S. K. TEWKSBURY, "Systems Analysis of a TDM-FDM Translator/Digital A-Type Channel Bank," *IEEE Trans. on Communication Tech.*, **COM-19**, 1050–1059, Dec., 1971.

5. C. F. KURTH, "SSB/FDM Utilizing TDM Digital Filters," *IEEE Trans. on Communication Tech.*, **COM-19**, 1, 63–70, Feb., 1971.

6. J. TIERNEY, C. M. RADER, AND B. GOLD, "A Digital Frequency Synthesizer," *IEEE Trans. on Audio and Electroacoustics*, **19**, No. 1, 48–58, 1971.

7. J. L. PERRY, R. W. SCHAFER, AND L. R. RABINER, "A Digital Hardware Realization of a Random Number Generator," *IEEE Trans. on Audio and Electroacoustics*, **AU-20**, No. 4, 236–240, Oct., 1972.

8. B. F. GREEN, J. E. SMITH, AND L. KLEM, "Empirical Tests of an Additive Random Number Generator," *J. Assn. Computer Machinery*, **6**, No. 4, 527–537, Oct., 1959.

9. M. D. MACLAREN AND G. MARSAGLIA, "Uniform Random Number Generators," *J. Assn. Computer Machinery*, **12**, 83–89, 1965.

10. C. M. RADER, L. R. RABINER, AND R. W. SCHAFER, "A Fast Method of Generating Digital Random Numbers," *Bell. Sys. Tech. J.*, **49**, 2303–2310, Nov., 1970.

11. C. M. RADER, "A New Method of Generating Gaussian Random Variables by Computer," *Lincoln Laboratory Technical Note*, 1969–49, 1969.

10

Special-Purpose Hardware for the FFT

10.1 Introduction

Chapter 6 included a fairly extensive introduction to the FFT algorithm as a necessary prelude to the techniques of digital spectrum analysis. This chapter is devoted exclusively to methods for implementing various FFT algorithms. For economic implementation, it behooves us to be aware of the many different FFT structures so that a structure matched to the particular problem can be chosen. Thus, we shall begin with a more complete survey of different FFT flow charts than given in Chapter 6.

For special purposes it may be true that algorithms other than radix 2 are preferred. In particular, radix 4 algorithms have received much attention because of their possible savings (compared to radix 2) in hardware. We shall study radix 4 systems in some detail and try to make general comparisons between radices 2 and 4.

For very high-speed systems, such as high performance radars, a pipeline FFT configuration is required. Such systems will be described, as well as other forms of parallelism in FFT hardware.

10.2 Review of FFT Fundamentals

As discussed in Chapter 6 FFT's can be done "in-place" and "not-in-place." The in-place algorithm stipulates that the DFT's of which the FFT is

573

comprised all be returned to the registers from where they came. For example, in the 16-point FFT of Fig. 10.1, the contents of registers 0 and 8 enter the two-point DFT represented by the top open circle in the first stage. Let us define the inputs as $f_0[= x(0)]$ and $f_1[= x(8)]$. Then the two-point output is F_0 and F_1; F_0 replaces f_0 and F_1 replaces f_1.

Scrambling of the result takes place for the in-place algorithm and as seen in Fig. 10.1, for a radix 2 FFT, the scrambling corresponds to a bit reversal of the frequency index. In Sec. 10.3, we shall discuss scrambling for higher radices.

As shown in Chapter 6, FFT structures may be decimation-in-time (DIT) or decimation-in-frequency (DIF). DIT corresponds to having the twiddles before the two-point DFT node points, while DIF corresponds to having the twiddles after the two-point DFT. Both sets of twiddles are shown in Fig. 10.1; depending on whether you use DIT or DIF, one or the other set is used. The numbers by the arrows correspond to multiplication by W^k where k is the number shown. The structure of Fig. 10.1 holds even if the input indexing is bit reversed; in this case the resulting output will be in natural order.

Figure 10.2 gives an alternate arrangement that yields an equally useful

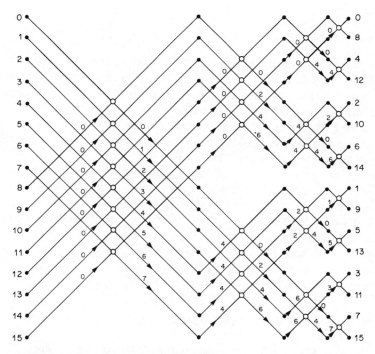

Fig. 10.1 In-place 16-point FFT with normally ordered inputs and bit-reversed outputs.

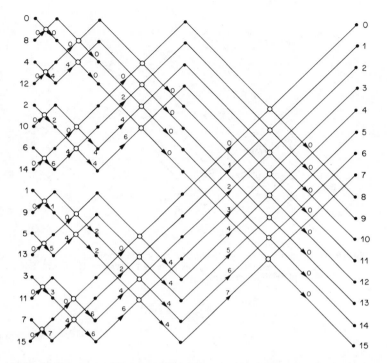

Fig. 10.2 In-place FFT with bit-reversed inputs and normally ordered outputs (DIT coefficients to left of nodes; DIF coefficients to right of nodes).

algorithm. Again both DIT and DIF algorithms are shown on the same diagram.

Unless otherwise stated, our convention will be to number the registers from top to bottom; these numbers will generally not be shown. The numbers at the inputs and outputs shown in Fig. 10.1 and 10.2 are the indices associated with the samples and the transformed results.

A 16-point radix 2 constant geometry algorithm is shown in Fig. 10.3. Here the butterfly outputs are not put back where they come from so it is a not-in-place algorithm. The indexing is kept constant from stage to stage, which can result in simplification in some programs or hardware designs. The inputs are normally ordered, while the outputs are bit-reversed. As a general rule, *in-place* means that N complex registers are needed, while $2N$ such registers are required for *not-in-place* algorithms. Figure 10.4 shows a constant geometry algorithm for bit-reversed inputs, leading to normally ordered outputs. The twiddle factors for both DIT and DIF can be found by looking at Fig. 10.3 *backward* and *reversing the arrows*. Figure 10.5 shows an algorithm for avoiding bit reversal, at the cost of doing a not-in-place

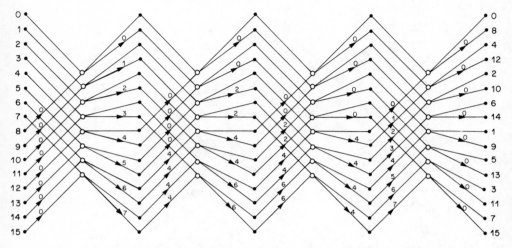

Fig. 10.3 Constant geometry algorithm, radix 2, 16 points, not-in-place, normally ordered inputs, bit-reversed outputs; (DIT and DIF multipliers both shown).

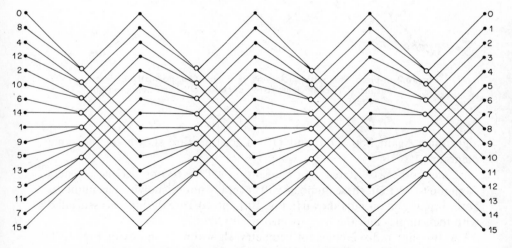

Fig. 10.4 Constant geometry algorithm, radix 2, 16 points, not-in-place, bit-reversed inputs, normally ordered outputs.

algorithm. Only the DIT version is shown but DIF can easily be done by placing twiddles after the nodes.

Figure 10.6 shows how an FFT algorithm can be designed that makes use of a high-speed scratch memory. We shall discuss the hardware implications of this structure later but for now we point out that nodes can be paired and two butterflies done on four input samples so that these samples pass through two FFT stages before the next four samples are handled. For example,

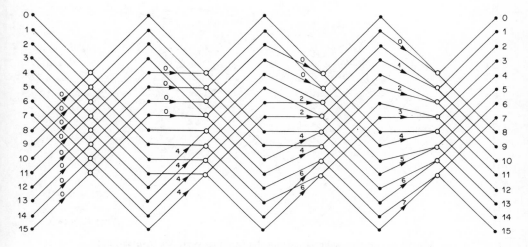

Fig. 10.5 Non-bit-reversed input and output; DIT; not-in-place, 16 point, radix 2 FFT.

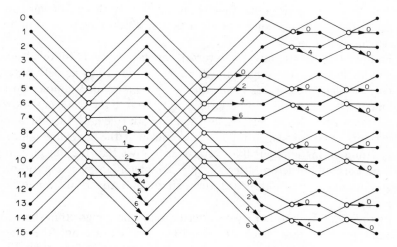

Fig. 10.6 A 16-point, radix 2, suitable for use with hardware that can handle two stages for four input samples with no need to access memory during the two butterflies, input in normal order, output is scrambled (not bit-reversed), DIF.

assume we enter samples 0 and 8 as a pair into node 0 of stage 0 and samples 4 and 12 into node 4 of stage 0. After doing these two butterflies we proceed to node 0, stage 1 and node 4, stage 1, with all the results winding up in the same four registers, i.e., 0, 4, 8, and 12. In the same way, we can enter registers 1, 5, 9, and 13 and again proceed through two stages. In this way,

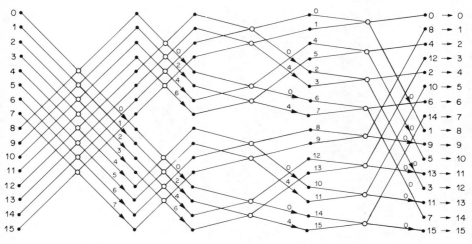

Fig. 10.7 Two-butterflies-at-a-time algorithm to avoid bit reversal, DIF.

half as many memory cycles are needed, provided that the arithmetic element can handle four samples rather than the usual two. In this particular version, bit reversal takes place since the algorithm can be thought of as in-place, two stages at a time.

Figure 10.7 shows how bit reversal can be avoided. In this 16-point algorithm the first two stages are done as in Fig. 10.1, namely, straight-forward in-place operation. From then on we enter four samples into our arithmetic element to do two butterflies and permute the results as shown in the last two stages. Notice that we are violating our rule not to assign numbers to registers, the purpose being to trace through the indexing. Since our register numbering emerges bit-reversed, it follows that the output samples must be normally ordered (after all, in a completely in-place algorithm, the register numbering is untouched throughout and the result emerges bit-reversed).

Figure 10.8 is another two-butterflies-at-a-time algorithm but for a different purpose than Fig. 10.7. In Fig. 10.8 we arrange the memory registers so as to be able to read (or write) two complex words at a time. Thus, for example, samples 0 and 8 are entered into the butterfly in parallel, saving a memory cycle. The table in the right-hand corner shows the desired matching up of samples as the FFT progresses. In order to achieve this match so that parallelism can be maintained, permutation of four output points at a time must be performed. For example, the samples 0, 8 and 4, 12 are permuted and entered as 0, 4 and 8, 12. Again we violate our rule and number the registers. The final results, i.e., the rightmost column of numbers, shows the scrambling of the output points, which is actually a combination of bit reversal and row–column permutation of an (8×2) array.

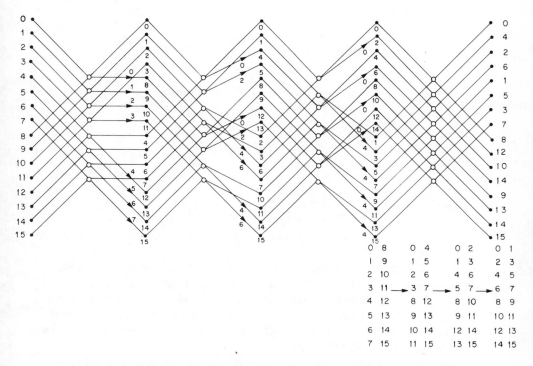

Fig. 10.8 Algorithm requiring two butterflies at a time and lending itself to two-fold parallelism.

10.3 FFT Indexing—Bit Reversal and Digit Reversal for Fixed Radices

We begin with a very simple example to show how the output points in an in-place FFT become scrambled; that is, $X(k)$ does not, in general, reside in register k even though $x(k)$ did reside in register k. By way of example, consider a four-point sequence $x(n)$. The DFT of $x(n)$ can be written as

$$X(0) = x(0) + x(1) + x(2) + x(3) = x(0) + x(1) + x(2) + x(3)$$

$$X(1) = x(0) + x(1)W^1 + x(2)W^2 + x(3)W^3 = x(0) + jx(1) - x(2) - jx(3)$$

$$X(2) = x(0) + x(1)W^2 + x(2)W^4 + x(3)W^6 = x(0) - x(1) + x(2) - x(3)$$

$$X(3) = x(0) + x(1)W^3 + x(2)W^6 + x(3)W^9 = x(0) - jx(1) - x(2) + jx(3)$$

Now, let us perform the same computation by means of a four-point FFT, according to the flow chart of Fig. 10.9. Notice that the positions of $X(1)$ and $X(2)$ have been reversed; that is, the result has been scrambled. This scrambling in FFT's can be explained generally by applying the ideas discussed in Chapter 6, which show how a one-dimensional DFT can be

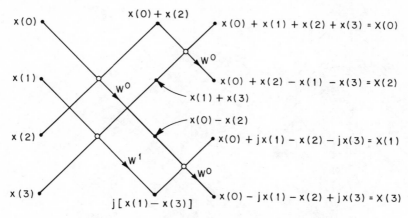

Fig. 10.9 Flow chart of a radix 2, four-point FFT, DIF, in-place.

performed as a two-dimensional DFT. The key result is in the row–column permutation of the indices that takes place as a result of the transform. If an FFT is conceived as a succession of "decimations" (representation of one-dimensional data in two dimensions), then the scrambling that takes place can be expressed as a succession of nested row–column permutations, as illustrated in Fig. 10.10 for a 16-point FFT.

For a fixed radix FFT, it turns out that the output data ordering can be very neatly expressed as a function of the input data ordering. The result, which we shall first state and then prove, goes by the name of *bit reversal* for radix 2 and, more generally, *digit reversal* for radix r, where the digits are numbers from a base-r number system. Thus, in radix 2, if we begin with

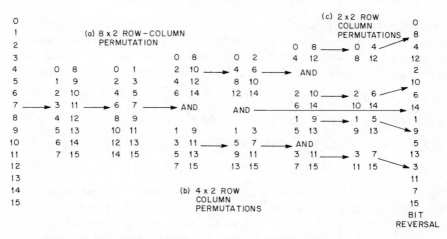

Fig. 10.10 Illustration of bit reversal in a 16-point FFT by successive row–column permutations.

$f(n)$ in register n and if n is represented as a binary number $x_5x_4x_3x_2x_1x_0$ (for a six-bit address where each x_j is zero or one), then the transform output $F(n)$ is stored in the register with the address $x_0x_1x_2x_3x_4x_5$; i.e., the order of the bits in the binary number are *reversed*. In radix 4 FFT's the same result holds *if* x_j is a base-4 number, which can take on only the values 0, 1, 2, 3; for base-8 numbers, x_j would be an *octal* number.

To prove digit reversal, we make use of a simple numerical relationship between the initial ordering of any two-dimensional array and the ordering of the same array after permutation. Consider the matrix

$$
\begin{matrix}
0 & 1 & 2 & \cdots & m-1 \\
m & m+1 & m+2 & \cdots & 2m-1 \\
2m & & & & \cdot \\
\cdot & & & & \cdot \\
\cdot & & & & \cdot \\
\cdot & & & & \cdot \\
(p-1)m & & \cdot & \cdot & N-1
\end{matrix}
$$

having p rows and m columns with $N = mp$. If we multiply each element in the matrix by $p \bmod (N-1)$ and take the result, we obtain the row–column permutation previously discussed. Thus, the transformed matrix will be

$$
\begin{matrix}
0 & p & 2p & \cdots & (m-1)p \\
1 & p+1 & 2p+1 & & (m-1)p+1 \\
2 & & & & \cdot \\
& & & & \cdot \\
& & & & \cdot \\
p-1 & \cdot & \cdot & \cdot & N-1
\end{matrix}
$$

Next we observe that if $p = 2$, then multiplication of a binary number by 2, taking the result mod $(N-1)$, is equivalent to clockwise circular rotation of the bits. Similarly, multiplication by 4, taking the result mod $(N-1)$, is a *two-bit* clockwise circular rotation or a *one-digit* (base 4) rotation; in general, multiplication by $r \bmod (N-1)$ corresponds to a *one-digit* (base r) clockwise rotation.

Now, notice what happens when we perform a succession of such rotations on a base-r number in the following way: Let the number be

$$x_5x_4x_3x_2x_1x_0$$

1.   Rotate all six digits

$$x_4x_3x_2x_1x_0x_5$$

2. Rotate five high level digits

$$x_3 x_2 x_1 x_0 x_4 x_5$$

3. Rotate four high level digits

$$x_2 x_1 x_0 x_3 x_4 x_5$$

4. Rotate three high level digits

$$x_1 x_0 x_2 x_3 x_4 x_5$$

5. Rotate two high level digits

$$x_0 x_1 x_2 x_3 x_4 x_5$$

Finally, to prove that the reordering of a radix r FFT can be described by digit reversal, begin with an $[(N/r) \times r]$ decomposition. Permutation of rows and columns due to such a transform corresponds to step 1. But now each of the r columns (of length N/r) must be further decomposed into $[(N/r^2) \times r]$ matrices. To transform each of the (N/r^2) matrices requires the rotation of all except the low level digit; as we go from stage to stage, we require successively higher level digits. It is perhaps best to try a few examples. Consider a 16-point radix 2 DFT, where the ordering of the input samples corresponds to the addresses of the registers. Choosing a decimation-in-frequency algorithm, we proceed (in our minds) to perform two-point row DFT's. If we had done all this, the new ordering would be as listed under step 1, namely, a row–column permutation of the entire matrix. Notice that each index in a given register is replaced by a new index obtained via step 1 of our digit-reversing procedure. But, as we know, each column under step 1 comes from transforming (4×2) arrays and so the indices are permuted according to step 2 of our procedure. Finally, treating each half column under step 2 as a (2×2) array we arrive at the reordering, which is an actual bit reversal. It is important to remember that in tracing through this example we assumed that the addresses of the registers correspond to the *geometric* location of the numbers and that they remain invariant throughout the transformation. Thus, for example, when the algorithm is completed, register 14 contains sample 7, register 5 contains sample 10, etc.

Radix 4 Algorithm

If N is a power of 4, it can be expressed as $(N/4) \times 4$; similarly $N/4 = (N/16) \times 4$, etc. In this way the original one-dimensional array can be broken down so that all the elementary computations are actually four-point DFT's. A simple example for $N = 16$ is shown in Fig. 10.11. Here only a single decimation is needed from a single dimension into a (4×4)

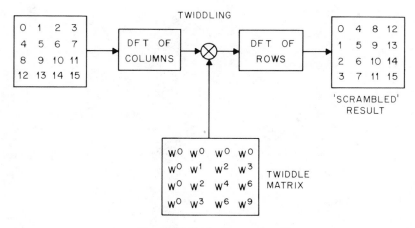

Fig. 10.11 A 16-point radix 4 FFT.

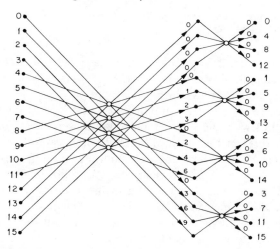

Fig. 10.12 Radix 4, 16-point DIF, normally ordered input, bit-reversed output.

array. Figure 10.11 can be put in the form of a flow diagram similar to the radix 2 flow diagrams, as shown in Fig. 10.12. The notation has been generalized (as already discussed in Chapter 6) so that an arithmetic node point (open circle) represents a K-point DFT, where K is the number of input (and output) lines. Thus, in Fig. 10.12, all open circles are four-point DFT's, with the twiddles being shown as arrows, as in the radix 2 diagrams. Memory in Fig. 10.12 is denoted in exactly the same way as in radix 2, with the memory node points (heavy dots) being labeled numerically from top to bottom and the register numbers not being explicitly shown.

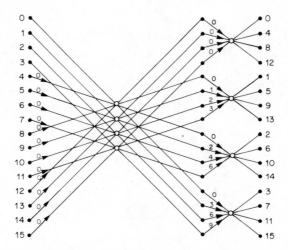

Fig. 10.13 In-place, 16-point, radix 4, DIT, FFT, normally ordered input, digit-reversed output.

By analogy with radix 2 diagrams, we can have radix 4 decimation-in-time (DIT) or decimation-in-frequency (DIF). The latter is shown in Fig. 10.12 and a DIT version is shown in Fig. 10.13.

It is instructive to compare the number of arithmetic computations of Figs. 10.12 and 10.13 with the number of computations in any of the 16-point radix 2 structures of Figs. 10.1 through 10.8. A careful count of the radix 2 case reveals 10 nontrivial complex multiplications (a trivial multiplication is by either ± 1 or $\pm j$). A similar count for radix 4 yields eight such multiplications. Thus, at least we have some insight into the fact that not all FFT's that yield the same result are equally efficient computationally. Whether or not radix 4 is advantageous in any given case will depend on many factors, too numerous to worry about now.

Figure 10.14 shows a 64-point radix 4 flow diagram. To see how this structure "decimates," consider that the first 16 samples represent the first row of a 16-column, 4-row matrix; the next 16 points are the second row, etc. Now, the first stage in Fig. 10.13 performs a total of 16 4-point DFT's on each of the columns, followed by twiddling, as shown by the exponents listed just before the second stage of memory. Now, each row is decimated into a (4×4) matrix and is transformed exactly as if we were doing 4 16-point radix 4 FFT's except that the twiddles are different.

We can consider Fig. 10.14 like a decimation-in-frequency algorithm because the twiddles take place after the DFT's. It is possible to derive a DIT algorithm for this case by building up from smaller to larger two-dimensional matrices rather than the other way around. An interesting problem for the reader is to find the twiddles for a DIT version of Fig. 10.14. How would this

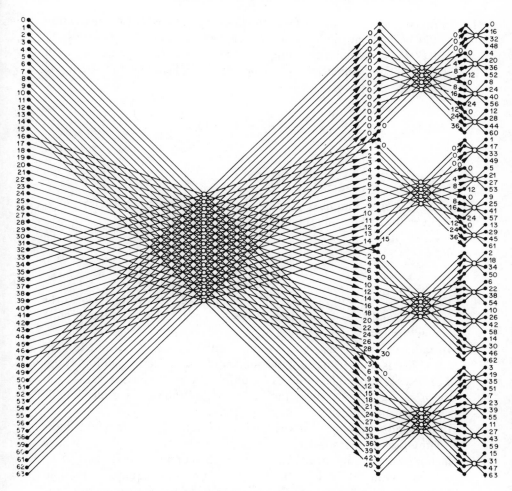

Fig. 10.14 A 64-point, radix 4, DIF, normally ordered inputs, digit-reversed outputs.

result be derived based on repeated reduction of the matrix into four-point DFT's?

10.4 Comparison of Computations for Radices 2, 4, and 8

Both radix 4 and radix 8 FFT algorithms are of theoretical and practical interest because they can lead to savings (relative to radix 2) for both software and hardware implementations of the FFT. To see this intuitively, let us choose $N = 64$ and do the FFT three ways, by radix 2, radix 4, and radix 8, and count the number of complex multiplications needed for each case.

1. In radix 2, if we are doing DIT, the first stage consists of multiplication by $W^0 = 1$, or no multiplications. The second stage has twiddles W^0 and $W^{16} = j$ so again no multiplications. The third stage has one-half of the full complement of $(N/2)$ complex multiplications and so on. As a formula, we can write

$$M_N = \frac{N}{2} \sum_{n=1}^{\log_2 N - 2} \frac{2^n - 1}{2^n} = \frac{N}{2}(\log_2 N - 2)$$

$$- \frac{N}{2}\left[\frac{1 - 2^{-(\log_2 N - 1)}}{1 - 2^{-1}}\right] + 1 \tag{10.1}$$

where M_N is the number of complex multiplications needed for an N-point radix 2 FFT. Simplifying Eq. (10.1), we find

$$M_N = \frac{N}{2} \log_2 N - \tfrac{3}{2}N + 2 \tag{10.2}$$

For $N = 64$, Eq. (10.2) gives $M_N = 98$. Notice that this formula works for $N = 2$ and $N = 4$, for which $M_N = 0$. The constraint on the formula is that N must be a power of 2.

2. In radix 4, we can derive a formula similar to Eq. (10.2) but for our example it is simpler to count the number of multiplications in Fig. 10.14. The first set of twiddles requires 4 multiplications; the next stage, 32; and the final stage, none; resulting in 76 multiplications—a quite notable improvement compared to the radix 2 case.

3. In both radix 2 and radix 4, the DFT part of the computation introduces no multiplications so that all multiplications are really the twiddle factors. Such is no longer the case for higher radices. In particular, an eight-point DFT, done by an FFT algorithm, requires 2 multiplications (by numbers of the form $\pm a \pm ja$), as we can see by applying Eq. (10.2) Thus, the multiplications in a radix 8 FFT are caused by both the DFT's and the twiddles. Figure 10.15 shows a radix 8, 64-point FFT. There are a total of 16 8-point DFT's, as can be seen by counting the open circles in Fig. 10.15, giving a total of 32 multiplications. There are 48 nontrivial twiddles so that we obtain a total of $(32 + 48 = 80)$ multiplications.[1] So, in some sense, it appears that, at least for a 64-point FFT, radix 4 is an "optimum." This optimum property should certainly *not* be taken literally to mean that one should always use radix

---

[1] It can be argued that since an 8-point DFT entails multiplications by the same coefficient, it corresponds to a single (rather than 2) complex multiplications. This would associate $(16 + 48 = 64)$ multiplications with radix 8, which would make radix 8 "better" than radix 4.

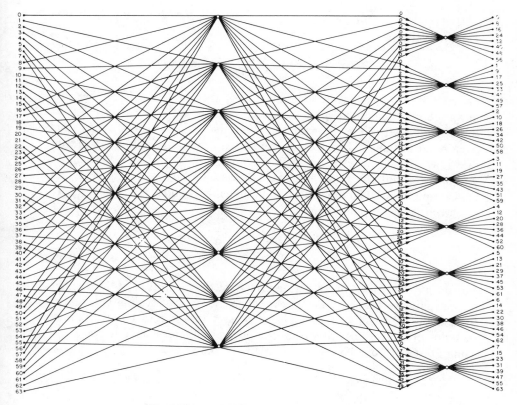

Fig. 10.15 Radix 8, 64-point FFT, in-place, DIT.

4 but it does mean that for any system that is big enough one should seriously look at all these radices.

10.5 Introduction to Quantization Effects in FFT Algorithms

When programming the FFT (or equivalently when building FFT hardware), one has to take into consideration the effects of representing the data and coefficients with finite precision. These effects include roundoff noise in truncating the result of a multiplication, scaling the data to prevent overflows, and inaccurate transformation due to errors in finite precision representation of the coefficients W^k. Extensive analysis has been done by Welch, among others, on the questions of roundoff noise and data scaling for radix 2 DIT and DIF algorithms using both fixed and floating point computations. The question of coefficient accuracy has not been studied so extensively and only qualitative results can be given here. We begin by considering fixed point realizations of radix 2 FFT algorithms.

If $\{x(n)\}$ is an N-point sequence with DFT coefficients $\{X(k)\}$, then by Parseval's theorem

$$\sum_{n=0}^{N-1} x^2(n) = \frac{1}{N} \sum_{k=0}^{N-1} |X(k)|^2 \tag{10.3}$$

or, equivalently, the mean square value of the transform is N times the mean square value of the input sequence. Thus the DFT of a sequence tends to be significantly larger in magnitude than the sequence itself. When using fixed point arithmetic one must therefore use scaling procedures to prevent overflow. To see how such overflows might occur, consider the computations for a butterfly at the mth stage of the FFT. If we let $f_m(i)$ and $f_m(j)$ be the input to a DIT butterfly with multiplier W^p and let $f_{m+1}(i)$ and $f_{m+1}(j)$ be the butterfly outputs, then

$$f_{m+1}(i) = f_m(i) + W^p f_m(j)$$
$$f_{m+1}(j) = f_m(i) - W^p f_m(j) \tag{10.4}$$

Equation (10.4) shows that from stage to stage the magnitude of the numbers in the sequence generally increases, which means that scaling can be performed by right shifts of the data. It is easy to show that the maximum modulus of the complex numbers is nondecreasing from stage to stage and, using Eq. (10.4), can be shown to satisfy the bound

$$\max \{|f_m(i)|, |f_m(j)|\} \leq \max \{|f_{m+1}(i)|, |f_{m+1}(j)|\}$$
$$\leq 2 \max \{|f_m(i)|, |f_m(j)|\} \tag{10.5}$$

Hence the signal level grows at a rate less than one bit per stage.

Based on the above, three reasonable scaling techniques are suggested. These include

1. Shifting right one bit at every iteration. If $|f_0(i)| < \frac{1}{2}$ for all i and the data is shifted one bit to the right after each iteration (except the last), there will be no overflows.
2. Controlling the sequence so that $|f_m(i)| < \frac{1}{2}$ for all i. At each stage of the FFT the array $f_m(i)$ is calculated and if any value exceeds one-half, the entire array is right shifted one bit.
3. Testing for an overflow. In this case the initial sequence is scaled so that $|\text{Re}\,\{f_0(i)\}| < 1$ and $|\text{Im}\,\{f_0(i)\}| < 1$ rather than one-half as in techniques 1 and 2. Whenever an overflow is detected during a butterfly, the entire sequence (including the results of those butterflies that have been successfully computed at this stage) is shifted right by one bit and the iteration is continued beginning with the butterfly in which the overflow occurred. More than one overflow may occur per stage but no more than two.

Technique 1, although fast and easy to program, is not accurate since scaling at each stage leads to unnecessary loss in accuracy when no scaling is required. The second method is both time-consuming (since the magnitudes of each element of the array must be computed at each stage) and slightly inaccurate since the sequences are always scaled to be less than one-half; thus one bit of the data word is never used to represent the data. The third technique is the most accurate one although it has the disadvantage that one must process the sequence an additional time each time an overflow is detected.

Given the last of the scaling procedures above, Welch has shown that it is reasonably straightforward to calculate bounds on the roundoff noise generated in the FFT. There are two sources of noise in the computation.

1. Roundoff error in calculating the product $W^p f_m(j)$ as in Eq. (10.4). If both W^p and $f_m(j)$ are represented by b-bit numbers [i.e., the real and imaginary parts of W^p and $f_m(j)$ are b-bit numbers], then rounding of each of the four real multiplications in $W^p f_m(j)$ to a b-bit number gives an error that is uniformly distributed between $-2^{-b}/2$ and $2^{-b}/2$, with zero mean and variance $2^{-2b}/12$.

2. If as a result of addition there is an overflow, then the sum must be shifted right one bit. If the low-order bit shifted out is a 0, there is no error. If it is a 1, an error of $\pm 2^{-b}$ is made depending on whether the number is positive or negative. The variance of this error is $2^{-2b}/2$.

An upper bound[2] on the ratio of rms (root-mean-square) roundoff noise to rms of the completed transform can be obtained by assuming that an overflow occurs at each stage of the algorithm, thus causing a need to rescale the data. If $f_k(j)$ is a typical element at the kth stage and the symbol $V(f_k)$ is the variance of f_k, defined as

$$V(f_k) = \frac{1}{N} \sum_{j=0}^{N-1} \text{Var}\,\{f_k(j)\} \tag{10.6}$$

where

$$V(f_0) = \frac{2^{-2b}}{2} = 6\Delta^2 \tag{10.7}$$

where $\Delta^2 = 2^{-2b}/12$ and $6\Delta^2 = 2^{-2b}/2$, Equation (10.7) reflects the assumption that the first stage gives an overflow; hence the zeroth stage must be truncated by one bit. The butterflies of the first stage involve multiplications only by ± 1 so no multiplication roundoff noise exists.

---

[2] It should be noted that the upper bound to be obtained is strictly an upper bound on an expectation that may be exceeded on an experimental measurement without contradicting the theory.

At the second stage we assume an overflow will occur; hence the $f_1(j)$ array must be rescaled. Thus the variance of f_1 is

$$V(f_1) = 2V(f_0) + (4 \times 6\Delta^2) = 36\Delta^2 \tag{10.8}$$

where the factor of 4 reflects the fact that the size of the error of the truncated signal due to scaling at the first stage is twice the size of the error at the zeroth stage; hence its variance is four times as great.

The second stage of the DIT algorithm involves multiplications only by ± 1 and $\pm j$; hence again no multiplication roundoff noise exists. Thus assuming scaling of the f_2 variables, the variance of the error at the second stage is

$$V(f_2) = 2V(f_1) + (4^2 \times 6\Delta^2) = 144\Delta^2 \tag{10.9}$$

where the factor of 4^2 reflects the doubling of the magnitude of the truncation error from the first to second stage.

All stages beyond the second stage involve nontrivial multiplications in most of the butterflies—i.e., in the third stage, half the butterflies have non-trivial multiplications; in the fourth stage, three-fourths of the butterflies have nontrivial multiplications; etc. As seen from Eq. (10.4), the general form for the real part of the output of a butterfly at the mth stage is (an equivalent expression can be written for the imaginary part)

$$\text{Re}\,[f_{m+1}(i)] = \text{Re}\,[f_m(i)] + \text{Re}\,[f_m(j)]\,\text{Re}\,(W^p) - \text{Im}\,[f_m(j)]\,\text{Im}\,(W^p) \tag{10.10}$$

For those butterflies where W^p is nontrivial, the variance of f_{m+1} (including roundoff noise for each of the products and scaling truncation noise) is

$$V(f_{m+1}) = V(f_m) + V(W^p)\{\overline{\text{Re}^2\,[f_m(j)]} + \overline{\text{Im}^2\,[f_m(j)]}\}$$
$$+ \{\text{Re}^2\,(W^p) + \text{Im}^2\,(W^p)\}V(f_m) + V(e^r_{m+1}) + V(e^s_{m+1}) \tag{10.11}$$

where the term $V(e^r_{m+1})$ stands for roundoff noise variance at the mth stage and $V(e^s_{m+1})$ stands for scaling noise variance at the mth stage; they are defined by the relations

$$V(e^r_{m+1}) = 4^m\Delta^2$$

$$V(e^s_{m+1}) = 4^m \times 6\Delta^2 \tag{10.12}$$

as discussed above. The third term in Eq. (10.11) involves the product of magnitude of W^p and $V(f_m)$. Since $|W^p| = 1$, this term is identical to the first term in Eq. (10.11). The second term involves a product of $V(W^p)$ and a term involving the average square modulus of the variables at the mth stage. The variance of W^p is Δ^2 (since W^p is inexactly represented by a b-bit word); however, to define the average squared modulus of the numbers at the mth stage we must define the number K, the average squared modulus of the input array, as

$$K = \overline{f_0(j)^2} = \frac{1}{N} \sum_{j=0}^{N-1} |f_0(j)|^2 \tag{10.13}$$

Recalling that the average squared modulus increases by a factor of 2 from stage to stage, we have the result that the average squared modulus at the mth stage output is $2^m K$. Thus the variance of f_{m+1} reduces to

$$V(f_{m+1}) = 2V(f_m) + 2^m K\Delta^2 + 4^{m+1}\Delta^2 + (4^{m+1} \times 6\Delta^2) \tag{10.14}$$

whenever *all* butterflies at the mth stage involve nontrivial multiplications. If α is the fraction of butterflies for which nontrivial multiplications occur $(0 \le \alpha \le 1)$, then the second and third terms in Eq. (10.14) are scaled by α. Thus if $\alpha = 0$ (as for the first and second stages), Eq. (10.14) reduces trivially to Eq. (10.8) (for $m = 0$) or Eq. (10.9) (for $m = 1$). If we assume $\alpha = \frac{1}{2}$ for the third stage, $\alpha = 1$ for all succeeding stages, and M is the index of the last stage, then [using Eq. (10.7) to (10.14)]

$$\begin{aligned}
V(f_m) &= 2^M(6\Delta^2) + 2^{M-1}(4 \times 6\Delta^2) + \ldots + 2(4^{M-1}6\Delta^2) \\
&\quad + 2^{M-2}K\Delta^2 + (M-3)2^{M-1}K\Delta^2 + 2^{M-4}(4^3\Delta^2) \\
&\quad + 2^{M-4}(4^4\Delta^2) + \ldots + (4^M\Delta^2) \\
&= (1.5)2^{M+2}\Delta^2(1 + 2 + \ldots + 2^{M-1}) + (M - 2.5)2^{M-1}K\Delta^2 \\
&\quad + 2^{M+2}\Delta^2 + 2^{M+4}(1 + 2 + \ldots + 2^{M-4}) \\
&\approx 2^{2M+3}\Delta^2 + (M - 2.5)2^{M-1}K\Delta^2 + 2^{M+2}\Delta^2 \tag{10.15}
\end{aligned}$$

Since the mean square of the absolute values of the output array $[f_M(j)]$ is $2^M K$, the mean square of the real part of the output (on the imaginary part) is $2^M K/2$. Thus a measure of the rms noise output to the rms signal output (in the limit of large M) is

$$\begin{aligned}
\frac{\text{rms (error)}}{\text{rms (signal)}} &\approx \frac{2^{(M+3)/2}\Delta}{\sqrt{K/2}} = \frac{2^{(M+3)/2}2^{-b}(0.3)}{\text{rms (input)}} \\
&= \frac{\sqrt{N}\,2^{-b}(0.3)\sqrt{8}}{\text{rms (input)}} \tag{10.16}
\end{aligned}$$

This upper bound of the error-to-signal ratio is seen to increase as $\sqrt{N}$, i.e., $\frac{1}{2}$ bit per stage.

A lower bound similar to Eq. (10.16) may be derived by assuming no overflows ever occur; hence all truncation errors due to scaling are removed from the derivation above. Hence up to the third stage there is no error at all.

Proceeding as above, the variance of f_M can be shown to be of the form

$$V(f_m) = 2^{M-2}K\Delta^2 + (M-3)2^{M-1}K\Delta^2 + 2^{M-3}\Delta^2$$
$$+ 2^{M-5}\Delta^2 + 2^{M-6}\Delta^2 + \ldots + \Delta^2$$
$$\approx (M-2.5)2^{M-1}K\Delta^2 + 2^{M-3}\Delta^2 + 2^{M-4}\Delta^2 \qquad (10.17)$$

Thus the ratio of rms (error) to rms (signal) for this lower bound in the case of large M assumes the form

$$\frac{\text{rms (error)}}{\text{rms (signal)}} \approx (M-2.5)^{1/2}(0.3)2^{-b} \qquad (10.18)$$

The model above has been experimentally validated by Welch in a wide variety of cases. Figure 10.16 shows the rms ratio of the error to the output signal when the input is random numbers with zero mean and uniform distribution over the interval $(-1, 1)$ with b, the number of bits, equal to 17. For this case the value of $\sqrt{K/2}$ was 0.58. As seen in this figure, the theoretical upper bound provides a good bound to the actual results, although the slope of the bound is somewhat steeper than the real data indicates it to be. The lower bound of Eq. (10.18) is not plotted because it is an overly optimistic bound and hence is not too practical.

Even though the DIT algorithm was used for the analysis above, extensions to other FFT algorithms are straightforward and results of the form of Eqs. (10.16) and (10.18) will generally be obtained. Thus guidelines for implementation of fixed point FFT algorithms are provided by Eq. (10.16). Knowing the maximum size of transform N and the desired ratio of rms error to rms signal, one can choose a word length (b bits) to attain the desired accuracy in the transform.

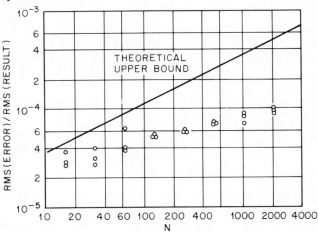

Fig. 10.16 Comparison between FFT noise bound and actual data (After Welch).

The analysis above deals primarily with errors produced by roundoff and signal truncation to prevent overflows. Another source of error in FFT computation is inexact representation of the multiplier coefficients W^p. Although a thorough analysis of this effect has not been made, Weinstein has used a random jitter model for the coefficients to obtain a simple expression for the ratio of mean square output error to mean square output signal of the form

$$\frac{\sigma_e^2}{\left[\dfrac{1}{N}\sum_{n=0}^{N-1}|f(n)|^2\right]} = \left(\frac{\log_2 N}{6}\right)2^{-2b} \tag{10.19}$$

where b is the number of bits used to represent the FFT coefficients. Equation (10.19) predicts that the variance of the error increases *very slowly* with N. Although Eq. (10.19) is very approximate, experimental measurements verify the main result—i.e., the error variance increases very slowly with N. Figure 10.17 shows the results Weinstein obtained using both fixed and floating

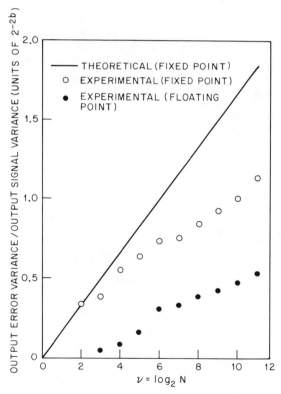

Fig. 10.17 Comparison between bound of FFT coefficient error and data (After Weinstein).

point computation. The main implication of the result above is that a fixed word length (of sufficient size to give a desired coefficient accuracy) is sufficient for a broad range of values of N. Thus FFT software or hardware can be applied for a broad spectrum of problems where N may vary over several orders of magnitude.

10.6 Some Hardware Considerations for Radix 2 Algorithms

We would now like to shift the emphasis to the actual performance of FFT hardware. It is worth pointing out that there is a substantial speed–flexibility tradeoff involved. The fact that computer A can do an FFT 10 times faster than computer B does not necessarily make it a better computer since, by specializing computer A architecture to gain FFT speed, other algorithms may be substantially slowed down, resulting in an overall performance decrease.

Let us begin with an example of a structure suited to radix 2 algorithms and for the moment restrict the set of possible structures to those with a single random access memory, a single arithmetic element (AE), and control, as shown in Fig. 10.18. For any butterfly with twiddle W^p, the computations can be expressed in either real or complex notations as follows.

I. Complex Notation

$$
\left.\begin{aligned}
f_{m+1}(i) &= f_m(i) + W^p f_m(j) \\
f_{m+1}(j) &= f_m(i) - W^p f_m(j)
\end{aligned}\right\} \quad \text{DIT} \qquad (10.20)
$$

$$
\left.\begin{aligned}
f_{m+1}(i) &= f_m(i) + f_m(j) \\
f_{m+1}(j) &= [f_m(i) - f_m(j)]W^p
\end{aligned}\right\} \quad \text{DIF} \qquad (10.21)
$$

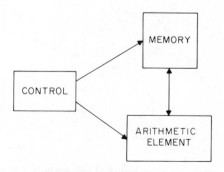

Fig. 10.18 Simplified FFT structure.

Notice that Eq. (10.20) is simply a copy of Eq. (10.4), which we repeat here for convenience.

2. Simplified Complex Notation

To show how a computer performing the FFT would behave, we can dispense with the various indices in Eqs. (10.20) and (10.21). Let us define

$$f_{m+1}(i) = A', \qquad f_m(i) = A$$
$$f_{m+1}(j) = C', \qquad f_m(j) = C$$

so that Eqs. (10.20) and (10.21) become

$$\left.\begin{aligned} A' &= A + W^p C \\ C' &= A - W^p C \end{aligned}\right\} \quad \text{DIT} \tag{10.22}$$

$$\left.\begin{aligned} A' &= A + C \\ C' &= (A - C)W^p \end{aligned}\right\} \quad \text{DIF} \tag{10.23}$$

3. Real Notation

A, A', C, C' are, in general, complex numbers; let us express them in terms of their real and imaginary components. Let $A = a + jb$, $A' = a' + jb'$, $C = c + jd$, $C' = c' + jd'$. Then we find

$$\left.\begin{aligned} a' &= a + c \cos \frac{2\pi p}{N} - d \sin \frac{2\pi p}{N} \\[2mm] c' &= a - c \cos \frac{2\pi p}{N} + d \sin \frac{2\pi p}{N} \\[2mm] b' &= b + c \sin \frac{2\pi p}{N} + d \cos \frac{2\pi p}{N} \\[2mm] d' &= b - c \sin \frac{2\pi p}{N} - d \cos \frac{2\pi p}{N} \end{aligned}\right\} \quad \text{DIT} \tag{10.24}$$

$$\left.\begin{aligned} a' &= a + c \\ b' &= b + d \\ c' &= (a - c) \cos \frac{2\pi p}{N} + (b - d) \sin \frac{2\pi p}{N} \\[2mm] d' &= (a - c) \sin \frac{2\pi p}{N} - (b - d) \cos \frac{2\pi p}{N} \end{aligned}\right\} \quad \text{DIF} \tag{10.25}$$

Now, from Fig. 10.18 and, for example, Eq. (10.24), we can determine the number of machine cycles required to perform a DIT butterfly. We assume that a word in memory contains only the real (or imaginary) part of a complex number and the arithmetic element contains a single real multiplier and a single real adder. Then, to perform Eq. (10.24), we first need six memory cycles to fetch a, b, c, d, $\cos(2\pi p/N)$, and $\sin(2\pi p/N)$ and four memory cycles to write back the results a', b', c', d'. Also there are four multiply operations and six addition (or subtraction) operations. The basic overhead on a computation such as Eq. (10.24) includes primarily extra temporary storage cycles and indexing cycles. Since the latter two depend so heavily on the details of the computer structure, we shall not include them in this analysis. Notice, by the way, that there is no basic computational difference between the DIF and DIT algorithms.

Now we can postulate a more expensive variation on the same basic structure; let us imagine a computer with two bytes per word, each byte being capable of holding a real (or imaginary) constituent of the complex number. Imagine also that the AE has two hardware adders and two hardware multipliers. With the proper control arrangement, the number of machine cycles should be reduced by half.

Some further reduction can be obtained by postulating a computer capable of one-cycle operations on complex numbers; then, from Eq. (10.22) or (10.23), we see that only two complex adds and one complex multiply are needed. The comparison among the three possibilities is listed in Table 10.1.

Table 10.1

COMPUTATIONAL TABLE

	Memory Cycles	Add Cycles	Multiply Cycles	Total
Single-word memory Real adder Real multiplier	10	6	4($\times$ 3)	28
Double-word memory Two real adders Two real multipliers	5	3	2($\times$ 3)	14
Complex-word memory Complex adder Complex multiplier	5	2	1 ($\times$ 3)	10

The *total* listed in the right-hand column is simply a count of the total cycles, with memory and add cycles having a weight of 1 and multiply cycles having a weight of 3.

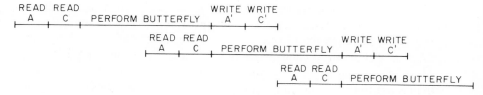

Fig. 10.19 Timing diagram for "optimum" radix 2 EC.

10.7 An "Optimum" Radix 2 Hardware Structure

The next logical question concerns the possibility of *overlapping* memory and arithmetic cycles since these cycles do derive from separate hardware. The first point to note is that the coefficients cos $(2\pi p/N)$ and sin $(2\pi p/N)$ need not be read from memory into the AE anew for each new butterfly because these coefficients remain constant over more than one cycle if the addressing is appropriately controlled. Hence, we can eliminate two memory cycles from the top row of the computational table and one memory cycle from the other two rows. Now the problem (for rows 2 and 3) becomes one of overlapping four memory cycles with arithmetic. This can be done as shown by the "optimum" timing diagram of Fig. 10.19.

In Fig. 10.19 the complex words A and C are fed into the AE; halfway through the computation, the inputs for the next butterfly are fed into the AE implying that buffer registers must be available within the AE. As soon as the first butterfly ends, the answers are written back and the AE continues without pause to perform the next butterfly. We see that this "matched" system requires that one butterfly computation time be equal to four memory cycle times. Since both memory and AE are working full time, it is fair to consider this control structure to be "optimum." That this timing creates the best match between memory and arithmetic can be seen by observing that a slowdown of either unit causes a slowdown in the result, whereas a speedup in either unit (but not the other) does *not* result in an overall speed increase.

As an example of the speed capability of this optimum structure, consider a memory time of $\tau = 100$ nsec, with $N = 1024$. The total time for such an FFT is $4\tau(N/2) \log_2 N \approx 2$ msec to process 1024 signal samples. For the system to be matched requires a butterfly time of $4\tau = 400$ nsec, which, with present technology, can readily be achieved.

Comparing the optimum result of this section with the tabulated results of Sec. 10.6, we see that a reduction from 28 to 4 cycles can be achieved in going from the simplest to the most sophisticated variation on the same basic structure. We might also comment that no attention has been paid to the time required to fetch instructions; if the program were to reside in the same memory as the data, much additional running time would be needed but it is true that many of the newer computers have separate memory banks with multiport facilities, making instruction–data overlap feasible and efficient. Finally, no attention has been paid to the "overhead" found in all FFT algorithms, consisting of setup time when changing coefficients and in moving from stage to stage. From our experience, the combined effects of all these peripheral activities (another one, not yet mentioned, is input–output time) can contribute from 20 to 70% of the total FFT running time.

10.8 Discussion of Parallel Processing to Speed Up FFT

Just as with other computer algorithms, parallel processing can lead to faster FFT's. The remainder of this chapter will be devoted primarily to discussions of forms of parallelism especially suited to more efficient FFT algorithms. In this section we shall try to establish some general categories of parallelism but the reader should keep in mind that any sophisticated FFT process design is usually a shrewd mix of several types of parallelism. Among the types of parallelism are

1. Time overlap of control arithmetic and memory functions. Figure 10.19 (which is already an example of this type of parallelism) explicitly shows memory and arithmetic overlap, and since no extra time has been allocated for memory addressing and other control functions, there must also be control overlap.
2. The addition of a small amount of high-speed memory. In a system containing a single arithmetic element, it may not be too expensive to make that element very fast but it could turn out to be very expensive to use comparably fast memory. As we shall show, a mix of two different memory speeds may be a cheap way to buy some speed.
3. Higher radix structures. A radix 2 butterfly consists of a single complex multiplication plus two complex additions, while a radix 4 butterfly consists of three complex multiplications and eight complex additions. It is straightforward to arrange radix 4 hardware so that three complex multiplications take place simultaneously, which should result in a 4:1 speed increase. This argument can be extended to higher radices, although beyond radix 8 it is doubtful if the extra parallelism is worth the cost.
4. Pipeline FFT's. An interesting special-purpose structure, employing $(\log_r N)$ arithmetic elements (where r is the radix), is useful for very high-speed applications and will be discussed in some detail in Secs. 10.12 and 10.13.
5. Superparallelism. For super-speed applications, efficient structures can be devices with as many as $N/2 \log_2 N$ parallel arithmetic elements; that

is, the entire FFT can be built as a single array. For even modest size N, this is a very large amount of hardware but, at least in the radar field, such devices are seriously contemplated.

To summarize, the beautiful symmetry of the FFT makes it possible to invent a host of structures containing almost the whole spectrum of possible parallelisms. In the following sections we shall enumerate and discuss this wide range of structures.

10.9 FFT Computation Using Fast Scratch Memory

In many cases, it is too costly to supply a large, fast memory capable of matching the speed of the arithmetic element. For example, in Sec. 10.6 (Fig. 10.19) we showed that, for a memory whose width encompasses a single complex datum, the best match will be a memory time of τ for a computation time of 4τ. Now, it is presently fairly economical to build a 400-nsec butterfly box but not very cheap to buy a 100-nsec memory, especially if you need a lot of it, say, 16,000 registers. By adding some fast scratch memory to the computational element, important speed savings can be obtained. One possible configuration is shown in Fig. 10.20.

To simplify the argument somewhat, let $M = 2\sqrt{N}$ (in Fig. 10.20). Now, given a scratch memory of $\sqrt{N}$, it is possible to go through half of the ($\log_2 N$) stages without involving a major memory cycle. For example, if $N = 16$, we can select samples 0 and 8, 4 and 12 from the major memory; then after performing a butterfly on each of these pairs, without leaving the scratch memory, we can proceed to do butterflies for 0 and 4, 8 and 12;

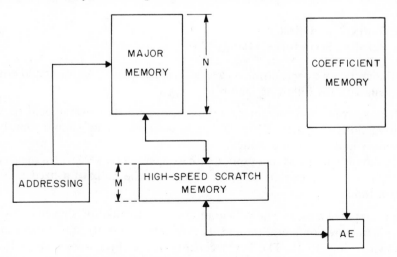

Fig. 10.20 Use of high-speed scratch memory to speed up FFT.

then we can select 1 and 9, 5 and 13 and repeat the procedure. In this way, to complete two stages of the FFT, we need to read in the 16-point data base once and write it back into major memory once; without the scratch memory this would need to be done twice. Table 10.2 compares the number of major

Table 10.2

N	With	Without
16	32	128
64	256	768
256	1,024	4,096
1,024	6,144	20,480
4,096	24,576	98,304

memory cycles with and without scratch memory for the case of a 16-register working scratch pad (which corresponds to $M = 32$ if we desire overlap in input–output and computation) for various values of N.

An interesting problem for the reader is as follows: Assume $\tau_b = 100$ nsec. Design a core memory–IC memory combination to give respectable performance for DFT sizes from 256 to 16,384. Choose some representative core memory speeds such as 1 and 0.5 μsec and find performance versus different sizes of IC memory.

Figure 10.6 is an example (for $N = 16$ and $M = 8$) of an FFT flow diagram that corresponds to the hardware configuration of Fig. 10.20. The scratch memory notion is a useful and economic compromise for achieving greater FFT speeds without undue expenses for large high-speed memory systems.

10.10   Radix 2 and Radix 4
Parallel Structures Using RAM's

In this section we concentrate on a specific computing structure employing parallelism with the following features.

1. Memory cycles and computation cycles are completely overlapped; such a structure is "matched" in that it is designed to be simultaneously memory and arithmetic limited.
2. The data is contained in a random access memory (RAM) with a register length of r complex data samples, where r is the radix of a fixed radix algorithm.

An example of a radix 2 flow diagram for such a structure is given in Fig. 10.8. A hardware configuration and its associated (partial) timing diagram are shown in Fig. 10.21. The RAM consists of $(N/2)$ registers each of the length of two complex data words. The numbers correspond to the RAM

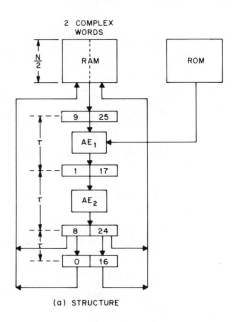

(a) STRUCTURE

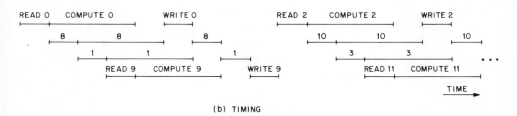

(b) TIMING

Fig. 10.21 Structure and timing for radix 2 FFT with double registers.

addresses. In order to realize a matched system, the second read cycle must, without pause, follow the first read cycle, etc. Thus, subsequent read cycles must be overlapped with the compute cycles. Furthermore, since writing must follow the completion of at least two compute cycles, we can see from the timing diagram in Fig. 10.21 that four successive read cycles are required before writing can occur. This means that the AE *must* contain pipelining; that is, a second butterfly begins before the first is finished.

If the AE is unable to perform its function in two memory cycles, a longer pipeline of six memory cycles will also result in a matched system; this timing is depicted in Fig. 10.22. The resulting structure will be a lengthened version of Fig. 10.21(a) with a total of six buffers entered between six partial arithmetic elements; if the computation can proceed more rapidly, some of these partial AE's will be *nulls*, simply lines between adjacent buffers.

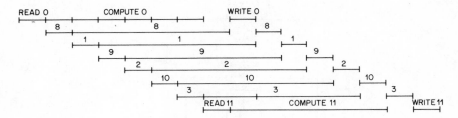

Fig. 10.22 Alternate timing diagram to Fig. 10.21.

An interesting problem for the reader is to construct a structure and determine the required timing to use the same AE to service two RAM's.

Figure 10.23 shows a radix 4 structure and its associated timing. Here the computation time is four units of memory time and eight reads are followed by eight writes. The pipeline culminates in a (4 × 4) permutation matrix, represented in Fig. 10.23 by four registers, each containing the result of a radix 4 butterfly.

The form of parallelism introduced in this section is based primarily on the notion of matching memory time to butterfly time. We have restricted ourselves to fixed radix systems and have assumed that we have r-fold parallelism for a radix r system. Now, in a radix r system, we have $(\log_r N)$ FFT stages. For each level, each of the (N/r) registers must be accessed twice, once to read the inputs to the butterfly and once to write back the answer. Thus, the number of computational units (or memory cycle times) needed to perform a complete FFT would be

$$C_r = \frac{2N}{r} \log_r N \qquad (10.26)$$

and the number of computational units per unit of sampling interval is

$$\frac{C_r}{N} = \frac{2}{r} \log_r N \qquad (10.27)$$

Equation (10.27) tells us the highest sampling rate that can be processed in real time given that we know the time per single computation (butterfly or memory). For example, for $N = 1024$ and $r = 2$, $C_r/N = 10$; thus, for a butterfly time of 100 nsec we can process a one-megasample signal.

10.11   General Discussion of the Pipeline FFT

If we go back to the flow diagrams of Figs. 10.1 through 10.8, we note that although the diagrams describe many properties of the algorithm, the precise sequence of butterflies in time is not specified. As a matter of fact, many such

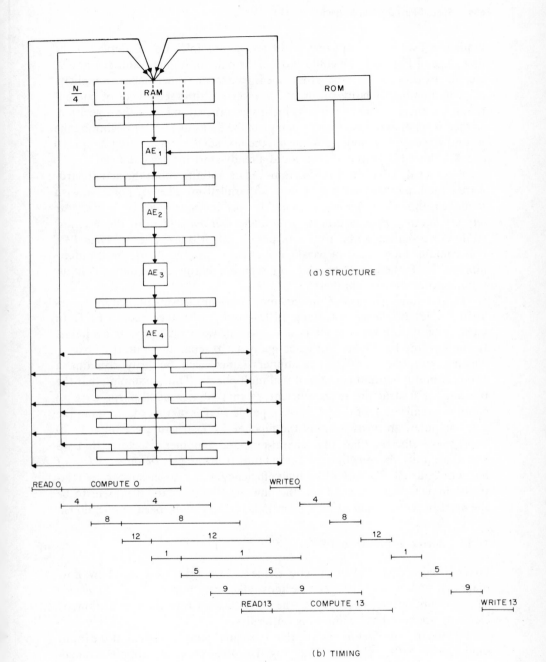

(a) STRUCTURE

(b) TIMING

Fig. 10.23 Radix 4 parallel structure and associated timing.

sequences leading to the same result are permissible. For example, in the first stage of Fig. 10.1, we could process the pairs of inputs 0 and 8, 1 and 9, etc., in any conceivable order; the same is true for the other stages. Simplicity of programming or hardware may favor certain time sequences of computation but there are no constraints intrinsic to the structure of the algorithm. In fact, it is not even necessary to complete the first stage before beginning the second stage; for example, if we began the first stage by processing samples 0 and 8 followed by 4 and 12, we could already start the second stage.

We note also that the flow diagrams tell us nothing about the actual hardware structure in terms of the amount of parallelism. The key point we wish to make is this: *Given hardware parallelism, definite constraints begin to appear on the allowable time sequences of the individual butterflies.* In the next few sections we shall describe a class of parallel algorithms called *pipeline FFT* that contains an amount of parallelism equal to $\log_r N$. Thus, for a radix r pipeline FFT there will be ($\log_r N$) separate hardware butterfly computations proceeding in parallel.

To give some perspective on the amount of parallelism entailed in a pipeline FFT, let us take as an example a 1024-point, or 10-stage, radix 2 FFT. In most general-purpose computers a single hardware multiplier is available. In the pipeline FFT there can be as many as 10 separate "butterfly boxes," which correspond to 40 real multipliers (since each butterfly contains a complex multiplier that contains 4 real multipliers). Thus, assuming that the pipeline FFT structure is as efficient as that of a general-purpose (g.p.) computer realization of the FFT, the pipeline FFT is 40 times faster than the g.p. computer. In turns out that the pipeline FFT structure is from 2 to 20 times more efficient than any general-purpose computer structures that we know of; thus the pipeline FFT structure is from *two to three orders of magnitude* faster. Because of its high efficiency and also because of a relatively simple control mechanism, the pipeline FFT appears at present to be the most important special FFT processor for very high-speed applications.

10.12   Radix 2 Pipeline FFT

Given ($\log_2 N$) parallel arithmetic elements, we first must ask how flow diagrams such as Fig. 10.1 can be most efficiently implemented. *Efficiency* can be quantitatively described as the percentage of time that the arithmetic elements are kept busy computing butterflies.

For the moment, let us assume that the signal samples appear at the input sequentially, $x(0)$, $x(1)$, etc. Then Fig. 10.24 shows a very simple arrangement for performing the first stage of an FFT corresponding, for example, to the flow diagram of Fig. 10.1. The first eight samples $x(0)$ through $x(7)$ are switched into the eight-stage delay element z^{-8}. The next eight samples are switched to the other input line to the system. Assuming that the butterfly

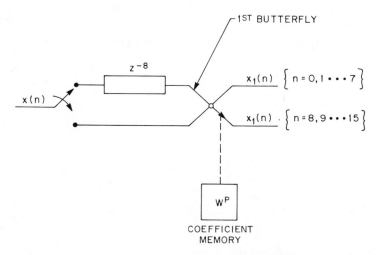

Fig. 10.24 First FFT pipeline stage.

computation time is *exactly* equal to the sampling interval, the entire first stage of the FFT is performed in the subsequent eight-sample intervals following the switching. Results of the first stage [which we have labeled $x_1(n)$] appear in parallel pairs at the butterfly output. Since the coefficient W^p changes from sample to sample, the coefficient memory must be entering its information to the butterfly at the same rate (the sampling rate) as the signal. We notice from Fig. 10.1 that the structural form of stage 1 is repeated twice in stage 2. Thus, we have to devise an arrangement that will process $x_1(n)$ $\{n = 0, 1, \ldots, 7\}$ and $x_1(n)$ $\{n = 8, 9, \ldots, 15\}$ in a manner similar to the way $x(n)$ $\{n = 0, 1, \ldots, 15\}$ was processed. This contrivance is shown in Fig. 10.25. We see that by means of appropriate delays and switching times, we line up the partly processed samples in exactly the way specified by Fig. 10.1. Thus, the "spacing" (difference between the samples in time) was eight time units for the first butterfly and four time units for the second. A complete 16-point pipeline FFT is shown in Fig. 10.26. Here we have an opportunity to observe the various symmetries and, by extrapolation, to construct pipeline FFT's with larger N. Let us make a few remarks about Fig. 10.26.

1. The delay elements in a given stage are half as long as that of the delay elements in an earlier stage.
2. The arithmetic elements are busy only half the time in the figures we have shown.
3. Each switch switches at double the rate of its predecessor.
4. The basic clocking interval of the whole system is naturally equal to the sampling rate.
5. The output is bit-reversed as a function of real time.

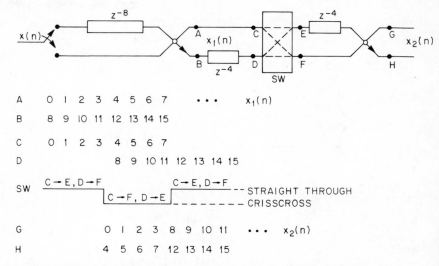

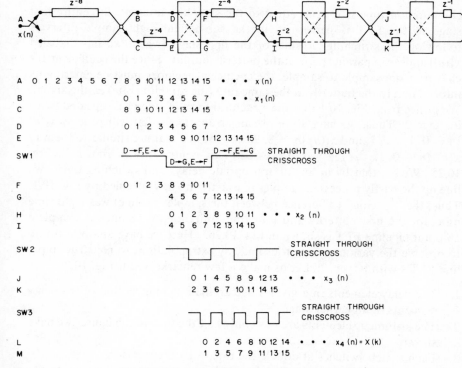

Fig. 10.25 First and second stage of 16-point pipeline FFT, radix 2, DIF.

Fig. 10.26 Complete 16-point, radix 2, pipeline FFT, DIF.

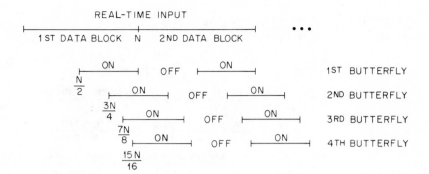

Fig. 10.27 On–off times for arithmetic elements processing contiguous
blocks of data.

To prove statement 5 we notice that the indices in Fig. 10.26 are in exact
correspondence with the (unlabeled) register numbers in Fig. 10.1. Since in
Fig. 10.1 the resultant output is bit-reversed, so is the output of Fig. 10.26.
More succinctly, Fig. 10.26 is a specific implementation of Fig. 10.1 and thus
possesses all the same properties plus timing properties not specified in Fig.
10.1. We must qualify this remark somewhat by observing that the pipeline
FFT structure has a two-port output so that two frequency samples at a time
are available. The important point is that the indices shown on the last two
lines of Fig. 10.26 are in actuality the bit-reversed indices of the output
frequency samples.

   With regard to statement 2, this is a rather tricky point and the on time
of the AE's is really dependent on how the input is interfaced with the
processor. For example, in Fig. 10.27 we chose a requirement that contiguous
data blocks be processed in real time. As we see from Figs. 10.24 through 10.26,
processing cannot begin until half the data block has entered the processor.
Then the first stage is completed in the next $(N/2)$ cycles. At this moment,
the first butterfly is turned off until the initial $(N/2)$ values of the next data
block have been gathered into the z^{-8} delay element. The other AE's follow
the same pattern with a delay. Therefore, the overall system efficiency is
50% since every AE is on exactly half the time. Figure 10.28 shows how
system efficiency can be made 100% by using the correct input buffering
scheme. After the first data block has been stored, ports (a) and (b) are
simultaneously played into the processor. Because of the parallelism of the
two ports, playout can be clocked at half the rate of the input sampling.
Thus, the first stage of the FFT is finished just when the second data block is
ready to be processed. The other stages perform the same way but with the
usual pipeline delays. The advantage of this scheme is that the computational
clock need be only half as fast as the input clock or, alternately, the same
system as that of Fig. 10.26 can handle double the data rate; the price paid is
extra input buffering and switching.

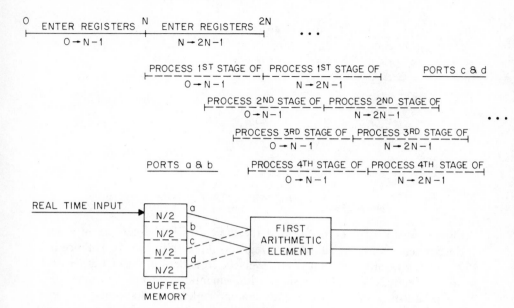

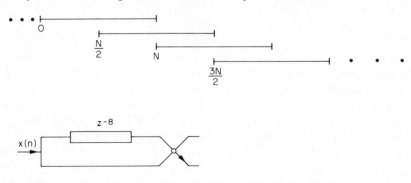

Fig. 10.28 Input buffer arrangement so that contiguous blocks of data can be processed 100% efficiently in real time.

In the special but interesting case of real-time processing with 2:1 overlap of the data blocks (as shown in Fig. 10.29), we simply connect the input to both the z^{-8} delay element and the first arithmetic element. As in Fig. 10.28, the system is 100% efficient in that all AE's are working full time. This special case fits a method of performing convolution by FFT; hence it is quite useful.

With some hindsight we can, in summary, adjust the remarks made with respect to Fig. 10.26. Remark 1 is generally true but alterations in the input buffering will influence the first stage delay; for example, in Fig. 10.28 this delay has been incorporated in the buffer system. Remark 2 need not be

Fig. 10.29 Input configuration for real-time processing of overlapped data blocks.

true since we have shown, via Figs. 10.28 and 10.29, how the AE's can be kept constantly busy. Remark 3 is again true with the first stage being a possible exception and, as seen in Fig. 10.28, the system clock can be slowed down compared to the sampling rate. In all our configurations thus far, the result is bit-reversed and always follows the flow diagram of Fig. 10.1. It appears that other possibilities exist in radix 2 and that pipeline FFT's can be devised from other flow diagrams but at this writing no other structure seems quite as compact and elegant.

A final remark on Fig. 10.26 is that no time was allotted for computation time of the AE's. Including such time does not in any way disturb the structures but it does insert extra delays within the system equal to the number of clock times needed to perform a butterfly. If this number is greater than 1, this implies some "staging" or "pipelining" *within* each AE.

10.13   Radix 4 Pipeline FFT

Beginning with Fig. 10.14 we can work out the structure of a radix 4, 64-point pipeline FFT. As our first exercise we consider the processing of a single data block of 64 samples arranged in normal order. It turns out that a radix 4 pipeline is blatantly inefficient for such an input because the AE's will be working only one-fourth of the time. Nevertheless, this exercise will allow us to analyze the entire structure such that many of the results are applicable for 100% efficient configurations. Making the system 100% efficient is really an input buffering problem that will then be discussed for a variety of input situations.

Figure 10.30 shows a block diagram of the radix 4 pipeline FFT. It is of the same general form as radix 2 but each of the basic elements (delay, commutators, and butterflies) are now geared to radix 4 operations. Thus, the butterfly, instead of performing a complex multiply and two complex adds (as in radix 2), now performs three complex multiplications and eight complex adds. The commutator is a four-input, four-output switch and there are delay elements in three out of the four parallel lines in the system.

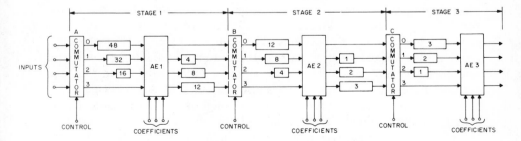

Fig. 10.30 Radix 4, 64-point, pipeline FFT.

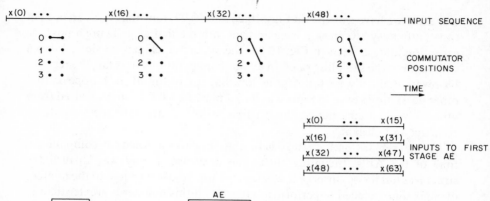

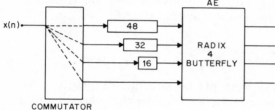

Fig. 10.31 Distribution of a block of 64 samples into four subblocks of 16 samples each to perform a radix 4 butterfly.

Referring to Fig. 10.31, we can see how the sequence $x(0), x(1), \ldots, x(63)$ is distributed among the four inputs to the first butterfly. The original sequence is shown on the top lines. Every 16 clock pulses, the commutator switches to the next position so that the subsequent subblock of 16 samples goes through a different delay. At clock time 48, samples $x(0), x(16), x(32)$, and $x(48)$ are "poised" to enter the first AE simultaneously and after clock time 63 the first stage of Fig. 10.14 has been completed. Note that the AE was used for only 16 of 64 cycles.

The rest of the algorithm can be traced, using Fig. 10.32. Each of the four parallel sequences is traced through all the processing elements. It is interesting to note that the commutator states follow a prescribed pattern. This is most easily seen for the outputs of commutator C, for which the switch positions are shown. A "switching cycle" really consists of a sequence of four different commutator positions. This sequence is cyclic. The end effects, which cause some of the commutator positions to be unused, can be otherwise interpreted as precisely the same sequence of four positions. In fact, the pattern of commutator B positions is also the same; in this particular case, end effects predominate. From all this, we might surmise that in a steady-state mode, where the AE's are being used full time, all commutators follow the switching cycle pattern indicated for commutator C, with each commutator switching four times as rapidly as its predecessor.

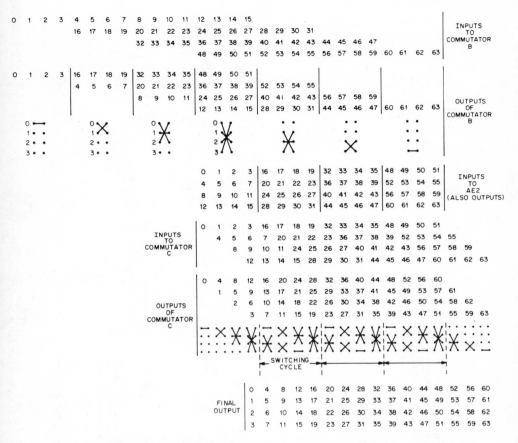

Fig. 10.32 Timing and switching of radix 4, 64-point, pipeline FFT of Fig. 10.14.

We also note that the *time* ordering of the outputs is effectively the same as the time ordering of the inputs, which means that the *frequency* ordering of the output is a base-4 *digit reversal* of the input, as described in Sec. 10.3.

The switching for a 4:1 overlapped sequence of FFT's is shown in Fig. 10.33. In the steady state, the single real-time input sequence is always connected to all four lines. To avoid end effects, the first three subsequences are switched in gradually. Once the steady state is reached, commutator *A* is really not needed; the input is merely connected to all four input lines all the time. Alternately, if, for convenience, we did want to switch commutator *A* in the standard way derived in Fig. 10.32, it would work, provided that the input sequence was connected directly to all four input lines of commutator *A*. While it seems natural that a radix 4 system should accommodate a 4:1 overlapped FFT with 100% efficiency, we might ask if it could remain 100% efficient in real-time FFT processing with 2:1 overlap. The answer is "yes,"

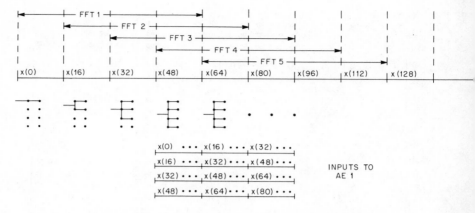

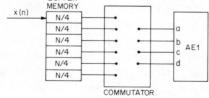

Fig. 10.33 Radix 4, pipeline FFT doing a 4:1 overlapped FFT.

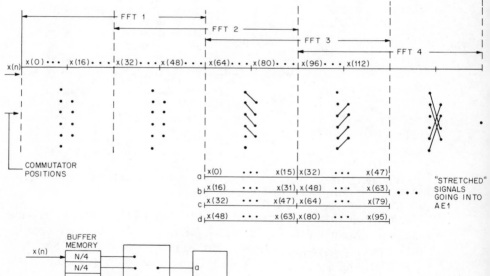

Fig. 10.34 Modification of input structure to accommodate 2:1 overlap processing with 100% efficiency.

although some modifications to the buffering and input switching are required. A configuration is shown in Fig. 10.34. Here the memory of the first stage in Fig. 10.30 is replaced by a memory of size $3N/2$ with a single input port and six output ports. During the first 64 clock cycles, the first N samples are entered sequentially into the buffer but no processing is done. Starting with sample number 64, the commutator switch is thrown so that the top four output memory ports are connected to AE1 and all actively supply data to AE1 but each at *half* the *rate* as the input sequence. Thus, during a single 32-clock interval, two new subblocks of memory are filled, while four already filled blocks are emptied into AE1. Therefore, 100% efficiency is maintained by virtue of a 2:1 "stretching" of the signals. All in all, as would be expected, twice as much processing per unit time can be accommodated compared to a 4:1 overlap. By way of exercise, the reader can figure out the input buffering when there is an overlap of only $N/4$; i.e., if the first FFT is from 0 to $N-1$, then the second FFT is $3N/4$ to $7N/4 - 1$ and a new FFT of size N is initiated every $(3N/4)$ clock pulses.

The radix 4 pipeline FFT can also perform FFT's on four distinct signals, $x_1(n)$, $x_2(n)$, $x_3(n)$, and $x_4(n)$. This is accomplished by having the first-stage commutator switch with the same pattern as the later-stage commutator. The arrangement is shown in Fig. 10.35; notice that each successive input is delayed by $N/4$ samples. Beginning with the fourth epoch of length $N/4$, the switching pattern goes into a steady-state mode, repeating every N samples. In a sense, the processing of four independent signals is the most natural mode for a radix 4 pipeline because the switching patterns of all commutators including the first commutator are now the same, with each successive commutator switching at four times the rate of its predecessor.

10.14 Comparison of Radix 2 and Radix 4 Pipeline FFT's

We have already shown that radix 4 uses fewer multipliers than radix 2 to perform the same FFT. Let us investigate this question for the particular case of a pipeline FFT. For a given N, radix 2 has ($\log_2 N$) stages, while radix 4 has ($\log_4 N$) stages. Thus radix 2 has twice the number of stages. But each radix 2 stage has two complex adders and one complex multiplier, whereas a single radix 4 stage has eight complex adders and three complex multipliers. Thus,

$$r_m = \frac{\text{No. of radix 4 multipliers}}{\text{No. of radix 2 multipliers}} = \frac{3}{2}$$

$$r_a = \frac{\text{No. of radix 4 adders}}{\text{No. of radix 2 adders}} = 2$$

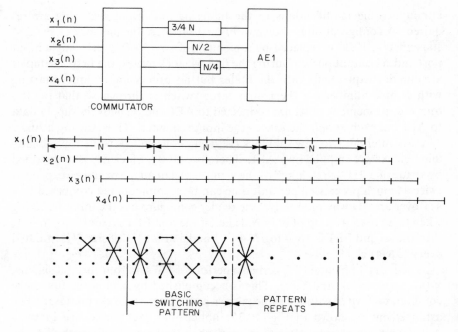

Fig. 10.35 Input configuration to process four independent signals with radix 4 pipeline.

So far it looks like radix 2 wins but we have not yet made a fair comparison because for the ratios above the radix 4 pipeline will process double the amount of data. Thus, if we normalize r_m and r_a to correspond to the same processing power for both cases, we find

$$\hat{r}_m \ (= \text{normalized } r_m) = \tfrac{3}{4}$$

$$\hat{r}_a = 1$$

Now we see that radix 4 does give the same capability for fewer multipliers. What we have *not* worried about are the comparisons of control problems one might face in designing two such systems.

10.15 Discussion of More Highly Parallel FFT Hardware Structures

We have established the criterion that a given FFT structure is 100% efficient if all hardware arithmetic elements are always busy computing their butterflies. The pipeline structures we have described can be made 100% efficient by appropriate input buffering matched to the specific signal or signals to be processed. By contrast, a general-purpose computer, being a

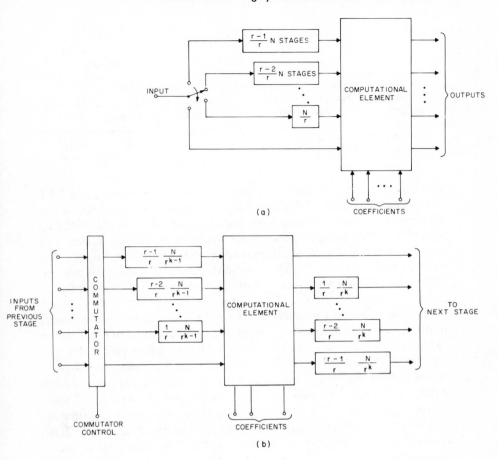

Fig. 10.36 Input stage and kth stage of radix r pipeline FFT.

highly sequential machine, will normally be much less efficient. In this section we shall describe FFT structures that remain 100% efficient even though the degree of parallelism is even greater than those already described.

1. A straightforward extension of the radix 2 and radix 4 pipeline FFT's is the radix r FFT. If r is a number like 8, for example, than a pipeline FFT, as shown in Fig. 10.36, can be constructed. This structure has r input lines, as shown in Fig. 10.36(a), and $(r-1)$ delay elements on each side of a computational element, as shown in Fig. 10.36(b). A computational element, or arithmetic element, is now a complete r-point DFT followed by $(r-1)$ twiddle factors. Now, let us look at what happens when we contemplate three FFT structures side by side, namely, radices 2, 4, and 8. Given that all three systems are clocked at the same rate,

Table 10.3

	Radix 2		Radix 4		Radix 8	
N	No. of Multi-plications	No. of Additions	No. of Multi-plications	No. of Additions	No. of Multi-plications	No. of Additions
64	16	32	24	60	44	118
4096	40	68	60	126	116	144

then since a radix r pipeline FFT handles r lines, the higher radix FFT will process more data. Table 10.3 shows the relative degree of parallelism involved in radices 2, 4, and 8 for some special cases. In this table we have counted the actual number of *real* multipliers and *real* adders. For example, a complex multiplication is counted as four real multiplications and two real additions. We also remind the reader that radix 4 is intrinsically twice as fast a processor as radix 2, while radix 8 is twice as fast as radix 4 (*provided that* the clock times are the same for all systems). (This means that, for example, in the radix 8 arithmetic element there must be sufficient hardware to perform an eight-point DFT plus seven twiddles in the same time that the radix 4 AE does a four-point DFT plus three twiddles and the radix AE does a two-point DFT plus one twiddle.)

2. Instead of having $(\log_2 N)$ parallel AE's or even $(\log_r N)$ parallel radix r AE's, we ought to be able to increase the parallelism to $(N/2)$ AE's and thereby gain even more processing capability. An efficient scheme for implementing such an algorithm is illustrated in Fig. 10.3, which shows a constant geometry 16-point FFT algorithm. The advantage of such a scheme lies in the small amount of switching needed in going from AE's to registers and back. Each AE is excited by two (complex) registers and unloads its results into two other registers; these connections remain constant through all the stages. The final result is bit-reversed; a hardware bit reversal can be wired in, at the cost of two extra connections per AE.

Probably the fastest and most economical way of entering the coefficients is to wire in the sequence for each AE. Thus, in Fig. 10.3, AE7 has the coefficients W^7, W^6, W^4, and W^0. Each AE has at most $(\log_2 N)$ coefficient registers so that $(N/2)$ AE's require less than $(N/2 \log_2 N)$ coefficient registers. Granted that this is appreciably more than the $(N/2)$ coefficients required, we benefit from the fact that no signal distribution from a central coefficient memory to a set of AE's is required.

To judge the speed of such a system, imagine that a butterfly can be done in time τ; then $\tau \log_2 N$ is the time needed to perform the FFT.

3. Given a complete set of ($N/2 \log_2 N$) AE's, an additional factor of $\log_2 N$ in speed can be bought by pipelining the configuration of Fig. 10.3. Thus, each of the open circles in Fig. 10.3 corresponds to a complete AE, whereas each of the nodes signifies a register. This is the ultimate because with array AE's and pipelining a complete FFT output is available every clock pulse.

An interesting fact about this configuration is that each AE requires a *single* wired-in coefficient and that many of these coefficients are either ± 1 or $\pm j$ so that no hardware multiplier is needed. Equation (10.2) shows how many hardware multiplications are really needed for an N-point FFT. Table 10.4 shows M_N, the number of hardware multipliers needed for a completely parallel and pipelined radix 2 array FFT.

Table 10.4

N	M_N
2	0
4	0
8	2
16	10
32	34
64	98

4. If one has a requirement for a large FFT, say, $N = 4096$, the parallelism described in (2) and (3) would result in overwhelmingly large amounts of hardware: 2048 AE's for (2) and 21,502 AE's for (3). On the other hand, the radix 2 pipeline FFT has 12 AE's. The radix 8 pipeline FFT discussed in (1) has the hardware equivalent of about 48 AE's for $N = 4096$ (by "AE's" we mean here the equivalent of a radix 2 AE). Another structure that lies somewhere between radix 2 and the ultimate is obtained by using a special memory structure, wherein the input data is arranged as a two-dimensional array and the same hardware processes (first) all elements of a row in parallel and (second) all elements of a column in parallel. For example, for $N = 4096$ we have a (64×64) array and 32 AE's operating in parallel so that we achieve nearly triple the parallelism of a radix 2 pipeline.

The two-dimensional memory structure can be implemented using commercially available IC memory elements. The concept is illustrated in Fig. 10.37 for a (4×4) array. If we label the data from 0 through 15 as shown in Fig. 10.37(a), then the data is distributed among four memory chips (each containing four elements) as shown on Fig. 10.37(b). If row j or column j is to be read or written, then the four required addresses are shown in the corresponding row or column label of Fig. 10.37(b); in parentheses are shown the corresponding accessed elements as defined in Fig. 10.37(a). Notice that

	0	1	2	3	COLUMNS
ROWS					
0	0	1	2	3	
1	4	5	6	7	
2	8	9	10	11	
3	12	13	14	15	

	CHIP 0	CHIP I	CHIP II	CHIP III
	0 13 10 7	4 1 14 11	8 5 2 15	12 9 6 3
ROW 0	0 (0)	1 (1)	2 (2)	3 (3)
1	3 (7)	0 (4)	1 (5)	2 (6)
2	2 (10)	3 (11)	0 (8)	1 (9)
3	1 (13)	2 (14)	3 (15)	0 (12)
COLUMN 0	0 (0)	0 (4)	0 (8)	0 (12)
1	1 (13)	1 (1)	1 (5)	1 (9)
2	2 (10)	2 (14)	2 (2)	2 (6)
3	3 (7)	3 (11)	3 (15)	3 (3)

Fig. 10.37 Bit arrangements on memory chips for a (4 × 4) array.

to read out a row the addresses of successive chips are incremented by 1, whereas to read out a column all addresses are equal. Notice also that successive rows all have the same sets of addresses but they are rotated.

A row, or column, of data is read out in parallel and enters a processor consisting of four AE's through a barrel switch. The barrel switch is equivalent to a circular switch and is needed so that the processor can be repetitive independent of the row or column it is processing. For example, referring to Fig. 10.37(b), when row 0 is to be processed, datum 0 goes to AE0, datum 1 to AE1, datum 2 to AE2, and datum 3 to AE3. When row 1 is to be processed, datum 4 (which comes from chip I) goes to AE0, datum 5 (chip II) goes to AE1, datum 6 (chip III) goes to AE2, and datum 7 (chip 0) goes to AE3. Figure 10.38 shows the connections required; an element of the jth column of Fig. 10.38 tells which chip to connect to the jth AE. To move from row to row or column to column of memory requires barrel switching. Notice that the barrel switch connections are the same for the ith column and the ith row in memory.

ROW 0	0	I	II	III
1	I	II	III	0
2	II	III	0	I
3	III	0	I	II
COLUMN 0	0	I	II	III
1	I	II	III	0
2	II	III	0	I
3	III	0	I	II

Fig. 10.38 Connections between AE's and memory.

The processor itself can be one of many types; for example, it might have the structure of any of the flow diagrams and it might be implemented as a complete pipeline array.

10.16 Overall Design Philosophy for Special FFT Processors

In this section we trace the train of judgments needed to specify and design a special-purpose processor that includes the FFT algorithm. There are many different FFT algorithms and one of the major decisions must be the choice of radix number, decimation-in-time or decimation-in-frequency, in-place or not-in-place, etc. These decisions, however, ought to follow from the more fundamental ones such as the choice of components, degree of parallelism, and memory structure, which in turn are based on the overall speed, size, and flexibility requirements imposed by the system specification.

While it is impossible to enumerate all possible system configurations within which an FFT processor might be imbedded, we can describe one typical example. In radar applications there is often need for a matched filter. The most important specifications encountered in such cases are (1) flexibility, in that the filter transfer function must be capable of change, and (2) speed, since practical bandwidths may be on the order of many megahertz. In addition, system requirements usually impose a set of time–bandwidth products that will ultimately determine the size of the FFT. For these cases, an FIR filter using FFT techniques with provisions for real-time input and output may be called for.

In cases such as described above, FFT processor speed is determined by the overall speed specification of the system. For example, if we consider convolution via FFT, it turns out that a single FFT processor embedded in such a system may have to work four times faster than the same size FFT processor embedded in a system where contiguous block spectral analyses were needed. Similarly, it is obvious that an FFT processor working on overlapped signals must work at a speed proportional to the degree of required overlap as well as the signal bandwidth.

Another common application for FFT processors is as attachments to general-purpose computer facilities. There are at least two reasons why such an attachment would be desirable.

1. For facilities where a large part of the work load is performing spectrum analysis. Computer time, hence money, would be saved.
2. For facilities where it is desired to do real-time processing for which only the lack of a high-speed FFT processor stands in the way.

In case 1, the precise speed specification is not as important as the ease of programming for the FFT processor, the cost of building the processor and its computer interface, and the flexibility of the processor.

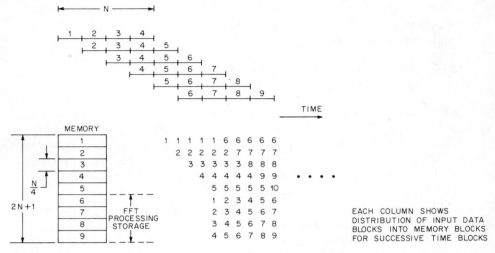

Fig. 10.39 A memory structure for a 4:1 overlapped real-time FFT.

10.17 Overlapped FFT with Random Access Memory

In many applications, it is desired to perform overlapped FFT's. We have already shown how a pipeline FFT can handle overlap. If we divide a data block for which the FFT is desired into four parts (for example, for a 4:1 overlap), then a complete N-point FFT must be performed for each new $(N/4)$ input point. A random access memory algorithm for handling this overlap is shown in Fig. 10.39. Assume the memory is divided into nine equal blocks, each of length $N/4$. While the fifth block is being read into memory, data blocks 1 through 4 are being transferred to the four bottom register blocks of the memory 6 through 9, whereupon an FFT is performed. For real-time operation, the sum of this transfer time plus the FFT processing time must be equal to the time needed to read $(N/4)$ input points. While data block 6 enters register block 1, thus erasing the no longer required data block 1, data blocks 2, 3, 4, and 5 are transferred to the same register blocks 6, 7, 8, and 9 for the second FFT. This procedure continues with the state of subsequent data blocks being indicated in successive columns.

10.18 Real-time Convolution via FFT Using a Single RAM and One AE

Let us assume we want to construct an experimental real-time FIR filter to perceptually test the effect of filtering on speech. For such an application, 20-kHz sampling of the speech signal (with 9 to 10 bits of quantization in the

A/D) seems appropriate. Based on the design criteria discussed in Chapter 3 we can guess at a reasonable impulse response length; assume that a 128-point impulse response suffices. This corresponds to 442 butterflies per FFT or $\frac{442}{128} \approx 3.45$ butterflies per sampling interval. Convolution via FFT can require (as we shall show) four times that speed; hence, we need 13.8 butterflies in one sampling interval of 50 μsec, which yields approximately 3.62 μsec required for each butterfly.

For these numbers it is quite clear that there is no need for hardware butterflies running in parallel since 3.62 μsec represents a moderate speed for even some of the slower logic lines. Nevertheless, the designer has a host of possible options open to him. For example, if he wants to use slow components, he may have to build a complete combinational array of all four multipliers and six adders. By using a faster line (such as TTL), however, only one hardware array multiplier (running at, say, 0.5 μsec per multiply) may be needed and the butterfly performed by proper sequencing of this hardware. Going to even faster logic may make it possible to design the multiplier as an add–shift combination, thus saving IC's but perhaps using more power and causing more grief to engineers who have never used such fast elements.

Let us now assume that the appropriate arithmetic system for the FFT convolutional filter is 8 bits for coefficients, 16 bits for data, and a fixed point 2's-complement number system and that these assumptions lead to the speeds we have already estimated. We now have the problem of configuring the system, which means (1) specifying the arithmetic, (2) specifying the memory element, (3) describing the memory addressing algorithm, (4) specifying the manner in which input and output interruptions take place, and (5) preparing a detailed timing chart of the various functions. These design steps ought to be performed in a detailed enough manner so that a complete logic design, with the chosen components, can be carried out. From the logic design we can proceed with board layouts, rack design, signal conditioning, and the whole host of construction problems.

There are two major timing structures in our system: The first one is the overall timing of the complete convolution (two FFT's plus a filtering function) and the second is the more detailed timing of the interaction between individual memory and arithmetic cycles. In order to deal with the first timing problem, we do not need to specify the precise memory or AE structure yet but simply the structure shown in Fig. 10.40.

To be specific, we have chosen to construct an FIR filter with an impulse response of length 256. In order to perform this convolution by FFT processing, we need to perform successive 512-point FFT's and save half of the output points. The overall timing begins by the entry of 256 complex samples of the input into memory addresses 0 to 225, which corresponds to one basic epoch and, as seen from Fig. 10.40, all timing has been divided in terms of this

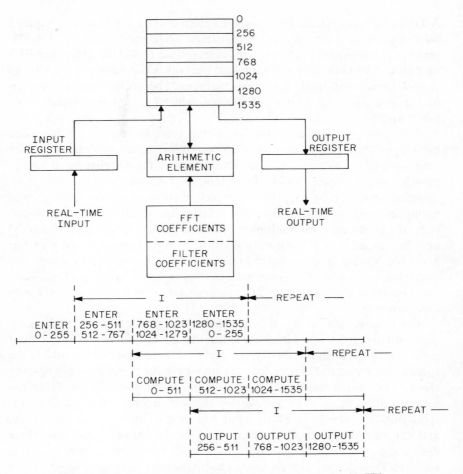

Fig. 10.40 Structure and overall timing for convolution via the FFT.

epoch. During the next epoch, signal samples 256 through 511 are entered
not only in memory registers 256 to 511 but also in registers 512 to 768. We
are now ready to perform the convolution that entails a 512-point FFT
followed by a multiplication of each FFT output by the filter coefficients
followed by an inverse FFT. For real-time operation, this computation must
be done no slower than in a basic epoch. Now, beginning with the fourth
epoch, input, output, and computation proceed in parallel and each of these
three functions has a three-epoch cycle associated with it, the cycles being
denoted by the Roman numerals. The coefficient memory proceeds in
parallel with both data memory and computation; if both FFT and filter
coefficients are fixed, the coefficients can be stored in read-only memories.
If it is desired to change coefficients, however, this memory must be writable

and at least a slow input (from a computer, paper tape, punched card, keyboard, etc.) has to be provided.

It is interesting to note in passing that although, in a typical general-purpose (non-real-time) computer program, performing a 512-point in-place convolution requires only 512 (complex) registers, when we go to real-time operation, we need thrice the memory. Also, the question arises,"Can this processor be used as a filter with an impulse response shorter than 256 without altering the constrained structure of Fig. 10.40?" The answer seems to be "yes." By choosing the appropriate values of the filter coefficients, any impulse response shorter than length 256 can be realized; but we should keep in mind that this statement is true only as long as finite coefficient register lengths are ignored.

10.19   10-MHz Pipeline Convolver

A proposed radar system wants to process 10-MHz bandwidth radar returns with filters matched to a variety of transmitted signals, such as linear FM and uniform and nonuniform bursts. Because of the flexibility desired, the matched filter transfer characteristic must be easily and arbitrarily manipulated. Let us assume that the longest signal duration is about 100 μsec so that the time–bandwidth product is 1000. This means, for example, that if the filter were realized as a digital tapped delay line, 1000 taps would be needed and 1000 complex multiplications would have to be performed every 100 nsec. It is not clear that the present level of component technology could handle such requirements. This, of course, is where the FFT method of convolution becomes an absolute requirement as the correct structure for a digital hardware realization. Using the FFT, we know from the previous example that two 2048-point FFT's need to be performed in 102.4 μsec. The total number of butterflies to be done in that time is 2048 $\times$ 11 or 22,528, which comes out to be exactly 22 butterflies every 100 nsec. Since present technology just about accommodates hardware that will perform a single complex multiply in about 100 nsec, the specifications call for 22 such boxes and this number corresponds exactly to the pipeline FFT structure described in Sec. 10.10. The convolutional processor is shown in Fig. 10.41 along with the timing. We see two pipeline structures working in parallel with each other and with the matched filter computation. Since the pipeline structure can handle two FFT's simultaneously, the 2:1 signal overlap exactly matches the pipeline timing and makes the arithmetic element perform 100% efficiently. For each block of 2048 input points, two sets of signals appear on the two output lines of the first pipeline. The second pipeline generates on its two output lines both the *good* half of a block plus the *bad* (first) half of the next block; thus, by accepting half of the output signal, a real-time filtered signal is generated by the overlap–save method.

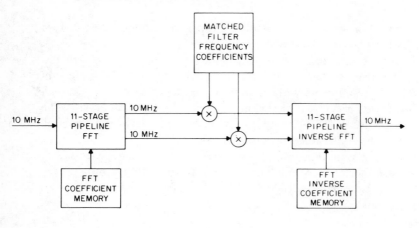

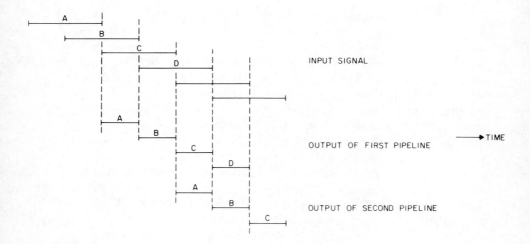

Fig. 10.41 Convolutional FFT processor.

For the pipeline structure, we are effectively employing $(2 \log_2 N)$ parallel arithmetic elements to match the speed requirements. In the previous example we used a single hardware arithmetic element. It is certainly reasonable to inquire as to attractive forms of parallelism for intermediate cases. One approach is to maintain the pipeline structure but decrease the number of arithmetic elements. For example, let us lower the input signal bandwidth from 10 to 0.8 MHz. In each of the two pipeline structures, replace the 11 arithmetic elements by buffer registers, all of which are multiplexed to a single high-speed AE. Thus, with two such AE's, one for each pipeline, the integrity of the overall structure is maintained but each of the pipeline stages shares the same hardware butterfly. If, instead, you desired a 3.2-MHz bandwidth, 4 AE's could be assigned to each pipeline, with 3 AE's each servicing three stages and the fourth one handling two stages. Thus, we see that the pipeline configuration remains efficient at a variety of data rates.

Given the overall structure above, detailed design questions can now be examined. In particular, there may be as many as five memory elements associated with Fig. 10.41. They are

1.   Memory internal to the pipelines.
2.   Coefficient memories for each of the pipelines.
3.   Filter memory.
4.   Input buffer memory.
5.   Output buffer memory.

In addition to these, the arithmetic number system must be specified and along with that the register lengths of all memory elements and the scaling and overflow strategy.

Convolution by FFT methods need not be done with a 2:1 overlap. In fact, by doing larger FFT's the convolver may be made more efficient. This follows from the fact that in FFT convolution the amount of information that is ultimately thrown away depends on both the convolution length and the FFT length.

Figure 10.42 illustrates this for three cases: Assume that in all cases the impulse response is of length N (represented by one unit in Fig. 10.42) but that for case 1 the FFT size is $2N$, for case 2 it is $4N$, and for case 3 it is $8N$.

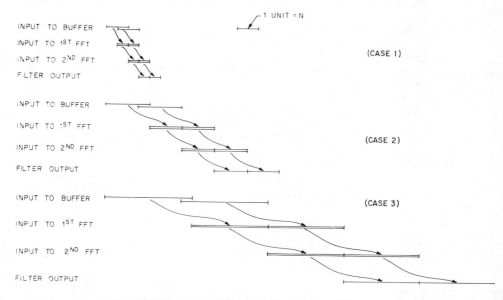

Fig. **10.42** Relative efficiencies of various size FFT's.

As we have previously shown, in case 1 for every pair of transforms (of size $2N$) we throw away half the result to obtain N good outputs (see the fourth line of case 1). In case 2 there are $3N$ good points for transforms of length $4N$. This means that we can allow a time proportional to $3N$ to perform transforms of size $4N$. Similarly, in case 3 we can allow a time proportional to $7N$ to perform transforms of size $8N$. This means that the *relative* efficiencies of cases 1, 2, and 3 are in the ratios $1:2, 3:4$, and $7:8$. For very high-speed systems, the hardware savings in going from $2N$ to $8N$ transforms may thus be substantial.

REFERENCES

1. G. D. BERGLAND, "Fast Fourier Transform Hardware Implementations—An Overview," *IEEE Trans. on Audio and Electroacoustics*, **AU-17**, 104–108, June, 1969.
2. H. L. GROGINSKY AND G. A. WORKS, "A Pipeline Fast Fourier Transform," *IEEE Trans. on Computers*, **C-19**, 1015–1019, Nov., 1970.
3. R. KLAHN, R. R. SHIVELY, E. GOMEZ, AND M. J. GILMARTIN, "The Time-Saver: FFT Hardware," *Electronics*, 92–97, June, 1968.
4. M. C. PEASE, "An Adaptation of the Fast Fourier Transform for Parallel Processing," *J.A.C.M.*, **15**, 252–264, Apr., 1968.
5. R. R. SHIVELY, "A Digital Processor to Generate Spectra in Real Time," *IEEE Trans. on Computers*, **C-17**, 485–491, May, 1968.
6. G. D. BERGLAND AND D. E. WILSON, "An FFT Algorithm for a Global, Highly-Parallel Processor," *IEEE Trans. on Audio and Electroacoustics*, **AU-17**, 125–127, June, 1969.
7. G. D. BERGLAND, "Fast Fourier Transform Hardware Implementations—A Survey," *IEEE Trans. on Audio and Electroacoustics*, **AU-17**, 109–119, June, 1969.
8. F. THEILHEIMER, "A Matrix Version of the Fast Fourier Transform," *IEEE Trans. on Audio and Electroacoustics*, **AU-17**, No. 2, 158–161, June, 1969.
9. G. C. O'LEARY, "Nonrecursive Digital Filtering Using Cascade Fast Fourier Transformers," *IEEE Trans. on Audio and Electroacoustics*, **AU-18**, No. 2, 177–183, June, 1970.
10. R. L. VEENKANT, "A Serial Minded FFT," *IEEE Trans. on Audio and Electroacoustics*, **AU-20**, No. 3, 180–184, Aug., 1972.
11. B. GOLD AND T. BIALLY, "Parallelism in Fast Fourier Transform Hardware," *IEEE Trans. on Audio and Electroacoustics*, **AU-21**, No. 1, 5–16, Feb., 1973.

Finite Word Length Effects in the FFT

1. A. V. OPPENHEIM AND C. J. WEINSTEIN, "Effects of Finite Register Length in Digital Filtering and the Fast Fourier Transform," *Proc. IEEE*, **60**, No. 8, 957–976, Aug., 1972.
2. P. D. WELCH, "A Fixed-Point Fast Fourier Transform Error Analysis," *IEEE Trans. Audio and Electroacoustics*, **AU-17**, No. 2, 151–157, June, 1969.
3. C. J. WEINSTEIN, "Roundoff Noise in Floating Point Fast Fourier Transform Computation," *IEEE Trans. on Audio and Electroacoustics*, **AU-17**, No. 3, 209–215, Sept., 1969.
4. T. KANEKO AND B. LIU, "Accumulation of Round-Off Errors in Fast Fourier Transforms," *J. Assn. Comp. Mach.*, **17**, No. 4, 637–654, Oct., 1970.

11

General-Purpose Hardware for Signal Processing Facilities

11.1 Introduction

For any specific computational task, special-purpose hardware should always be more economical than general-purpose (g.p.) computers. A great variety of situations, however, require hardware flexibility to which g.p. computers are the most useful answer. This chapter is devoted to a specialized class of such applications, namely, the design of general-purpose facilities for the study of signal processing techniques in the fields of speech, radar, sonar, communications, and seismology. Such facilities are proving to be more valuable than large time-shared or batch processing central facilities for an increasing number of workers in the fields above and is therefore worthy of our attention. A primary purpose of such a facility is the development of appropriate algorithms for simulation of proposed hardware devices.

At the present time, the computer field is still a rapidly evolving field. To a large degree, the appropriateness of a given computer structure is dependent on the characteristics of the available components. For example, at the time when magnetic core memory was the prime form of computer memory, arithmetic elements were just graduating from vacuum tubes to transistors and even a single flip-flop was large and expensive. Under these conditions, memory external to core was at a premium and designers tried very hard to minimize the amount of such hardware. Later, when core still reigned supreme but arithmetic and flip-flop hardware became cheap and powerful, such concepts as general registers and high-speed scratch pads became popular. The advent of MSI (medium scale integration) has led to a rash of parallel processing concepts, many of which are presently highly controversial. The rapidly changing components field has spawned a large number of ideas and, thus far, a few computers that take advantage of these components.

In this chapter, we shall first choose as an example a simple g.p. computer and show its signal processing capability. Then we discuss various methods of improving the speed of g.p. computers for signal processing. Finally we describe in some detail the FDP (Fast Digital Processor) and the LSP2 (Lincoln Signal Processor 2) as examples of both the structure of signal processing g.p. computers and the use of such computers in signal processing facilities.

11.2   Special- and General-Purpose Computers

Implementation of signal processing algorithms by digital means implies the use of either a special-purpose or a general-purpose computer. The distinction between these nomenclatures has always been vague and seems to be growing more vague so some discussion of our usage of these terms seems warranted.

General purpose implies flexibility, adaptability, and programmability. In order to attain these often-desired features, we expect to pay for them with money, time, size, power, etc. It turns out, however, that often it is better to buy and program a commercial computer to perform the task simply because it will do the job, it is already in production, and it is therefore cheap (usually cheaper than the development cost of even a relatively simple piece of digital hardware). Assuming that this is *not* so, we can point to at least one clear-cut case where specialized hardware should be used, namely, when the required processing speed is so great that specialized structures are called for to meet specifications. Perhaps the most obvious distinction between a general and special computer structure is that for any given algorithm performed by the g.p. computer, special hardware can be built that will perform the same algorithm faster. A corollary of this distinction is that, to implement a given algorithm with special hardware, parallel processing techniques can often be employed quite gracefully. For g.p. computers, parallelism is of course possible and has been used occasionally but experience indicates that too much of it tends to be inefficient, causes severe communications problems among the various components, and leads to awkwardness in programming. It seems to us that a sequential machine is in general more suitable when great variety and flexibility is desired. Thus, in designing digital hardware for many applications, be it a central facility, a facility dedicated to a single set of problems (such as radar), or even more highly specialized airborne equipment, the speed–flexibility tradeoff has perhaps the major impact on the nature of the subsequent design and construction effort.

11.3   How to Describe a Computer

Any computer, special or general, can be defined by specifying (1) the individual computational blocks, such as adders, multipliers, and logic nets; (2) the memory structure; (3) all the paths between the computational blocks and memories; (4) the sequence of operations; and (5) the connection to devices considered to be *outside* the computer. Special-purpose computers are designed to handle a limited number of algorithms, with each algorithm always being implemented via the same sequence of operations. The g.p. computer has a program that resides in one or more of the computer memories, with each "line" of the program corresponding to an instruction that exercises the system (usually) in a simple way—the entire algorithm being composed of an instruction sequence designed by the programmer.

Figure 11.1 shows the structure of a simple general-purpose computer. The central element is the random access memory, which holds both the program and the data to be processed. In this simplified scheme, an instruction consists of an operation (OP) and an address. The machine has a single index register whose contents (optionally) can be added to the address portion of the instruction to form the true memory address. The arithmetic portion of the computer consists of an adder and an accumulator register.

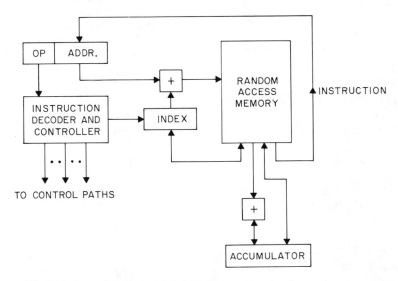

Fig. 11.1 Structure of a simplified general-purpose (g.p.) computer.

The arrowed lines show the possible communication paths. An important factor in the cost of a computer is the number of paths. For every path one must design gating or switching circuits so that the controller can indeed control which paths are active for a given instruction. In this section, we defer the question of how the information (program and data) was put into the memory; this question is dependent on the computer's input–output system, which we shall subsequently describe.

Let us now discuss several typical computer instructions and see if we can write a program to compute one of the very simplest signal processing algorithms, a running sum.

Instruction Name	Description
1 Load Y	Transfer the contents of memory register Y into accumulator.
2 Store Y	Transfer the contents of the accumulator into memory register Y.
3 Add Y	Add the contents of memory register Y to the number already existing in the accumulator and leave the sum in the accumulator.
4 Sub Y	Same as 3 except the operation is subtract.
5 JNX Y	Jump to the instruction located at Y if index register X is less than zero; otherwise go to the next instruction. In either case increment X by 1 after condition is satisfied.
6 YIX Y	Enter the number in the address portion of the OP code into the index register X.
7 Clear Y	Enter zero into memory register Y.
8 Halt	

11.4 Running Sum Program

By *running sum* we refer to the computation given by

$$y(n) = \sum_{m=0}^{k-1} x(n - m) \tag{11.1}$$

To increase the running speed (for $k \geq 3$) of the program, it is advantageous to rewrite Eq. (11.1) as a recursion:

$$y(n) = y(n - 1) + x(n) - x(n - k) \tag{11.2}$$

Figure 11.2 shows the realization of Eq. (11.1) as being simply the sum of the outputs of a digital delay network, while Fig. 11.3 shows the realization of the equivalent algorithm as being a transversal filter in cascade with a one-pole

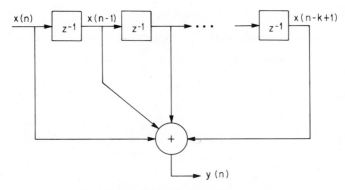

Fig. 11.2 Digital realization of a running sum.

recursive filter. Table 11.1 shows a convenient format for implementing Eq. (11.2) on our simple g.p. computer.

In order to avoid extra programming to account for end effects, the data, $x(0)$ through $x(999)$, are surrounded by zeros on either side; the number of zeros being one less than the width of the running window. To be specific, we have chosen a window size (k) of 26. The labels preceding the arrows in Table 11.1 are memories for addresses that will appear in the program given below, which implements the running sum.

			Comments
	YIX	-1025	Enter the number -1023 into X
	CLEAR	YN	Enter zero into YN
LOOP $\rightarrow$	ADD$_x$	DATA $+$ 1051	Add $x(n)$ to $y(n-1)$
	SUB$_x$	DATA $+$ 1025	$x(n) - x(n-k) + y(n-1)$ into accumulator
	STORE$_x$	YN $+$ 1025	Store $y(n)$
	JNX	LOOP	Go around loop 1024 times
	HALT		Stop

The signal processing lesson to be learned from the program above is that four instructions must be performed for each processed data point. The time required depends, of course, on the speed of the computer circuits. For

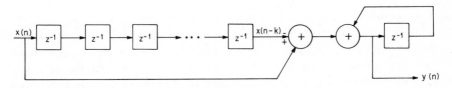

Fig. 11.3 Recursive realization of a running sum.

Table 11.1

DATA FORMAT FOR RUNNING SUM PROGRAM

DATA	$\rightarrow$	0
		0
		.
		.
		.
		0
DATA $+$ 26	$\rightarrow$	$x(0)$
		$x(1)$
		.
		.
DATA $+$ 1025	$\rightarrow$	$x(999)$
		0
		0
		0
		.
		.
DATA $+$ 1051	$\rightarrow$	0
		.
		.
		.
YN	$\rightarrow$	0
		.
		.
		.

convenience, let us use a reference time of 100 nsec for each memory cycle. Now, the top three of the four inner loop instructions require a minimum of *two* memory cycles, one to fetch the instruction and the next to execute it. Thus, the total time per processed point is 600 nsec plus the JNX time plus any arithmetic computation time that couldn't be conveniently buried in the memory cycles; in round numbers, about 1 μsec.

11.5 Input–Output Problems for Real-Time Processing

If the computer is required to perform real-time signal processing, many new problems arise. In the non-real-time case discussed in Sec. 11.4, we assumed that 1000 input samples were to be processed, that these samples were already available for processing in the computer memory, and that the processed samples would simply be stored in memory. For non-real-time processing, we can imagine a completely independent routine which accepts 1000 samples as input and another routine which outputs the stored result to some other device. Obviously, the time needed to run these routines must be added to the computation time already obtained but no difficulties or programming intricacies arise. In a real-time situation, however, there is a more complex interleaving of the in–out and computational routines that often lead to an appreciable slowdown in running time.

To illuminate these problems, let us conceive of a very simple input–output system attached to the basic configuration of Fig. 11.1. This system consists of an additional register B, which has a path to memory and also to the outside world. Whenever the instruction INPUT or OUTPUT is given, B becomes the recipient (for INPUT) of information from the peripheral device. Also, the INPUT or OUTPUT instruction causes the computer to hang up and wait for B to either receive or deliver the datum. Once this is done, the computer continues to the next instruction and it is up to the program to fit the addition (or removal) of this new datum into its format. Obviously, such a rudimentary input–output system is unsatisfactory (as is our rudimentary computer) and does not allow for automatic interrupts or interleaving of numerical and input–output programs but it should illustrate some of the additional programming complexity and time required to accommodate a real-time situation.

We now demonstrate the effects of the simple input–output system described above on the computation of a running sum. Let us design our situation so that the input data stream is made to circulate in computer memory as shown in Fig. 11.4. The computation strategy is as follows: Each new input from B is $x(n)$ and is first stored in a temporary register, TEMP. The address in the memory format of Fig. 11.4 points at $x(n - k)$ (we have changed the value of k to 6 in Fig. 11.4 to aid in depicting this simple input–output system), there being precisely k registers in Fig. 11.4. The computation $x(n) - x(n - k)$ is performed and then $y(n - 1)$, which is stored in another temporary register YN, is added and the result stored back in YN. To complete the loop, $x(n)$ replaces $x(n - k)$, the index is incremented, and an output instruction is given.

The following program is a simple implementation of the real-time running sum. (The value of k is 26 again.)

$x(0)$ →	$x(6)$	$x(6)$	$x(6)$	$x(6)$	$x(6)$	$x(6)$ →	$x(12)$ ···
$x(1)$	$x(1)$ →	$x(7)$	$x(7)$	$x(7)$	$x(7)$	$x(7)$	$x(7)$
$x(2)$	$x(2)$	$x(2)$ →	$x(8)$	$x(8)$	$x(8)$	$x(8)$	$x(8)$ ···
$x(3)$	$x(3)$	$x(3)$	$x(3)$ →	$x(9)$	$x(9)$	$x(9$	$x(9)$
$x(4)$	$x(4)$	$x(4)$	$x(4)$	$x(4)$ →	$x(10)$	$x(10)$	$x(10)$
→ $x(5)$	$x(5)$	$x(5)$	$x(5)$	$x(5)$	$x(5)$ →	$x(11)$	$x(11)$

Fig. 11.4 Illustration of memory format for a real-time running sum program ($k = 6$).

	YIX	−25	Store −25 in index register.
	CLEAR	YN	
1.	INPUT		
2.	STORB	TEMP	Store contents of in–out register [new $x(n)$] in a temporary register.
3.	SUB$_x$	XN + 25	Subtract $x(n - k)$ from $x(n)$.
4.	ADD	YN	Add $y(n - 1)$ thus computing $y(n)$.
5.	STORE	YN	Put $y(n)$ into register YN to be used for next iteration.
6.	LOAD	TEMP	
7.	STORE$_x$	XN + 25	
8.	JNX	10.	
9.	YIX	−25	Reset index register.
10.	LOADB	YN	Stick $y(n)$ into in–out register
11.	OUTPUT		
	JMP	1.	

To determine the rate at which this program can be run, we assume that after the INPUT is given, the computer waits for the input device's clock to arrive; then instructions 1 through 11 are executed and the computer again hangs up, waiting for the output clock. In order to operate at the maximum rate it is necessary for the delay between input and output clocks to be carefully adjusted. This means that should the program change, readjustment is needed. If the two clocks are not phase-synchronized, the maximum rate must be lowered to accommodate longer hang-up times after the INPUT or OUTPUT. But even with the adjustment, we see that the real-time program requires about twice as much processing time as the non-real-time version.

11.6   Methods of Increasing Computer Speed

Many variations of a few basic principles have been introduced into computer structures to increase speed. One set of ideas hinges around the use of "cache" memories, which usually consist of relatively small but very fast memories somehow imbedded into the overall computer structure; the trick in using such memories is to write a program in such a way (if possible) that most of the time-consuming operations are run in the cache. Another set of ideas for increasing speed of computation is associated with parallelism of arithmetic operations. This notion has a great many variations since an overall design will change depending on the degree of parallelism. In addition, parallelism of control, in–out memory, and arithmetic operations can help reduce the number of machine cycles needed to perform a given algorithm. Included in this type of parallelism are the overlapping of instruction cycles with data cycles and the wired subroutines. *Pipelining* is another form of parallelism in which the data is partially processed in an assembly line manner so that new data can enter the pipe as soon as the first operation is finished. In this way, data can enter and leave the pipeline much more rapidly.

Underlying the structural ideas above is the use of components of appropriate speed and cost in the right places. Circuit speeds matched to the structure are a necessity for efficient computer design.

11.7 Cache Memories

The cache memory may be incorporated into a computer in various ways.

1. By connecting it as a logical part of the arithmetic element (AE).
2. By connecting it as an input–output device.
3. By making it the main computer memory and connecting a larger slower memory as an input–output device.
4. By connecting it logically into the main memory structure and using a specified portion of the address field.

Cache memory in the computer AE is quite prevalent in modern high-speed computers. Typical examples are the CSP-30,[1] which has 32 such registers, and the META-4[2] and LX-1,[3] which have 16 registers each. The minicomputer PDP-11[4] has 8 registers and the Nova series[5] has 4. Usually, these memory elements are referred to as general registers since they can be used for a variety of purposes such as arithmetic, input–output, and indexing.

Item 2 in the list above is limited to hard-wired peripherals such as special FFT processors, as discussed in Chapter 10. In this configuration, the main computer can set up a dedicated controller to perform certain specialized functions at a high rate. Data is transmitted from main memory to cache memory, processed via the special AE, and the result sent back to main memory for further processing or inspection.

If the cache memory is large enough, it may be economical to structure the main computer elements around it. In this case, the large memory is used primarily as a storage buffer somewhat like a drum, while the cache memory does all the work. Such a system requires high-speed control and arithmetic circuitry but is logically closer to the goals postulated above, namely, that of having a general-purpose high performance signal processing computer.

[1] Computer Signal Processors Inc., Burlington, Mass.
[2] Digital Scientific Co., San Diego, Calif.
[3] M.I.T. Lincoln Laboratory, Lexington, Mass.
[4] Digital Equipment Corp., Maynard, Mass.
[5] Data General Corp., Southboro, Mass.

Finally, we can integrate high- and low-speed memories into a single logical structure. For efficiency, it is expected that the controller will know which memory is being used and will appropriately adjust its timing. In order for such a structure to be efficient, we again need substantial speeds for both AE and control circuitry.

Needless to say, all the methods above have their tradeoffs, both functionally and economically. Also the different ideas presented above can be varied and combined in different ways. Later in the chapter we shall tell how different computers use such notions.

11.8 Arithmetic Parallelism

It is apparent that duplication of computational hardware can lead to speedups but it is not trivial actually to attain the potential implied by such parallelism. An old but quite effective method of introducing optional arithmetic parallelism is by defining a computer register in terms of bytes. Thus, for example, the Lincoln Laboratory TX-2 computer has a 36-bit word arranged as four bytes, each of length 9 bits. The arithmetic hardware is also arranged by bytes so that, for example, with a single instruction, the computer can perform a single 36-bit by 36-bit multiplication, two (18×18)-bit multiplications, or four (9×9)-bit multiplications. With such a structure, and some tricky programming, substantial speedup may be attained if one can afford the smaller word lengths.

A greater degree of parallelism would involve multiple arithmetic elements each having full word lengths. [The Lincoln Laboratory *Fast Digital Processor* (FDP) is an example of such a machine, having four 18-bit AE's and two 18-bit memories.] With such parallelism, structuring each program to utilize this power fully can become an exercise of appreciable complexity. The basic problem is one of setting up and flushing out; before all AE's can be efficiently employed, they must all be set up, which could take an appreciable percentage of the time. Following the parallel computation, the answers must be stored in a convenient place, which could also be time-consuming. An additional problem arises when the different AE's need to communicate; the matrix of possible connections for n elements blows up like $n!$ so that the amount of gating gets out of hand unless relatively few of the possible connections are actually implemented. This necessary restriction could make communications awkward and time-consuming.

Some examples of arithmetic parallelism oriented toward efficient FFT algorithms will now be described. Since the FFT manipulates complex numbers, we might ask how a "complex" arithmetic element would enhance performance; such a computational scheme is shown in Fig. 11.5. We assume that R_1, R_2, R_3, and R_4 are double length, or complex, registers and that the adder and multiplier are also complex. A program for performing a

FROM MEMORY M

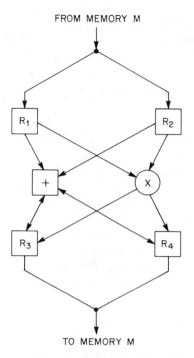

TO MEMORY M

Fig. 11.5 Complex arithmetic element.

single butterfly $A' = A + CW^k$, $C' = A - CW^k$ is shown below where we assume that initially R_2 contains W^k.

	Program	*Interpretation*
1.	$M \rightarrow R_1$	$C \rightarrow R_1$
2.	$R_1 \times R_2 \rightarrow R_4$	$CW^k \rightarrow R_4$
3.	$M \rightarrow R_1$	$A \rightarrow R_1$
4.	$R_1 + R_4 \rightarrow R_3$	$A' = A + CW^k \rightarrow R_3$
5.	$R_1 - R_4 \rightarrow R_4$	$C' = A - CW^k \rightarrow R_4$
6.	$R_3 \rightarrow M$	Store A'
7.	$R_4 \rightarrow M$	Store C'

The program above includes the memory transfers and arithmetic computations required for a decimation-in-time radix 2 butterfly. The indexing is missing and could add, say, two extra instructions, or perhaps be paralleled with the arithmetic and memory cycles.

At the cost of additional control complexity, the inner loop above can be tightened by permitting a degree of parallelism between memory and arithmetic cycles, as follows on page 638.

Program	*Interpretation*
1. $M \rightarrow R_1$	$C \rightarrow R_1$
2. $R_1 \times R_2 \rightarrow R_4$ $M \rightarrow R_1$	$CW^k \rightarrow R_4$ $A \rightarrow R_1$
3. $R_1 + R_4 \rightarrow R_3$	$A' = A + CW^k \rightarrow R_3$
4. $R_1 - R_4 \rightarrow R_4$ $R_3 \rightarrow M$	$C' = A - CW^k \rightarrow R_4$, store A'
5. $R_4 \rightarrow M$	Store C'

Note that this reduction in number of cycles requires no additional logical complexity in the AE and memory structure. For example, in instruction 2, there is no hardware difficulty in using R_1 to both pass information to the multiplier and accept information from memory; this apparent simultaneity works because the register flip-flops have built-in delays between transmitting and receiving paths.

The complex AE of Fig. 11.5 contains four (real) multipliers and four adders (two adders associated with the complex multiply and two for the complex addition). The question arises: Is it worth providing that much hardware to speed up the FFT butterfly? For purposes of comparison, Fig. 11.6 shows a two-adder two-multiplier AE but with as many registers (eight) as Fig. 11.5.

Exercise   Assuming that the memory and arithmetic of Fig. 11.6 can run in parallel, how many instructions are needed to do a radix 2 decimation-in-time butterfly? Design your own instructions.

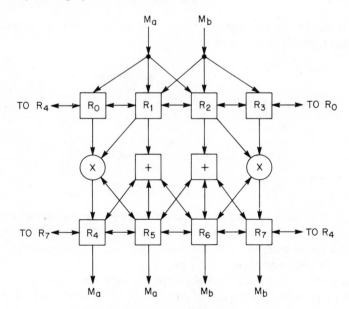

Fig. 11.6 Eight-register arithmetic structure.

11.9   Parallel Operation of Memories, Arithmetic, Control, and Instruction Fetches

Each computer instruction which is fetched controls a collection of gates and flip-flops which puts the computer through a series of operations. Many of these operations can be done in parallel without disturbing the logical flow of the program. The price of this resultant speedup is primarily in the increased complexity of the controller. For example, a jump, load, and add could be embodied in a single instruction. No logical hangups occur in this case but the instruction word becomes quite long. For example, let's associate a 12-bit address with both data and instruction memory and assume that all instruction operation codes are 6 bits long; this would result in a 42-bit word length for the three-operation, two-address instruction described above.

Additional parallelism (which can also be construed as pipelining) is obtained by operating the instruction and data memories in parallel so that new instructions start being executed before old ones are completed. This creates the danger that two instructions vie for control of the same hardware at the same time to perform different functions. Another danger is the failure of a conditional instruction (such as a jump) to sense in time what the actual condition will be, since the previous instruction, which may alter this condition, has not yet been finished.

There are several ways to avoid the dangers above. One way is via an interlock scheme that senses these possible dangers and holds up new fetches or new executions accordingly. Another way is by careful design and synchronization of the timing of each instruction, plus a collection of programming rules; this latter technique will be illustrated in Sec. 11.10 during the discussion of the FDP.

11.10   The Lincoln Laboratory Fast Digital Processor (FDP)

The original motivation for building the FDP came from the speech-compression work then going on at Lincoln Laboratory. It had already been established that computer simulation of vocoder systems[6] was a useful preliminary to the actual construction of the system. The main disadvantage of computer simulation was its incapability of performing real-time processing. In listening to speech or watching a radar scope or a sonar display, the judgment of the person who receives such information is greatly influenced by how he receives it. Without real-time operation of the systems, much of the valuable intuition of the human operator regarding the quality of his system is lost.

[6] See Chapter 12 on applications of digital signal processing to speech.

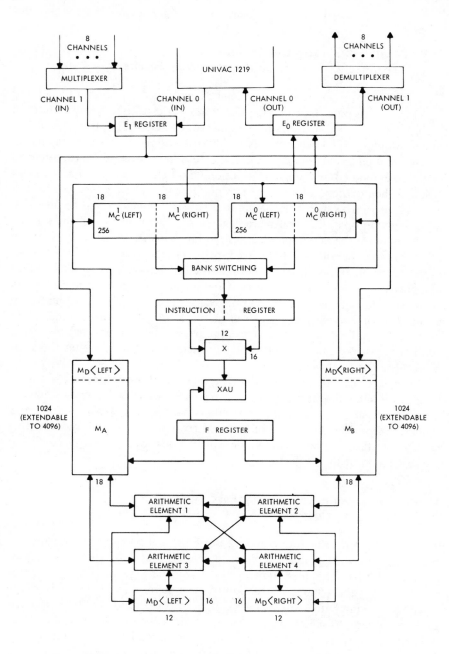

WORD LENGTHS APPEAR ALONG
HORIZONTAL EDGES OF MEMORY.

NUMBER OF WORDS APPEAR ALONG
VERTICAL EDGES OF MEMORY.

Fig. 11.7 Overall FDP structure.

By using the cache memory idea and high-speed ECL logic, it was felt that a speedup of about 10:1 was realizable. In order to program the desired speech processing system in real-time, however, an additional 10:1 factor was needed (these numbers are relative to the Univac 1219 minicomputer that was used at the time). This meant that a variety of tricks were required, which included

1. Arithmetic parallelism.
2. Memory parallelism.
3. Data-instruction cycle overlap.
4. Multipurpose instructions.
5. Parallel control of addressing, arithmetic, and memory.
6. Buffered multiplication.

Figure 11.7 shows the overall FDP structure. Some points to note are

1. There are two independently addressable 150-nsec 4096-register IC (integrated circuit) memories. Bulk memory storage is in the Univac 1219 minicomputer.
2. There are four independently controllable AE's each with a buffered multiplier, an adder, and three programmable registers.
3. The 512-register program memory can generate a new 36-bit instruction every 150 nsec.
4. A fifth but smaller arithmetic element, the XAU (index arithmetic unit), services the index registers. Communication between this and the four large AE's is via the data memories M_A and M_B and the F register.

Addressing is accomplished as shown in Fig. 11.8. Instructions that address M_A and M_B must be fetched from M_C (left), while instructions that address M_C must be fetched from M_C (right). Four bits of IR_A (left instruction register), address M_D, fetching the base addresses for M_A and M_B. Index memory X_A modifies the M_A base address, while index memory X_B modifies the M_B base address.

Addressing of M_C is done via the eight-bit address Y of IR_B. For certain jump instructions, the contents of X_C are added to Y. While all three index memories can be read independently, their addresses must be equal for writing. Thus, from the programmer's view, there is only a single X memory that can be addressed from three fields in a pair of instructions for purposes of M_B and M_C address modification.

11.11 Structure of the AE's

Figure 11.9 shows the transfer paths in one of the four identical arithmetic elements. The registers that are directly under program control are I, Q, and R. AE1 is connected to AE2 and 4; AE2, to AE3 and 1; AE3, to AE4 and 2;

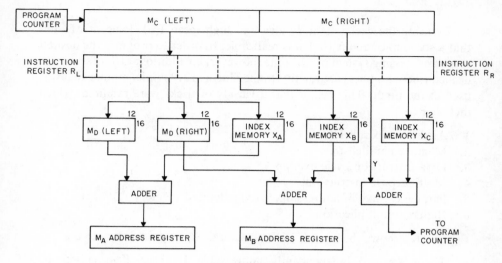

Fig. 11.8 Addressing for the FDP.

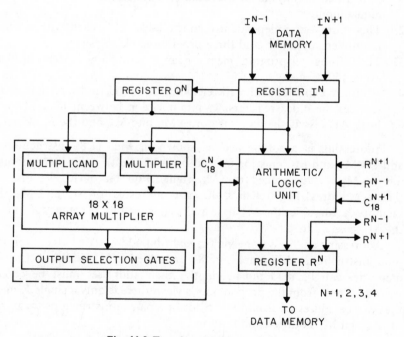

Fig. 11.9 Transfer paths in the FDP.

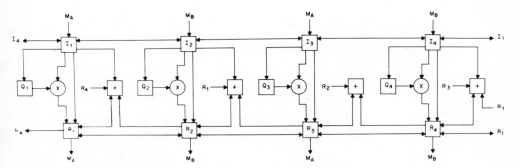

Fig. 11.10 Programmable transfer paths for all four AE's.

and AE4, to AE1 and 3. Thus the AE's are connected in a ring with $n = 4$ and $R^{n+1} = R^1$. R^n can be considered the accumulator since it receives the result of both addition and multiplication.

Important AE features are

1. Each AE can be programmed parallel to, and independently from, all the other AE's. Thus a single 18-bit instruction can cause a multiply in AE1, a transfer from I^2 to R^2 in AE2, etc.

2. The multiply instruction is a *transfer* instruction that transfers I^n and Q^n to their respective images in the multiplier, leaving the programmable registers I^n, Q^n, and R^n free so that other instructions can be performed within AEn during the multiply time, which is presently four instructions including the initial MUL. Thus, each multiplier acts as a special attachment to its associated AE and does not hang up in the AE during the extent of the multiply. Figure 11.10 shows programmable transfer paths for all four AE's.

11.12 Timing

Timing of instructions is shown in Fig. 11.11. Each instruction requires 400 nsec to complete but because of instruction overlap the effective rate is one instruction every 150 nsec. At any time, some aspect of three different consecutive instructions is being done. The three intervals corresponding to an instruction are

1. Fetch instruction (from M_C).
2. Decode instruction, index register operations, jump decisions.
3. Perform transfers, memory access, arithmetic, and logical operations.

During each 150-nsec epoch (Fig. 11.11) *all* the three classes of operation are performed but on three distinct instructions. Since a new fetch is made

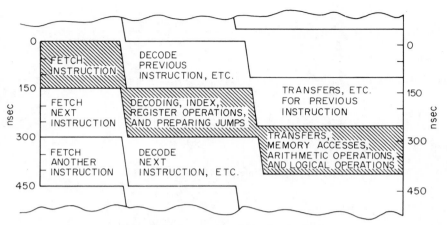

Fig. 11.11 Instruction timing for the FDP.

every 150 nsec, the *effective* instruction rate is 150 nsec, although the actual time expended for each instruction is 400 nsec. (Some special instructions require more than 400 nsec but the basic timing is as described.)

A consequence of this overlap is that a jump-type instruction executed under the timing constraints can only change the instruction sequence after the instruction immediately following the jump has already been fetched. For purposes of efficiency, the instruction following the jump is performed and it is thus the *second* instruction after the jump that is taken from the specified new M_C address. Thus, a jump that closes a loop should be the next to the last instruction rather than the last.

Skip-type instructions cause the instruction following to be interpreted as a no-op (no operation performed) so that the time taken by the instruction following skips is independent of the outcome of the skips. A special (left-half) instruction that has no memory function is SOJ (skip on jump), which, if the right-half instruction accompanying it is a jump, will cause the next instruction to be a no-op if the jump is done. Thus, SOJ with a jump behaves like a normal computer's jump instruction, except that it takes 300 nsec to jump and 150 nsec not to jump.

11.13 Summary of FDP Speedup Features

Let us now summarize how the FDP saves cycle times in the execution of a program.

1. The parallelism of the data memories M_A and M_B allows half the number of memory cycles. For example, in an FFT inner loop, two complex data points must be entered into the AE's that compute the butterfly and then

return two complex results to the memory. In the FDP this costs four memory cycles, whereas in a computer lacking this memory parallelism eight cycles would be needed.

2. Instruction pipelining allows a new instruction fetch to begin 150 nsec after the previous fetch. With the exception of the multiply instruction, each instruction takes 400 nsec from initiation of fetch to end of execution. Thus, a $2\frac{2}{3}$ speed savings occurs relative to completely nonoverlapped instructions. The philosophy adopted in the FDP was that instruction pipelining was mandatory so that all program speeds are more easily computed; an alternate approach would have been to make the pipelining dependent on the instruction sequence and the data by means of hardware interlocks. A disadvantage of such an approach is the greater difficulty of predicting how fast a program will run, thus making real-time programming more tricky.

3. The multipurpose instruction saves cycles because, for example, a memory cycle and a conditional jump may be fetched and performed simultaneously. Of course, the conditional jump depends on what transpired in previous instructions and is independent of the effect created by the instruction it is mated with. Figure 11.12 shows which right-half–left-half combinations of the 36-bit instruction word are permissible.

4. The fourfold arithmetic parallelism also saves cycles although, in view of the other time-saving features we have enumerated, it is rather difficult to assess how much is gained quantitatively by this parallelism. The straightforward way of making such an assessment would be to pick some typical programs and write them using one AE, two AE's, or four AE's. As a very simple illustration, let us consider the problem of accumulating a sum of a block of N numbers. First, let us assume that only AE1 and a single memory (M_A) is available. Then the program would be, provided that R was initialized to zero,

$$M_A \to I \quad ; \quad \text{NO-OP}$$

$$\text{BACK} \to M_A \to I \quad ; \quad \text{JNX} \quad \text{BACK}$$

$$\text{NO-OP} \quad ; \quad I + R \to R$$

$$\text{HALT}$$

The JNX mnemonic is a conditional jump (and increment by 1) when an index register X is negative; otherwise, don't jump. The most interesting aspect of this simple routine is that the third line of instruction is performed before the condition takes effect. The inner loop is thus two instructions, or 300 nsec, long.

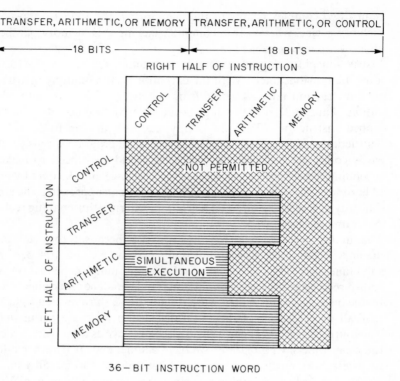

36 – BIT INSTRUCTION WORD

Fig. 11.12 Permissible instruction combinations.

Let us now assume that both memories M_A and M_B are available as are AE1 and AE2. This time both R_1 and R_2 are set to zero. Furtgermore, the block of length N is distributed equally in M_A and M_B, each memory holding $N/2$ samples.

$$M_A \rightarrow I_1,\ M_B \rightarrow I_2 \quad ; \quad \text{NO-OP}$$

$$BACK \rightarrow M_A \rightarrow I_1,\ M_B \rightarrow I_2 \quad ; \quad \text{JNX}\quad \text{BACK}$$

$$\text{NO-OP} \quad\quad\quad ; \quad I_1 + R_1 \rightarrow R_1,\ I_2 + R_2 \rightarrow R_2$$

$$\text{HALT}$$

Again we have an inner loop of two instructions but since AE1 and AE2 are running in parallel, we go around the loop only half as often, simply doing a final summation $(R_1 + R_2)$ after finishing the loop. Thus, we have almost exactly doubled the speed by doubling the arithmetic and memory.

Finally, we bring all four AE's into the picture.

$$M_A \rightarrow I_1, M_B \rightarrow I_2 \quad ; \quad \text{NO-OP}$$
$$M_A \rightarrow I_3, M_B \rightarrow I_4 \quad ; \quad \text{NO-OP}$$
$$M_A \rightarrow I_1, M_B \rightarrow I_2 \quad ; \quad R_n + I_n \rightarrow R_n \text{ for all values of } n$$
$$\text{BACK} \rightarrow M_A \rightarrow I_3, M_B \rightarrow I_4 \quad ; \quad \text{JNX} \quad \text{BACK}$$
$$M_A \rightarrow I_1, M_B \rightarrow I_2 \quad ; \quad R_n + I_n \rightarrow R_n$$
$$\text{HALT}$$

Remarkably, we still have only two inner loop instructions so that this routine is again twice as fast as the two-AE routine and four times as fast as the single-AE routine. From these simple examples we surmise that the FDP instruction format can make efficient use of its multiple arithmetic capacity.

In all the routines above, the memory addressing and the setting up of the correct indexing condition have been omitted. For such straightforward addressing, however, no additional instructions are needed so the inner loop numbers above are legitimate.

11.14 Doing the FFT in the FDP

One of the major problems in the FDP design was to try to keep the machine very general but at the same time to make it as efficient as possible for programming the FFT. It is interesting to consider all the required FDP operations that go into an FFT inner loop and then to see how compactly such a routine can be written with the FDP repertoire of instructions. To begin with, if we assume that the coefficients reside in the Q registers, then we require two read cycles to enter the data into the AE's and two memory write cycles to return the result to memory; so far, there are four instructions. A single multiply instruction suffices to perform all four multiplications but the execution of a multiply takes three (double) instruction cycles. A total of six additions are needed, two to complete the complex multiply and four more to do two complex additions, yielding two more instructions. It turns out that three index manipulation instructions, one conditional jump instruction, and one overflow check instruction are needed. Tabulating, we obtain

Type of Instruction	Number of Instructions
Memory	4
Multiply	6
Addition	2
Indexing	3
Conditional jump	1
Overflow check	1
Total	17

It turns out that an FDP inner loop can be written in 8 lines of code (without overflow checks it can be done in 5 lines of code), or 16 instruction places. This seems impossible since we have tabulated a total of 17 instructions or $8\frac{1}{2}$ lines. If we recall that the multiplier is buffered so that the AE's can be used while the multiply is proceeding, however, we then see that some of the coding for the next butterfly can be done while the present butterfly is being performed.

How much faster can a machine with an FDP-like structure perform a butterfly? We see that the memory structure constrains the number of cycles to be no less than four (two for reading in the data and two for writing the answer back into memory). Thus, the FDP loses *only a factor 2 in speed* compared to a special-purpose processor of the same components and architecture. As we shall see later in this chapter (in the discussion of the LSP2), faster structures for both special- and general-purpose computation can be constructed.

Another useful signal processing program is, of course, recursive digital filtering. The FDP, being a programmable computer, can perform any such algorithm. A good guess at the average number of instructions for a two-pole digital resonator is eight instructions or 1.2 μsec. It is interesting to note that the coupled form takes no longer than the direct or canonic form even though there are twice as many multiplications. This is because of the convenience of using the four multipliers simultaneously.

11.15   Floating Point Routines

The FDP is a fixed point 18-bit machine but floating point routines can and have been programmed on it. An estimate of a floating add is 6.5 μsec, taking the floating negative of a floating point number takes 1.5 μsec, floating multiplication is about 2.0 μsec, and floating divide is about 10 μsec. These times are approximate since the program running time is a function of the data.

The FDP data memory sizes are 4096 registers each for M_A and M_B. This means that FFT's for values of N larger than 2048 cannot be performed by the FDP without access to a larger memory system. In the FDP, this larger memory is M_L, which is connected to an FDP input–output channel. Let us consider the method of performing an FFT on a large number of samples stored in M_L. From Chapter 6 we note that the data can be represented as a two-dimensional ($L \times M$) array, with $N = LM$. Then the FFT algorithm can be performed by first doing FFT's on the rows, then doing the twiddling, and finally doing FFT's on the columns. Since there are N twiddle factors, they must be kept in M_L along with the data and when a data row is read into FDP memory, a row of twiddles must accompany it. After twiddling, the row is sent back to M_L. Thus, to complete all the rows

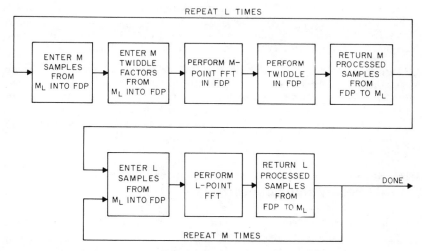

Fig. 11.13 Large FFT done by using the FDP's large core memory.

of the array, including twiddling, requires $3N$ access times of M_L. As seen in Fig. 11.13, another $2N$ cycles are required to do FFT's on all columns and return the result to M_L. Assuming that a twiddle takes as much FDP time as a butterfly, denoting this time by τ_b and the FDP input–output time needed to enter and return M_L information by T_m, we arrive at the formula for the total FFT time T (exclusive of overhead) to be

$$T = \frac{N\tau_b}{2} \log_2 N + 2 + 5NT_m \qquad (11.3)$$

For example, if $N = 65{,}536 = 2^{16}$, $\tau_b = 1.2$ usec, then $T = 0.87$ sec, most of the time being taken up by the butterfly operations that are needed in any case; if we counted only $(N/2 \log_2 N)$ butterfly times (which would be the case if the FDP memory could accommodate the full data base), we find $T = 0.629$ sec. We thus can conclude that the size of the fast FDP memory does not appreciably slow down large FFT's where the data are located in a large random access memory connected as an input–output device. We must be careful, however, to note that if the memory access time T_m of the large memory is too slow, such that $5NT_m$ is bigger than the first part of Eq. (11.3), then this time becomes the constraining factor. For example, in the FDP, T_m can be either 1.6 or 3.2 μsec, depending on the operating mode. For $T_m = 1.6 \ \mu$sec and $N = 65{,}536$, $5NT_m = 0.52$ sec, which is less than the computation time; thus, this time is totally buried. For $T_m = 3.2 \ \mu$sec, $5NT_m = 1.04$ sec and this time now becomes the constraining speed factor on the program.

Exercise Assume a 1024-FDP memory size and try to do a 1024-point FFT that requires the use of M_L. Using the numbers $\tau_b = 1.2$ μsec, $\tau_m = 0.5$ μsec, and $T_m = 3.2$ μsec, determine the constraining time. Compare with the time required to do the same FFT in an expanded FDP fast memory.

Exercise A slightly complicated variant of the exercise above is programming a real-time large FFT. Try to enumerate the basic steps required (and associated times) to structure this program. When you're through, the system should look like a one-input–one-output port system with signal going in and spectrum coming out sequentially in real time. Estimate the data rate that can be handled by the FDP.

11.16 Summary of Special Features on the FDP Caused by the Parallelism

Since the FDP has instruction, memory, and arithmetic parallelism, some of the standard computer instructions must be modified and some special instructions introduced to take this into account. We have already mentioned the problems of jumps due to instruction pipelining and the solution of permitting the instruction following the jump to take place before the jump condition is invoked. Such a scheme works well for an isolated jump but difficulties arise when a succession of jumps appear. In such cases the degree of data dependence of this succession may become too complex for the programmer to follow. A solution to this problem is the SOJ (skip on jump) instruction that appears in the left 18 bits when a jump appears in the right 18 bits. Although the next instruction is in the process of being fetched while the jump condition is being inspected, it is not yet being executed. Thus, it is possible to "kill" that instruction if the inspection says that a jump will indeed take place. Thus an SOJ transforms an FDP jump into a "normal" computer jump but results in the loss of the wasted 150-nsec fetch time of the killed instruction.

Most computers have jumps on various arithmetic conditions, such as the accumulator being zero, positive, or overflow. Since the FDP has four AE's, the conditions are more complicated. The format below shows how this problem is handled.

	6		8				
Arithmetic jump		Y	E_1	E_2	E_3	E_4

The four right bits are "activity" bits; if any E_i is 1, AE_i is active and the jump will take place if the specified arithmetic conditions are met. If more than one E_i is 1, a jump will take place on the OR of these conditions. Thus, if $E_1 = E_2 = E_3 = E_4 = 1$ and a JPR (jump if R is positive), the jump will occur if any of the four R's are positive.

Another method of handling conditional situations in the FDP is the nullify (NUL) instruction. The concept here is that given four AE's running in parallel, each may want to behave differently depending on the data. NUL allows one to make any combinations of AE's inactive so that their states won't change even though instructions are supplied to them. The format is as follows:

6	3	3	3	2
NUL	AE1	AE2	AE3	AE4

There are eight nullify conditions for each AE, as follows.

000	do *not* nullify
001	nullify if $R = 0$
010	nullify if $R \neq 0$
011	nullify if $R > 0$
100	nullify if $R \geq 0$
101	nullify if $R < 0$
110	nullify if $R \leq 0$
111	nullify unconditionally

There is also an accompanying activity instruction (ACT) that conditionally (or unconditionally) reactivates the AE's. Both NUL and ACT have a one-instruction time delay that allows the AE to perform two instructions after the NUL condition is tested and one after the ACT condition is tested.

Another problem that arises is caused by the separation of program and data memories plus the synchronous pipelining of M_C. This pipelining would be upset if we were permitted to *write* as well as read M_C; with no writing permitted, all programs cannot be modified and this is quite inconvenient. For example, an FFT and an inverse FFT program are nearly alike, differing in perhaps five or six instructions. Also, programs may be longer than the 512 registers available in M_C and, in order not to slow the overall performance drastically, it is necessary to keep the excess of the programs in M_A and M_B and to provide a fast path from these memories to M_C. This is accomplished in the FDP via a block transfer instruction. What happens is that the control of the machine is switched to a special piece of hardware that is responsible solely for transferring instructions from M_A and M_B into M_C, the number of transfers and the assigned memory blocks being determined by the block transfer instruction.

We have referred to the FDP as an 18-bit machine but it is worth remembering that instruction words are 36 bits. Since both M_A and M_B can simultaneously read or write, we have access to a 36-bit data word. Furthermore, using four AE's implies 72 bits of arithmetic word length. Realizing all this, the designers endeavored to add multiple precision capabilities to the FDP via the link (LNK) instruction. Invoking LNK allows the carry bit from AE_{n+1} to propagate into AE_n and allows shift instructions to move data between AE_{n+1} and AE_n. Thus, for example, a 36-bit add is done via LNK, doing the low-order 18-bit add followed by the high-order 18-bit add. If all four LNK bits are set, then it is possible, with a single 18-bit instruction, to rotate the entire 72 bits of the four R registers right or left by 1 bit.

11.17   The LSP2 (Lincoln Signal Processor 2)

The FDP design was, to a great extent, an exercise in parallelism as applied to signal processing. On the whole, it was successful because the notion that this mix of speed and flexibility was, on the average, more useful than (1) much speed, no flexibility or (2) much flexibility, little speed, proved to be correct. In addition, a number of lessons were learned that allowed us to think of more efficient signal processing structures. To understand this, let us first analyze the FDP structure. The first point to notice is that the FDP, like most conventional g.p. computers, spends a lot of time doing memory, indexing, and jump instructions. The second point is that making really efficient use of the four parallel AE's often requires very tricky and time-consuming programming. (This objection tends to lose its validity with time through the accumulation of good software). Third, the FDP was built using less highly integrated ECL circuits that were only about half as fast as the improved ECL circuits available at the current time. Some study indicated that an appreciably more powerful processor, yet cheaper and more compact and also more easily programmable than the FDP, could be built. Fourth, an important objection to the FDP design was its overly simple input–output structure, which, for real-time applications, proved to be troublesome. Fifth, FDP memory size was limited by the machine architecture. Sixth, FDP structure paid little attention to many numerical analysis programs that could conceivably require more precision or floating point arithmetic routines.

The LSP2 design was an attempt to overcome some of the difficulties above by making use of new component technology. The overall structure is shown in Fig. 11.14. The central element is a three-rail busing system. Connected to this bus is a small high-speed memory M_R, various function boxes, and a larger memory M_s that communicates with peripherals and also with the program memory M_P.

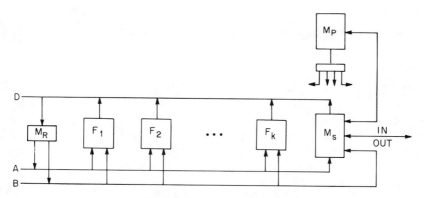

Fig. 11.14 Structure of the LSP2.

Some of the motivations behind LSP2 architecture will now be enumerated.

1. The high-speed memory M_R embodies an attempt to strip away many memory and indexing operations from the required mathematical manipulations. To understand this, we first define the M_R instruction format

OP	A	B	D

The three fields A, B, and D are addresses in M_R and the OP selects a given function box. Thus, in response to such an instruction, two data points from the A and B address fields will be read from M_R onto the output buses A and B and will appear at the input to the chosen function box. These function boxes will respond to putting the result (of, say, an ADD or MULTIPLY) onto the D bus, which is then written back into the M_R register with address D. All this transpires in a single instruction cycle; the time for the execution of such a cycle varies from about 60 to about 150 nsec, depending on the function. Figure 11.15 shows roughly how the hardware control of these instructions works to make this possible. The example chosen involves three successive ADD's. After the 60-nsec fetch, M_R is read and distributed to the bus in 30 nsec. To attain this speed requires two parallel versions of M_R so that both A and B addresses can be simultaneously accessed. Following this, 20 nsec is allotted for the ADD itself. The result, rather than being returned to M_R, is saved in an external buffer and written back during the execution time of the *second* instruction: This trick allows the M_R write cycle to be buried. No conditional jump problems arise since the result of the ADD is available in the buffer and if a conditional jump is called for, the buffer can supply the datum.

 The advantages of the high-speed memory M_R derive from the ability of the programmer to process a substantial chunk of data without

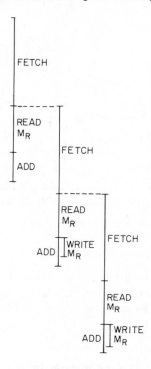

Fig. 11.15 LSP2 Instruction–data timing.

memory references and even without indexing. A good example is a small FFT; let's say M_R contained 64 registers and we were doing a 32-point FFT. After entering all 32 data samples into M_R, the entire FFT routine could be executed without a single reference to M_s or a single jump instruction. Under these conditions, we can show that an FFT in the LSP2 can run about six times faster than it can in the FDP. Considering that LSP2 is a substantially *simpler* computer (although about twice as fast), this is a considerable architectural improvement.

2. The function box concept takes advantage of advances in high-speed circuit integration techniques by the practice of employing more circuitry for purely arithmetic functions and avoiding the use of the same circuits to perform a multitude of functions. Although it might appear to be a more costly approach, the savings in control hardware, the simplification of structure, and the improved modularity make this approach useful. For example, the arithmetic part of an FFT butterfly may take six instructions given a two-multiplier function box. If, however, we incorporate a complex multiplier function box, this butterfly can be done in three instructions. Similarly, it is simple to add function boxes for special applications; thus some versions of the LSP2 may have a special divide function box or square root function box to speed up the computer for

special applications. In most computer structures, addition of special hardware function boxes are costly to implement because of interfacing difficulties, whereas in the LSP2 these units are essentially pluggable.

3. The main memory of LSP2 is M_s. The length of the address word can be either 12 or 16 bits, depending on the chosen computer size, but provision can be made to extend the 12-bit address to 16 bits by suitable bank switching hardware. Addressing of M_s is either direct, via the instruction word, or indirect, via the M_R registers. This means that M_R doubles as an AE memory and an index memory and no parallel hardware is needed for these operations. It also means that indexing and arithmetic cannot be done in parallel, as in the FDP, and this can cost instructions in LSP2. Programming should be simpler because of the necessarily sequential nature of the LSP2.

4. M_s is the input–output link of the LSP2. There are two input and two output data paths and six control paths from M_s to other devices. Because so much can be done in M_R without using M_s and because the input–output control is parallel hardware, very efficient real-time programming is feasible in the LSP2. This means that both the computational and input–output portions of the programs run efficiently. In particular, the LSP2 system has a priority interrupt and direct memory access (to M_s) capability, which should allow several LSP2's to run in parallel and still maintain efficiency.

At present, the LSP2 "exists" as a design study and before its features can be well evaluated, construction programming experience is necessary. The brief discussion above was included here because the LSP2 design was a natural consequence of the FDP experience and the continuing improvement in high-speed component technology.

11.18   A Laboratory Computer Facility for Digital Signal Processing

In this chapter we have primarily been concerned with ways of speeding up the processing of signals with general-purpose hardware (i.e., a high-speed number crunching computer) that is attached to a general-purpose laboratory minicomputer. Although the bulk of the computation is performed in the fast processor attachment, it is generally the job of the minicomputer facility to provide sufficient input–output capability to handle a variety of signals. In this section we discuss an existing minicomputer facility with regard to its input–output capability.

Figure 11.16 shows a block diagram of the Bell Laboratories Acoustics Research Department interactive computer facility. The facility is built around a Honeywell DDP-516 minicomputer. The DDP-516 is an integrated

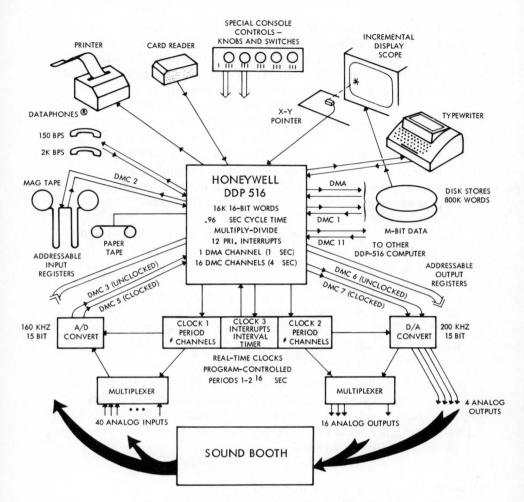

Fig. 11.16 Facility for digital signal processing.

circuit machine with a 0.96-μsec cycle time and a 16-bit word length. As seen in Fig. 11.16, the computer has 16K of core memory, hardware multiply and divide, direct multiplex control (DMC) with 16 data channels (0.25 MHz each), and a direct memory access (DMA) channel that operates up to 1 MHz. The software supplied with the machine included a FORTRAN IV compiler, an assembler, a mathematical subroutine library, a relocatable loader, and various utility software such as a debugger.

In addition to the main frame minicomputer and its associated hardware and software, a variety of peripheral devices have been found to be highly desirable for interactive work in signal processing. At the Bell Laboratories facility, the following peripherals have been interfaced to the minicomputer.

1. Two fixed head disks, each providing 394K words of storage with a 33-msec maximum access time and a 180-kHz word transfer rate.
2. Four independent D/A converter channels with program controlled clocks for rates from dc to 180 kHz.
3. A 180-kHz, 15-bit A/D converter and a 40-channel 100-kHz multiplexer.
4. An incremental display scope driven from a track on the disk. This peripheral is almost indispensable for visual examination of waveforms, spectra, etc.
5. A 300-per-minute card reader and keypunch for off-line program editing.
6. A hard copy output device for listings and debugging.
7. A magnetic tape drive for mass storage of signals and programs.
8. A paper tape reader for inputting programs supplied by the computer manufacturer.

The operating system is a disk operating system that resides permanently in 48 K write protected words of the disk. A 15-word core resident bootstrap load enables the user to access a variety of utility programs conveniently from the disk. A fairly sophisticated graphics package makes the incremental scope useful for display of text as well as waveforms or general graphical material. Hard copy graphical output is obtained from a central computer facility via a dump onto magnetic tape.

As stated previously, in order to use a high-speed processor of the type described earlier in this chapter in an efficient manner, a good facility must already be available to handle the routine input–output operations that are necessary in the course of debugging and experimenting with programs. Thus a large variety of peripheral equipment, of the type described above, has been found useful in this regard.

REFERENCES

1. B. Gold, I. L. Lebow, P. G. McHugh, and C. M. Rader, "The FDP, A Fast Programmable Signal Processor," *IEEE Trans. on Computers*, C-20, 33–38, Jan., 1971.
2. G. D. Hornbuckle and E. I. Ancona, "The LX-1 Microprocessor and its Application to Real-Time Signal Processing," *IEEE Trans. on Computers*, Aug., 1970.
3. H. W. Gschwind, *Design of Digital Computers*, pp. 235–243, Springer-Verlag, Austria, 1967.
4. P. Blankenship, B. Gold, P. McHugh, and C. J. Weinstein, "Design Study of the Advanced Signal Processor," *Lincoln Lab. Tech. Note*, 1972–17, 1972.

12

Applications of Digital Signal Processing to Speech

12.1 Introduction

Some of the most important applications of digital signal processing techniques have been in the area of speech processing. In fact, a large percentage of the theoretical background of digital signal processing has been derived from speech studies and by speech researchers. As we shall see, digital processing has been applied to a wide range of problems in speech including spectrum analysis, channel vocoders, homomorphic processing systems, speech synthesizers, linear prediction systems, and computer voice response systems. In this chapter we present fairly complete discussions of several representative speech systems where digital signal processing has played an important role in the realization of the system. Included will be examples of both software and hardware realizations of speech processing systems. Before proceeding to specific examples, we shall first present a comprehensive summary of models of how speech is produced.

12.2 Model of Speech Production

Figure 12.1 shows a schematic diagram of the human speech production mechanism. In normal speech production, the chest cavity expands and contracts to force air from the lungs out through the trachea past the glottis.

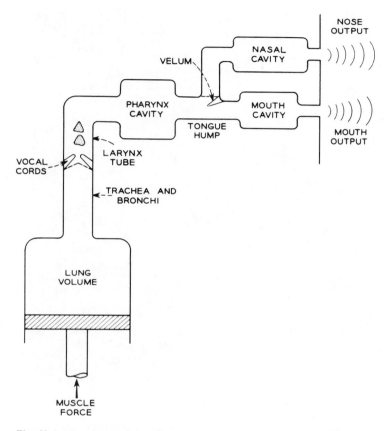

Fig. 12.1 Schematic diagram of the human speech production mechanism
(After Flanagan).

If the vocal cords are tensed, as for voiced sounds like vowels, they will
vibrate in the mode of a relaxation oscillator, modulating the air into discrete
puffs or pulses. If the cords are spread apart, the air stream passes through
the glottis and is unaffected. The air stream passes through the pharynx
cavity; past the tongue; and, depending on the position of the trap door
velum, through the mouth and/or nasal cavity. The air stream is expelled
at either the mouth or nose, or both, and is perceived as speech. In the case of
unvoiced sounds like *s* as in snow or *p* as in pit, the vocal cords are spread
apart (no voicing) and one of two conditions prevails. Either a turbulent
flow is produced as the air passes a narrow constriction in the vocal tract (as
in *s*) or a brief transient excitation occurs following a build up of pressure
behind a point of total closure along the tract (as in *p*). As the various articulators (e.g., lips, tongue, jaw, velum) change position during continuous
speech, the shapes of the various cavities change drastically. Figure 12.2

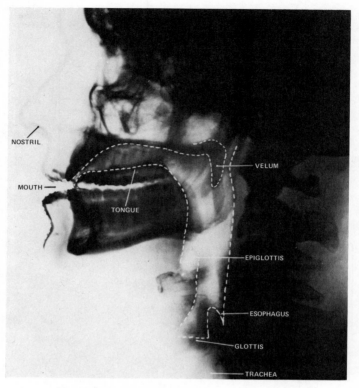

NOSTRIL

VELUM

MOUTH

TONGUE

EPIGLOTTIS

ESOPHAGUS

GLOTTIS

TRACHEA

Fig. 12.2 X-ray of a male vocal tract (After Flanagan).

shows an X ray of a male vocal tract indicating the positions of several of the articulators and the sketched outline of the various cavities.

The human vocal tract is a nonuniform acoustical tube that extends from the glottis to the lips. It is about 17 cm long in an adult male and therefore its first quarter-wave resonance occurs at a frequency given by

$$F_1 = \frac{1}{4}\frac{c}{l} = \frac{1}{4}\frac{34,000 \text{ cm/sec}}{17 \text{ cm}} = 500 \text{ Hz} \tag{12.1}$$

Its nonuniform cross-sectional area depends strongly on the position of the articulators and varies from 0 cm² at closure to about 20 cm². The vocal tract has certain normal resonant modes of vibration, called *formants*, that depend heavily on the exact position of the articulators. Figure 12.3 shows schematic vocal tract profiles for several vowels and gives typical values of the frequencies of the first three formants in hertz.

Figure 12.4 shows the frequency transmission characteristics of these vowels. The nature of the resonances is clearly seen in these plots. It is useful to note at this point that perceptually only the first three formants are

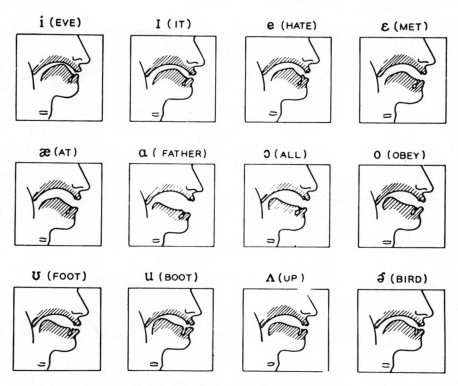

FORMANT FREQUENCIES FOR THE VOWELS

Typewritten Symbol for Vowel	IPA Symbol	Typical Word	F_1	F_2	F_3
IY	i	(beet)	270	2290	3010
I	I	(bit)	390	1990	2550
E	ε	(bet)	530	1840	2480
AE	æ	(bat)	660	1720	2410
UH	ʌ	(but)	520	1190	2390
A	a	(hot)	730	1090	2440
OW	ɔ	(bought)	570	840	2410
U	u	(food)	440	1020	2240
OO	μ	(boot)	300	870	2240
ER	ɝ	(bird)	490	1350	1690

Fig. 12.3 Schematic vocal tract profiles and table of formant frequencies for several vowels (After Flanagan).

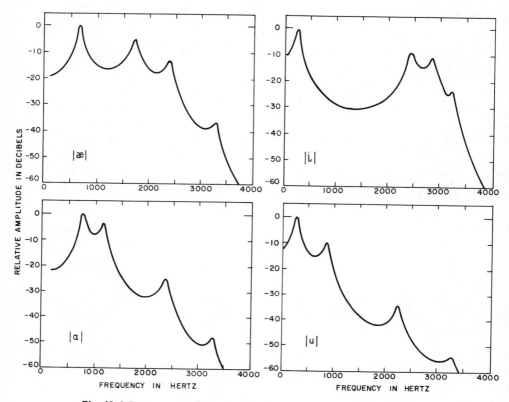

Fig. 12.4 Frequency transmission characteristics of several vowels (After Flanagan).

important in determining the sound that is heard, although the higher formants are necessary to produce sounds of acceptable quality. This finding is the basis for several speech bandwidth compression systems to be discussed later.

As mentioned earlier, there are three primary modes for exciting the vocal tract system. For voiced sounds the source is at the glottis and consists of broad-band quasi-periodic puffs of air produced by the vibrating vocal cords. For unvoiced sounds like *s*, the source is at the point of constriction and consists of turbulent quasi-random airflow. Finally, for unvoiced sounds like *p* as in pop, the source is at the point of closure and consists of a rapid release of the air pressure built up behind the total constriction.

The basic assumption of almost all speech processing systems is that the source of excitation and the vocal tract system are independent. It is this source–system independence that allows us to discuss the transmission function of the vocal tract and to let it be excited by any of the possible sources. The validity of the assumption above is quite good for the majority of cases

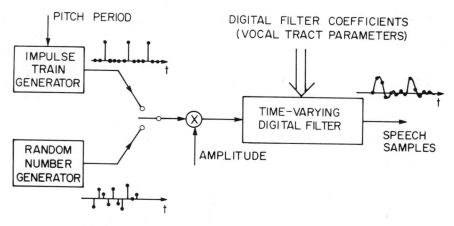

Fig. 12.5 Digital model of speech production (After Schafer).

of interest. There are some cases, however, when the assumption is invalid and the basic model breaks down such as during transient sounds like *p* in pot. For the most part in this chapter, we shall assume its validity. Based on the ideas above, a simple digital model of speech production is shown in Fig. 12.5. The sources of excitation are an impulse generator (controlled from the outside world by the pitch-period signal) and a random number generator. The impulse generator produces an impulse (corresponding to the initiation of a puff of air) once every N_0 samples. This duration is referred to as the pitch period and its reciprocal is the pitch frequency or rate of oscillation of the vocal cords. The random number generator output simulates both the quasi-random turbulence and the pressure buildup waveform for unvoiced sounds.

Either or both of these sources may be applied as input to a linear, time-varying digital filter. This filter simulates the vocal tract system and thus the filter coefficients specify, in some manner, the vocal tract as a function of time during continuous speech. Once every 10 msec, on average, the filter coefficients are varied, indicating a new vocal tract configuration.

Finally a gain control between the source and system allows a certain flexibility in acoustic level of the output. The digital waveform at the output of the filter corresponds to the final speech output, sampled at the appropriate rate.

To control the model above requires a knowledge of the appropriate parameters (pitch period, switch position, amplitude, and filter coefficients) as a function of time. This is the goal of almost all speech analysis systems—i.e., to estimate the appropriate model parameters from real speech. The goal of most speech synthesis systems is to use these parameters, obtained in any reasonable manner, to derive a synthetic speech signal that is indistinguishable perceptually from the original signal. Speech analysis–synthesis systems combine the two problems with the twin goals of *efficiency*, i.e., to lower the bit rate of the synthesis system below that required for conventional waveform representations, and *flexibility*, i.e., to be able to modify and alter the speech in some desired manner through manipulation of the model parameters. In the remaining sections of this chapter, we discuss various aspects of several systems designed with the considerations above in mind.

12.3 Short-Time Spectrum Analysis

The Fourier transform of a discrete-time signal, $x(nT)$, $-\infty < n < \infty$, is defined as

$$X(e^{j\omega T}) = \sum_{n=-\infty}^{\infty} x(nT)e^{-j\omega nT} \tag{12.2}$$

As discussed in Chapter 6, for a time-varying signal, such as speech, the Fourier transform is not too meaningful as the spectrum of speech changes with time. A more useful measure of the energy content of a speech waveform is the short-time Fourier transform defined as

$$X(\omega, nT) = \sum_{r=-\infty}^{n} x(rT)h(nT - rT)e^{-j\omega rT} \tag{12.3}$$

Equation (12.3) may be viewed as measuring the infinite-time Fourier transform of the speech signal, at time nT, seen through a window with response $h(nT)$, as shown in Fig. 12.6, and may alternatively be obtained via the convolution relation

$$X(\omega, nT) = [x(nT)e^{-j\omega nT}] * h(nT) \tag{12.4}$$

Equation (12.3) can be rewritten in the form

$$X(\omega, nT) = a(\omega, nT) - jb(\omega, nT) \tag{12.5}$$

where $a(\omega, nT)$ and $b(\omega, nT)$ are the real and imaginary parts of the short-time transform and may be obtained as

$$a(\omega, nT) = \sum_{r=-\infty}^{n} x(rT)h(nT - rT) \cos(\omega rT) \tag{12.6a}$$

$$b(\omega, nT) = \sum_{r=-\infty}^{n} x(rT)h(nT - rT) \sin(\omega rT) \tag{12.6b}$$

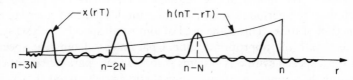

Fig. 12.6 Representation of short-time spectrum analysis.

ANALYSIS

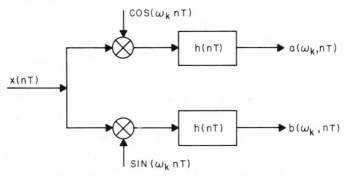

Fig. 12.7 Simple techniques for analyzing speech based on short-time spectrum analysis.

Equations (12.6a) and (12.6b) suggest a simple technique for measuring the short-time transform that is illustrated in Fig. 12.7. Generally $H(e^{j\omega T})$, the Fourier transform of $h(nT)$, is chosen to approximate the ideal lowpass filter with cutoff frequency ω_c, as shown in Fig. 12.8. Thus $X(\omega, nT)$ is the energy of the speech waveform at time nT and at frequency ω. The energy measurement reflects the speech energy in the band from $(\omega - \omega_c)$ to $(\omega + \omega_c)$.

In most speech spectrum analysis systems it is desired to measure the short-time transform at a finite set of N frequencies, spaced (often uniformly) over the band $0 \leq \omega T \leq 2\pi$. These measurements are accomplished by iterating the measurement technique above for each of the N frequencies. In the case where $h(nT)$ is the impulse response of an FIR filter and where the analysis frequencies are uniformly spaced, the FFT algorithm can be used to simultaneously make all the desired measurements in an extremely efficient manner. To see this, let $h(nT)$ be nonzero for $0 \leq n \leq M - 1$ and let the center frequencies for analysis ω_k be chosen as

$$\omega_k = \frac{2\pi}{NT} k \qquad k = 0, 1, \ldots, N - 1 \tag{12.7}$$

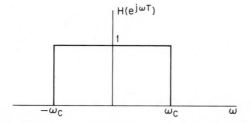

Fig. 12.8 Ideal lowpass filter for short-time spectrum analysis.

Then Eq. (12.3) can be written as

$$X(\omega_k, nT) = \sum_{r=n-M+1}^{n} x(rT)h(nT - rT)e^{-j\omega_k rT} \tag{12.8a}$$

$$= \sum_{m=0}^{[M/N]+1} \sum_{r=n-(m+1)N+1}^{n-mN} x(rT)h(nT - rT)e^{-j\omega_k rT} \tag{12.8b}$$

where $[M/N]$ stands for the greatest integer less than or equal to M/N. If we let $l = n - mN - r$,

$$X(\omega_k, nT)$$

$$= \sum_{m=0}^{[M/N]+1} \sum_{l=0}^{N-1} x(nT - rT - mNT)h(lT + mNT)e^{j\omega_k(l-n+mN)T} \tag{12.9}$$

Substituting Eq. (12.7) for ω_k gives

$$X(\omega_k, nT) = e^{-j(2\pi/N)kn}$$

$$\times \sum_{l=0}^{N-1} \left[\sum_{m=0}^{[M/N]+1} x(nT - lT - mNT)h(lT + mNT)e^{j(2\pi/N)kl} \right] \tag{12.10}$$

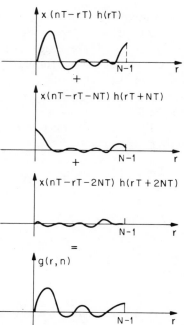

Fig. 12.9 The construction of $g(r, n)$ from $x(nT)$ and $h(nT)$.

where we have replaced $e^{j2\pi m}$ by 1. Equation (12.10) can now be written as

$$X(\omega_k, nT) = e^{-j(2\pi/N)kn} \sum_{l=0}^{N-1} g(l, n)e^{j(2\pi/N)lk} \qquad (12.11)$$

$$\underbrace{\phantom{\sum_{l=0}^{N-1} g(l, n)e^{j(2\pi/N)lk}}}_{\text{DFT}}$$

where

$$g(l, n) = \sum_{m=0}^{[M/N]+1} x(nT - lT - mNT)h(lT + mNT) \qquad (12.12)$$

Equation (12.11) shows that $X(\omega_k, nT)$ may be obtained as the product of the sequence $e^{-j(2\pi/N)kn}$ and the DFT of the sequence $g(l, n)$. Figure 12.9 illustrates how the sequence $g(r, n)$ is obtained term by term from the individual sequences $x(rT)$ and $h(rT)$.

Thus short-time Fourier analysis of speech is readily performed either directly using a bank of digital filters and modulators or indirectly using the FFT.

12.4 Speech Analysis–Synthesis System Based on Short-Time Spectrum Analysis

The principles of measuring the short-time spectrum of speech may be applied to an entire analysis–synthesis system. The basic idea is to measure the outputs of a bank of M bandpass filters and reconstruct the speech from these M signals. A simplified schematic of this system is shown in Fig. 12.10. The input speech is $x(nT)$ and the reconstructed synthetic waveform is $y(nT)$. The M individual bandpass filters have impulse responses $h_k(nT)$, $k = 1$, $2, \ldots, M$. The bandpass outputs are labeled $y_k(nT)$, $k = 1, 2, \ldots, M$. If the bandpass filter impulse responses are restricted to be of the form

$$h_k(nT) = h(nT) \cos (\omega_k nT) \qquad (12.13)$$

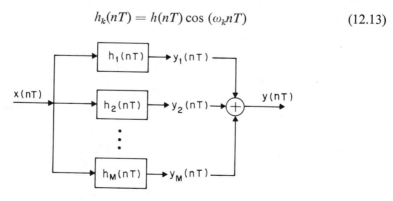

Fig. 12.10 Schematic of an analysis–synthesis system based on short-time spectrum analysis.

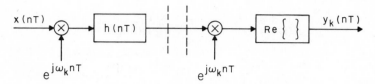

Fig. 12.11 The operations for the kth channel.

where $h(nT)$ is the impulse response of a lowpass filter (i.e., the bandpass impulse response is a modulated lowpass impulse response), then the bandpass outputs $y_k(nT)$ can be written as

$$y_k(nT) = \sum_{r=-\infty}^{n} x(rT)h(nT - rT) \cos [\omega_k(nT - rT)] \qquad (12.14)$$

$$= \text{Re}\, [e^{j\omega_k nT} X(\omega_k, nT)] \qquad (12.15)$$

where $X(\omega_k, nT)$ is as defined in Eq. (12.3). Thus each channel of the system can be obtained in the manner shown in Fig. 12.11. Since $X(\omega_k, nT)$ can be written in terms of real and imaginary components [Eq. (12.5)], Eq. (12.15) can be put in the form

$$y_k(nT) = a(\omega_k, nT) \cos (\omega_k nT) + b(\omega_k, nT) \sin (\omega_k nT) \qquad (12.16)$$

which is realized as shown in Fig. 12.12. The dashed lines in Figs. 12.11 and 12.12 indicate points of transmission and reception when the system is implemented as a speech bandwidth compression system. The straight path between the dotted lines represents the communications channel (assumed error-free here). The transmitted parameters $a(\omega_k, nT)$ and $b(\omega_k, nT)$ would have to be sampled to a lower rate than the speech transmission rate and quantized to achieve any significant bandwidth reduction. Further discussion on this topic is given in Sec. 12.6.

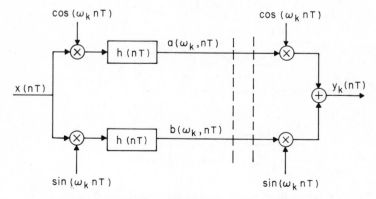

Fig. 12.12 The operations for the kth channel using real processing.

12.5   Analysis Considerations

The quality with which this system can represent speech is dependent on the extent to which the bank of M filters adequately represents the speech spectrum. One simple way of measuring this is to determine the overall impulse response of the system and examine its Fourier transform. If we denote the impulse response of the composite bank as $\tilde{h}(nT)$, then

$$\tilde{h}(nT) = \sum_{k=1}^{M} h_k(nT) = h(nT) \sum_{k=1}^{M} \cos{(\omega_k nT)} \qquad (12.17)$$

If we denote by $d(nT)$ the summation over $\cos{(\omega_k nT)}$ in Eq. (12.17), i.e.,

$$d(nT) = \sum_{k=1}^{M} \cos{(\omega_k nT)} \qquad (12.18)$$

then

$$\tilde{h}(nT) = h(nT)\, d(nT) \qquad (12.19)$$

i.e., the composite filter bank impulse response is the *product* of the prototype lowpass impulse response and a response dependent entirely on the number of filters M and their center frequencies ω_k.

To see how well $\tilde{h}(nT)$ approximates a digital impulse (perhaps with some delay), we can examine the behavior of either $\tilde{h}(nT)$ or its Fourier transform. In the special case of a uniform filter bank with

$$\omega_k = \Delta\omega \cdot k \qquad \text{(uniform spacing)} \qquad (12.20)$$

($\Delta\omega$ is a constant), then $d(nT)$ may be solved by using Eq. (12.18) to give

$$d(nT) = \sum_{k=-M}^{M} e^{jk\Delta\omega nT} - 1 \qquad (12.21)$$

$$d(nT) = \frac{\sin{[(M + \tfrac{1}{2})\Delta\omega nT]}}{\sin{[(\Delta\omega/2)nT]}} - 1 \qquad (12.22)$$

If $\Delta\omega = 2\pi/(NT)$, with N an integer, then the sequence $d(nT)$ is periodic with period N samples. If $2\pi/(\Delta\omega T)$ is not an integer, the sequence $d(nT)$ is not periodic but still has peaks at intervals of NT seconds.

A particularly interesting choice of parameters is as follows: Let N be an odd integer (similar results can be derived for N even) and $M = (N - 1)/2$. For $\Delta\omega = 2\pi/(NT)$, it can easily be seen that this corresponds to evaluating the short-time Fourier transform at equally spaced frequencies in the range $0 < \omega < \pi/T$. If, in addition, we include a channel centered on zero frequency, it can be shown that

$$
\begin{aligned}
d(nT) &= \frac{\sin{(\pi n)}}{\sin{(\pi n/N)}} \\
&= N \qquad n = 0, \pm N, \pm 2N, \ldots \\
&= 0 \qquad \text{elsewhere}
\end{aligned}
\qquad (12.23)
$$

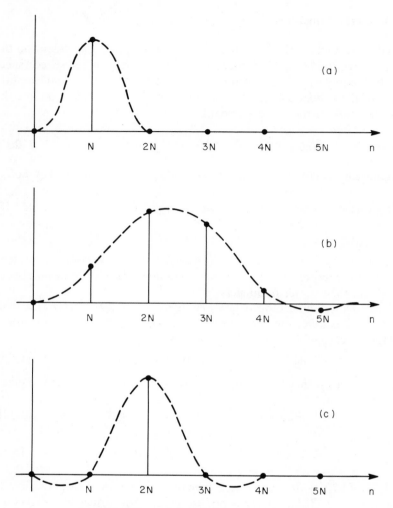

Fig. 12.13 Time versus frequency resolution tradeoffs.

Thus, for these conditions $d(nT)$ is a periodic train of impulses, with a period NT that is inversely proportional to the frequency spacing between channels. Since $\tilde{h}(nT) = h(nT) d(nT)$, it is clear that the composite impulse response will also be an impulse train. Since the ideal composite impulse response is a delayed impulse, we must choose the prototype lowpass impulse response $h(nT)$ so as to eliminate all but one of the impulses in $d(nT)$. Suppose we fix T and N, corresponding to fixed frequency spacing $\Delta\omega$. Then if we choose a very narrow impulse response, e.g., of duration less than $2N$, the composite impulse response will appear as in Fig. 12.13(a). Here we have shown the prototype lowpass response or data window as a dotted curve superimposed on the impulse train that represents the composite response. Clearly there is only one impulse; however, such a narrow impulse response $h(nT)$ corresponds to a rather wideband lowpass filter that would not give satisfactory frequency resolution. If we use a narrower bandwidth filter, the impulse response will become proportionately greater in duration, as in Fig. 12.13(b), where we note that the composite impulse response consists of several impulses that would give rise to a reverberant quality in the output speech. Thus we see that good frequency resolution, i.e., narrowband channels, seems to be at odds with low reverberation. Figure 12.13(c) suggests one way in which, at least theoretically, the output can match the input exactly, however. Here we have used a wider filter but have constrained the values of $h(nT)$ to be zero at integer multiples of the period N. In this case the composite response is a single impulse delayed by $2N$. Thus the output is a delayed and scaled replica of the input. Such a data window can be designed. Therefore, the short-time Fourier transform can theoretically represent the speech signal exactly.

In many practical systems it will not be convenient to choose the parameters so that the composite response is as depicted in Fig. 12.13(c). The analysis and synthesis equations can be modified to effect further improvements, however, even if the optimum response cannot be achieved. The example of Fig. 12.14 illustrates this point for an analysis–synthesis system of 30 channels spaced at 100-Hz intervals, with a sampling rate of 10 kHz. The dotted curve on the left in Fig. 12.14(a) shows the lowpass impulse response $h(nT)$ (sixth-order Bessel filter) and the solid curve shows the composite impulse response $\tilde{h}(nT)$. From this latter curve it is possible to visualize the periodic pulse-like character of the sequence $d(nT)$ in the case when not all the channels are used in synthesis. We see that, in addition to the main pulse at 10 msec, there is a significant echo at 20 msec. [The period of $d(nT)$ is 10 msec.] This echo manifests itself in the composite frequency response as an amplitude and phase ripple [Fig. 12.14(b) and (c) on the left] and perceptually as a reverberant quality in the synthesized output. This example, together with the fact that $\tilde{h}(nT)$ is the product of $d(nT)$ and $h(nT)$, suggests two ways of improving the composite response. As we have noted, for a given frequency spacing we could widen the bandwidth of the lowpass filter thereby reducing the duration of $h(nT)$. As seen from Fig. 12.14(a) on the left, this would have the effect of increasing the amplitude of the first pulse and decreasing the amplitude of the second pulse. This means, however, that we must effectively sacrifice frequency resolution. An alternative approach is suggested if we note that if $d(nT)$ could be shifted to the right relative to $h(nT)$ (the dotted curve), then the main pulse would grow in amplitude and the echo would become smaller. At the same time, however, the pulse in $d(nT)$ at $nT = 0$, which was completely suppressed by $h(nT)$, grows in amplitude as $d(nT)$ moves to the right. This is shown in Fig. 12.14(a) on the right. Thus for a

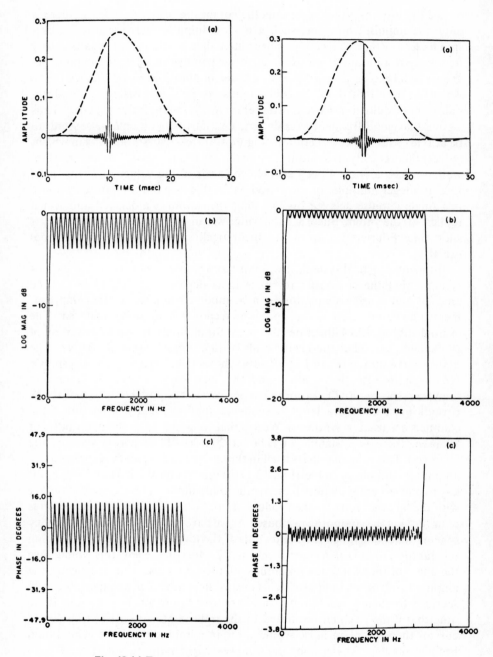

Fig. 12.14 Time and frequency responses of two filter banks.

given frequency resolution there is an optimum delay of $d(nT)$ relative to $h(nT)$ for which $\tilde{h}(nT)$ consists of a large central pulse and two small pulses of equal size, one on each side of the main pulse. It can be shown that this condition gives minimum amplitude and phase ripple for a given frequency resolution.

The mechanism for delaying $d(nT)$ relative to $h(nT)$ is available in either the analysis or the synthesis stage. If we change Eq. (12.6a) and (12.6b) to

$$a(\omega_k, nT) = \sum_{r=-\infty}^{n} h(nT - rT)x(rT) \cos [\omega_k(rT - n_aT)] \qquad (12.24)$$

and

$$b(\omega_k, nT) = \sum_{r=-\infty}^{n} h(nT - rT)x(rT) \sin [\omega_k(rT - n_aT)] \qquad (12.25)$$

where $n_a = n_0$, the desired delay in samples, and we use Eq. (12.16) for synthesis, the effective impulse response of the kth channel will be

$$h_k(nT) = h(nT) \cos [\omega_k(nT - n_0T)] \qquad (12.26)$$

and the composite impulse response will be

$$\tilde{h}(nT) = h(nT) \cdot d(nT - n_0T) \qquad (12.27)$$

Alternatively, the same channel response will be obtained if we use Eqs. (12.6a) and (12.6b) for analysis and substitute for Eq. (12.16).

$$y_k(nT) = a(\omega_k, nT) \cos [\omega_k(nT - n_sT)] + b(\omega_k, nT) \sin [\omega_k(nT - n_sT)] \qquad (12.28)$$

where $n_s = n_0$. As a third possibility we can use Eq. (12.24) and (12.25) for analysis and Eq. (12.28) for synthesis if $n_a + n_s = n_0$. An interactive design program facilitates the choice of the parameters of such systems so as to obtain composite responses similar to the right-hand side of Fig. 12.14.

12.6 Overall System

Based on the theoretical considerations of Sec. 12.5, a complete analysis–synthesis system can be simulated and tested. Figure 12.15 shows all the processing required for the kth channel. Each of the M channels requires similar processing. Figure 12.15 is conveniently segmented into three parts: an analysis section, a section for bit-rate reduction, and a synthesis section. The analysis section works as described in the previous section, computing $a(\omega_k, nT)$ and $b(\omega_k, nT)$ for each channel. To achieve any bit-rate reduction (bandwidth compression), these signals must be sampled at a lower rate (i.e., once every T_1 seconds) and quantized to a smaller number of bits. These are the functions of the sampler and quantizer of the bit-rate reduction section. Appropriate values of T_1 and number of bits per sample must be

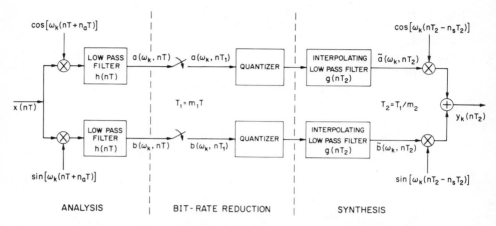

Fig. 12.15 Overall analysis-synthesis system including sampling and quantization.

obtained from speech perception experiments. The synthesis stage is similar to the one described in Sec. 12.5 with the exception of the interpolating low-pass filters which interpolate the received values of $a(\omega_k, nT)$ and $b(\omega_k, nT)$ to the appropriate synthesis rate, T_2 seconds, which need not be identical to the analysis rate.

The details of simulation experiments with this analysis–synthesis system are given by Schafer and Rabiner. It is sufficient to note that reasonably good

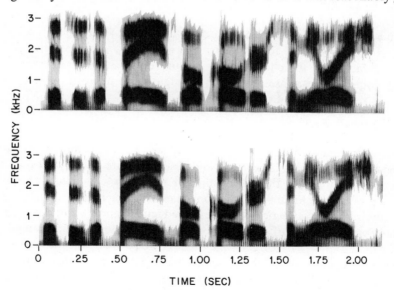

Fig. 12.16 Spectrogram comparison between natural and synthetic utterance.

quality speech may be obtained at a bit rate of about 15 kilobits/sec (kbps), or about one-quarter the bit rate of 7 log-bit, 8-kHz pulse code modulation (PCM) speech. A spectrogram comparison of the input speech and synthetic reconstructed speech is shown in Fig. 12.16 in the case where the analysis parameters were unquantized. It is difficult to tell the input from the output in this case.

12.7 Channel Vocoder

The channel vocoder (voice coder) is an analysis–synthesis system based on our knowledge of the speech production and perception mechanism in man. In particular the channel vocoder exploits the insensitivity of the hearing mechanism to phase and only attempts to reproduce the short-time power spectrum of the speech waveform. (This is equivalent to preserving the magnitude of the short-time Fourier transform and disregarding the phase.) The spectral envelope of the speech is measured with a bank of filters and ascribed to the vocal tract filter, while the excitation is estimated to be either a quasi-periodic pulse train or noise. (Thus the vocoder is a direct application of the assumed source–system independence model.) There are several methods of combining these extracted parameters to reconstruct the speech. In this section we describe several vocoder configurations and discuss the factors that affect specification of design parameters.

Figure 12.17 shows a block diagram of a typical channel vocoder. The incoming speech $x(n)$ is analyzed by a bank of bandpass filters (16 in this case) that nonuniformly cover the frequency range of interest (generally 0 to 3 kHz). Considerations in the design of these filters will be discussed later in this section. The outputs of the bandpass filters are rectified and lowpass filtered to give the signals $y_k(n)$ that more or less represent the spectral envelope of the speech signal. Parameters representing the nature of the excitation are obtained by the voiced–unvoiced detector that determines if the input speech is voiced (i.e., the vocal cords are vibrating) or unvoiced. If the speech is voiced, a pitch extractor determines the fundamental frequency of vibration, F_0.

The 16 lowpass filtered channel signals and the voiced–unvoiced and fundamental frequency signals are coded and transmitted over a channel to the receiver. Assuming error-free transmission, the job of the receiver is to reconstruct the speech from the transmitted parameters. The excitation is created from either a pulse generator whose fundamental frequency is controlled by the F_0 signal or a random noise generator. The voiced–unvoiced signal controls a switch to determine which source will excite a bank of bandpass filters that are identical to those used in the analysis. The lowpass spectrum envelope signals modulate the outputs of the bandpass filters to control the speech power in each of the frequency bands. The resultant synthetic speech is obtained by summing the modulated bandpass outputs.

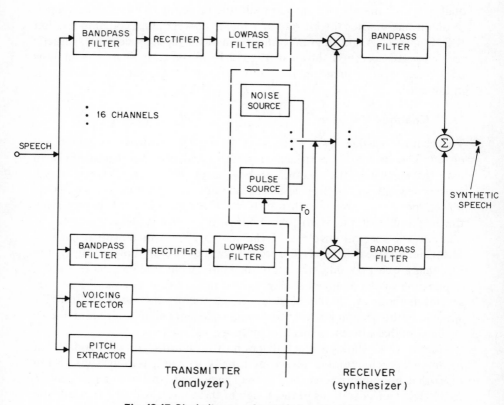

Fig. 12.17 Block diagram of typical channel vocoder.

One of the goals of the channel vocoder is low bit-rate transmission of reasonable quality speech. A large number of studies have been made as to how many bits are required to adequately represent channel vocoder parameters. Although different vocoder configurations require different bit rates, an overall rate of from 2400 to 9600 bits/sec (bps) has generally been found adequate for channel vocoders. The quality of the speech is generally monotonically related to the bit rate; thus 9600-bps vocoders sound best, whereas most 2400-bps vocoders are barely acceptable to most listeners.

12.8 Vocoder Analyzers—Signal Processing Considerations

The use of a fixed filter bank to measure the speech spectrum is complicated by the fact that speech resembles either noise or a periodic signal of widely varying fundamental frequency. Additional complication arises from the time-varying nature of the speech spectrum. From steady-state frequency

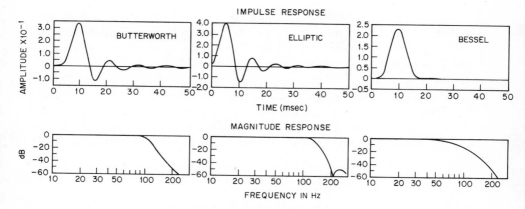

Fig. 12.18 Time and frequency responses of Butterworth, elliptic, and Bessel lowpass filters (After Golden).

response analysis, one would assume that sharp cutoff filters would result in the best spectral measurements; in practice such filters have relatively long impulse response durations. Thus their use would result in a smearing, in time, of rapid spectral changes and, as previously discussed, subsequent reverberation effects on the synthetic speech. To illustrate this effect the sequence of three pairs of plots in Fig. 12.18 shows the impulse and magnitude response of eighth-order, 100-Hz cutoff Butterworth, elliptic (transition ratio = 0.5), and Bessel filters. The Butterworth and elliptic filters have sharper cutoffs than the Bessel filter, but the impulse response of both of these filters lasts almost 50 msec, whereas the Bessel impulse response lasts less than 20 msec. Thus, for a channel vocoder, the Bessel filter is the only one of these three that would be considered acceptable.

A second manifestation of disturbing temporal effects arises if the individual bandpass filters have unequal time delays, such as might occur if different channels used different bandwidths. In a practical situation the bandwidths of the high-frequency channels increase to give a frequency resolution characteristic similar to that of the human ear. For example, a reasonable bandwidth for the low-frequency channels is about 125 Hz, whereas a bandwidth of about 400 Hz is often used for the highest channel employed. There are several ways of equalizing the delays of differing bandwidth filters. One simple technique is to increase the order of the filter in direct proportion to the increase in bandwidth, paying careful attention to the magnitude and phase at the crossover frequencies of adjacent channels to preserve good overall frequency response of the filter bank. It should be noted that the use of nonuniform filter bandwidths guarantees that, at least for the higher-frequency channels, more than one harmonic of the speech will be measured by the wide bandwidth filters. Thus a type of spectral distortion is introduced into the synthetic speech to which listeners have been found to be rather insensitive.

The design considerations for the fixed lowpass filters in the analyzer are relatively straightforward. The lowpass filter needs to be well attenuated by 50 Hz (the lowest fundamental expected from male speech) to eliminate any voicing component present in the rectifier output. Since the speech spectrum is slowly varying, the lowpass cutoff need be no greater than about 25 Hz. With the additional constraint that the lowpass filters have very little ringing to prevent reverberation effects, it is seen that the design is quite constrained. Filters approximating linear phase characteristics such as Bessel or Lerner IIR filters have generally been used, although linear phase FIR filters are good candidates for this filter.

12.9 Vocoder Synthesizers—Signal Processing Considerations

The design of the various components in the synthesizer is straightforward. Generally the synthesizer bandpass filters are identical to those in the analyzer for the same reasons as discussed above. An alternate synthesizer channel is often used, however, in place of the standard channel. Figure 12.19 shows a comparison between the standard channel and the modified channel. The modified channel is often referred to as a spectral flattening channel. The introduction of an infinite clipper or hard limiter with the property that its output is either $+1$ or -1 depending on the sign of the input creates, at the input to all the modulators, an array of square waves that all have equal power.

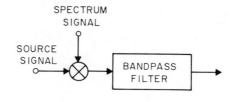

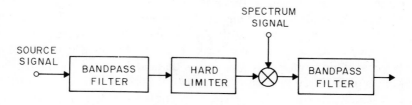

Fig. 12.19 Comparison between standard and modified vocoder channel.

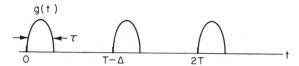

Fig. 12.20 Excitation pulses for voiced speech.

A steady-state voiced excitation signal results in a set of square waves of frequencies that are approximately harmonics of the fundamental. Thus, the behavior of the spectrally flattened channel appears to be the same as that of the normal channel in the simple case when only a single harmonic passes through the first bandpass filter; the third and higher harmonics produced by the hard limiter are filtered out by the final bandpass filter. These two configurations do not yield the same sound, however, even in the case of high fundamental frequencies where the approximation of no more than a single harmonic per filter is valid.

To explain this apparent discrepancy, it seems necessary to deal with situations wherein the glottal source spectrum is not constant but fluctuates with time. This fluctuation is caused both by variations in the glottal area from one period to the next and by changes in the vocal cord vibration rate. In a vocoder, only the vocal cord vibration period, not the wave shape, is extracted. The evidence that synthetic speech quality is strongly affected by spectral flattening implies that spectral fluctuations in the speech are at least partly induced by the time variations between successive laryngeal pulses.

Short-time spectral fluctuations caused by variations in the excitation pulse periods can be examined quantitatively by computing the continuous power spectrum of three pulses as shown in Fig. 12.20. This computation gives

$$G(\omega) \approx \frac{\sin^2(\omega\tau/2)}{(\omega\tau/2)^2} [3 + 4\cos(\Delta\omega)\cos(\omega T) + 2\cos(2\omega T)] \quad (12.29)$$

If $\tau/T \ll 1$ (i.e., small pulse width compared to the pulse spacing), then $\sin^2(\omega\tau/2)/(\omega\tau/2)^2$ is approximately 1 for the frequency range of interest. A plot of $G(\omega)$ for $\Delta = 0$ and for $\Delta = 0.1T$ is shown in Fig. 12.21.

If the preceding argument is accepted, then pitch-induced spectral fluctuations would be carried in both pitch and spectral analysis of the original speech. Spectral flattening effectively wipes out (for the case of no more than a single harmonic per filter) spectral variations in frequency carried by the excitation signal.

When the fundamental frequency is low enough to allow two or more harmonics to pass through a filter, the spectral flattening configuration no longer completely flattens the excitation spectrum. This follows from the

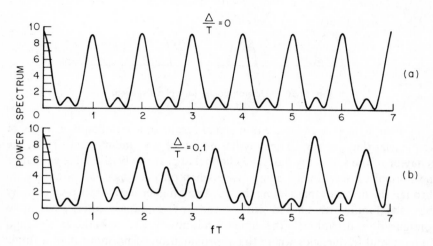

Fig. 12.21 Spectra of two different pulse trains.

effect of the limiter on two harmonics; often the smaller harmonic is further suppressed. These difficulties can be substantially avoided by replacement of the limiter with an automatic volume control (AVC) of the type shown in Fig. 12.22. The delayed narrowband signal is divided by its own average intensity. The delay τ of the lowpass filter is matched to the delay of the signal. As τ approaches zero, the output of the divider becomes proportional to the signal divided by its own magnitude, thus the limiting case of the AVC is the hard limiter. For moderate delays on the order of 10 msec, very little spectral distortion occurs even if more than a single harmonic is present. In short, the AVC form of spectral flattening preserves intrafilter amplitude relations, while removing interfilter spectral fluctuations. Since little or no third harmonic distortion is created, the possibility exists that the final bandpass filter in each channel may not be needed.

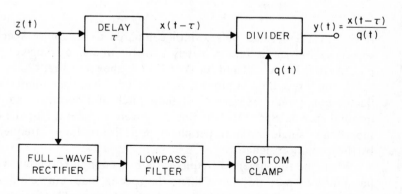

Fig. 12.22 Automatic volume control for vocoder.

12.10   Other Vocoder Configurations

Many other variations on the channel vocoder analyzer and synthesizer have been proposed and studied in great detail. The reader is referred to the many excellent reports available in the literature for specific details on particular systems.

12.11   Pitch Detection and Voiced–Unvoiced Decisions

To complete these sections on the channel vocoder, it is appropriate to mention the problems of pitch detection and voiced–unvoiced decision making in the analyzer. There exists a wide variety of algorithms for estimating the pitch period. For the sake of illustration we shall discuss a particularly efficient algorithm that works in the time domain and uses parallel processing techniques to make its final decisions. The problems of pitch detection and voiced–unvoiced decisions is really a combination of signal processing and feature extraction. Since pitch detectors are embedded in a large number of speech processing systems, however, it is worthwhile discussing them here. In Sec. 12.12 a completely different kind of algorithm for pitch detection is discussed in conjunction with homomorphic speech processing.

A block diagram of a pitch-period estimation algorithm is shown in Fig. 12.23. The algorithm is conveniently segmented into four distinct processing or decision-making parts including

1. Lowpass filtering of the speech signal.
2. Generation of six functions of the peaks of the filtered speech signal.
3. Six identical simple pitch-period estimators, each working on one of the six functions above.
4. Final pitch-period computation, based on examination of the results from each simple pitch period estimator.

The primary purpose of the lowpass filter is to filter out higher harmonics of the speech waveform. A lowpass filter with a cutoff of about 600 Hz works well.

The second part of the algorithm generates pulses at various peaks in the lowpass filtered waveform as illustrated in Fig. 12.24. Pulses of height m_1, m_2, and m_3 are generated at every positive peak, while pulses of height m_4, m_5, and m_6 are generated at each negative peak. Measurements m_1 and m_4 are simple peak (positive and negative) measurements, whereas m_2 and m_5 are peak-to-valley and valley-to-peak measurements and m_3 and m_6 are peak-to-previous-peak and valley-to-previous-valley measurements. All the m's are converted into positive pulse trains. Thus if a current peak (valley) is not so large as the previous peak (valley), measurement m_3 (m_6) is set to zero.

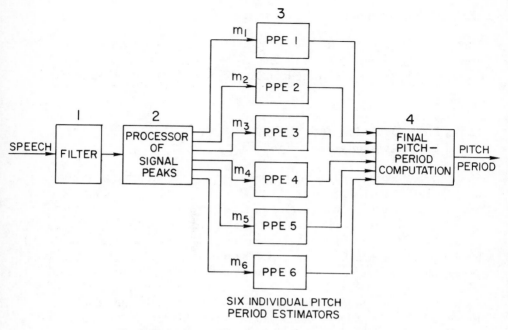

Fig. 12.23 Block diagram of pitch-period estimation algorithm.

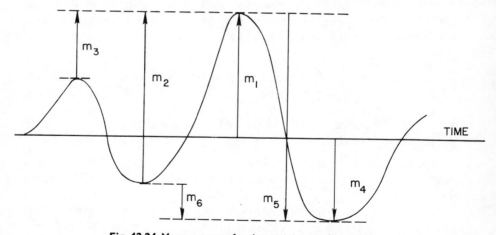

Fig. 12.24 Measurements for determining pitch period.

The choice of this particular set of measurements was based on consideration of two extreme cases as shown in Fig. 12.25. For the case when only the fundamental is present (as on the left), measurements m_3 and m_6 fail but measurements m_1, m_2, m_4, and m_5 provide strong indications of the period. For the case when a very strong second harmonic and some fundamental are

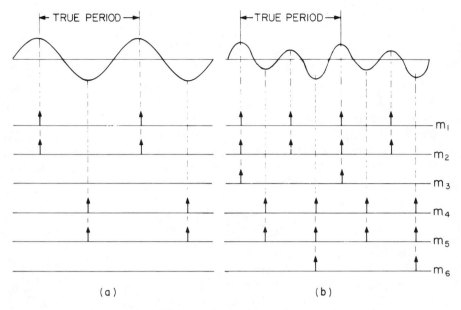

Fig. 12.25 Pitch measurements for two extreme conditions.

present (as on the right), measurements m_3 and m_6 will be correct and all others will fail. In this case, although four of the six measurements may fail, it will be shown below how the final computation has high probability of being correct.

The six sets of pulse trains are applied to six identical pitch detectors, each of which operates as shown in Fig. 12.26. In essence each pitch-period estimator is a peak-detecting rundown circuit. Following each detected pulse there is a blanking interval (during which no pulses can be detected) followed by a simple exponential decay. Whenever a pulse exceeds the level of the run-down circuit (during the decay), it is detected and an exponential run-down circuit is reset. The run-down time constant and the blanking time of each detector are made to be functions of the smoothed estimate of pitch period

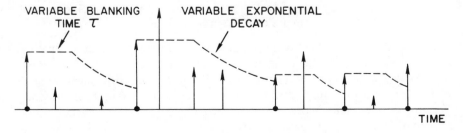

Fig. 12.26 Operations of each pitch detector.

P_{av} of that detector. P_{av} is derived from the iteration

$$P_{av}(n) = \frac{P_{av}(n-1) + P_{new}}{2} \qquad (12.30)$$

where P_{new} is the most recent estimate of pitch period; $P_{av}(n)$ is the current smoothed estimate of pitch period; $P_{av}(n-1)$ is the previous smoothed estimate of pitch period. Each time a new peak is detected, P_{av} is updated according to the iteration. To prevent extremes of values of blanking time or run-down time constant, P_{av} is limited to be greater than 4 msec and less than 10 msec. Within these limits, the dependence of blanking time τ and run-down time constant β on P_{av} is given by

$$\tau = 0.4 P_{av}, \qquad \beta = \frac{P_{av}}{0.695} \qquad (12.31)$$

The final computation of pitch period is performed by block 4 (of Fig. 12.23), which may be thought of as a special-purpose computer with a memory, an arithmetic algorithm, and control hardware to steer all the incoming signals. At any time t_0 an estimate of pitch period is made by

1. Forming a (6×6) matrix of estimates of pitch period. The columns of the matrix represent the individual detectors and the rows are estimates of period. The first three rows are the three most recent estimates of period. The fourth row is a sum of the first and second rows; the fifth row is the sum of the second and third rows; and the sixth row is a sum of the first three rows. The technique for forming the matrix is illustrated in Fig. 12.27.

    The reason for the last three rows of the matrix is that sometimes the individual detectors will indicate second or third harmonic rather than fundamental and it will be entries in the last three rows that are correct rather than the three most recent estimates of pitch period.

2. Comparing each of the entries in the first row of the matrix to the other 35 entries of the matrix and counting the number of coincidences. (A precise definition of coincidence is given below.) That particular P_{i1} $(i = 1, 2, 3, 4, 5, 6)$ that is most popular (greatest number of coincidences) is used as the final estimate of pitch period.

At this point it is well to describe *coincidence*. First, to determine whether two pitch-period estimates "coincide," it seems more appropriate to observe their ratios rather than their differences. However, the ratio measurement can be very approximate to avoid the need of a divide computation. Second, because during many parts of the speech there are sizable variations of successive pitch-period measurements, it is useful to include several threshold values to define coincidence and then to try to select, for each overall

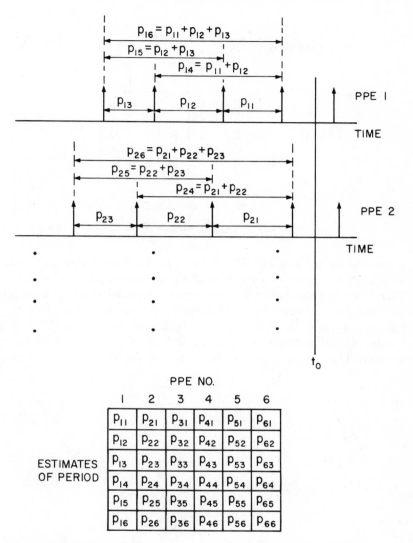

Fig. 12.27 Technique for forming matrix of estimates of pitch periods.

pitch period computation, the threshold that yields the most consistent answer. With this explanation, we now define the computation of block 4.

Figure 12.28 shows a table of 16 coincidence window widths. As indicated in Fig. 12.27, only the most recent estimated pitch period from a given detector is a candidate for final choice. This candidate is thus one of six possible choices for the correct pitch period. To determine the "winner," each candidate is numerically compared with all the remaining 35 pitch numbers. This comparison is repeated four times, corresponding to each

PITCH–PERIOD RANGE (msec)	BIAS			
	1	2	5	7
1.6 – 3.1	1	2	3	4
3.1 – 6.3	2	4	6	8
6.3 – 12.7	4	8	12	16
12.7 – 25.5	8	16	24	32

COINCIDENCE WINDOW WIDTH IN
HUNDREDS OF MICROSECONDS

Fig. 12.28 Table of coincidence window widths.

column in the table of Fig. 12.28. From each column, the appropriate window width is chosen as a function of the estimate associated with the candidate. Thus, if this estimate, for example, were 4 msec, coincidence between the candidate and any compared interval would mean that their difference was less than or equal to $\pm 200 \ \mu sec$ at a sampling rate of 10 kHz. After the number of coincidences is tabulated, a bias of 1 is subtracted from that number. The measurement is then repeated for the second column; this

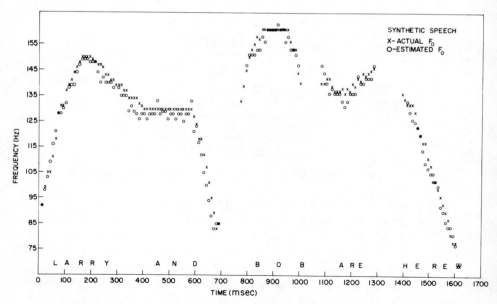

Fig. 12.29 Comparison between synthetic pitch and that generated by the algorithm.

time the windows are wider and this increases the probability of coincidence but, in compensation, a bias of 2 is subtracted from the compilation. After the computation has been repeated in this way for all four columns, the largest biased number is used as the number of coincidences that represents that particular pitch-period estimate. The entire procedure is now repeated for the remaining five candidates and the winner is chosen to be that number with the greatest number of biased coincidences. In the course of this computation, a total of $(6 \times 4 \times 35)$ coincidence measurements (comparison of the magnitude of a difference with a fixed number) have to be made. Repetition of the complete computation every 5 msec suffices to follow even rapid pitch-period variations.

To demonstrate typical results obtained with this algorithm, Fig. 12.29 shows a comparison between fundamental frequency estimates obtained by the method above and the true values as used in generating the synthetic utterance used in the test. The algorithm clearly works very well in this case.

12.12 Voiced–Unvoiced (Buzz–Hiss) Detection

The pitch-period estimation algorithm described above can readily be converted to give voiced–unvoiced estimates. Whenever the speech is unvoiced, the number of coincidences observed by the individual detectors will be small. Quantitative measurements can be made to set thresholds for the appropriate decisions. In addition, if the energy measurement out of the pitch detection lowpass filter is below a fixed threshold, this strongly indicates either silence or unvoiced speech. By combining this with functions of the pitch detector output, as indicated above, a voice–unvoiced algorithm can be implemented as discussed by Gold.

12.13 Homomorphic Processing of Speech

The term *homomorphic processing* is generally applied to a class of systems that obey a generalized principle of superposition. This generalized superposition can be stated as follows. If $x_1(n)$ and $x_2(n)$ are inputs to a homomorphic system, $y_1(n)$ and $y_2(n)$ are the respective outputs, and c is any scalar, then if

$$y_1(n) = \phi[x_1(n)]$$
$$y_2(n) = \phi[x_2(n)]$$

$$\phi[x_1(n) \, \triangle \, x_2(n)] = \phi[x_1(n)] \; \square \; \phi[x_2(n)] \tag{12.33}$$

and

$$\phi[c \, \lozenge \, x_1(n)] = c \bigcirc y_1(n) \tag{12.34}$$

where $\triangle$, $\square$, $\lozenge$, and $\bigcirc$ correspond to unspecified mathematical operations such as multiplication, addition, and convolution.

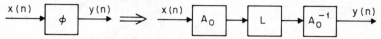

Fig. 12.30 A homomorphic processing system.

The importance of this type of processing lies in the fact that the operation ϕ of the homomorphic system can be decomposed into a cascade of operations as shown in Fig. 12.30. The systems A_0 and A_0^{-1} are inverse systems. The system L is a linear time-invariant system, i.e., a simple filter. Thus, with the decomposition as shown in Fig. 12.30, one can process the output of system A_0 using standard techniques and perform processing in a relatively straightforward manner. Systems A_0 and A_0^{-1} are readily determined from ϕ as we will show in the case of speech.

As discussed earlier, the speech waveform is modeled as the convolution of three components—a train of impulses representing the pitch, the excitation pulse, and the vocal tract impulse response. (The effects of radiation from the mouth also enter into the model but these effects are generally combined with the excitation pulse.) Using the notation $p(n)$ to represent the train of pitch impulses, $e(n)$ to represent the excitation pulse, $u(n)$ to represent the vocal tract impulse response, and $w(n)$ to represent a time-limited window through which the speech waveform $x(n)$ is viewed, we find

$$x(n) = [p(n) * e(n) * u(n)]w(n) \tag{12.35}$$

Since $w(n)$ is generally a smooth sequence, Eq. (12.35) can be simplified to the approximate form

$$x(n) \approx [p(n) \cdot w(n) * e(n) * u(n)] \tag{12.36}$$

$$x(n) \cong \hat{p}(n) * e(n) * u(n) \tag{12.37}$$

Equation (12.37) shows $x(n)$ to be a triple convolution. This convolution can readily be converted to a summation by Fourier transforming Eq. (12.37) (giving a triple product) and then taking the logarithm of the result. The resulting waveform may then be processed by a linear, time-invariant system to process each of the components of $x(n)$ in some desired manner. To recover a processed waveform, the inverse system A_0^{-1} consists of an exponentiator and an inverse Fourier transformation. Thus the homomorphic system for processing speech is as shown in Fig. 12.31.

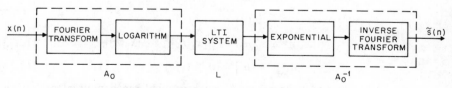

Fig. 12.31 A homomorphic system for processing speech (After Oppenheim, Schafer, and Stockham).

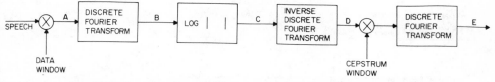

Fig. 12.32 Homomorphic processing for speech (After Oppenheim, Schafer and Stockham).

Depending on the specific application in mind, several variations on the system above have been used to process speech. For example, the system shown in Fig. 12.32 has been used to estimate parameters of both the vocal tract transmission function and the excitation function. In this case, the excitation is considered to be $\hat{p}(n) * e(n)$ and the vocal tract impulse response is $u(n)$. Thus $x(n)$ is a simple discrete convolution

$$x(n) = u(n) * s(n) \tag{12.38}$$

where $s(n)$ is the excitation signal. In this case $x(n)$ is the signal at point A in Fig. 12.32. The application of the discrete Fourier transform gives a signal at point B that is the product of the discrete Fourier transforms of $u(n)$ and $s(n)$. The next block takes the log magnitude of the signal at point B giving a signal at point C that is the sum of the log magnitudes of the DFT's of $s(n)$ and $u(n)$. The sequence of blocks following point C (an inverse discrete Fourier transform, a windowing, and a discrete Fourier transform) is readily seen to be a linear filtering of the signal at point C. The filtering is carried out in the transform domain (as a multiplicative operation) for reasons to be discussed below. Since the inverse discrete Fourier transform is linear, the signal at point D (called the *cepstrum* of the signal at point A) is the sum of the cepstra of the excitation and the vocal tract impulse response.

It can be argued that the cepstrum at point D serves to separate the excitation from the vocal tract impulse response in the following manner. The excitation signal can be viewed as a sequence of quasi-periodic pulses whose Fourier transform consists approximately of a line spectrum where the lines are spaced at harmonics of the fundamental frequency. The process of taking the log magnitude does not affect the general characteristics of the excitation spectrum. The IDFT operation yields another quasi-periodic waveform with pulses spaced at the fundamental period. Thus the cepstrum of the excitation signal should consist of pulses around $n = 0, T, 2T, \ldots$, where T is the pitch period. The vocal tract impulse response is a sequence that generally is non-negligible for about 20 to 30 msec. Its Fourier transform is a slowly varying function of frequency, as shown earlier in this chapter. The process of taking the log magnitude and IDFT yields a sequence that is nonnegligible for only a small number of samples (generally less than the number of samples in a pitch period). It can be shown that for a sequence that decays as $1/n$, its cepstrum

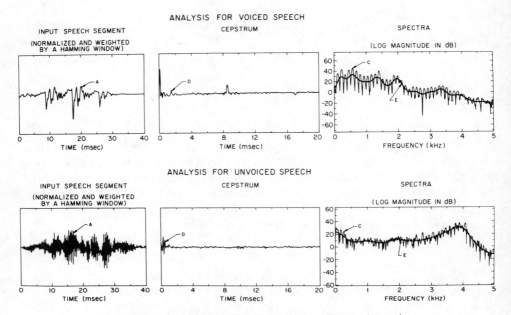

Fig. 12.33 Homomorphic analysis for voiced and unvoiced speech.

decays as $1/n^2$. Thus the cepstrum serves to differentiate the excitation information from the vocal tract impulse response information.

Figure 12.33 illustrates this type of processing on both voiced and unvoiced speech. The upper half of this figure shows typical waveforms obtained at points A to E for a voiced sequence, the lower half for an unvoiced sequence. The upper left shows a Hamming window weighted voiced sequence. On the order of three distinct pitch periods are included in this analysis. The wiggly curve at the upper right shows the log magnitude of the transform of this sequence, which consists of a rapidly varying periodic component (due to the excitation) and a slowly varying component due to the vocal tract transmission. The upper middle shows the resulting cepstrum. The strong peak at about 9 msec shows the pitch period. The low-time portion corresponds to the cepstrum of the vocal tract impulse response. Application of a low-time cepstral window (to eliminate excitation information) and discrete Fourier transformation yields the slowly varying curve in the upper right. Based on the peaks in the resulting spectrum at point E, efficient algorithms exist for estimating the formant resonances corresponding to the particular vocal tract transmission function.

For unvoiced speech the excitation is a random input rather than a quasi-periodic pulse train. In this case the waveform at point A is as shown in the lower left of Fig. 12.33. The random nature of the input is evident from this plot. The log magnitude of the DFT is as shown in the rapidly varying curve

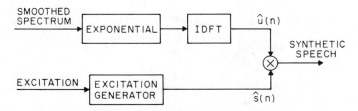

Fig. 12.34 Block diagram of a homomorphic vocoder (After Oppenheim).

in the lower right-hand corner of Fig. 12.33. The curve may again be modeled as the linear combination of a random component due to the source and a slowly varying component due to the vocal tract. The resulting cepstrum, shown in the lower middle, displays no strong peak, indicating the absence of voicing. The result of using a cepstrum window and the DFT is shown as the slowly varying curve in the lower right of the figure. This curve represents the transmission of the vocal tract. Generally both poles and zeros are used to represent the shape of the unvoiced spectrum. Further discussion of this point will be given in Sec. 12.15.

Thus the system of Fig. 12.32 is capable of separating the various components that comprise the speech waveform, even though they are combined in a convolutional manner.

12.14 Homomorphic Vocoder

The analysis scheme above may be readily combined with a synthesizer of the type shown in Fig. 12.34 to comprise an entire vocoder system. Instead of coding the vocal tract impulse response spectrum into either formants or a pole–zero representation, it is preserved and put through an inverse system to the original nonlinear processing. This inverse system consists of an exponentiator (to undo the logarithm) and an inverse discrete Fourier transformation (to undo the DFT) to give $\hat{u}(n)$, an estimate of the vocal tract impulse response. The excitation period (as obtained from a cepstral measurement) is used to create either a quasi-periodic pulse train or a random train of impulses to act as an estimate of the true excitation. These two sequences are convolved to give the synthetic speech. Figure 12.35 shows a spectrographic comparison (due to Oppenheim) of an original and a synthetic utterance as processed by a homomorphic vocoder. Clearly the similarity between spectrograms is quite good.

12.15 Formant Synthesis

One of the most important speech research problems concerns techniques for synthesizing speech from appropriate excitation parameters. Speech synthesis applications include several types of computer voice response systems

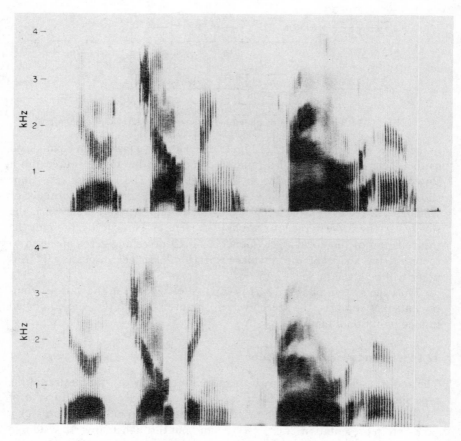

Fig. 12.35 Spectrogram comparison between natural and synthetic utterances for a homomorphic vocoder (After Oppenheim).

and provide important insight into the basic mechanism of speech production and perception. One of the most basic sets of such parameters is the set of formant frequencies as a function of time. We shall see later in this chapter how this representation of speech lends a considerable degree of flexibility and efficiency to the various applications of synthetic speech. In this section we present some of the signal processing problems associated with synthesizing speech from formant data. It is assumed that an analysis system, e.g., of the type discussed in Sec. 12.14, is available for deriving the formant data from natural speech.

Figure 12.36 shows a schematic block diagram of a general-purpose formant synthesizer of the type used in several computer voice response studies. There are two excitation sources: an externally controllable impulse generator (the source for voiced sounds), whose output consists of a unit

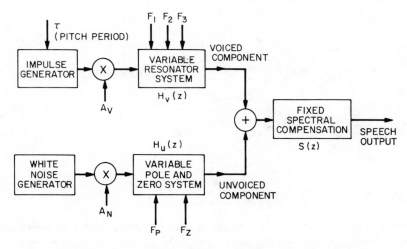

Fig. 12.36 Schematic block diagram of a formant synthesizer.

pulse once every pitch period (P samples), and a pseudo-random uniform number generator (the source for unvoiced sounds), whose output approximates a white-noise generator.

There are two basic signal processing paths in the synthesizer. The upper path consists of an intensity modulator (A_v) and a time-varying digital filter consisting of a cascade of L variable resonators (poles). The transfer function of this filter (under steady-state conditions) is

$$H_v(z) = \prod_{k=1}^{L} \left[\frac{1 - \exp(-\alpha_k T) 2 \cos(b_k T) + \exp(-2\alpha_k T)}{1 - \exp(-\alpha_k T) 2 \cos(b_k T) z^{-1} + \exp(-2\alpha_k T) z^{-2}} \right] \quad (12.39)$$

where α_k is the radian bandwidth of the kth pole, b_k is the radian center frequency of the kth pole, and T is the sampling period. A typical z-plane plot of the pole locations for a vowel ($L = 5$) is shown in Fig. 12.37. Although all the pole center frequencies and bandwidths can be controlled, generally only the lowest three center frequencies are varied as shown by the control signal inputs (F_1, F_2, F_3) to the variable resonator system in Fig. 12.36. The variable resonator system accounts for the effects of the time-varying shape of the vocal tract on the speech spectrum.

The effects of radiation of sound from the mouth (or nose) into air and glottal excitation pulse shape must be accounted for. This is the function of the fixed spectral compensation network whose transfer function is of the form

$$S(z) = \frac{[1 - \exp(-\alpha T)][1 + \exp(-bT)]}{[1 - \exp(-\alpha T) z^{-1}][1 + \exp(-bT) z^{-1}]} \quad (12.40)$$

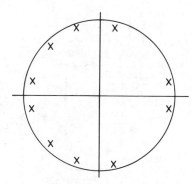

Fig. 12.37 Pole locations for a typical vowel.

This network consists of two real axis poles (one in the right-half z plane, one the left-half z plane), which approximates the desired transfer function. The z-plane plot of pole locations for this network is as shown in Fig. 12.38.

The lower path in Fig. 12.36 consists of a modulator (A_N) that controls the variance of the noise generator output and another time-varying digital filter consisting of a cascade of a pole and zero. Its transfer function is of the form

$$H_u(z) = \frac{H_1(1)H_2(z)}{H_1(z)H_2(1)} \tag{12.41}$$

where

$$H_1(z) = 1 - 2e^{-aT} \cos{(bT)}z^{-1} + e^{-2aT}z^{-2}$$

and

$$H_2(z) = 1 - 2e^{-cT} \cos{(dT)}z^{-1} + e^{-2cT}z^{-2}$$

where a, b, c, and d are the radian bandwidths and center frequencies of the time-varying pole and zero. Generally, the bandwidths of the pole and zero

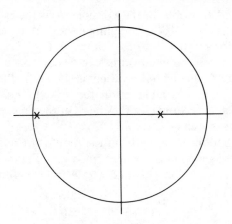

Fig. 12.38 Pole locations for the source function.

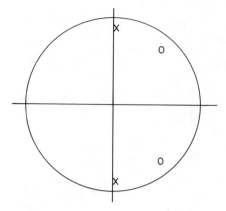

Fig. 12.39 Pole–zero locations of a typical unvoiced sound.

are fixed and only the center frequencies vary as shown by the control signal inputs F_p and F_z to the variable pole and zero system in Fig. 12.36. The z-plane pole–zero plot for a typical unvoiced sound is given in Fig. 12.39. The output of this system is passed to the fixed spectral compensation system to provide the final unvoiced speech output.

It should be noted that each of the transfer functions [Eqs. (12.39) to (12.41)] of the synthesizer has the property that at zero frequency the transfer function is unity independent of the center frequencies and bandwidths of any pole or zero. This property is essential to account for the unity transmission of the vocal tract at zero frequency and is achieved by using resonators that are individually normalized to have this property.

The synthesizer configuration above is incomplete in its ability to synthesize the sounds of speech in several aspects that are desirable in a general-purpose synthesizer. For example, there is no provision for a network to produce the nasal consonants n and m or a network to produce the voiced fricatives z (as in zoo), zh (as in azure), v (as in very), and th (as in there). To synthesize nasal consonants, a network consisting of a time-varying pole and zero must be placed in cascade with the variable resonator system of Fig. 12.36. To synthesize voiced fricatives adequately, a network that modulates the noise generator output by the voiced path output is necessary. Also, for additional flexibility in the synthesizer, provision should be made to allow the noise generator output to excite the voiced processing path in order to produce whispered speech.

To remedy these problems, Fig. 12.40 shows a more versatile synthesizer configuration that has been both simulated and built in digital hardware. The synthesizer derives its time-varying control parameters (indicated as external inputs to each of the signal processing blocks) synchronously—i.e., it changes all parameters once per pitch period, at the beginning of the

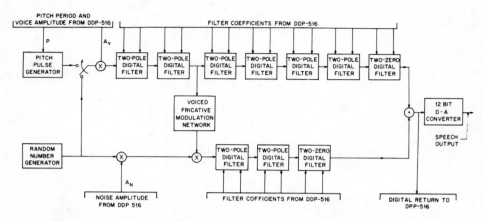

Fig. 12.40 Block diagram of hardware synthesizer.

period. At this time, each of the filters has minimum energy and the adverse effects of any large change in any control parameter are minimized. The control parameters are supplied to the hardware from a control computer (indicated as a Honeywell DDP-516 computer in Fig. 12.40).

This synthesizer is similar in concept to the one discussed earlier but differs slightly in details. Specifically the upper signal processing path consists of six two-pole digital filters [$L = 6$ in Eq. (12.39)] and one two-zero filter, where the bandwidth and center frequency of each filter is controllable. The sixth two-pole filter and the two-zero filter account for a nasal pole and zero and cancel each other during nonnasal sounds. (Exact cancellation of a pole by a zero is easily accomplished in a digital system.) Four of the two-pole filters (or possibly five during nonnasal sounds) are used to represent the time-varying vocal tract transfer function $H_v(z)$ and the last two-pole filter provides the desired spectral compensation $S(z)$.

The unvoiced signal processing path consists of two two-pole filters and one two-zero filter. Again the bandwidths and center frequencies of each of the filters can be varied externally. One two-pole and one two-zero filter are used to represent $H_u(z)$ and the remaining two-pole filter is used to provide the necessary spectral compensation $S(z)$. In this synthesizer, for added flexibility, the voiced and unvoiced spectral compensation networks may be different since they are included separately in each path of the synthesizer.

12.16   Voiced Fricative Excitation Network

The voiced fricative excitation network connects the output at one point in the voicing path to the unvoiced path. It is used to model the production of the unvoiced component of voiced fricatives. Figure 12.41 shows the relevant

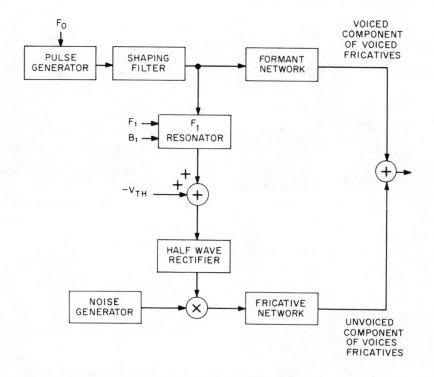

EXCITATION NETWORK FOR VOICED FRICATIVES

Fig. 12.41 Excitation network for voiced fricatives.

networks used to synthesize the entire voiced fricative. The unvoiced excitation is produced as follows. The pitch pulses excite a resonator tuned to the first formant of the voiced component of the fricative. This resonator is the first-order approximation to the transfer function of volume velocity from the glottis through the point of constriction of the vocal tract. A threshold level (V_{TH}) is subtracted from the output of the resonator and the result is half-wave rectified. These operations model the physical observation that turbulence is not produced until the volume velocity of the airflow exceeds a threshold value.

The output of the half-wave rectifier modulates the output of the noise generator, producing a pitch-synchronous excitation for the unvoiced component of the fricative. The final unvoiced component is produced by feeding this excitation into the fricative network (i.e., the lower branch of the synthesizer). The voiced component is produced by exciting the formant network in the usual manner.

Figure 12.42 shows spectrograms of synthetic and natural versions of the voiced fricatives |zh| and |z|. A careful examination of these spectrograms

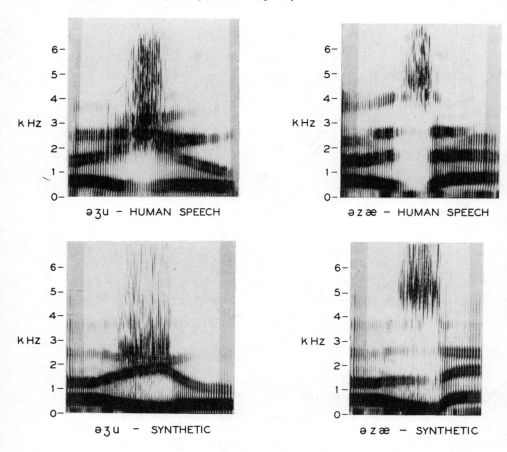

Fig. 12.42 Spectrograms of synthetic and natural voiced fricatives.

shows the effects of the pitch-synchronous modulation during the fricatives for both the synthetic and the natural speech.

12.17   Random Number Generator

To generate the pseudo-random numbers needed as the source for unvoiced sounds, any of a large number of available algorithms could be used. For the hardware realization, the specific pseudo-random number generator used is a 16-bit maximal length shift register sequence. This algorithm generates a random bit from mod-2 sums of the previous 16 bits, shifts out the bit generated 16 clock pulses earlier, and shifts in the new bit. The algorithm used to generate the current bit is

$$X_n = X_{n-1} \oplus X_{n-12} \oplus X_{n-14} \oplus X_{n-15} \qquad n = 1, 2, 3, \ldots \quad (12.42)$$

where each X is either 1 or 0 and a 1 physically corresponds to a positive excitation pulse and a 0 to a negative excitation pulse. Thus the noise generator output consists of a random succession of positive and negative pulses. The spectrum of the noise generator output is flat.

12.18 Principles of Digital Operation

The basic principle behind the digital hardware is the multiplexing of a single arithmetic unit among all the two-pole filters and the two-zero filters. The arithmetic operations required to realize a two-pole filter, for example, are two additions, two subtractions, and two multiplications for each output sample. High-speed integrated circuits are currently capable of doing about 25 times this number of arithmetic operations in the time between output samples (100 μsec at a 10-kHz sampling rate). Thus the notion of sharing a single arithmetic unit among many filters attains practical significance in the synthesizer. By providing storage for the filter coefficients and the delayed outputs of the filters and by dynamically controlling which inputs go into the arithmetic unit and where the outputs go, a single arithmetic unit can service the entire synthesizer.

A schematic block diagram of the digital logic used to realize the synthesizer is shown in Fig. 12.43. The arithmetic unit consists of a three-input adder, a shift register delay (which holds the delayed filter variables), a subtracter, and a multiplier. The length of the shift register delay is 480 bits (20 delayed variables times 24 bits per variable). Another shift register memory of 320 bits (20 filter coefficients times 16 bits per coefficient) holds the multipliers for each of the filter sections. This arithmetic unit can perform a multiplication, an addition, and a subtraction simultaneously in about 3.9 μsec; therefore

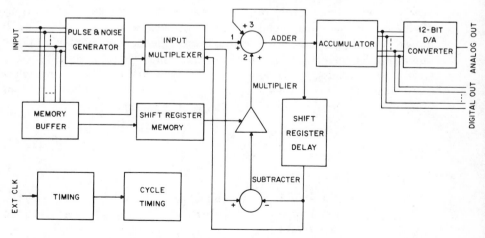

Fig. 12.43 Block diagram of logic in hardware synthesizer.

each filter section requires about 7.8 μsec per iteration. In this manner the 10 filter sections of the synthesizer require about 78 μsec. Thus the synthesizer can operate at sampling frequencies up to 12.8 kHz.

The remainder of the logic diagram is straightforward. The synthesizer control signals come from the computer output line to the input of the synthesizer. A memory buffer transfers the gain coefficients and pitch period to the pulse and noise generator and to the input multiplexer. The memory buffer shifts the filter coefficients to the shift register memory. The pulse and noise generator provides excitation to the arithmetic unit via the input multiplexer. An accumulator sums the voiced and unvoiced outputs and sends the 16 most significant bits back to the computer, simultaneously converting the 12 most significant bits to analog form. The switching and timing logic is determined from timing logic, which uses an externally supplied clock to determine the basic synthesizer sampling rate. The sampling rate is thus easily changed without any internal modifications to the synthesizer.

12.19   Linear Prediction of Speech

The basic idea of formant analysis and synthesis is that speech production is well modeled by exciting a cascade of linear time-varying second-order section digital filters (formant resonators) with either quasi-periodic pulses or noise. The major difficulty with this idea lies in assigning computed formants to specific second-order sections. Formants seem to disappear during certain sounds and additional formants seem to be present during other sounds. A large number of errors of either of these types can quickly render the synthetic output unintelligible or at best make its quality unacceptable. Such errors are generally not uncommon across sentence length utterances.

To remedy these problems, the basic speech synthesis (production) model can be modified slightly to the form shown in Fig. 12.44. The L individual second-order systems of the formant model are combined to give one Pth-order linear system (where $P \geq 2L$). This system accounts for the vocal tract transmission, the source pulse shape, and the radiation characteristics. The input $\delta(n)$ is either a stream of digital impulses or a quasi-random input. The transfer function of the filter is of the form

$$H(z) = \frac{X(z)}{\delta(z)} = \frac{1}{1 - \sum_{k=1}^{p} a_k z^{-k}} \tag{12.43}$$

The analysis of speech to determine pitch and the voiced–unvoiced decision is performed as for any other system using a pitch detector of the type shown earlier in this chapter or any other algorithm that is desired. The predictor coefficients $\{a_k, k = 1, 2, \ldots, p\}$ are determined from a minimum

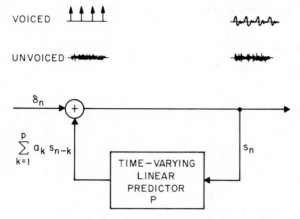

Fig. 12.44 Linear prediction model of speech production (After Atal and Hanauer).

mean square error analysis. The difference equation characterizing the system is of the form

$$s(n) = \sum_{k=1}^{p} a_k s(n - k) + \delta(n) \qquad (12.44)$$

For voiced speech, $\delta(n)$ is zero except for one sample at the beginning of every pitch period. Thus, except at this sample when $\delta(n)$ is nonzero, Eq. (12.44) becomes

$$s(n) = \sum_{k=1}^{p} a_k s(n - k) \qquad (12.45)$$

Thus theoretically if the model is perfect, the speech samples $s(n)$ are completely predictable from Eq. (12.45). Since the speech waveform does not fit the model perfectly, it is possible to define an error between $s(n)$, the true value at sample n, and $\hat{s}(n)$, the value predicted by Eq. (12.45). Let $E(n)$ be the error; i.e.,

$$E(n) = s(n) - \hat{s}(n) = s(n) - \sum_{k=1}^{p} a_k s(n - k) \qquad (12.46)$$

The predictor coefficients are chosen so as to minimize the mean square prediction error $\langle E(n)^2 \rangle$, averaged over all n.

The expression for the mean square error can be put into the form

$$\langle E(n)^2 \rangle = \sum_{n=1}^{\infty} \left[s(n) - \sum_{k=1}^{p} a_k s(n - k) \right]^2 \qquad (12.47)$$

To solve for the predictor coefficients, Eq. (12.47) is differentiated with respect to $a_j, j = 1, 2, \ldots, p$, and the result is set to zero giving the set of equations

$$\sum_{k=1}^{p} a_k \sum_{n=1}^{\infty} s(n - k) s(n - j) = \sum_{n=1}^{\infty} s(n) s(n - j) \qquad j = 1, 2, \ldots, p \quad (12.48)$$

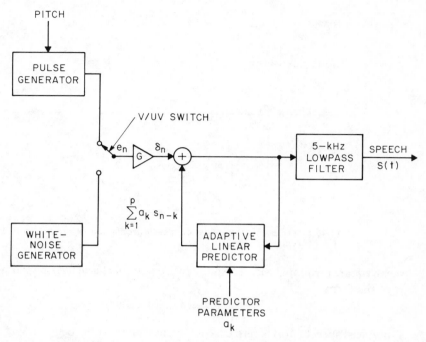

Fig. 12.45 Linear prediction synthesizer (After Atal and Hanauer).

In matrix formulation the set of equations given by Eq. (12.48) can be written as

$$\Phi \underline{a} = \psi \tag{12.49}$$

where

$$\varphi_{ij} = \sum_{n=1}^{\infty} s(n - i)s(n - j) \tag{12.50}$$

and

$$\psi_j = \varphi_{0j} \tag{12.51}$$

Thus Φ is a matrix of autocorrelations and ψ is a vector of autocorrelations. Since Φ is symmetric and positive definite, there exist several efficient methods of solving the set of equations implied by Eq. (12.48). Thus the analysis required by linear prediction is relatively straightforward.

For synthesis the system shown in Fig. 12.45 is used to give a high quality representation of the natural signal. The distinctions between this synthesizer and the formant synthesizer of Sec. 12.18 are worth noting. The most important difference is the use of a single pth-order recursive filter in place of the cascade of second-order filters. In the time-invariant case, e.g., a steady vowel, these two models are exactly equivalent. In the time-varying case, e.g., most of the time during speech, these two configurations are not equivalent. In the formant synthesizer it is essential that each resonance be assigned to

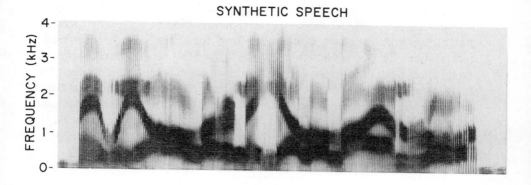

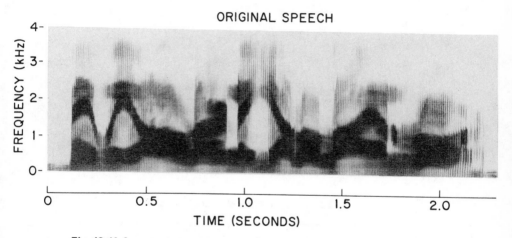

Fig. 12.46 Spectrogram comparison of natural and synthetic utterances (After Atal and Hanauer).

the proper formant or improper operation results. This is not necessary for the linear prediction case as all formants are synthesized by one recursive filter. A further important difference is that the amplitude of the pitch pulses, as well as the white noise, is adjusted by the gain network G to provide the correct rms value of the synthetic speech samples. No such adjustment is generally made for synthetic speech.

The spectrograms in Fig. 12.46 show how well linear prediction works. The upper spectrogram is of synthetic speech, while the lower one is of real speech. It is difficult to tell them apart.

12.20 A Computer Voice Response System

As discussed earlier, the representation of speech in parametric form, e.g., in terms of pitch and formants, has two important advantages in terms of its utility in computer voice response systems. First, since the formants change at rates comparable to the motions of the vocal tract, they can be sampled and quantized to low bit rates. Hence representation of speech by formant parameters constitutes an economical form for digital storage of speech information. The second advantage of the formant representation of speech is its inherent flexibility. Since contextual information is contained in the formant data, and prosodic data (e.g., inflection, rate of speaking, etc.) is contained in the pitch data and the timing information, the formant representation enables you to separate "what is said," from "the manner in which it is said." This flexibility and economy form the basis for a simple computer

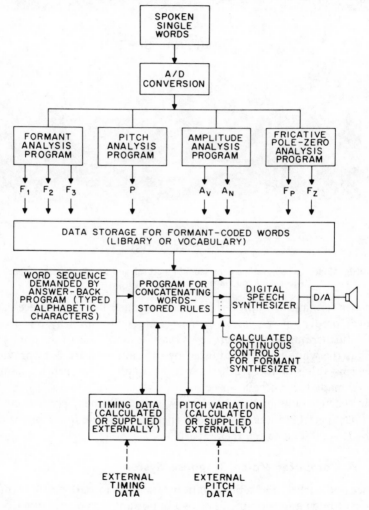

Fig. 12.47 Block diagram of concatenation voice response system.

voice response system where isolated vocabulary elements are smoothly assembled into connected speech.

A block diagram of the system used for synthesis of connected speech from a vocabulary of formant-coded words is shown in Fig. 12.47. Naturally spoken, isolated words (or phrases) are analyzed by a formant analyzer to give three formants (F_1, F_2, F_3), voiced and unvoiced amplitude (A_V, A_N), pitch period (P), and unvoiced pole and zero (F_p, F_z) once every 10 msec. These control parameters are smoothed by programmed digital filters, sampled at their Nyquist rates (typically $33\frac{1}{3}$ sec^{-1}), quantized, and stored in the word catalog as the reference library. The typical bit rate used for storage of these data is 700 bps when the pitch signal is saved. When pitch is not saved (the usual situation since it is normally calculated by the concatenation program), the bit rate for the stored data is 533 bps. Table 12.1 shows a breakdown of how

Table **12.1**

CODING OF FORMANT PARAMETERS

Parameter	No. Bits per Frame	No. Frames (sec)	No. Bits (sec)
F_1 or F_p	3	$33\frac{1}{3}$	100
F_2 or F_z	4	$33\frac{1}{3}$	$133\frac{1}{3}$
F_3	3	$33\frac{1}{3}$	100
P	5	$33\frac{1}{3}$	$166\frac{1}{3}$
A_V or A_N	3	$33\frac{1}{3}$	100
V/U	1	100	100
		Total	700
		Pitch	$-166\frac{2}{3}$
	Data rate for synthesis using calculated pitch data		$533\frac{1}{3}$

these bit rates are achieved. The data in this table were derived from experimental investigation of the effects of smoothing and quantization on the perception of the synthetic output.

As shown in Table 12.1, at every 10-msec interval the speech is classified as voiced or unvoiced (V/U) by a 1-bit signal. Thus, for each frame, storage is required for either voiced parameters or unvoiced parameters but not for both. It should be noted that the control parameter frame rate ($33\frac{1}{3}$ sec^{-1}) is one-third the rate of the V/U signal.

Once input words and phrases are coded in terms of the formant representation, they can easily be modified for use with the synthesis program. Words can be lengthened or shortened; formants can be changed easily; and a pitch contour, different from the one originally spoken, can be superimposed on the data. Thus the vocal resonance data is available to the synthesis program in a form flexible enough to conform to the timing and pitch generated by the concatenation program.

The lower portion of Fig. 12.47 shows how the system assembles a synthetic message composed of words and phrases from the reference library. First, the answer-back program requests the word sequence for a specific message. The word concatenation program first determines timing data for the message from an auxiliary program. The timing data is in the form of a word duration for each word in the output message. The concatenation program then accesses, in sequence, the control parameters for each of the words in the string. A duration modification adjustment on each word is first made so the word duration in context matches the duration specified by the timing rules. Next the concatenation program smoothly interpolates the formant control parameters when the final part of any word and the initial part of the following word are both voiced. An interpolation algorithm designed to produce realistic formant transitions is used. Finally a continuous function for pitch variation is produced for the whole message. All computed control parameters are outputted to a hardware digital speech synthesizer. Digital-to-analog conversion produces a continuous synthetic speech output.

The computer voice response system described above has been applied to several specific problems including automatic generation of telephone numbers and computer-aided voice wiring. A spectrogram comparison of a typical synthetic telephone number and a natural version of the same number is shown in Fig. 12.48. From this figure it is seen that the timing and formant data of the synthesized utterance are reasonably good matches to those of the natural utterance.

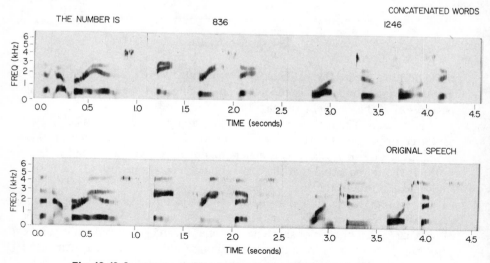

Fig. 12.48 Spectrogram of synthetic and natural telephone numbers.

12.21 Summary

As seen throughout this chapter, the concepts of digital signal processing and speech research seem to go hand in hand. Progress toward practical solutions of most speech analysis and synthesis problems has been made primarily with the advent of signal processing computers and algorithms. It should be reiterated that the specific problems discussed in this chapter are merely representative problems in the field of speech research and do not constitute, by any means, a complete documentation of the vast amount of research in this field. Furthermore, the specific systems discussed are not necessarily the optimum systems for solving the particular problems for which they are used. Their selection was based primarily on the familiarity of the authors with the basic approaches involved, and their connection to digital signal processing techniques.

REFERENCES

General References

1. J. L. FLANAGAN, *Speech Analysis, Synthesis and Perception*, 2nd ed., Springer-Verlag, New York, 1972.
2. R. W. SCHAFER, "A Survey of Digital Speech Processing Techniques," *IEEE Trans. on Audio and Electroacoustics*, **AU-20**, No. 4, 28–35, Mar., 1972.
3. J. L. FLANAGAN, C. H. COKER, L. R. RABINER, R. W. SCHAFER, AND N. UMEDA, "Synthetic Voices for Computers," *IEEE Spectrum*, **7**, No. 10, 22–45, 1970.

Short-Time Spectrum Analysis

1. J. L. FLANAGAN AND R. M. GOLDEN, "Phase Vocoder," *Bell Syst. Tech. J.*, **45**, 1493–1509, 1966.
2. R. W. SCHAFER AND L. R. RABINER, "Design of Digital Filter Banks for Speech Analysis," *Bell Sys. Tech. J.*, **50**, No. 10, 3097–3115, Dec., 1971.
3. R. W. SCHAFER AND L. R. RABINER, "Design and Simulation of a Speech Analysis–Synthesis System Based on Short-Time Fourier Analysis," *IEEE Trans. on Audio and Electroacoustics*, **AU-21**, 165–174, June 1973.

Channel Vocoder

1. M. R. SCHROEDER, "Vocoders: Analysis and Synthesis of Speech," *Proc. IEEE* **54**, 720–734, 1966.
2. B. GOLD AND C. M. RADER, "Systems for Compressing the Bandwidth of Speech," *IEEE Trans. on Audio and Electroacoustics*, **AU-15**, No. 3, 131–135, Sept., 1967.
3. R. GOLDEN, "Vocoder Filter Design: Practical Considerations," *J. Acoust. Soc. Am.*, **43**, 803–810, Apr., 1968.
4. B. GOLD AND C. M. RADER, "The Channel Vocoder," *IEEE Trans. on Audio and Electroacoustics*, **AU-15**, No. 4, 148–160, Dec., 1967.

Pitch Detection

1. B. GOLD, "Computer Program for Pitch Extraction," *J. Acoust. Soc. Am.*, **34**, 916–921, 1962.

2. B. GOLD, "Description of a Computer Program for Pitch Detection," *Proc. Int. Cong. Acoustics,* 4th, Copenhagen, Paper G34, 1962.

3. B. GOLD, "Note on Buzz–Hiss Detection," *J. Acoust. Soc. Am.,* **36,** 1659–1661, 1964.

4. B. GOLD AND L. R. RABINER, "Parallel Processing Techniques for Estimating Pitch Periods of Speech in the Time Domain," *J. Acoust. Soc. Am.,* **46,** No. 2, 442–449, Aug., 1969.

5. A. M. NOLL, "Cepstral Pitch Determination," *J. Acoust. Soc. Am.,* **41,** 293–309, 1967.

Homomorphic Processing of Speech

1. A. V. OPPENHEIM, R. W. SCHAFER, AND T. G. STOCKHAM, "Nonlinear Filtering of Multiplied and Convolved Signals," *Proc. IEEE,* **56,** 1264–1291, 1968.

2. A. V. OPPENHEIM AND R. W. SCHAFER, "Homomorphic Analysis of Speech,' *IEEE Trans. on Audio and Electroacoustics,* **AU-16,** 221–226, 1968.

3. A. V. OPPENHEIM, "Speech Analysis–Synthesis System Based on Homomorphic Filtering," *J. Acoust. Soc. Am.,* **45,** 459–462, 1969.

4. R. W. SCHAFER AND L. R. RABINER, "System for Automatic Analysis of Voiced Speech," *J. Acoust. Soc. Am.,* **47,** Part 2, 634–648, 1970.

Formant Synthesis

1. L. R. RABINER, "Digital-Formant Synthesizer for Speech Synthesis Studies," *J. Acoust. Soc. Am.,* **43,** 822–828, 1968.

2. B. GOLD AND L. R. RABINER, "Analysis of Digital and Analog Formant Synthesizers," *IEEE Trans. on Audio and Electroacoustics,* **AU-16,** 81–94, Mar., 1968.

3. L. R. RABINER, L. B. JACKSON, R. W. SCHAFER, AND C. H. COKER, "Digital Hardware for Speech Synthesis," *IEEE Trans. on Communication Tech.,* **COM-19,** 1016–1020, 1971.

Linear Prediction of Speech

1. B. S. ATAL AND S. L. HANAUER, "Speech Analysis and Synthesis by Linear Prediction of the Speech Wave," *J. Acoust. Soc. Am.,* **50,** 637–655, 1971.

2. F. ITAKURA AND S. SAITO, "An Analysis–Synthesis Telephony System Based on Maximum Likelihood Method," *Electronics and Communication in Japan,* **53A,** 36–43, 1970.

3. J. I. MAKHOUL, AND J. J. WOLF, "Linear Prediction and the Spectral Analysis of Speech," Bott, Beranek, and Newman Report 2304, Aug. 1972.

4. J. D. MARKEL, A. H. GRAY JR., AND H. WAKITA, "Linear Prediction of Speech-Theory and Practice," Speech Communication Research Lab. Monograph No. 10, Sept. 1973.

Computer Voice Response Systems

1. L. R. RABINER, R. W. SCHAFER, AND J. L. FLANAGAN, "Computer Synthesis of Speech by Concatenation of Formant-Coded Words," *Bell Sys. Tech. J.,* **50,** No. 5, 1541–1558, May–June, 1971.

2. J. L. FLANAGAN, L. R. RABINER, R. W. SCHAFER, AND J. DENMAN, "Wiring Telephone Apparatus from Computer-Generated Speech," *Bell Sys. Tech. J.,* **51,** No. 2, 391–397, Feb., 1972.

13

Applications to Radar

13.1 Introductory Discussion of Radar Principles and Applications

Our attention in this chapter will be directed toward but a small segment of the subject of radar, namely, the improvement in signal processing made possible by digital technology. We know that the development of digital computer technology has led to great sophistication of radar tracking algorithms. In addition, computers in conjunction with electronically steerable phased array antennas have led to refined methods of scheduling of the radar's repertoire of transmitted signals. We anticipate that future radars will incorporate high-speed digital hardware to perform the desired filtering and thresholding algorithms. Several radars familiar to us have incorporated or are incorporating digital signal processing hardware. There is no doubt that these processors are substantially more flexible than their analog counterparts.

Radar Applications. During World War II, the British used radars to detect incoming German bombers. Beginning in the early 1950's guided missiles for defense against bombers were developed; these missiles required compact and reliable homing radar systems. At the same time the SAGE (semiautomatic ground environment) global air-defense system against intercontinental bombing was developed; this gigantic system was revolutionary in combining advanced radar, computers, communications, and display technology. By the time this equipment was made operational, the ICBM had surpassed the bombers as the number one threat; this led to new efforts in radar technology to detect small but highly lethal targets at long range under adverse conditions in minimum time. Debates as to the strategic success of such efforts are bound to continue for a long time but it is certainly true that these efforts have led to substantial technological advances. Meanwhile, radars have found applications in such diverse subjects as air traffic control, weather monitoring, accurate ground mapping, and radar astronomy (moon mapping). Following is a brief discussion of some of these applications.

An air traffic control (ATC) system at a major airport is a large complex of electronic equipment run by a team of highly skilled controllers. The "brains" of the system are in a room with many display terminals. Each terminal is handled by a controller who knows the flight schedules and flight paths, communicates by voice with airborne personnel and other controllers, observes on the display scope both radar and beacon[1] information, and has at his disposal a variety of controls to edit the displays.

In a typical ATC radar, the antenna rotates mechanically, sweeping out a full 360° every 4 to 12 sec. Azimuth beam resolution is 1 to 2° and the vertical antenna pattern is a fan beam of 30 to 45° angle. Thus, as the antenna sweeps by a target aircraft, there will be a succession of "hits"; i.e., radar returns at the pulse repetition period after which no information is obtained from that target until the antenna has made a complete revolution. Based on this information, the system must track up to several hundred aircraft within its field of view (typically 40 mi) and display these tracks to the controller who can then correlate and use all this information.

During the Vietnam war, a major problem was the detection of enemy supply lines. Technologically, the problem was one of determining the presence of men and trucks in heavy foliage. The problem is essentially a subset of the problem of finding interesting objects amidst dense clutter. Other types of problems are the monitoring of aircraft on the ground, searching for people lost in the woods, and guarding of property such as airbases.

The chief characteristic of these problems is the presence of ground clutter radar returns that may be 80 to 100 dB stronger than the signal return. The only known mechanism for detecting such small signals in large clutter is via the Doppler effect, which allows the measurement of the frequency displacement of moving targets.

---

[1] For large commercial aircraft, beacons have supplanted radars as the major aircraft trackers. All commercial United States aircraft contain transponders that "answer" a coded transmitter signal. Thus, at the transmitter site, much greater signal return is available for processing. At present, many private aircraft have no transponders so that radar or voice communications must be used for aircraft location. Hence, for the next 20 years, radar should remain in use for ATC; beyond that the future is doubtful.

Perhaps the most demanding radar task is long-range missile detection and the consequent tracking of multiple targets with the aim of discriminating among missiles, decoys, tank fragments, and chaff clouds. During the detection period, a portion of space is repetitively searched using matched filtering techniques to obtain both long range and good range resolution. During the track mode, the radar tries to discard, by means of velocity differences, uninteresting targets. The final phase of discrimination requires analysis of the missile's wake properties and that will not be discussed here.

Radar Environments. The central problem in most radars is background clutter, which appears in a wide variety of forms; in every radar design the clutter problem must be well understood before an adequate design is possible. For example, in ATC problems if the antenna is not tilted vertically upward, the radar beam will intersect the ground at small ranges so that returns from near-in aircraft will be buried in large ground clutter returns. Also, rainclouds will cause clutter that creates difficulties in tracking aircraft during bad weather. Even flocks of birds cause clutter. Rain and bird returns, although not so large as ground returns, have a velocity component that complicates Doppler discrimination techniques.

Detection of moving objects on the ground is greatly inhibited by the huge ground clutter returns. Detection of moving objects in foliage is further complicated by a noticeable frequency spread imparted to the clutter return by wind-induced foliage motion. "Discretes," namely, large returns from big stationary objects such as water towers, can usually be dealt with by the radar operator who eventually will memorize the radar "map" of the surrounding terrain.

A guided missile reentering the atmosphere is subject to great temperature gradients. Soon after takeoff the missile is separated from its surrounding tank. Since both the missile and the jettisoned tank are above the atmosphere, they will follow the same trajectory until reentry at which time the tank will slow down and break into many fragments that appear as moving targets on the radar screen. In addition, a missile can carry its own penetration aids such as decoys (streamlined but nonlethal bodies) and chaff (a multitude of tiny but highly reflective dipoles).

13.2 Radar System and Parameter Considerations

The major components of a radar are the antenna, the tracking computer, and the signal processor. Associated with the antenna are the transmitter and modulator and the receiver hardware. The tracking computer is the "brains" of the system, scheduling the appropriate antenna positions and transmitted signals as a function of time, keeping track of important targets and running the display system. The major traditional functions of the signal

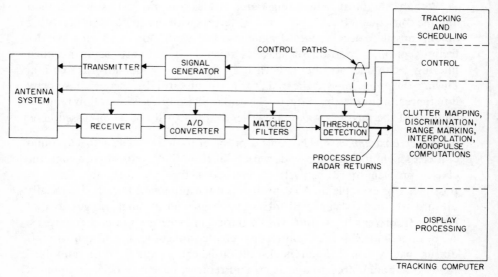

Fig. 13.1 Block diagram of a modern radar system.

processor are matched filtering and removal of useless information by threshold detection. A key element in the design of the overall radar is signal design. Transmitted radar signals may vary from simple pulse trains to high bandwidth chirps or LFM signals (linear frequency modulated signals), bursts of pulses or chirps, nonuniform bursts, or polyphase codes.

Figure 13.1 shows the major blocks of a modern radar system. Notice that the tracking computer controls all functions. There is a control path from it to the antenna system; if the antenna beam is electronically steerable, the tracking computer can control the beam position on a pulse-to-pulse basis and can determine whether monopulse information[2] (for better angle resolution) is called for. The tracking computer also controls the signal generator and coordinates the transmitted signal and the matched filter configuration. If the tasks of the tracking computer become too great, the jobs may be apportioned among several computers, as indicated in Fig. 13.1.

Before we can proceed with a useful signal processing discussion we must digest a certain amount of radar jargon plus some simple but important physical phenomena; in brief, we must describe the important radar parameters.

---

[2] A monopulse antenna system is in essence a multiantenna configuration whereby signal returns due to a single transmitted signal are received simultaneously from two antenna beams that are spatially offset. The result is that reliable angle information can often be obtained from a single transmitted signal.

1. Antenna Aperture and Wavelength

The formula for antenna beamwidth is given by

$$b \propto \frac{\lambda}{D} \qquad (13.1)$$

where b is beamwidth, λ is wavelength, and D is antenna width. If the antenna geometry is symmetric, as in a parabolic reflector, then b is the same in both horizontal (azimuth) and vertical (elevation) dimensions;[3] this corresponds to a pencil beam. In many cases, such as the air traffic control case, a fan beam is required to obtain full coverage in a reasonable time. A fan beam antenna is built with large horizontal and small vertical aperture to obtain a beam that is narrow in azimuth and wide in elevation.

2. Range and Range Resolution

The maximum unambiguous range R_{max} is given by,

$$R_{max} = \frac{cT}{2} \qquad (13.2)$$

where c is the velocity of light ($\simeq 3 \times 10^8$ m/sec) and T is the pulse repetition interval. For example, if $T = 10^{-3}$ sec, $R_{max} = 1.5 \times 10^2$ km. If T is decreased, targets at ranges greater than R_{max} will appear after the next radar pulse has been transmitted, causing an ambiguity in the interpretation of measured range. Of course, receiver noise and clutter may clamp a much lower limit than R_{max} on the maximum range at which target detection probabilities are good.

The range resolution ΔR is a measure of how well two targets that are near each other can be resolved by the radar. If the transmitted signal is a pulse of constant carrier frequency, then the minimum ΔR is determined by the pulse width p; if the range difference ΔR between two targets in the same beam is less than $pc/2$, the two received radar signals will interfere and the two targets may be mistaken for a single target. Narrowing the pulse width improves range resolution but diminishes the maximum range by decreasing the average power; this loss cannot be compensated because radars are generally peak-power limited. As we shall see later, radar signals can be much more complicated than sine waves and the fundamental limitation on range resolution is signal bandwidth rather than pulse narrowness.

---

[3] A spherical coordinate system is normally used for radar antennas. The radius of the sphere corresponds to range (distance from antenna), while azimuth is the angular dimension parallel to the earth; elevation is the orthogonal angular dimension perpendicular to the earth.

3. Doppler Filtering

Doppler filtering of received target information is based on the Doppler effect. For our purpose, the use of Doppler is succinctly explained by examination of the equation

$$\Delta f = \frac{2v}{c} f_0 = \frac{2v}{\lambda} \tag{13.3}$$

where f_0 is the carrier frequency, v is the target velocity, λ is the wavelength ($c = f_0 \lambda$), and Δf is the resultant shift in the received (relative to the transmitted) frequency. Equation (13.3) is the result when a continuous sine wave is transmitted for a constant velocity target. However, continuous signals or, practically speaking, signals of long duration would result in poor range resolution. To obtain both range and velocity resolution it becomes necessary to deal with pulsed Doppler signals. We shall now show that detection of such signals corresponds to sampling a sinusoid of frequency Δf at the pulse repetition period; assume that each pulse of carrier frequency f_0 begins at exactly the same phase; then at some given range a return signal is received. Since, during the pulse duration p, the aircraft can be considered stationary, there is no measurable Doppler shift of the received signal. After T seconds, however, the aircraft has moved slightly and if the next return is sampled at the same range, a phase shift relative to the first return is discernible. The amount of phase shift will be determined by the proportion of a wavelength moved during T; that is,

$$\varphi = 2\pi \frac{vT}{\lambda} \tag{13.4}$$

If the aircraft maintains constant radial velocity, there will be an additional phase shift φ for each repetition interval T. Thus, any signal return from a given range can be represented as

$$s(n, t) = a(n)e^{j[2\pi f_0(t-T)+n\varphi]} \tag{13.5}$$

where $a(n)$ is the amplitude modulation caused by the motion of the antenna beam as it sweeps by the target. For an electronically steerable antenna, it is possible to stop the antenna beam so that $a(n)$ can be unity.

In the radar receiver we assume that the returned complex exponential is multiplied by a coherent local oscillator source with an arbitrary but fixed phase ψ. Calling this signal $s_r(t)$, we obtain the demodulated result

$$f(n, t) = s(n, t)s_r(t) = a(n)e^{j[2\pi f_0(t-T)+n\varphi]}e^{-j(2\pi f_0 t+\psi)}$$

$$= a(n)e^{-j\omega_0 T}e^{-j\psi}e^{jn2\pi(vT/\lambda)} \tag{13.6}$$

The exponentials $e^{-j\omega_0 T}$ and $e^{-j\psi}$ are constants of unity amplitude and can be ignored. The variable part is simply an oscillation of frequency v/λ, which is the Doppler frequency.

The development above implies that the transmitted carrier is coherent from pulse to pulse so that the phase of the carrier is the same every time the radar transmits, that the carrier frequency is very stable, and that there are two orthogonal demodulators and associated well-matched amplifiers so that a complex signal can be processed. All this assumes a high degree of accuracy in the analog portion of the radar receiver. Presently deployed ATC radars do not, for example, fulfil these conditions. In many existing radars the transmitting tubes do not have pulse-to-pulse coherence. To make good use of Doppler processing will require the updating of many radars.

13.3   Signal Design and Ambiguity Functions

Radar signal design is directed toward achieving the best range and velocity measurement on one or more targets. From simple physical reasoning we know that a transmitted narrow pulse results in good range but poor velocity measurement, while a wide pulse of a single frequency yields good velocity but bad range information. From this, we can conjecture that signal design will turn out to be a compromise between range and velocity measurement. In the process of explaining why and how this conjecture turns out to be true, we shall make use of the *ambiguity function* of the two variables, range and velocity. The ambiguity function is at the center of analog radar signal design using analog matched filtering and the emphasis in this chapter will be on digital ambiguity functions. The interested reader can find adequate literature on analog ambiguity functions; thus, we shall base our development entirely on the assumption that the signal processor is digital.

The ambiguity function is an idealized mathematical model of the system shown in Fig. 13.2. We assume that the signal is generated digitally but must pass through an analog filter on its way to the transmitter. Ideally the analog signal return $s(t - \tau)e^{j2\pi f(t-\tau)}$ is a delayed and frequency shifted version of the transmitted signal $s(t)$. These effects, due to target displacement and velocity, are assumed to be carried undisturbed through the receiver analog filter and the A/D converter so that the input to the matched filter is the digital signal

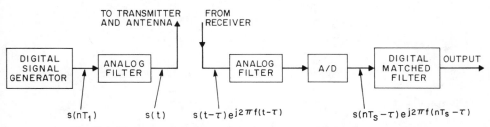

Fig. 13.2 Block diagram of the radar model leading to the ambiguity function.

$s(nT_s- \tau)e^{j2\pi f(nT_s-\tau)}$. Note that this signal is a function of two continuous parameters, τ and f (range and Doppler).

We now must inquire as to the nature of the digital matched filter. To preserve radar power it is desirable to make the signal duration large; but to preserve range resolution the output from a given range bin must be compact in time. This apparent contradiction is resolved by designing long duration signals with short duration correlation functions so that when the received signal passes through the appropriate matched filter, the output will be a very sharp pulse. Hence, if we think of a digital filter matched to the signal return for zero range and zero Doppler, this filter must have the impulse response $s^*(-nT_s)$; the digital matched filter output, defined as the digital ambiguity function, is also in actuality the cross-correlation function between the signal and the matched filter impulse response, which we know to be

$$A(\tau, f) = \sum_{n=-\infty}^{\infty} s(nT_s + \tau)s^*(nT_s)e^{j2\pi f(nT_s+\tau)} \tag{13.7}$$

where T_s is the sampling rate at the A/D in Fig. 13.2. If the matched filter is matched to a given range and Doppler, this corresponds to a two-dimensional displacement of $A(\tau, f)$ so that Eq. (13.7) is a perfectly general formulation.

The major importance of the ambiguity function (the magnitude of $A(\tau, f)$) is its bringing into focus a basic constraint of radar signal design, namely, that a signal cannot be designed that gives high performance everywhere in the range–velocity plane. The mathematical manifestation of this constraint, as we shall show, is that the total volume under the squared magnitude of the ambiguity function is independent of signal wave shape. Thus pushing $A(\tau, f)$ down anywhere in the (τ, f) plane will cause it to pop up elsewhere. To prove this, we begin with Eq. (13.7) and form

$$V = \int_{-\infty}^{\infty} \int_{-F_s/2}^{F_s/2} A(\tau, f)A^*(\tau, f) \, df \, d\tau \tag{13.8}$$

where F_s is the sampling frequency ($F_s = 1/T_s$).

Substituting Eq. (13.7) into Eq. (13.8) and integrating first with respect to f,

$$V = \sum_{n=-\infty}^{\infty} \sum_{m=-\infty}^{\infty} \int_{-\infty}^{\infty} s(nT_s + \tau)s^*(nT_s)s^*(mT_s + \tau)s(mT_s)$$

$$\times \frac{\sin [\pi F_s T_s(n - m)]}{\pi T_s(n - m)} \, d\tau \tag{13.9}$$

Noting that

$$R[(n - m)T_s] = \int_{-\infty}^{\infty} s(nT_s + \tau)s^*(mT_s + \tau) \, d\tau \tag{13.10}$$

is the autocorrelation function of the signal, Eq. (13.9) becomes

$$V = \sum_{n=-\infty}^{\infty} \sum_{m=-\infty}^{\infty} s^*(nT_s)s(mT_s)R[(n-m)T_s]\frac{\sin\,[\pi F_s T_s(n-m)]}{\pi T_s(n-m)} \quad (13.11)$$

Since $F_s T_s = 1$,

$$\frac{\sin\,[\pi F_s T_s(n-m)]}{\pi T_s(n-m)} = \begin{cases} 0 & \text{when } n \neq m \\ F_s & \text{when } n = m \end{cases} \quad (13.12)$$

Therefore, the double sum of Eq. (13.11) can be replaced with a single sum

$$V = F_s R(0) \sum_{n=-\infty}^{\infty} |s(nT_s)|^2 \quad (13.13)$$

We see that V, the volume under the ambiguity function, is dependent only on the total signal energy and not in any way on the signal shape. Thus for a given amount of signal energy, we have demonstrated that any decrease of the ambiguity function must result in an increase somewhere else in the (τ, f) plane.

1. Ambiguity Functions of Chirps and Sinusoidal Pulses

A chirp waveform is a linear frequency modulated (LFM) signal which combines some of the useful properties of both long and short pulses of a single sinusoidal carrier. For mathematical convenience we first find the ambiguity function for the chirp and then obtain that of the continuous wave (CW) pulse as a special case.

From a practical point of view we are interested in signals and filters with finite duration; this means that we have to establish some convention on the limits in Eq. (13.7) that takes into account the precise way that $s(nT_s + \tau)$ and $s^*(nT_s)$ overlap.

Let us agree on the following convention, as illustrated in Fig. 13.3; let the width of the signal T be exactly equal to MT_s where M is the number of samples in the matched filter. Thus for $-T_s < t \leq 0$, the overlap between signal and impulse response is perfect. For $0 < \tau \leq T_s$, they are misaligned by one sample. Now let us define $I(\tau)$ as the nearest rounded *up* integer of the ratio $|\tau|/T_s$. Thus, when $0 < \tau \leq T_s$, $I(\tau) = 1$; when $T_s < \tau \leq 2T_s$, $I(\tau) = 2$, etc. With this nomenclature, we can see from Fig. 13.3 that for τ negative, the limits on Eq. (13.7) become

$$A(\tau, f) = \sum_{n=I(\tau)}^{M-1} s(nT_s + \tau)s^*(nT_s)e^{j2\pi fnT_s} \quad (13.14)$$

Similarly, for positive τ, we obtain

$$A(\tau, f) = \sum_{n=0}^{M-1-I(\tau)} s(nT_s + \tau)s^*(nT_s)e^{j2\pi fnT_s} \quad (13.15)$$

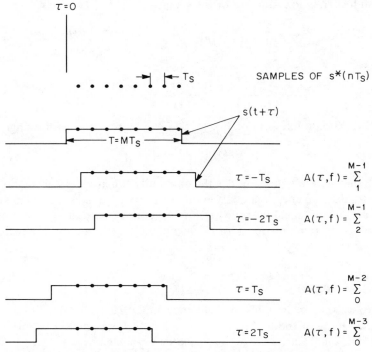

Fig. 13.3 Synchronization conventions between digital matched filter and sampled signal.

If $s(nT_s)$ is of the form of an exponential, then Eq. (13.14) can be changed into Eq. (13.15) by the simple change of variable $k = n - I(\tau)$, the inclusion of $I(\tau)$ in the arguments of the three functions inside the sum of Eq. (13.14) has been factored out to become a phase term that can be discarded. Thus, any computation can be done using Eq. (13.15) and $A(\tau, f)$ is really a function of $|I(\tau)|$, the magnitude of $I(\tau)$. Now we are in a position to develop Eq. (13.15) for the special case of a chirp.

$$s(t) = e^{j(\pi W/T)t^2} \tag{13.16}$$

In Eq. (13.16), W is the swept bandwidth of the chirp and T is the total signal time duration. We replace the parameter W by N/T so that N is the time–bandwidth product of the signal and, for the case when T_s is the Nyquist sampling interval ($= 1/W$), we find $N = T/T_s = M$. In the following we shall derive the result keeping M and N as independent variables so that our answer will be valid for any sampling rate. From Eqs. (13.15) and (13.16), neglecting phase terms, we have

$$|A(\tau, f)| = \sum_{n=0}^{M-1-I(\tau)} e^{j[2\pi(W/T)\tau nT_s + 2\pi f nT_s]} \tag{13.17}$$

To normalize the variables τ and f, we introduce

$$\gamma = \frac{\tau}{T} \quad \text{and} \quad \nu = fT \tag{13.18}$$

and after some further manipulation we obtain

$$|A(\gamma, \nu)| = \frac{\sin\,[\pi(N\gamma + \nu)][1 - (|I(\gamma T)|/M)]}{\sin\,[(\pi/M)(N\gamma + \nu)]} \tag{13.19}$$

Equation (13.19) is our fundamental result. It is of interest to see what happens when γ or ν are zero. When $\nu = 0$, we find

$$|A(\gamma, 0)| = \frac{\sin\,(\pi N\gamma)[1 - (|I(\gamma T)|/M)]}{\sin\,[(\pi N/M)\gamma]} \tag{13.20}$$

When $\gamma = 0$, we find

$$|A(0, \nu)| = \frac{\sin\,(\pi\nu)[1 - (|I(0)|/M)]}{\sin\,(\pi\nu/M)} \tag{13.21}$$

The important point to note is that both Eqs. (13.20) and (13.21) look like sharp pulses (about either $\gamma = 0$ or $\nu = 0$) of about the same width, given that M is a reasonably large integer. The term $1 - (|I(\gamma T)|/M)$ has the effect of reducing the frequency of the side lobe ripples at large range offsets. Because the term $N\gamma + \nu$ appears as an entity in both arguments of Eq. (13.19), *there is no way to separate the effects of range and velocity offsets.* This is illustrated in Fig. 13.4, which shows two range cross sections of the ambiguity

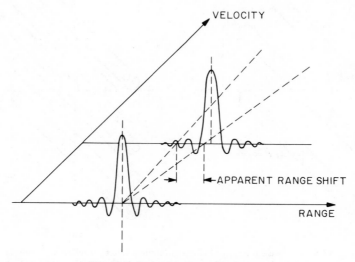

Fig. 13.4 Ambiguity function of chirp.

function at different Dopplers. Assume the chirp signal is used to track two targets in the same range bin traveling at different velocities; the result will be target returns apparently displaced in range.

Since the chirp may be of long duration, it illuminates a target with substantial energy, thus increasing range. In addition, since the matched filter response is always a sharp pulse, range resolution is also obtained. In this sense, the chirp has several good properties. One must not, however, lose sight of the fact that range–Doppler coupling makes it impossible to separate range from velocity measurements; thus the chirp is not a good signal for velocity measurements. Fortunately, in many practical problems, the range offset caused by Doppler is very small so that the signal return can also be used as a range measure.

Along the line $N\gamma + v = 0$ in the (γ, v) plane, Eq. (13.19) reduces to

$$|A(\gamma, v)| = \left(1 - \frac{|I(\gamma T)|}{M}\right)M \qquad (13.22)$$

Thus, in three dimensions one can imagine a ridge along the line $N\gamma + v = 0$ with a triangular decrease in the height of the ridge with increasing range offset. Due to the stepwise nature of $|I(\gamma T)|$, the top of the ridge actually decreases in discrete steps. Figure 13.5(a) is a sketch of Eq. (13.19), showing where $|A(\gamma, v)|$ (or $|A(\tau, f)|$) has substantial value (shaded region) and where

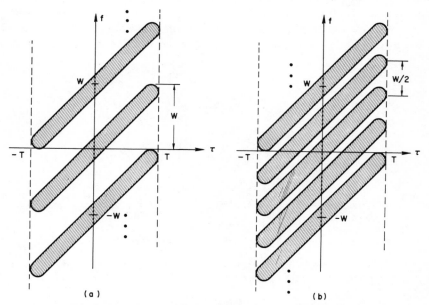

Fig. 13.5 (a) Dense portion of chirp ambiguity function when sampling at the Nyquist rate. (b) Dense portion of chirp ambiguity function when sampling at half the Nyquist rate.

it doesn't, for the case $M = N$ or $T_s = 1/W$, which corresponds to Nyquist sampling. Figure 13.5(b) shows what happens when the sampling rate is halved so that $T_s = 1/(2W)$; extra shaded regions due to aliasing appear. By comparing Eq. (13.7) with the well-known formula for the analog ambiguity function

$$a(\tau, f) = \int_{-\infty}^{\infty} s(t + \tau)s^*(t)e^{j2\pi ft}\, dt \qquad (13.23)$$

we can derive the relationship between $a(\tau, f)$ and $A(\tau, f)$. From Chapter 2 we know that the relationships between the (continuous) Fourier transform of an analog signal and the Fourier transform of the same analog signal are as follows; if

$$G_a(f) = \int_{-\infty}^{\infty} g(t)e^{-j2\pi ft}\, dt$$

and

$$G(e^{j2\pi f}) = \sum_{n=-\infty}^{\infty} g(nT_s)e^{-j2\pi fnT_s}$$

then

$$G(e^{j2\pi f}) = \sum_{n=-\infty}^{\infty} G_a\left(f + \frac{n}{T_s}\right) \qquad (13.24)$$

Inspection of Eqs. (13.7) and (13.23) shows that Eq. (13.23) is the Fourier transform of the "signal" $s(t + \tau)s^*(t)$ and that Eq. (13.7) is the Fourier transform of the sampled version; hence

$$A(\tau, f) = \sum_{n=-\infty}^{\infty} a\left(\tau, f + \frac{n}{T_s}\right) \qquad (13.25)$$

From the aliasing argument we conclude that the digital ambiguity function is periodic in frequency but not in time. Figure 13.6 shows plots of $|A(\tau, 0)|$ for $TW = 512 = M$. These plots were obtained by a program that passed the signal through a simulated matched filter. Notice that the side lobes appear very low in Fig. 13.6(a) but are more like the familiar -13 dB that characterizes an unweighted analog chirp in Fig. 13.6(b). This is simply an artifice of the display, which sampled $|A(\tau, 0)|$ every T_s seconds. For Fig. 13.6(a) these samples fall close to the zeros of the (continuous) ambiguity function ($\rho = 0$), whereas in Fig. 13.6(b), the points were shifted half a sampling interval ($\rho = 0.5$) to near the peak side lobe levels.

2. Ambiguity Function of a CW Pulse

From Eq. (13.19), if we set $W = 0$ so that $N = 0$, this corresponds to a CW pulse and we find

$$|A(\gamma, \nu)| = \frac{\sin(\pi\nu)[1 - (|I(\gamma T)|/M)]}{\sin(\pi\nu/M)} \qquad (13.26)$$

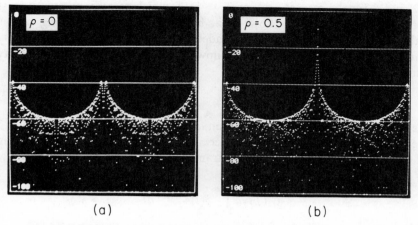

Fig. 13.6 (a) Chirp ambiguity function for zero doppler for $\rho = 0.0$. (b) Chirp ambiguity function for zero doppler for $\rho = 0.5$.

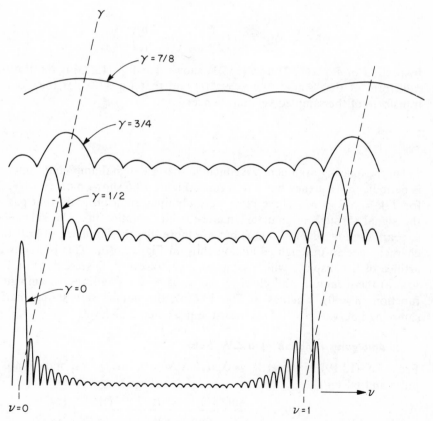

Fig. 13.7 Cross sections of the ambiguity function of a CW pulse.

Figure 13.7 shows a few cross sections versus v for fixed γ's. Notice that as the range offset increases, not only does the main lobe decrease in size but it widens so that velocity resolution is lost. If we were to take cross-section cuts versus γ for fixed v, poor results would be obtained. For $v = 0$, $|A(\gamma, 0)|$ has a triangular shape; for other values of v, the shape is sinusoidal with peaks occurring in rather arbitrary places.

3. Ambiguity Function of a Burst

While the chirp signal results in good signal detectability, it leaves unanswered the numerical values of velocity for a given target. Thus, although the chirp is a popular radar signal to help increase range, precise velocity measurements require additional radar signals.

As explained in Sec. 13.2, an increased unambiguous range window is obtained through transmitting at a low repetition rate, while an increased unambiguous velocity window requires increasing the pulse repetition rate. In either case the signal being transmitted is a succession of pulses, for which we now derive and discuss the ambiguity function. Since the burst is a pulse sequence, the analog and digital ambiguity exhibit the same basic features and there is no new property imparted to the signal by virtue of its digitization. For example, Fig. 13.8 is a sketch of the ambiguity function of a 10-pulse burst with uniform spacing. This is the well-known "bed of nails." The parameter Δ is the spacing between pulses. Figure 13.8 displays both range and velocity ambiguity. Since the periods in both τ and f are functions of only Δ, we see that increasing Δ lessens range ambiguity and increases velocity ambiguity and vice versa. Thus, a burst most usefully yields range

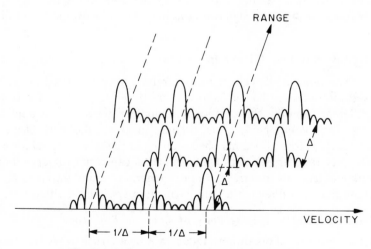

Fig. 13.8 Ambiguity function cross section of a burst.

and velocity measurements, provided there is already rough but reliable information on either range or velocity or both.

Exercise Derive the ambiguity function of a burst and see if you can justify Fig. 13.8 from the mathematical derivation.

4. Other Signals

In some instances it may be useful during a radar search to try to measure both range and velocity. This can be done by designing a signal with an ambiguity function that approximates a thumb tack, with a substantial peak of energy concentrated in a small section of the range–Doppler plane. Many such signals have been invented, such as up–down chirps, Barker and polyphase codes, and shift register codes. Little would be gained by essentially repeating the analyses that are available in contemporary books.

The reader might have noticed the similarity (at least philosophically) between the design of filters and the design of signals having desired ambiguity functions. There are several reasons why a formal signal synthesis procedure is quite difficult. First of all, one is faced with a criterion for approximating a two-dimensional ideal function, which is appreciably more difficult than the filter design counterpart of approximating either a (one-dimensional) spectrum or impulse response. Second, radar signal design is greatly influenced by many other radar design parameters, involving the antenna, transmitter, and receiver tracking computer, which are not necessarily under the control of the radar signal processor. Third, and perhaps most important, clutter environments are often very difficult to model and the "right" ambiguity function depends on the clutter properties. This probably means that a radar system will not be optimized until the radar has undergone much field testing and leads to the conclusion that the extra flexibility inherent in digital signal processing should be exploited as fully as possible.

13.4 Digital Matched Filters for Radar Signals

For any given range-angle cell, the signal return from a desired target may be masked by undesired background clutter. If the target velocity is sufficiently different from the clutter velocity, Doppler filtering can be applied to extract the signal component. Thus, broadly speaking, the work of a radar signal processor is to perform matched filtering of the return signal for every range–Doppler (and angle) cell of interest. This can result in very complex equipment and leads us to retreat from the general case and inquire as to the appropriate matched filter for a variety of more specialized situations.

I. Filter Matched to a Long Pulse of Constant Frequency

Consider the problem of obtaining a radar track on a satellite. We assume that the approximate angular position of the satellite relative to the radar is known

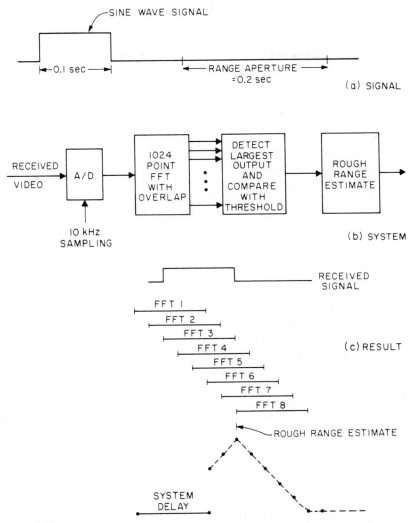

Fig. 13.9 System for detection and crude range estimation at very long ranges.

but both range and velocity are relatively unknown. We know that a long pulse results in poor range resolution but the major problem is detection because of the large distance and small size of the target. In this case it behooves us to construct a bank of filters, each filter being tuned to a different presumed Doppler frequency. Assume the pulse is on for 0.1 sec as shown in Fig. 13.9 and that the search interval encompasses 0.2 sec, which corresponds to an unambiguous range aperture of 18,600 mi. Let the maximum satellite velocity be 1000 mph and choose an S-band radar, with $\lambda \approx 10$ cm.

Then, according to Eq. (13.3), the maximum Doppler frequency is about 10,000 Hz. The velocity resolution is inversely proportional to the signal "on" time and is therefore 10 Hz. Hence, a bank of 1000 Doppler filters and a sampling rate of 10,000 Hz is called for.

We know from Chapter 6 that a sliding DFT is equivalent to a bank of filters. It is, in fact, true that the sliding, unweighted FFT corresponds exactly to a bank of "matched" filters since the impulse responses of the filters comprising the sliding FFT correspond exactly to the received signal. A sliding 1024-point FFT at a 10-kHz sampling rate is a more expensive item than is really needed, however, because each FFT filter is only 10-Hz wide and need not be sampled at 10 kHz. If we replace the sliding FFT by a hopping FFT, performing the FFT at, say, 50 times per second, this is equivalent to sampling the sliding FFT output 50 times per second. This results in a 200:1 reduction in required computation speed. As shown in Fig. 13.9(c), the detected returns from the successive hopping FFT's can be utilized to yield a rough range marking.

Exercise How many butterflies per second are needed to perform the 1024-point FFT 50 times per second? Is it reasonable to use a minicomputer as the FFT processor? Consider both computation capability and memory size in your response.

2. Matched Filters for General Signals

Since the "right" radar signal is often a function of the environment (which can be time-varying), it is useful to contemplate the design of digital matched filters with arbitrary impulse response and inquire as to how to design and implement them. We can postulate two different implementations of such a filter; one is a straightforward direct form FIR filter and the other an FFT realization of the same filter. In the first case, flexibility is attained by changing the impulse response, which means changing the filter coefficients. In the second case, the DFT of this impulse response is made flexible. The relative desirability of these two methods depends primarily on the time–bandwidth product TW; the larger TW, the more one should favor the FFT. Time–bandwidth product is an extremely important concept for matched filtering. Bandwidth determines the best range resolution possible with a given signal, and time determines the best possible velocity resolution. Furthermore, in digital systems the total "on" time T of a signal is equal to NT_s, where T_s is the sampling interval and N is the total number of signal samples. But, in a digital system that samples the received analog signal at the Nyquist rate, T_s is precisely $1/W$. Thus,

$$N = TW \tag{13.27}$$

that is, the time–bandwidth product is equal to the number of signal samples. Since, in general, a filter matched to a given signal has the same number of samples in the impulse response as does the signal, N also determines the

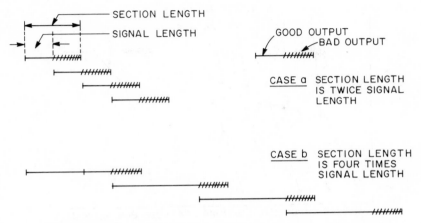

Fig. 13.10 Comparison of information thrown away for two different section lengths, with the same signal lengths.

complexity of the digital matched filter. As we shall see in the later examples, time–bandwidth products vary over a very wide range; yet in most instances the FFT version of digital matched filtering seems most appropriate.

In any given radar application, time–bandwidth product is determined by radar requirements; hence the number of samples in the matched filter impulse response, being equal to the number of signal samples, is a design condition more or less imposed by system considerations. The section size (and thus the FFT size) is an additional parameter, however, that can be chosen to minimize the hardware cost. Two possibilities are shown in Fig. 13.10; in case (a) the section length is twice the signal length (and also twice the impulse length), whereas in case (b) the section length is made four times as big as the signal and impulse length. The advantage of case (a) is the use of a smaller FFT size, while the advantage of case (b) is the throwing away of fewer points. This tradeoff efficiency can be given by the formula

$$E = \left(\frac{L_s - L_h}{L_s}\right)\left(\frac{1 + \log L_h}{\log L_s}\right) \qquad L_s \geq 2L_h \qquad (13.28)$$

where L_h is the impulse response length (number of samples in the impulse response) and L_s is the section length. We are assuming equal signal and impulse response lengths so that the smallest section length must be at least twice the impulse response length. The first term in parentheses in Eq. (13.28) expresses the increased efficiency (in terms of the fraction of good samples obtained from the evaluation) and the second term in parentheses expresses the decreased efficiency resulting from the logarithmic increase in multiplication power needed to perform a larger FFT. E is therefore efficiency of computation and can be thought of as the multiplication power per processed

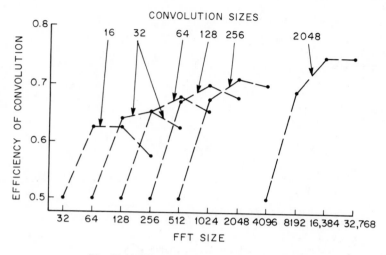

Fig. 13.11 Efficiency of convolution versus FFT size.

datum. As a norm, we take $L_s = 2L_h$, which gives $E = \frac{1}{2}$ for $L_H \gg 1$. We can now plot E versus L_s with L_h as a parameter in Fig. 13.11. We see that improvements in multiplication efficiency are possible with larger sections but that the curves are not monotonic; i.e., there is a "best" section length for each value of L_h and, in fact, this best length, for the cases we have shown, turns out to be $L_s = 8L_h$. Keep in mind, however, that only the multiplication cost has been computed; for example, the memory size increases linearly with FFT size so that the gain in multiplication efficiency may be more than offset by the increased memory hardware. As a practical matter, in most cases there is little justification for deviating from the case $L_s = 2L_h$.

3. Weighting to Reduce Matched Filter Side Lobes

The matched filter for a typical radar signal such as a chirp is designed to yield the greatest signal-to-noise ratio at a single instant of time; this usually means that the matched filter output will look like a sharp pulse. For most systems this desirable main lobe of the filtered signal is accompanied by side lobes of fairly high amplitude. This can be distressing when the radar is processing signals from several targets of differing cross section because the main lobe of the smaller target can be masked by a side lobe of the large target.

At the cost of both range resolution and signal-to-noise ratio, the side lobes can be reduced by windowing or weighting. This can be done in the time domain by passing the matched filter output through another filter or in the frequency domain by appropriate spectral weighting sandwiched between the forward and inverse FFT's. In fact, referring to Fig. 13.2, weighting can be accomplished by appropriate design of the analog filter prior to the A/D

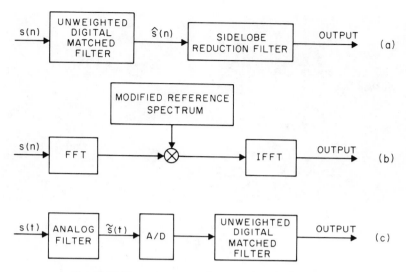

Fig. 13.12 Three ways to reduce matched filter synthesis.

converter. Figure 13.12 shows the three different ways that weighting can be inserted. In the case of Fig. 13.12(a) this means quite a bit of extra computation if r is made too large where r is the filter length in samples. For this reason, Hamming and hanning windows are very useful since they correspond to $r = 3$ and still reduce the side lobe levels to more than 40 dB below the main lobe. In the case of Fig. 13.12(b) the situation is a bit more subtle. Since the output signal is always $2L_h$ long (where L_h is both signal and impulse response length), any spectral weighting equivalent to an r-point FIR filter must be taken away from the matched filter duration. As an example, let the signal consist of 256 samples; then we do 512-point forward and inverse FFT's. With or without spectral weighting, the number of points generated by the inverse FFT is 512 but since r of these samples are associated with weighting, the effective time–bandwidth product shrinks slightly but this loss is minor compared to mismatch loss. In the case of Fig. 13.12(c), we note that if the A/D converter is ignored, we have a linear system and the weighting filter can be the presampling analog filter. The danger in this form of weighting is the fact that this analog filter serves another role, namely, that of removing noise mixed with the signal that could be aliased in with the signal due to sampling. At the present writing, no comprehensive study of this last arrangement [Fig. 13.12(c)] has been undertaken.

Figure 13.13 shows the ambiguity function of a chirp for zero Doppler when Hamming weighting is used for $\rho = 0$ and $\rho = 0.5$. The two large responses on either side of the peak are clear evidence of a broadening of the main response. The reduction of the near-in side lobe level to -40 dB is also seen.

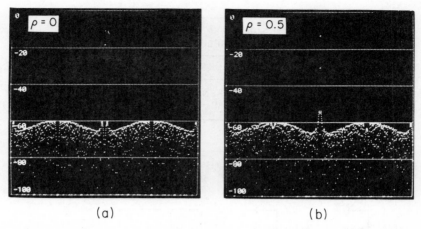

Fig. 13.13 Weighted pulse compression for a chirp for $\rho = 0$ and $\rho = 0.5$.

Other crucial design questions arise in digital matched filters for radar signals but these depend greatly on the particular problem so further discussion will be deferred until the examples.

4. Matched Filter for a Burst

When trying first to detect a target at long range and then track the target as the range decreases, a standard procedure is first to transmit a chirp followed by a burst. The chirp is useful for detection but, as pointed out in Sec. 13.3, one cannot distinguish between range and velocity changes. To measure both velocity and range unambiguously, the radar signal generator will send out a burst. This signal may consist of a succession of short pulses or, to increase signal strength, it may consist of a succession of subpulses each of which is a chirp. In the latter case, first passing the signal through a filter matched to these subpulses produces a succession of very narrow pulses with greater peak amplitude. Thus, in either case we are interested in the filter or filters matched to presumed target velocity or velocities for a burst of sharp pulses.

Figure 13.14 shows one method of matched filtering that corresponds to a sliding FFT on N signal samples separated in time by the interpulse interval. Assuming that one of the N FFT outputs is tuned to the correct Doppler, we would expect that output to be a pulse train with a triangular envelope that corresponds to the matched filter output of a pulse train with rectangular envelope.

This type of burst processing is inherently wasteful because K samples, later $N - 1$ of the original N outputs at the delay line taps, will be reprocessed, where K is the interpulse spacing. Assume that the sampling interval is T_s; then every T_s seconds a complete N-point FFT must be performed. For

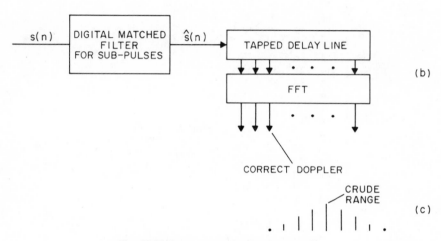

Fig. 13.14 Burst processing by sliding FFT.

example, in a 10-megasample system processing 16 Doppler channels, a 16-point FFT must be performed every 100 nsec.

The efficiency shortcoming of the scheme above can be overcome through the use of a permute memory structure and a bank of digital filters, as shown in Fig. 13.15; that is one of the rare cases for which a digital filter system is

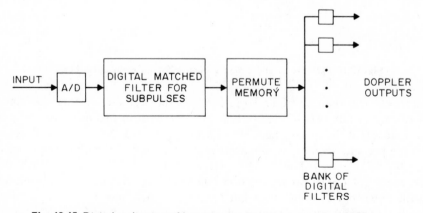

Fig. 13.15 Digital realization of burst processor using a permute memory.

0	0	0	0
1	5	25	8
2	10	11	16
3	15	36	24
4	20	22	32
5	25	8	1
6	30	33	9
7	35	19	17
8	1	5	25
9	6	30	33
10	11	16	2
11	16	2	10
12	21	27	18
13	26	13	26
14	31	38	34
15	36	24	3
16	2	10	11
17	7	35	19
18	12	21	27
19	17	7	35
20	22	32	4
21	27	18	12
22	32	4	20
23	37	29	28
24	3	15	36
25	8	1	5
26	13	26	13
27	18	12	21
28	23	37	29
29	28	23	37
30	33	9	6
31	38	34	14
32	4	20	22
33	9	6	30
34	14	31	38
35	19	17	7
36	24	3	15
37	29	28	23
38	34	14	31
39	39	39	39

SEQUENCES REPEAT
→

Fig. 13.16 Example of addressing sequence for an eight-pulse burst and five range bins.

more efficient than an FFT. To understand the permute memory, imagine a sequence of numbers labeled $0, 1, 2, 3 \ldots$ as input to this memory. Assume as before that the spacing between adjacent pulses is K samples. The function of the permute memory is to rearrange the ordering of the input sequence to be $0, K, 2K, \ldots, 1, K+1, 2K+1, \ldots, 2, K+2, 2K+2, \ldots$. With this reordering, Doppler processing now takes place on each range bin, beginning with range bin 0, then range bin 1, etc. For each new input sample, if we wish to process all N Doppler channels, we have to perform N complex multiplications, compared to $(N/2) \log_2 N$ complex multiplications using the scheme of Fig. 13.14. The relationships between FFT filters and recursive realizations of such filters have been analyzed in Chapter 6. A further saving in the filter bank implementation is possible if not all N Doppler channels are needed; in the FFT implementation, all channels are computed whether needed or not.

Implementation of the permute memory is accomplished using a random access memory and a special addressing algorithm. By means of this algorithm, real-time permutation of the data can be accomplished with no extra memory beyond that needed to hold the original data base. The algorithm is illustrated in Fig. 13.16 for a system where five range bins are to be processed for an eight-pulse burst. Each column represents the permuting of all 40 data samples. For the first 40 inputs, the memory addressing is sequential, 0, 1, 2, 3, etc. The Doppler filter bank must receive samples in the order 0, 5, 10, . . . , 1, 6, 11, etc., however, and this can be done by using the address sequence in the second column. To prevent extra buffering, each new input datum must replace the most recent output datum. Thus, for each new set of 40 data points, entry and exit from the memory follows a new addressing pattern, as shown in the succeeding columns.

Exercise   Write a formula defining the addressing pattern of Fig. 13.15 for the general case of L input samples and an interpulse spacing of K samples.

13.5   Airborne Surveillance Radar for Air Traffic Control—Doppler Processing to Combat Clutter Problems

In present air traffic control radars, the antenna rotates mechanically, sweeping out a full 360° every 4 to 12 sec; azimuth beam resolution is about 1° to 2° and the vertical antenna pattern is a fan beam, usually having a 30° to 45° width. Thus, as the antenna sweeps by a target aircraft, there will be a succession of "hits" as seen in Fig. 13.17; i.e., radar returns at the pulse

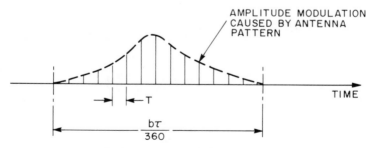

Fig. 13.17 Radar returns from a target at a given range as antenna sweeps by.

repetition period after which no information is obtained from that particular target until the antenna has made a complete revolution. Based on this information, the system must track up to perhaps 50 aircraft within its field of view (typically 25 to 40 mi) and display these tracks in a useful manner to the air traffic controller who can then correlate this information with his knowledge of the traffic schedule, the planned flight paths of the various commercial and private aircraft, and his audio communication links with the aircraft and other control towers.

Present airport surveillance radars operate at S band ($\lambda \approx 10$ cm). At this wavelength, weather clutter, caused by radar reflections from rain, can be quite troublesome but perhaps the most important disturbing effect is ground clutter, which is picked up by the antenna. Such effects can be alleviated through better signal processing, in particular, by Doppler filtering that permits the relatively fast aircraft targets to be discriminated from the ground and weather returns. It is worth noting that ground clutter, which we would normally think of as having only a dc spectral component, actually has a spectral spread induced by the antenna motion; this makes Doppler processing less effective and could be avoided by other antenna designs (such as a phased array), but antenna redesign for so many existing systems is an expensive proposition. Lincoln Laboratory had the task of demonstrating signal processing techniques which, hopefully, could lead to relatively inexpensive additions to existing systems which would greatly alleviate some of the problems above.

A sketch of the experimental setup is shown in Fig. 13.18. The antenna and associated radar equipment was about $\frac{1}{2}$ mi away from the FDP facility

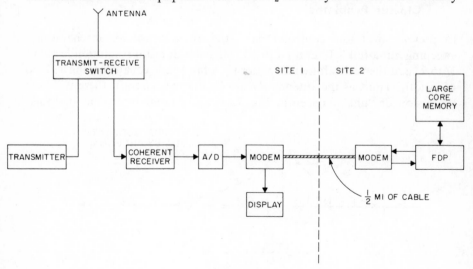

Fig. 13.18 Experimental setup for ASR signal processing.

described in Chapter 11. Two communication links were used; one to relay the received radar signal to the FDP for signal processing and threshold detection and the other to relay back the processed information to the radar display. The radar used was a coherent S-band radar with a $1°$ fan beam antenna having a width of 5.25 m, a rotation rate of 1.36 rad/sec and a wavelength of 10.7 cm. The pulse repetition frequency (PRF) was 1000 pulses per second and these pulses were about 1 μsec wide, corresponding to a 1-MHz bandwidth. Given the antenna beam width and PRF, the number of hits as the antenna scanned by the target was 15. This corresponds to the burst signal described in Sec. 13.3; for any range bin in the range aperture, the problem would be to design a matched filter to the transmitted signal, taking into account some reasonable model of the clutter. Before doing this, however, let us make use of the system parameters to obtain an idea of the range–velocity ambiguity question.

First, the repetition period of 1 msec is needed to give the required unambiguous range coverage of 60 mi. For an S-band radar, the range of Doppler frequencies, corresponding to a velocity range of $\frac{1}{3}$km/sec is about 3300 Hz. *But*, since the PRF is 1000 Hz, the Doppler spectrum will be periodic with 1000-Hz period so that, for example, the system will not be able to distinguish among, say, a 400-Hz, 1400-Hz, or 2400-Hz Doppler. Furthermore, since the Doppler frequency of ground clutter is close to dc, the 1000-Hz sampling will introduce large clutter components at 1000, 2000, 3000, etc., Hz; these frequencies correspond to "blind speeds" at which airplanes will be lost in the clutter. Thus, an important aspect of a complete signal processing system is *disambiguation;* one way to accomplish this is to transmit at two different PRF's. The scheme devised for the present experiment was to transmit eight pulses at one PRF, switch to another PRF for the next eight pulses, and then switch back again, etc. This will allow a target whose radial velocity corresponds to a blind speed at one PRF to become detectable at a different PRF. Thus, the matched filter for our system becomes an eight-pulse processor. Hofstetter and Delong have studied problems of this sort of assumed clutter models and have derived formulas for the optimum linear processor for any given Doppler frequency. Clearly, a bank of such processors is needed for many Doppler frequencies since there is no a priori way of knowing the target's radial velocity. It is not our intention here to go into the details of optimum processing;[4] from a computational point of view, what needs to be done is, for each range bin, to perform a weighted sum of eight signal returns. (Both weights and signal are assumed to be complex numbers.) If N Doppler bins are to be examined, then $8N$ multiplications

---

[4] With no clutter and just a white-noise background, an optimum processor would simply be a filter matched to the return signal in the sense described in Sec. 13.4. With clutter, the optimum processor is, in effect, a filter that "prewhitens" the clutter and is then matched to the modified signal.

must be performed; for real-time operation, since each range gate is 1 μsec wide (the width of the transmitted pulse), these multiplications must be performed in an 8-μsec interval. For example, if $N = 8$, 64 complex multiplications must be performed every 8 μsec or 1 complex multiplication every 125 nsec. The question naturally arises as to how the algorithm can be altered to maintain close to optimum performance while reducing the computation load. Intuition says that a filter that greatly attenuates the large dc clutter component followed by a bank of filters tuned to the various Dopplers of interest should yield good results. Since an FFT resembles a filter bank and since FFT's tend to be computationally efficient, a suboptimal signal processor using a three-pulse cancellor,[5] followed by an FFT was designed and simulated and compared with a simulated optimum processor. Figures 13.19,

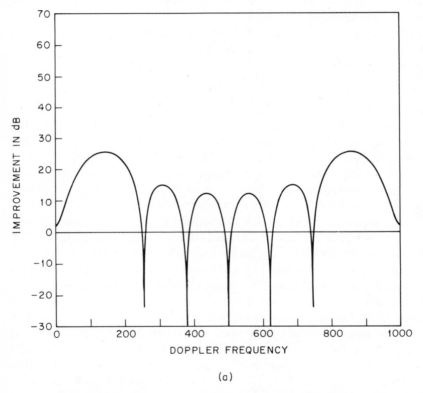

(a)

Fig. 13.19 (a) SIR improvement of optimal eight-pulse processor optimized at 0 Hz.

[5] In radar jargon, a three-pulse cancellor is an FIR third-order digital filter with impulse response coefficients $-\frac{1}{2}$, 1, $-\frac{1}{2}$. Similarly, a two-pulse cancellor is a second-order FIR filter with coefficients $+1$, -1. Both filters have a zero at $z = 1$ (dc); hence ground clutter tends to be cancelled—i.e., greatly attenuated.

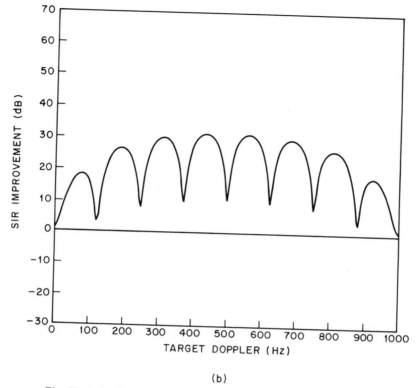

(b)

Fig. 13.19 (b) SIR improvement of suboptimal processor three-pulse cancellor and unweighted DFT filter tuned to 0 Hz.

13.20, and 13.21 show comparisons of the two types of processors for 0, 125, and 500 Hz, respectively. We have plotted, in all cases, signal-to-interference (SIR) ratio improvement due to the filtering as a function of the Doppler, in each case optimizing for a new Doppler. For 0-Hz optimization, we note that the best improvement for both cases is not at 0 Hz. The peak improvement for the optimum processor occurs at about 130 Hz and is about 26 dB, whereas the peak improvement for the suboptimum processor is at about 80 Hz but is only about 18 dB. For the 500-Hz case, both curves peak at 500 Hz. It appears that the suboptimal peak is actually bigger than the optimal but this is misleading because to perform three-point clutter cancellation filtering followed by an eight-point FFT really requires 10 input samples, whereas the optimum processor uses 8 samples. To be fair we should have computed a 10-pulse optimum processor. Despite this, however, the consensus seemed to be that the suboptimal processor was close enough so that it would be worth implementing provided that it led to an appreciably simpler piece of hardware. From Chapter 11,

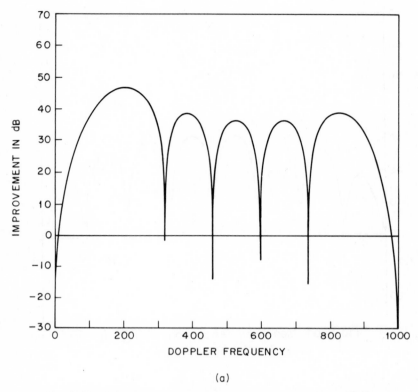

(a)

Fig. 13.20 (a) SIR improvement of optimal eight-pulse processor optimized at 125 Hz.

it was shown that 2 multiplications are needed to perform an eight-point FFT, compared to 64 for the optimal processor. Without delving into the details of the two designs it can be argued that the clutter cancellation filter plus eight-point DFT are the correct filters to build. Actually, benefits result from weighting of the signal (after passing through the clutter cancellation filter) prior to the FFT. This adds 4 complex multiplications; the benefits are reduction in the side lobes of the curves of Figs. 13.19 to 13.21. Figure 13.22 shows a complete suboptimal filter block diagram, consisting of a third-order FIR digital filter for pulse cancellation, weighting, and an eight-point FFT.

Given ground clutter suppression, we are still faced with the very severe problem of signal detection. Remember that Figs. 13.19 to 13.21 represent an improvement factor; since the intensity of the clutter varies greatly with the terrain, the curves of these figures will be raised or lowered as a function of the range-angle sector. In order to obtain reasonably constant false alarm probabilities for all the illuminated space, we need a "clutter map" as a reference, that is, an averaged clutter intensity for each range-angle sector.

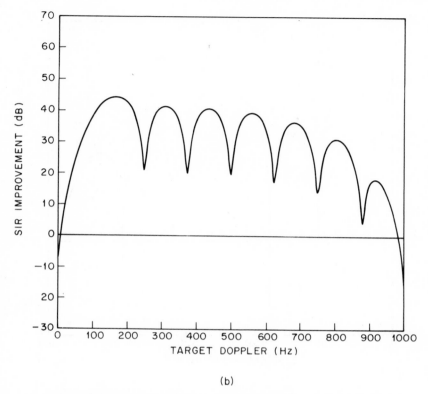

(b)

Fig. 13.20 (b) SIR improvement of suboptimized processor three-pulse cancellor and unweighted DFT filter tuned to 125 Hz.

This is accomplished by means of a scan-to-scan averaging of the dc component of the signal. Letting x_j be that component due to the jth scan, we can then prescribe the clutter map intensity to be

$$y_j = (y_{j-1} - x_j)\alpha + x_i \qquad (13.29)$$

For any range-angle sector, threshold detection is accomplished by comparing the largest output of the FFT with y_j. α will determine the length of time to build up a reliable clutter map indication.

We now turn our attention to some details of the FDP simulation of the signal processing and clutter mapping algorithms. To begin with, the FDP is not a fast enough processor to process all information coming to the antenna in a 60-mi radius. This means that, in order to approximate real-time simulation, the data from a given section of space must be buffered and then processed while the information from the next scan is being accumulated; the situation is shown in Fig. 13.23. The FDP facility proved to be useful for this processing for two reasons. First, its high speed meant that a large enough sector could be examined so that a useful number of targets are displayed;

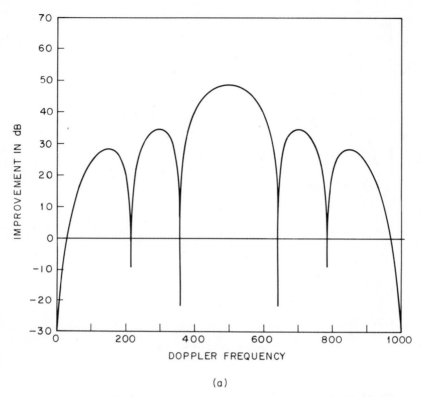

(a)

Fig. 13.21 (a) SIR improvement of optimal eight-pulse processor optimized at 500 Hz.

second, the large core memory connected as input–output device to the FDP (160,000 18-bit registers) allowed the buffering of raw video from a reasonably large sector. To determine the buffer storage requirements for the full 60 mi and 360°, assume 600 range bins are needed for 60 mi (at 1 μsec per range bin and 10 μsec per mi); this means 600 numbers must be stored in 1 msec or 600,000 numbers per second and 2,700,000 numbers for a full 4.5-sec scan. Given the buffer memory size roundoff to 150,000, this means that at most one-eighteenth of the total space can be processed in real time (with a one-scan delay before display). Thus, for example, the shaded sector could be 10 mi by 120° or 30 mi by 40°. It turns out that the buffer size is the real limitation in our case since the FDP can do real-time processing of sectors two or three times these sizes. For display purposes the sector was rectangularized. Figures 13.24 and 13.25 show some typical results. The aircraft tracks are clearly visible through ground clutter and the false alarms are quite few. The photograph taken was a long exposure lasting over perhaps 20 scans or well over a minute. Preliminary results thus far encourage us to believe

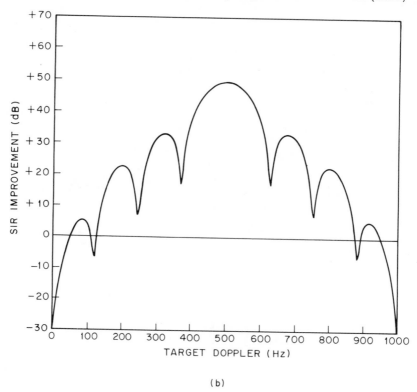

(b)

Fig. 13.21 (b) SIR improvement of suboptimal processor three-pulse cancellor and unweighted DFT filter tuned to 500 Hz.

that good Doppler processing and clutter mapping using special-purpose digital signal processing hardware could be a useful addition to ASR (Airport Surveillance Radar) systems.

13.6 Long-Range Demonstration Radar (LRDR)

In 1967 to 1971, a radar system using the FDP as a real-time simulator was built at the Lincoln Laboratory. Its purpose was the detection of moving objects in large amounts of ground clutter background. The clutter spectrum consisted of a very large dc return, plus an ac component caused by foliage motion. The moving object, which could be either a vehicle, such as a car or a truck or a taxiing airplane, or else an animal, such as a crocodile or a human lost in the woods or an enemy soldier, returned a signal level about 80 dB below the steady-state clutter and 60 dB below the fluctuating background caused by foliage motion. The radar antenna was a phased array UHF antenna with a 1.5° horizontal beam width and a vertical fan beam. This experimental antenna consisted of several thousand array elements

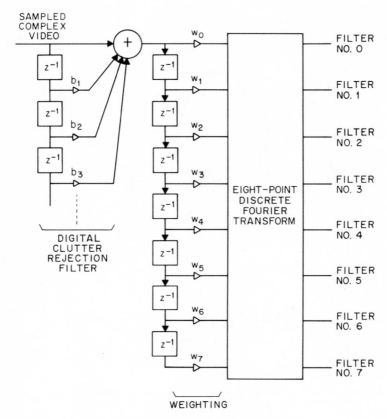

Fig. 13.22 Suboptimal processor.

arranged on half a cylinder, which permitted illumination of a 45° sector; a full cylindrical array would of course cover a full 360° sector.

Because of the very high clutter-to-signal ratio, a variety of signal processing tricks were necessary to obtain good signal detection. Included in these tricks were pulse compression using complementary coding, presumming, Doppler processing, and post detection integration; we shall try to describe the ideas and implementation of each of these techniques briefly. First, however, we present a diagram of the complete experimental setup. Figure 13.26 illustrates the system and the extent to which digitization has taken over. Prior to the A/D converter, the received radar signals are amplified and heterodyned. Sensitivity time control (STC), which is a range-dependent gain control, takes place in a digitally controlled stepped attenuator.[6] Demodulation is performed in quadrature so that the video return is a

---

[6] All radars need STC because the signal return is inversely proportional to the fourth power of range, which would lead to overwhelming dynamic range problems without STC.

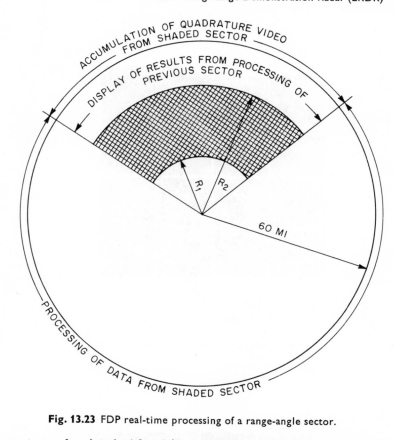

Fig. 13.23 FDP real-time processing of a range-angle sector.

coherent complex signal. After A/D conversion at 10 megasamples and seven bits, the video is pulse compressed, details of which we defer for now. This radar is assumed to survey 2048 range bins for each beam position; since there are 30 beam widths in 45°, a total of $2048 \times 30 = 61{,}440$ range-azimuth bins are illuminated.

Let us try, in a heuristic way, to show how the parameters of this particular radar were chosen. Since one of the main purposes of the project was to study the possibility of detecting moving targets in foliage, the wavelength had to be large enough to penetrate foliage;[7] this led to a choice of UHF (a wavelength of about 5 m). Targets of interest moved as slowly as several tenths of a meter per second; using Eq. (13.3) leads to a frequency resolution requirement of about 0.5 Hz. This meant that the total integration time before target detection was about 2 sec; the interesting design questions revolve around the signal processing techniques used during this 2-sec epoch. If we assume that

---

[7] As wavelength decreases, radar penetrability decreases. At S band, signal-to-clutter ratio would have been too low. The disadvantage of a long wavelength is the need for a larger antenna.

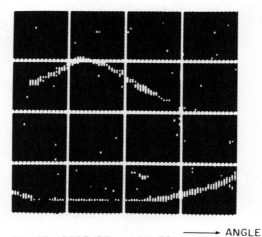

RANGE APERTURE ≈ 4 MILES
ANGULAR COVERAGE = 26°

Fig. 13.24 Airplane tracks for ASR experiment.

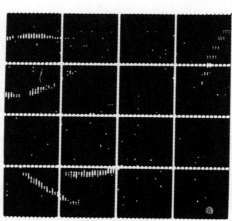

RANGE APERTURE ≈ 4 MILES
ANGULAR COVERAGE = 26°

Fig. 13.25 More airplane tracks for ASR experiment.

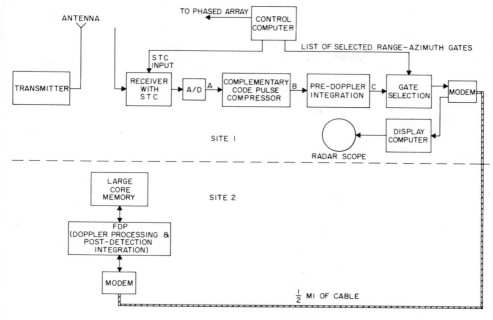

Fig. 13.26 Block diagram of LRDR.

the maximum Doppler we desire to detect is about 32 Hz, then we need 64 velocity bins. This does not necessarily mean that only 32 transmissions are sent per second, however, but rather that the information in the received signal is collected and put in a "bundle" 32 times per second and 64 of these bundles are eventually velocity filtered. These Doppler requirements determine the lowest permissible repetition period. The highest permissible repetition period is determined by range ambiguity. The largest unambiguous range was assumed to be 22.4 nmi (nautical miles), which led to a radar PRF of 3600 per second. Thus, approximately 113 radar return signals at the PRF rate can be combined to create a single bundle. The bandwidth of 10 MHz was chosen to give the required 50-ft range resolution.

As seen in Fig. 13.26, the succession of signal processing devices also includes a range-azimuth gate selector, which was necessitated by the inability of the FDP to process the entire field of view in the basic 2-sec coherent time interval. The FDP was able to process 2048 bins in that time; thus the function of the gate selector was the selection of 2048 out of the possible 61,440 bins for each 2-sec interval.

Following gate selection the partially processed radar data is sent, via a modem and appropriate digital interfaces, to the large core memory that resides next to the FDP. The selected data enters the memory sequentially with respect to range bins; the memory addressing algorithm permutes the

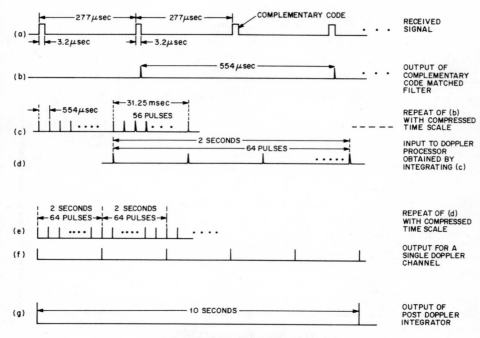

Fig. 13.27 Time epochs in LRDR.

data so that 64 successive inputs from a single range bin are sequentially entered into the FDP. This reordered data can now be processed by the FDP that performs FFT's for Doppler discrimination. Then the FDP in conjunction with the large memory perform postdetection integration on each Doppler bin by adding five successive FFT magnitudes at that Doppler frequency. Following this, various statistical decision algorithms (not to be discussed here) are implemented by FDP programs, leading to detection of targets. Finally, this processed information is sent back to the radar site, entering a g.p. computer for formating and display purposes. The other g.p. computer at the radar site acts primarily as an input–output processor, controlling the antenna beam, the STC, the radar timing, and the selection pattern for use by the gate selection processor. Figure 13.27 summarizes the various time epochs in the system.

Having briefly described the sequence of major blocks we now discuss some details of A/D conversion and the implementation of the pulse compressor algorithm.

I. A/D Conversion

In the presence of large ground clutter, a minute target return rides within a huge clutter signal. Eventually, the target detection will depend on the Doppler discrimination properties of the radar signal processor; meanwhile there is a concern lest the minuscule signal gets wiped out by nonlinear and noise effects of quantization beginning with the analog-to-digital converter. It is really not necessary for the target return strength to be as big as a complete quantum step. As long as the clutter fluctuates and does not "get stuck" between two quantum levels at a precise value that would wipe out the target, one can expect that target presence will not be wiped out. For example, in the present radar, seven bits are used and the target is presumed to be about 12 bits below clutter level, yet enough signal strength is present to make it useful to perform further processing to raise signal-to-noise ratio.

2. Pulse Compressor

The transmitted signal consists of a 32-bit code and its complement sent out at alternate repetition intervals. As we shall show, the matched filter response to this pair of signals is a sharp pulse with *no* side lobes for zero Doppler. For this particular radar, if the requirement is imposed that only very slowly moving targets be detected, the theoretical result is quite valid for a noiseless signal.

To define complementary codes we begin with a very simple example; let $s_1(n)$ be the sequence $+1$, $+1$ and $s_2(n)$ be the sequence $+1$, -1. The matched filters at zero Dopplers will yield the correlation sequences; for $s_1(n)$ the correlation sequence is $+1$, $+2$, $+1$, while for $s_2(n)$ we find -1, $+2$, -1. Adding the two matched filter outputs yields the final result 0, 4, 0. Larger codes can be generated from $s_1(n)$ and $s_2(n)$ by the following algorithm.

1. Define a new signal $s_3(n)$ as $s_1(n)$ followed by $s_2(n)$, namely, $+1$, $+1$, $+1$, -1. Define another new signal $s_4(n)$ as $s_1(n)$ followed by $\bar{s}_2(n)$ [$s_2(n)$ with all signs reversed], or $+1$, $+1$, -1, $+1$. These two new sequences form a complementary code pair of length 4 and have the matched filter outputs -1, 0, 1, 4, 1, 0, -1 and 1, 0, -1, 4, -1, 0, 1, which when added yield the side-lobeless sequence 0, 0, 0, 8, 0, 0, 0.
2. The procedure above can be iterated to double the length of succeeding complementary codes. Thus, the pair would be $+1$, $+1$, $+1$, -1, $+1$, $+1$, -1, $+1$ and $+1$, $+1$, $+1$, -1, -1, -1, -1, $+1$, -1. Figure 13.28 lists the complementary code pairs through length 32. The LRDR transmits a pair of length 32 (or 3.2 μsec, since the bandwidth is 10 MHz and the subsequent sampling rate is 10 megasamples).

Exercise How would you compute the matched filter output for one of the code pairs if the effect of the Doppler were included in the calculations?

Implementation of the matched filter was a digital FIR filter with the interesting property that all the 32 multiplications per iteration were by $+1$ or -1. If not for this, 32 multiplications per 100 nsec would have been necessary. Instead, 32 additions per 100 nsec are required, a much less formidable task although certainly nontrivial. The implementation algorithm

$$
\left.\begin{array}{l}
+\ + \\
+\ -
\end{array}\right\}2
$$

$$
\left.\begin{array}{l}
+\ +\ +\ - \\
+\ +\ -\ +
\end{array}\right\}4
$$

$$
\left.\begin{array}{l}
+\ +\ +\ -\ +\ +\ -\ + \\
+\ +\ +\ -\ -\ -\ +\ -
\end{array}\right\}8
$$

$$
\left.\begin{array}{l}
+\ +\ +\ -\ +\ +\ -\ +\ +\ +\ +\ -\ -\ -\ +\ + \\
+\ +\ +\ -\ +\ +\ -\ +\ -\ -\ -\ +\ +\ +\ -\ +
\end{array}\right\}16
$$

$$
\left.\begin{array}{l}
+\ +\ +\ -\ +\ +\ -\ +\ +\ +\ +\ -\ -\ -\ +\ -\ +\ +\ +\ -\ +\ +\ -\ +\ -\ -\ -\ +\ +\ +\ -\ + \\
+\ +\ +\ -\ +\ +\ -\ +\ +\ +\ +\ -\ -\ -\ +\ -\ -\ -\ -\ +\ -\ -\ +\ -\ +\ +\ +\ -\ -\ -\ +\ -
\end{array}\right\}32
$$

Fig. 13.28 List of complementary code pairs from 2 to 32.

is illustrated in Fig. 13.29 for a code of length 4. The hardware was implemented with 32 hardware adders operating in parallel, using TTL adder logic and TTL shift register memory for the delays. Thus, one clock cycle corresponded to a complete iteration of the digital matched filter.

13.7 Digital Matched Filter for a High Performance Radar

The specification for this example included a phased array antenna with scheduling activities controlled by a large g.p. computer and a digital signal processor capable of handling bandwidths somewhat higher than 10 MHz with time–bandwidth (TW) products of about 2000. The signals to be handled included chirps, bursts with chirp subpulses, and nonuniform bursts. The part of this large system that we shall treat is the digital matched filter.

First of all, a digital filter matched to a 10-MHz signal with $TW = 2000$ must be a two-thousandth-order FIR filter. If one attempted to implement

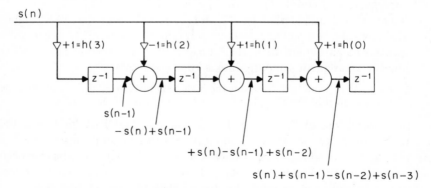

Fig. 13.29 Filter matched to one of a length 4 complementary code pair.

such a filter by non-FFT methods, then 2000 complex multiplications would have to be performed every 100 nsec. A system with 8000 100-nsec multipliers has a very good chance of never working (with present component technology). Use of the FFT cuts this requirement considerably. One would have to perform a 4096-point forward FFT followed by multiplication of the output by a reference spectrum (that of the matched filter) followed by a 4096-point inverse FFT. Half of the processed result would be the correct matched filter output, while the other half would be worthless results of circularly convolved information. Thus, the total number of multiplications for a radix 2 FFT's to process 2048 points is $(N/2) \log_2 N + N + (N/2) \log_2 N = N(1 + \log_2 N) = 4096 \times 13$ or a total of 26 complex multiplications every 100 nsec. This is nearly two orders of magnitude reduction compared to non-FFT convolution. It also fits very neatly into the pipeline architecture described in Chapter 10. In fact, after some thought, it became clear that the radar specifications demanded a pipeline convolver. The next step was to determine the most suitable radix number. Only the radix 2 and radix 4 cases were considered. For radix 4, we have six stages in the forward FFT and six stages in the inverse FFT. Each stage consists of three complex multipliers, for a total of 36 plus 1 for the reference spectrum, making 37. This is greater than the 26 attributed to radix 2 but, for the same clock rate, the radix 4 can process at twice the data rate of the radix 2 since it has four parallel paths compared to two. This advantage can be used either by designing a more powerful system or by slowing the clock rate so that the multiplication speed requirements are eased. For reasons that are fairly complicated but, it seems, quite valid, the radix 4 structure was chosen. Since the speed requirements of this matched filter are so severe and since it was hoped that this radar would lead the way to even more powerful radars, we looked for the fastest logic family with sufficient versatility and this turned out to be 2-nsec ECL. Important design decisions dealing with the precision and the overall strategy of computation could only be answered by extensive computer simulation experiments. Here are some of the problems that had to be studied.

1. Register length of coefficients (W^{nk}).
2. Register length of samples of the reference spectrum.
3. Register length of the data.
4. Type of arithmetic; fixed or floating point or a hybrid scheme.
5. Scaling strategy.
6. Rounding or truncation.

The following simulation experiments were used to help evaluate the problems above.

1. Computation of the matched filter output for a single LFM (linearly frequency modulated) or chirp signal with Hamming weighting to determine whether any spurious side lobe peaks appear.

2. Computation of the matched filter output for the sum of a large LFM signal and a small LFM signal close in range to the large signal to see if the small signal is lost in a side lobe of the large one.

3. A low level LFM signal is processed in the presence of simulated receiver noise. The purpose of this experiment is to ascertain whether the processor has altered the statistics of the random signal component thereby affecting signal-to-noise ratio.

The first simulation experiments were run with fixed point arithmetic. For LFM signals, the signal amplitude as it passes through the successive FFT stages builds up quite rapidly, essentially one bit per stage. Thus, to avoid using very long register lengths, the result after each butterfly was right shifted by one bit. For this strategy in both forward and inverse FFT it was found that coefficient register lengths of eight bits and data register lengths of 16 bits were required.[8]

Next, the system was altered to do floating point arithmetic. As we know from Chapter 5, some decrease in the number of bits needed could be expected. Simulation showed this to be true; it was found that a nine-bit mantissa and four-bit exponent was adequate for all registers in the system. Despite the decrease in memory size, however, arithmetic complexity increase and arithmetic speed decrease made this system less attractive than the fixed point case.

The system that was simulated most extensively used a hybrid arithmetic scheme that was a cross between fixed and floating point. This scheme had the following features.

1. Coefficients were fixed point nine-bit fractions.
2. The complex data word had two mantissas (one for the real part and one for the imaginary part) and a single exponent that served both mantissas.
3. Mantissas were never left justified (normalized). On overflow of either mantissa of the complex datum, however, both mantissas were right shifted one bit and the exponent incremented by one.
4. The exponent is assumed to be a positive number; mantissas are signed, 2's-complement numbers.

Figure 13.30 shows such an arithmetic system for one state of a radix r pipeline FFT processor, using the decimation-in-time algorithm.

Processing begins by immediately performing the $(r - 1)$ complex multiplications. These are simply fixed point operations, and overflow is prevented in the cross product combination by carrying an extended sign bit. In parallel with the multiplications, the largest of the r exponents is determined and transmitted to the output of the stage for later use. Before

[8] It is possible to construct a system where the register length increases for successive FFT stages but this seems awkward for hardware implementation.

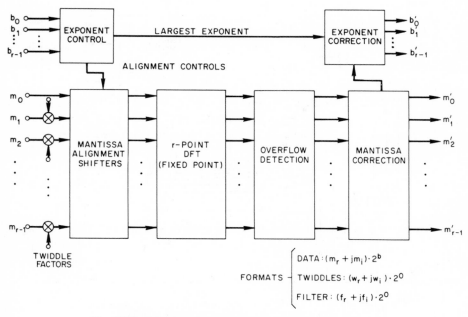

Fig. 13.30 Hybrid arithmetic scheme for radix r pipeline FFT.

proceeding with the r-point DFT it is necessary to align the twiddled mantissa pairs to have the same exponents. The r-alignment shifter pairs are controlled by the difference between the maximum exponent and the individual exponents of each mantissa pair. Following this alignment, all operations are fixed point. Notice that the twiddles and the search for the largest exponent, which are both time-consuming operations, are performed in parallel; this was possible because, in the DIT algorithm, multiplication precedes the r-point DFT (which, in the radix 4 case, consists only of addition).

Depending on r, several further extended sign bits are carried through the fixed point DFT computations to avoid possible overflow. At the DFT output, each complex mantissa pair is inspected to see if these extended bits have been used. Special logic examines these bits and determines the number of right shifts needed to bring the mantissas back into range so that the length of the memory registers need not include these extra bits. A more detailed sketch of this overflow correction logic is shown in Fig. 13.31.

Simulation results convinced us that the scheme above was highly competitive in performance with either the fixed or floating point schemes previously discussed. From a hardware point of view, this scheme was faster and cheaper than the full floating point and yet overcame the crucial problem of dynamic range that made the fixed point system worrisome.

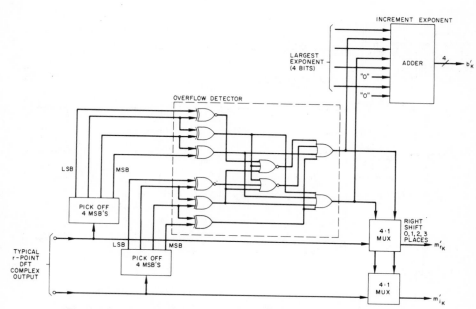

Fig. 13.31 Details of overflow correction for hybrid arithmetic scheme.

Finally, the question of truncation versus rounding was addressed. Various simulation experiments were performed to compare the two methods. It appears that rounding to 9 bits is equivalent in performance to truncating to 12 bits so despite the added hardware needed to do rounding, it was the recommended procedure.

13.8 Summary

The radar professional reading this chapter will quickly sense that a very restricted portion of the radar field has been included. We believe, however, that this chapter is the most thorough one in print on the application of digital signal processing to radar problems. At this writing, for the first time since World War II, radar technology appears to be in a decline. On the one hand, the air traffic control agencies are aiming toward beacon systems; on the other hand, the effects of the SALT talks may result in curtailment of high performance ballistic missile defense (BMD) radars. Despite this, there seems little doubt that radar technology is a permanent and important aspect of research and development in electronics. Just as surely, digital technology, including digital signal processing, is bound to find increasing use in radar.

REFERENCES

1. *Radar Handbook* (M. Skolnik, ed.), McGraw-Hill Book Co., New York, 1970.

2. B. GOLD AND C. E. MUEHE, "Digital Signal Processing for Range-Gated Pulse Doppler Radars," *Agard Conference Proceedings*, No. 66, 1970.

3. P. BLANKENSHIP AND E. HOFSTETTER, "Digital Pulse Compression via Fast Convolution," submitted to IEEE Transactions on Acoustics, Speech, and Signal Processing.

4. C. E. MUEHE, ET AL., "New Techniques Applied to Air Traffic Control Radars," *Proc. IEEE*, 62, No. 6, 716–723, June 1974.

5. D. F. DELONG, JR. AND E. M. HOFSTETTER, "On the Design of Optimum Radar Waveforms for Clutter Rejection," *IEEE Trans. on Information Theory*, **IT-13**, No. 3, 454–463, July, 1967.

6. J. R. KLAUDER, A. C. PRICE, S. DARLINGTON, AND W. J. ALBERSHEIN, "The Theory and Design of Chirp Radars," *Bell Sys. Tech. J.*, **39**, 1–76, No. 4, July, 1960.

7. C. E. COOK AND M. BERNFELD, *Radar Signals*, Academic Press, New York, 1967.

8. A. M. RIHACZEK, *Principles of High-Resolution Radar*, McGraw-Hill Book Co., New York, 1969.

Index

A

Add-shift multiplier, 515–16
Airborne surveillance radar (ASR), 733–41
Air traffic control (ATC) radar, 710
Aliasing, 26–28, 218
All-pass equalizer, 288
All-pass network, 206
Alternation theorem, 127
ALU (arithmetic-logic unit), 507–8
Ambiguity function, 715–24
Ambiguity function of a burst, 723–24
Ambiguity functions of chirps and
 sinusoidal pulses, 717–21
Ambiguity function of a CW pulse, 721–23
Analog lowpass filter design techniques,
 226–38
Analog-to-digital (A/D) conversion,
 296–300
Analog-to-digital conversion noise, 296
Analytic signal, 72
AND function, 491
Antenna aperture and wavelength, 713
Aperiodic convolution, 61–62

B

Approximation problem:
 FIR filters, 75–204
 IIR filters, 205–94
Array multipliers, 518–24, 533–36
Autocorrelation, 401
Automatic volume control (AVC), 680

Backward differences, 212–14
Bandpass filters:
 frequency sampling, 119
Bandpass sampling, 72–73
Bandpass transformation, 257–63
Bessel filters, 228–30
Bessel function program, 103
Bilinear transformation, 219–24
Bipolar logic, 496–501
Bit-reversal, 363–66, 575, 579–85
Bit-reversed counter, 364–66
Blackman window, 105
Block floating point arithmetic, 305–6
Bluestein's algorithm, 392–93